The Architecture of Science

The Architecture of Science

EDITED BY

Peter Galison and

Emily Thompson

The MIT Press
Cambridge, Massachusetts
London, England

© 1999 Massachusetts Institute of Technology

All rights reserved. No part of this book may be reproduced in any form by any electronic or mechanical means (including photocopying, recording, or information storage and retrieval) without permission in writing from the publisher.

This book was set in Adobe Garamond by Graphic Composition, Inc., Athens, Georgia.

Printed and bound in the United States of America.

Library of Congress Cataloging-in-Publication Data

The architecture of science / edited by Peter Galison and Emily Thompson.
    p.   cm.
    Includes bibliographical references and index.
    ISBN 0-262-07190-8 (hc : alk. paper)
    1. Architecture and science.   2. Architecture and technology.   3. Laboratories.   I. Galison, Peter Louis.   II. Thompson, Emily Ann.
NA2543.S35A73   1999
720—dc21
                                                                               98-43966
                                                                                 CIP

Contents

Acknowledgments   xi

Notes on Contributors   xiii

1   Buildings and the Subject of Science   1
Peter Galison

I   OF SECRECY AND OPENNESS: SCIENCE AND ARCHITECTURE IN EARLY MODERN EUROPE

2   Masculine Prerogatives: Gender, Space, and Knowledge in the Early Modern Museum   29
Paula Findlen

3   Alchemical Symbolism and Concealment: The Chemical House of Libavius   59
William R. Newman

4   Openness and Empiricism: Values and Meaning in Early Architectural Writings and in Seventeenth-Century Experimental Philosophy   79
Pamela O. Long

II   DISPLAYING AND CONCEALING TECHNICS IN THE NINETEENTH CENTURY

5   Architectures for Steam   107
M. Norton Wise

6   Illuminating the Opacity of Achromatic Lens Production: Joseph von Fraunhofer's Use of Monastic Architecture and Space as a Laboratory   141
Myles W. Jackson

7   The Spaces of Cultural Representation, circa 1887 and 1969: Reflections on Museum Arrangement and Anthropological Theory in the Boasian and Evolutionary Traditions   165
George W. Stocking Jr.

8   Bricks and Bones: Architecture and Science in Victorian Britain   181
Sophie Forgan

III MODERN SPACE

9   "Spatial Mechanics": Scientific Metaphors in Architecture   213
Adrian Forty

10 Diagramming the New World, or Hannes Meyer's "Scientization" of Architecture   233
K. Michael Hays

11 Listening to/for Modernity: Architectural Acoustics and the Development of Modern Spaces in America   253
Emily Thompson

12 Of Beds and Benches: Building the Modern American Hospital   281
Allan M. Brandt and David C. Sloane

IV IS ARCHITECTURE SCIENCE?

13 Architecture, Science, and Technology   309
Antoine Picon

14 Architecture *as* Science: Analogy or Disjunction?   337
Alberto Pérez-Gómez

15 The Mutual Limits of Architecture and Science   353
Kenneth Frampton

16   The Hounding of the Snark   375
Denise Scott Brown

V   PRINCETON AFTER MODERNISM: THE LEWIS THOMAS LABORATORY FOR MOLECULAR BIOLOGY

---

17   Thoughts on the Architecture of the Scientific Workplace: Community, Change, and Continuity   385
Robert Venturi

18   The Design Process for the Human Workplace   399
James Collins Jr.

19   Life in the Lewis Thomas Laboratory   413
Arnold J. Levine

20   Two Faces on Science: Building Identities for Molecular Biology and Biotechnology   423
Thomas F. Gieryn

VI   CENTERS, CITIES, AND COLLIDERS

---

21   Architecture at Fermilab   459
Robert R. Wilson

22   The Architecture of Science: From D'Arcy Thompson to the SSC   475
Moshe Safdie

23   Factory, Laboratory, Studio: Dispersing Sites of Production   497
Peter Galison and Caroline A. Jones

Index   541

Acknowledgments

We would like to thank the Mellon Foundation, the Harvard University Provost's Fund for Educational Innovation, the Faculty of Arts and Sciences at Harvard, Peter Rowe and the Graduate School of Design at Harvard, and Suzanne Rauffenbart and the Harvard Medical School, all of whom supported the conference from which this volume originated. Thanks also to Jean Titilah and Shelly Brener, who helped prepare the manuscript. The Clark Fund/The Cooke Fund at Harvard and the Research Foundation at the University of Pennsylvania provided additional support during manuscript preparation.

Most of the papers contained in this volume were presented at "The Architecture of Science," a conference that met 21–22 May 1994 at Harvard University. Questions and suggestions by session commentators and by the audience were a great help to all of the authors in revising their papers for publication. Finally, many thanks to the editors and referees at The MIT Press, especially Larry Cohen.

Notes on Contributors

Allan M. Brandt is the Amalie Moses Kass Professor of the History of Medicine at Harvard Medical School and Professor of the History of Science at Harvard University. Brandt is the author of *No Magic Bullet: A Social History of Venereal Disease in the United States Since 1880* (Oxford University Press, 1985). He has written on the social history of epidemic disease, the history of public health, and the history of human subject research. He is currently writing a book on the social and cultural history of cigarette smoking in the United States.

James Collins Jr. joined Payette Associates, Inc., in Boston in 1979, became principal in 1986 and is now the firm's president and chief executive officer. As a winner of Rensselaer Polytechnic Institute's Ricketts Prize in Architecture, Collins graduated from RPI in 1978 with bachelor's degrees in Architecture and in Science, and a master's degree in Business Administration. He has designed a significant number of academic research buildings across the country as well as written and lectured extensively on laboratory design.

Paula Findlen teaches history as well as the history and philosophy of science at Stanford University and is the coeditor of *Configurations,* a journal of literature, science, and technology. She is the author of *Possessing Nature: Museums, Collecting, and Scientific Culture in Early Modern Italy* (University of California Press, 1994) and recipient of the 1996 Pfizer Prize in History of Science. Her essay in this volume is part of a recently completed book, *A Fragmentary Past: Museums and the Renaissance* (Stanford University Press, forthcoming).

Sophie Forgan studied history at London University and worked on the history of the Royal Institution for her doctorate. She teaches history of architecture and design in the School of Law, Arts and Humanities, University of Teesside, United Kingdom. Her research interests cross the disciplinary divide between architecture and science, and she is particularly concerned with the ways that buildings function in practice, and how they may shape, constrain, and reflect ideas about scientific practice and knowledge. She has published on the history and architecture of learned societies, universities, museums and exhibitions.

Adrian Forty is Senior Lecturer in the History of Architecture at The Bartlett, University College London. He is the author of *Objects of Desire: Design and Society Since 1750* (Thames and Hudson, 1986), and he is at present writing a book on the history of the relationship between architecture and language, to be published by Thames and Hudson in 1999.

Kenneth Frampton holds the position of Ware Professor of Architecture at Columbia University. He is an architect and architectural historian. He was educated at the Architectural Association in London and has worked as an architect in England, Israel, and the United States. He has taught at Princeton University and the Royal College of Art, as well as at Columbia. Frampton has been the recipient of numerous awards, including the National Honors Award of the American Institute of Architects, and the Gold Medal of l'Academie d'Architecture. His publications include *Modern Architecture: A Critical History* (Thames and Hudson, 1980), and most recently, *Studies in Tectonic Culture* (The MIT Press, 1996).

Peter Galison is the Mallinckrodt Professor of the History of Science and of Physics at Harvard University. Author of *How Experiments End* (University of Chicago Press, 1987), and *Image and Logic: A Material Culture of Microphysics* (University of Chicago Press, 1997), he has coedited *Big Science* with Bruce Hevly (Stanford University Press), *The Disunity of Science* with David Stump (Stanford University Press), and *Picturing Science, Producing Art* with Caroline Jones (Routledge, 1998). He was awarded a MacArthur Fellowship in 1997.

Thomas F. Gieryn is Professor of Sociology at Indiana University (Bloomington). His new book, *Cultural Boundaries of Science: Credibility on the Line,* was due to be published by the University of Chicago Press in late 1998. His current research on the design of research facilities for the biological sciences has been supported by The Andrew W. Mellon Foundation.

K. Michael Hays is Professor of Architectural Theory at the Graduate School of Design at Harvard University, where he is also director of the Advanced Independent Studies Program and chair of the doctoral program. He is founder and editor of *Assemblage,* a critical journal of architecture and design culture. Hays' published work focuses on ideological issues in the history of the avant-garde and on current debates in architectural and critical theory. His books include *Modernism and the Posthumanist Subject* (1992) and *Architecture Theory Since 1968* (1998), both published by The MIT Press.

Myles W. Jackson received his Ph.D. from the Department of History and Philosophy of Science of the University of Cambridge in 1991. He was a Walther Rathenau Fellow of the Technical University of Berlin and a Mellon and National Science Foundation Fellow of the Department of the History of Science at Harvard University. He has just completed a study on German and British optics during the nineteenth century, entitled *Spectrum of Belief: Fraunhofer's Artisanal Knowledge and Precision Optics* (University of Chicago Press, 1999). He is currently Assistant Professor of the History of Science at Willamette University.

Caroline A. Jones teaches contemporary art and theory in the Art History Department at Boston University, where she is also director of museum studies. Author of *Modern Art at Harvard* (Abbeville, 1985), *Bay Area Figurative Art* (University of California Press, 1990), and *Machine in the Studio* (University of Chicago Press, 1996), she is also the curator-essayist of "Painting Machines" (Boston University Art Gallery, 1997) and co-editor of *Picturing Science, Producing Art* (Routledge, 1998). Her forthcoming book will focus on the art critic Clement Greenberg.

Arnold J. Levine is President of Rockefeller University. He was formerly the Harry C. Weiss Professor in the Life Sciences and former chairman of the Department of Molecular Biology at Princeton University. He teaches and carries out research in the areas of virology and cancer biology. He has received honorary degrees from the University of Pennsylvania and the University Pierre and Marie Curie in Paris. Elected to the Institute of Medicine of the National Academy of Sciences, he has also received the first Charles Rodolphe Brupbacher Foundation Award from Zurich for his work on the p53 gene; the Thomas A. Edison Science Award from the State of New Jersey; the Memorial Sloane-Kettering Katharine Berkan Judd Award; the Josef Steiner Cancer Foundation Prize from Berne, and the seventeenth annual Bristol-Myers Squibb Award for Distinguished Achievement in Cancer Research.

Pamela O. Long is a historian of late medieval/early modern science and technology. She has taught most recently in the Department of the History of Science, Medicine, and Technology at Johns Hopkins University. She has published on the sixteenth-century Vitruvian commentary tradition, on humanism and science, on the ideal of openness in sixteenth-century mining traditions, and on the origins of patents. The descriptive title of her book in progress is *Openness, Secrecy, Authorship, Ownership: Studies in the Technical, Practical, and Knowledge Traditions of Premodern and Early Modern Europe*.

William R. Newman is Professor of the History and Philosophy of Science at Indiana University. He has published two books, *Gehennical Fire: The Lives of George Starkey, An American Alchemist in the Scientific Revolution* (Harvard University Press, 1994), and *The Summa Perfectionis of Pseudo-Geber* (Brill, 1991). His interests include the history of alchemy and the occult sciences, matter theory, chemistry up to Dalton, natural philosophy, and early technology. He is currently working on the collaboration of George Starkey and Robert Boyle, on early modern corpuscular theory, and on a reevaluation of the concept of the occult sciences.

Alberto Pérez-Gómez has taught at the Universities of Mexico, Houston, Syracuse, Toronto, and Carleton, and is now the Saidye Rosner Bronfman Professor of the History of Architecture at McGill University, where he has been director of the history and theory of architecture graduate program since 1987. He was also the director of the *Institute de recherche en histoire de l'architecture* in Montreal. Pérez-Gómez is the author of *Polyphilo or the Dark Forest Revisited* (The MIT Press, 1992), an erotic narrative/theory of architecture based upon a kindred text from late fifteenth-century Venice. He is coauthor, with Louise Pelletier, of *Architectural Representation and the Perspective Hinge* (MIT Press, 1997), and editor of the series *Chora: Intervals in the Philosophy of Architecture* (McGill-Queen's University Press). His first book, *Architecture and the Crisis of Modern Science* (MIT Press, 1983), won the Alice Davis Hitchcock Award for architectural history in 1984.

Antoine Picon is *Ingénieur des Ponts et Chaussées,* architect and doctor of history. A graduate of the *Ecole des Hautes Etudes en Sciences Sociales,* he is currently Professor at the *Ecole des Ponts et Chaussées* and director of research at the University of Paris I–Sorbonne. He has published numerous articles and books on the history of engineering and on the relations among architecture, science, and technology, including *Claude Perrault, 1613–1688, ou, La curiosité d'un classique* (Picard, 1988), *L'Invention de l'ingénieur moderne. L'Ecole des Ponts et Chaussées 1747–1851* (Presses de l'Ecole Nationale des Ponts et Chausées, 1992), and *French Architects and Engineers in the Age of Enlightenment* (Cambridge University Press, 1992).

Moshe Safdie established Moshe Safdie and Associates Inc. in 1967 to design and supervise the construction of Habitat '67 in Montreal. With offices currently in Boston, Jerusalem, and Toronto, he has designed numerous influential and award-winning projects ranging from public institutions including museums, performing arts centers, and university campuses to airports, housing, mixed-use complexes, and new communities. Safdie served as the Ian Woodner Professor of Architecture and Urbanism and the director of the Urban Design Program at the Harvard University Graduate School of Design from 1978 to 1990 and has also taught at Yale, McGill, and Ben Gurion Universities. In addition, he is the recipient of many honorary degrees and awards, including the Order of

Canada and the Governor General's Gold Medal. He has published several articles and books including *Beyond Habitat, Form and Purpose, Jerusalem: The Future of the Past,* and most recently, *The City After the Automobile* (Basic Books and Stoddart, 1997). In 1996, Academy Editions published a comprehensive monograph on the work of Safdie's firm for the past thirty years.

Denise Scott Brown, a principal of Venturi, Scott Brown and Associates, has been a world leader in architecture for more than thirty years. Her projects in architecture and planning bring artistry and imagination to hard social and civic problems. She has taught at the Universities of Pennsylvania, Yale, California and Harvard and has published, inter alia, *Learning from Las Vegas* (with R. Venturi and S. Izenour, The MIT Press, 1972) and *Urban Concepts* (Academy Editions, 1990). As a theorist, practitioner, educator, and writer, she has influenced several generations of architects and urban designers, helping them relate their work to the city and understand their roles in its public and private realms. Her awards include the Chicago Architecture Award, the U.S. President's National Medal of Arts, and the AIA-ACSA Topaz Medallion.

David C. Sloane is Associate Professor of Planning and Urban History in the School of Policy, Planning and Development at the University of Southern California. Educated at the University of Wisconsin-Madison and Syracuse University, Sloane previously taught at Dartmouth College and Medical School. He is the author of *The Last Great Necessity: Cemeteries in American History* (Johns Hopkins University Press, 1991) and is currently completing a book on the influence of modern retail concepts on the recent transformation of the American hospital and health care system. He is additionally investigating the history of health care in Southern California.

George W. Stocking Jr. is Stein-Freiler Distinguished Service Professor of Anthropology and the Conceptual Foundations of Science at the University of Chicago. He received his Ph.D. in American Civilization from the University of Pennsylvania in 1960 and the Huxley Medal of the Royal Anthropological Institute in 1993. In addition to his own numerous publications on the history of anthropology in the United States and Britain, he was the founding editor in 1983 of the series *History of Anthropology* and continues to serve on its editorial board.

Emily Thompson is Assistant Professor of History and Sociology of Science at the University of Pennsylvania, where she teaches the history of technology. She received her Ph.D. from Princeton in 1992 and later studied at Harvard as a Mellon Fellow in history of science. She is completing a book on architectural acoustics and the culture of listening in America, and is about to begin another on the development of the science, art, and craft of sound motion pictures. She is also working on a cultural history of noise and of campaigns for noise abatement.

Robert Venturi, a principal of Venturi, Scott Brown and Associates, has been a world leader in architecture for more than thirty years. He has collaborated in the design and construction of eight laboratory buildings since 1983. He was educated at Princeton University and has taught at the University of Pennsylvania and Yale University. He was a fellow of the American Academy in Rome and has lectured widely in America, Europe, and Japan. His awards include the Arnold W. Brunner Memorial Prize in Architecture, the AIA Medal for *Complexity and Contradiction in Architecture* (New York Museum of Modern Art, 1966), the U.S. President's National Medal of Arts, and the Pritzker Architecture Prize.

Robert R. Wilson studied physics with Ernest O. Lawrence at the University of California-Berkeley, and received his Ph.D. in 1940. An expert in cyclotron design and operation, he initially taught at Princeton, then joined the scientific war effort and was named head of the Experimental Nuclear Physics Division at the Los Alamos Laboratory in 1944. He opposed the use of atomic weapons against Japan, and shortly after the bombs were dropped, gave up his national security clearance in protest. Wilson was Professor of Physics at Cornell University from 1947 to 1967, taking a sabbatical in 1960 to study sculpture at the Academia Belli Arte in Rome. He served as director of the Fermi National Accelerator Laboratory in Batavia, Illinois, from 1967 to 1978. He is currently Professor of Physics Emeritus at Cornell and at Columbia University.

M. Norton Wise is Professor of History at Princeton University. Coauthor with Crosbie Smith of *Energy and Empire: A Biographical Study of Lord Kelvin* (Cambridge University Press, 1989), he has also concentrated on the relation of natural philosophy to political economy in Britain, and on the cultural foundations of quantum mechanics in central Europe. He is preparing a book on the mediating technologies which, in particular cultural "moments," have grounded scientific explanation.

1

Buildings and the Subject of Science

Peter Galison

The Architecture of Science

The Architecture of Science means many things to the authors whose work is assembled here. We are (collectively) trying to understand the alternately harmonious and tense, but always compelling relationship between alchemists, natural philosophers, and scientists on one side, and those charged with the centuries-old task of constructing the built world on the other. How do the buildings of science literally and figuratively configure the identity of the scientist and scientific fields? Conversely, how do the sciences procedurally and metaphorically structure the identity of the architect and the practice of architecture?

To emphasize just how diversely science has been architecturally sited and architecture has been "scientized," we have quite deliberately mixed historians of science, historians of architecture, historians of art, architectural theorists, sociologists, architects, and scientists—you have before you papers diverse in method, style, period, and subject. In a sense, it could hardly be otherwise; our object of inquiry is too multiform. There is, for example, no single transtemporal, transcultural entity that is "the laboratory" that would include all spaces from the alchemist's secretive basement array of furnaces through the clinical research hospital to the $10 billion Large Hadron Collider outside Geneva.[1] There is no single way in which science is invoked within architectural theory and practice, and no single method by which architecture facilitates, appropriates, displays, and gives identity to the scientific processes contained within its walls.[2] Historically, spaces for the production of knowledge about nature have ranged in type from castles to industrial factories, university groves to corporate headquarters. Even within a given genre of buildings, there is no simple unity. Some hospitals evoke a church or alms house, whereas others call to mind sprawling suburban malls. Some laboratories shoot up with modernist fervor ten, twenty stories into the air; others branch horizontally in quasi-historical reference to nineteenth-century mill towns.

Just as the category of the scientist alters across time, so too does that of the architect. Within the history of architectural practice, there has been, as several of our authors tell us, a long history of "scientism"—a fiercely sought (and equally fiercely contested) ambition to make architects themselves into mirror images of the science their buildings sometimes contain: architecture as physics, biology, or ecological science. But if architecture is not a subcategory of the natural sciences, architecture and architects do appropriate scientific practices, materials, theories, and values in a constant reworking of disciplinary identity.

There is no reductionist message, then, to the essays in this book, nor, given the heterogeneity of our subject, could there be. In a hypertext introduction, there would surely be many different ways to group these essays: One could, for example, sort them by discipline, segregating the work of architectural historians from that of architectural theorists, or putting the architects together and isolating them from the historians or sociologists of science. Intentionally, however, we have chosen to group the essays thematically, very deliberately straddling disciplinary fences. Their variegated authorial backgrounds notwithstanding, I aim, in this introduction, to view these essays through the prism of the subject: What can architecture tell us about the changing identity of the scientist? And conversely how can science inform us about the shifting identity of the architect?

I find especially helpful the work of Francesco Dal Co as he explores the relation of the modern subject to the notion of "home" and the city. For theorists like Ferdinand Tönnies, Oswald Spengler, and to a certain extent Werner Sombart, place and dwelling held a preexisting attachment to the world, an organic connectedness of the home to the landscape that was irrevocably separated by the city. Against this nostalgic position, Dal Co pits Emmanuel Levinas and Martin Heidegger's argument that dwelling always presupposes a difference, a separation from the surroundings. It was *never* otherwise. Our sense of "place," even before the city, is not an outgrowth of some "natural" space. Rather, it is the other way around, as Dal Co puts it: "Without dwelling there can be no place; it is construction that evokes the place and transforms the space."[3]

Extending and transforming this reasoning about home and city to the places of science, we can put aside the idea that there was once a "natural," purely individual subject position for the scientist that was destroyed by the expansion of the laboratory into its large-scale, industrial form. Instead, we can ask of each age: "What kind of scientist, doctor, or viewer of science does a certain kind of laboratory, hospital, or museum architecture presuppose?" There is an analytic, design side to this question: Are the "scientific dwellings" (so to speak) constituted such that they are appropriate for individual, small group, or mass work? Are their spatial divisions permanent or flexible? Are the symbolic elements of architecture such that they link the space to libraries, offices, factories, homes, or churches? There is also a psychological-sociological side of the question: How does the architecture act as a guiding, daily reminder to practitioners of who they are and where they stand? By exposing the adjacency and distance between disciplines, between theorists and experimenters, between patients and doctors, between the scientist and the public, the built world helps define how scientists see themselves. Architecture can therefore help us position the scientist in cultural space; buildings serve both as active agents in the transformation of scientific identity and as evidence for these changes.

Conversely, several of our contributors not only explore whether architects are, were, or ought to be "scientists," but more probingly, they demand that we reassess the image of scientific practice against which that question is posed.[4] The architects in this volume are looking to design in ways that renounce a too functional, scientistic conception of laboratory design; the architectural theorists reject the idea that architecture can or ought be assimilated to a particular scientific "method"; and our architectural historians distance themselves from a narrow definition of technology. Examining the history of architects' vision of science can tell us a great deal about the evolving sense of architectural self-identity.

In different ways, and in strikingly different registers, the following twenty-two chapters explore the means by which architecture and science define one another through their encounter. We are after the scientific subject inhabiting the scientific edifice and presupposed by it; and we are after the architectural subject that has itself been drawn in the confrontation with science.[5] Who is it (and who ought it be) who steps into the alchemical chamber designed after the secret alchemical hieroglyph? What identity attaches to the patient who walks into a medical center qua mall to purchase genetic tests? What kind of scientist comes to work on the swing shift of the data analysis team at Fermilab? When architects structure a molecular biology laboratory with contiguity in space between mouse-based genetics and virology, their decision facilitates certain moves in biological research and contributes to the identity of a field. When electrical engineers' workspaces join those of physicists, or when Renaissance women are excluded from a natural history collection, space is manipulated to concentrate the meanings crystallized around the science of a given time. It is through appropriation, adjacency, display, and symbolic allusion that space, knowledge, and the construction of the architectural and scientific subject are deeply intertwined. These studies—"Of Secrecy and Openness: Science and Architecture in Early Modern Europe"; "Displaying and Concealing Technics in the Nineteenth Century"; "Modern Space"; "Is Architecture Science?"; "Princeton after Modernism: The Lewis Thomas Laboratory for Molecular Biology"; and "Centers, Cities, and Colliders"—move through a general chronological framework, yet at the same time raise common issues and questions that span time and thus serve to link the different sections of the book.

PART I: OF SECRECY AND OPENNESS

Architecture begins at the door, and that door, as Georg Simmel argued a century ago, is both a statement of connection and of separation.[6] Where "modern" science happens is often (too simply) assumed to be open space, doors ajar and ideas available to all. Medieval science was a closed affair (so the story goes) until suddenly the sunlight radiating from Galileo, Descartes, and Newton flooded the scene, collapsing the shadowy secrecy of the alchemist and replacing it with the published, daylight knowledge of modern science. But neither architectural nor scientific knowledge changed quite that way. Among others, Steven Shapin has shown us the more ambivalent nature of this openness: how Gresham House, site of Boyle's early experiments, was not by any means open to any comer. Gentlemen—and the term had a multiply restrictive meaning—would pass through where others were excluded.[7] The essays by our first group of authors show further that the doors of science swung open selectively, and the passages revealed upon entrance were by no means public streets.

Paula Findlen begins with the natural history collection of the Renaissance naturalist Ulisse Aldrovandi; she asks, Who came? Who saw? Not women. Of the many who inscribed their names in the guest book, almost all were men. The "hairy girl" from Spain came, but more as an exhibit than a visitor.

Two issues immediately come to mind. The first is the matter of simple exclusion: At the most elemental level, if women were not allowed to view (much less maintain) natural history collections, the modality of their exclusion from natural philosophy in the early Renaissance becomes clearer. Second, and more subtly, the very act of excluding women from certain sites of knowledge impressed a symbolic set of associations on the natural history collections themselves and so conditioned the gendered patterns through which natural historical knowledge was understood. Alberti's writing (as Findlen so remarkably shows) made this transparent by codifying gender-defined access to domestic space. Alberti's ideal patrician household contained adjacent bedrooms for husband and wife (with a "secret door" to allow conjugal visits). Off the husband's bedroom was his study; off the wife's bedroom, her dressing room. The ethos underlying such sentiments was pervasive, even absorbed by some women, as they strove to locate spaces and subjects appropriate to themselves. Moderata Fonte's imaginary female academy was set in a (properly feminine) garden, not in the masculine location of the museum. In so dividing architectural space, the ever more feminized domestic sphere acquired a specifically gendered identity for the prosecution of natural historical knowledge.[8]

Like Findlen, William Newman is not at all convinced that the shift from secrecy to openness came easily—if at all—into Renaissance culture. In a pathbreaking article in 1986, Owen Hannaway contrasted the plans for a chemical house of Andreas Libavius with the secret, "elitist" alchemical laboratory of Tycho Brahe.[9] To Hannaway, every architectural nook and cranny of Tycho's castle, isolated on a small island in the bay of

Hven, spoke to a traditional princely power and privileged knowledge, and as such stood for the older order of science. Where Tycho's island fortress excluded the public, Libavius's laboratory stood in the heart of the city and welcomed visitors. Where Tycho saw the laboratory as an extension of the textual obfuscations that impeded the entrance of the uninitiated and unworthy to the realm of alchemical knowledge, Libavius (Hannaway contended) wanted openness. Tycho became a recluse; Libavius's plan was to act as philosopher and paterfamilias, as a citizen in town and an inquirer into nature. In the architecture itself, Hannaway's argument went, one could see the end of the alchemist and the beginning of the chemist.

Newman views this architecture differently, and he shows persuasively that far from being a harbinger of everything modern, Libavius saw himself as restoring faith in authority against the rebellious, chaotic interventions of the Paracelsans. Indeed, Libavius's self-identification with the older alchemical tradition becomes manifest when we observe that, as Newman demonstrates, the alchemical monad (the master alchemical symbol) actually stood as the fundamental design element of Libavius's laboratory. Far from instantiating civic duty, on Newman's reading, Libavius's laboratory was built to mark in the walls themselves the most secret transmutational scheme of all: the hypersymbolized drive to the philosopher's stone. In the sanctum sanctorum of Libavius's lab, certain furnaces were to be separated "from the crowd." There were sites to which no one is admitted "except one quite close to perfection." Nor did such ambivalence towards openness perish at the end of the Renaissance: As Shapin has shown, even seventeenth-century laboratories were not open to just anyone. Simmel's "open" and "closed" spaces turn out to be nuanced concepts indeed, each containing elements of the other.[10]

The conjoint issues of architectural, epistemic, and social access Findlen, Newman, Shapin, and others raise throw into question the very notion of "openness" in post-Renaissance science. Yet, the rhetoric of openness was clearly present at the origins of (sixteenth-century) modern science. Where did this rhetoric come from, and how was it tied to the idea of empiricism? These are the questions that Pamela Long addresses, and to respond to them she takes us back to classical uses of "openness" in the world of architecture. Long shows us that to the Roman architect Vitruvius (20s B.C.E.), architecture was not distinct from other practices that later would have been classified variously as technical or engineering: clock making, machine operation, and the rational and practical aspects of construction. Vitruvius favored openness in print and experimentation (in the sense of trying out); he detested stolen writings and tirelessly defended the need to make precepts explicit. All this sounds, on the surface, quite "modern," but one needs to listen carefully, as Long does, to the context in which Vitruvius was writing. In the politically unstable last decades before the common era, Augustus was working furiously to bolster the legitimacy of his rule, and his temple-building program was designed to secure the authority of priests and ancestor worship, as well as his own claim to power. For Vitruvius, stealing text from past authors was a breach of etiquette against ancestors and authority; openness in the use of sources stood as an act of filial devotion.

Long tells us that the belief that architecture was bound to the conduct of governance was new in the Renaissance. Architectural symbolism went further: As Vitruvius saw it, each type of ruler (tyrant, magistrate, and so on) ought to build according to different styles. This architectural knowledge, said fifteenth-century architect Leon Battista Alberti, could come only from a just combination of study of the past, innovation in the present, and the "limpid, clear, and expeditious" development of argument. Unlike the higher-born Alberti, Filarete was trained as an artisan, and he had a different case to make for openness. In his "ideal city" of Sforzinda, Filarete insisted that artisanal be joined to "learned" knowledge—he published guides to everything from blast furnaces to stainless plaster. But openness here was tied to an abandonment of guild secrecy in favor of princely patronage. For some architects, openness was indeed linked to empirical work, on the surface an ethos usually associated with early modern science. Looked at carefully, Long concludes, natural philosophers of the sixteenth century were indeed struggling to define themselves through openness, but in doing so they were building on a long tradition of architectural proclamations of openness from antiquity through the Renaissance.[11]

Part II: Displaying and Concealing Technics

What happens when architecture builds around—literally encloses—the technical? Norton Wise explores this concretely in his tour of the English-style gardens that appeared in nineteenth-century Germany. After the Prussian defeat of Napoleon, architects and statesmen urgently sought a symbolic, cultural representation of their new victorious identity. One prominent celebratory structure was an ostentatious use of the steam engine to create an "English" landscape architecture by powered waterfalls, artificial streams, and forced irrigation. Disguised engines both hid and displayed industrial might.

But Wise's Prussian-English garden was also a hidden, subtle aesthetic, one that self-consciously sought to avoid the British "Coketowns" while alluding to both material and spiritual elevation. Cloaked in the exterior of mosques and Italianate villas, the steam engines were technology, but technology tamed by culture. Here was where a Hermann von Helmholtz could "grow up"; indeed, a generation of scientists and industrialists built on their experience of this landscape to forge a new science of work. To walk in the German English steam garden was to feel oneself both at one with nature and in command of it; it was to reinscribe in nature the self-fashioned identity of an older Prussian culture and to join to it a contemporary identity of modernity and power.[12]

Nineteenth-century German science was not all sited amid steam engines and pseudoclassical ruins, however. As Myles Jackson shows in his exploration of glassmaking in the Benedictine monastery of Benediktbeuern, the identity of the German optician could be both antique and modern, secret and open, industrial and craft centered at the same instant. Joseph von Fraunhofer, known to every physics student for the last century and a half for his study of the physical properties of light, made his prisms far from the famous urban physical institutes. Deep in the Black Forest, the Benediktbeuern monastery where Fraunhofer worked had both natural and human resources to support his science: wood, quartz, and workers (religious and secular) well skilled in the production and use of precision optics. Additionally, the physical architecture of the cloister itself was vital.

But the pairing of secrecy and openness reinforced a particular identity for Fraunhofer's optics. Making a great display of his techniques for measuring and calibrating prisms, Fraunhofer swung the doors open for important scientific visitors to witness the dark lines of the spectrum and the attainment of achromatism. Only then could Fraunhofer claim to be a "scientist" rather than a craftsman. At the same time, other doors slammed shut: No one—under any conditions—was going to be ushered into the quasi-sacred site of the glass fabrication house itself. Covert guild practices fused with proprietary as well as national secrecy and a transformed monastic silence. Architecture permitted this separation by the disposition of closed and open spaces, but more importantly, the symbolic history of the cloister naturalized such a distinction. Like Newman's alchemical laboratory or Findlen's natural history collection, Fraunhofer's "lab" was ostentatiously open *and* shut—a scientific showcase with a hidden back room.

Display continued in natural history museums, of course, the museums that over the course of the eighteenth and nineteenth centuries came to replace the curiosity cabinets of the Renaissance. Two of the most prominent types of museums that came into existence were natural history museums and anthropological museums, and our next two essays turn to these. In the first, George Stocking examines the often acrimonious debate between Otis Mason, curator of anthropology at the United States National Museum, and Franz Boas, then an immigrant scholar from Germany. At stake was what one might call the meaning of adjacency, the conceptual significance of artifact arrangement. Mason believed that similar artifacts were found around the world because similar circumstances gave rise to similar responses. Given such an outlook it made sense, for example, that in the National Museum, all rattles were grouped together: Why separate them? Boas argued vehemently against the view that artifacts could stand as universal constituents of a universal culture; instead, he contended that practices could be grasped only in separate, culturally "complete" collections.

One can see in Stocking's depiction of these two visions of the architecture of artifacts two very different cultural contexts. On the one side was an American vision of a world in which identity could be forged around universalist notions of human capacity. On the other stood the powerfully inscribed German concern with *Rasse* and *Kultur,* even if in Boas's hands such categories were molded into a relativistic anthropology. But Stocking's story continues. An ambitious new museum building was planned at Oxford in the 1960s, a museum to house the collection of some 14,000 objects donated by General Pitt Rivers, a contemporary of Mason and Boas. The architects chose an ideological pastiche: Circumferential rows would display objects of a certain type, while radial sections would exhibit diverse objects from a specific geographical area—although Stocking might not put it this way, it was Boasian in r and Masonian in θ, a never realized British compromise between American amalgamation and German tribal purity. In this decades-long architectural struggle, we encounter not only attempts to formulate tribal and human development, but also the traces of the anthropologists' own battle over the purposes and identity of their profession.

Like Stocking, Sophie Forgan excavates a fascinating comparative story by examining two nineteenth-century museums: Richard Owen's Natural History Museum, and the Museum of Practical Geology in Jermyn Street (London).[13] The Jermyn Street Museum was designed to educate the public not only about the "natural" geographic riches of the British island, but also about the economic benefits that resulted from the commercial exploitation of these natural resources. Museum rooms exposed the local riches of different districts, and even the walls spoke of British origins, through facades of Yorkshire dolomite or red Peterhead granite. Some of the great disputes of British stratigraphic geology became displayed fact as the museum presented layers of minerals divided according to one or another theory. Forgan shows us how the galleries became a horizontal representation of a geological section: One could walk through geological time, from the Lingula Flags or Cambrian through the Permian, Jurassic, Eocene, and Oligocene epochs. The space of the museum, marked by fossils, became the symbolic space of geological history and a microcosmic representation of Britain's national existence. Here was the museum as map and flag.

In contrast to the Jermyn Street Museum, the Natural History Museum was designed not as a map, but as an architectural representation of the natural historical convictions of its first director, anatomist Richard Owen. Current zoology stood to the west, paleontology and extinct species to the east; the "spine" of the building bore vertebrates, and the skeletal design of the whole reflected the Cuvierian conception of the order of nature. According to contemporaries, the Natural History Museum was a well-ordered "biography" of nature and nation, writ in stone. Each museum reflected the cultural and scientific preoccupations of its designers and directors; but once built, each served as a powerful guide in the construction of the identity of natural history as a discipline. The inescapable gap between a national geology and the supposedly transcendent character of science is, to a certain degree, imaginatively bridged by the solid existence of national museums of natural history.

From their origins in private rooms of contemplation designed to bolster Renaissance aristocratic privilege, national museums became, in the nineteenth century, spaces that conveyed ancient permanence and the modernism of national power. Walking through the natural history museums, the museum-goer and scientist alike surveyed an orderly world acquired by empire. Natural history gave order to the diversity of that world, and in the architectural materiality of museums, scientific displays helped define what science was, and where it was going.

Part III: Modern Space

If scientists use architecture to fashion and refashion their identity, the converse is also true. As Sophie Forgan argues elsewhere in her essay, nineteenth-century British architects enlisted scientists not only for their contribution to building principles, but also as symbolic allies in the architects' own struggle for an identity distinct from that of engineers and builders. But the mutual positioning of symbolic orders as complex as science and architecture could never remain a purely strategic alliance. Adrian Forty makes it clear that the metaphorical vocabulary of science enters into architecture in ways so familiar to us as to be practically invisible—and yet these scientific tropes express the way architecture is conceptualized and received.

Think, Forty directs us, of the metaphor of "circulation." William Harvey first developed the concept in his physiological studies of the heart in the seventeenth century. Yet the usage of "circulation" within architecture did not occur for centuries: Suddenly, in the latter part of the nineteenth century, buildings came, in the architectural imaginary, to resemble physiological systems, and passages in buildings transmuted into pulsating arteries and veins. Whereas it is easy to see buildings as, in fact, like the body, Forty asks us to step back, arguing instead that nineteenth-century architects like Eugène Viollet-le-Duc wanted this enclosed self-sufficiency, and the physiological circulation metaphor bolstered their desired autonomy. (They could have chosen an interactive metaphor like respiration.) Similarly, one can, and Forty does, pursue architects' deep attachment to mechanical metaphors (used not as structural dynamics but as a means of expressing form): compression, stretch, tension, torsion, shear, equilibrium, along with centrifugal and centripetal forces. Traceable to the aesthetics of Hegel and Schopenhauer, mechanical metaphors came into widespread use with Heinrich Wölfflin. Later critics, including Colin Rowe (writing on Le Corbusier) or, more recently, Peter Eisenman boosted such mechanical metaphors to new heights.

Perhaps, Forty suggests (following Alberto Pérez-Gómez), the interwoven premodern identity of architecture and science before the 1700s made the analogic function of scientific metaphors superfluous. On this reading, the very existence of scientific mechanico-circulatory metaphors signals the separation of architecture from science. Modern architecture therefore revealed its identity through the scientific language on which it drew for expression, an identity simultaneously turning toward science and acknowledging its separation from it.

To restrict "science" to the natural-scientific sphere of biology, chemistry, and physics, however, would be to miss entirely the significance of the notion for one of the leading modern architects of the Bauhaus, its often neglected post-Gropius leader, Hannes Meyer. Michael Hays makes this clear in his essay, showing that for Meyer, the German sense of *Wissenschaft* embraced far more than the natural sciences. In particular, Meyer's scientized architecture linked the practice of the discipline to the aspirations of scientific Marxism. True, architecture would embrace the calculational physics and physiology of the day, but science itself folded into a social and historical network of forces, and history itself unfolded into the socialist future. Hays argues that in Meyer's design of the Swiss Petersschule, Meyer was striving to eliminate the architect's personality and the traditional art of facture in favor of a scientized engineering. Meyer's was, in the strongest possible sense, a modernist intervention, an attempt to reorder people's daily experience of the world. By producing communal spaces, spaces that referred only to the rational calculus of a new way of life, Meyer wanted architecture to reshape us and so the order of society. Architecture went all the way down, so to speak: Architecture was the physical world that rebuilt the psychological and social subject who inhabited that world. The result was the new, "modern man." For Hays, this modern man held a new subject position within Meyer's architecture, that of a rationalized life, and it was precisely a rationality embedded in the order of scientific Marxism.

As I have argued elsewhere, the "unaesthetic aesthetics" of the Dessau Bauhaus went hand in hand with a technocratic Marxism, an "unpolitical politics," and a new scientific "unphilosophical philosophy." Rudolf Carnap, Otto Neurath, Philip Frank—the left wing of the Vienna Circle—made common cause with left-leaning Bauhäusler, seeing in their logico-empirical epistemology a form of *Neue Sachlichkeit* that they hoped would supplant traditional philosophy. Radiating outward from Vienna and Berlin, this movement held together by offering the architects a form of scientific justification, while providing the new "antiphilosophy" a grounding in the wider modernist movement.[14]

It is a commonplace in architectural history to contrast the enthusiastically scientizing worldview of the hard-edged Bauhaus with the recalcitrant early-twentieth-century American architects, who were far less willing to pursue the aesthetic possibilities of the modern machine age. More specifically, the cliché is this: Americans, in their own unselfconscious, pragmatic, and unreflective way, had produced a vast technical infrastructure from grain silos to machines of mass production. But only in European hands, the cliché continues, could this engineering know-how be converted into the cultural modernism that transformed the meaning of architecture and society. Emily Thompson dissents, offering a very different picture of the way science joins the modern American building. She is interested less in building exteriors than in the interior uses of technology: modernist scientific architecture from the inside out, not the outside in.

Thompson's focus is on sound.[15] In the nineteenth century, there were many scientific principles of acoustics, but not much success in applying them to the practical art of controlling sound in buildings. Around 1900, Harvard physicist Wallace Sabine was enlisted as an acoustic consultant to help create the sound for a concert hall for the Boston Symphony Orchestra. He developed a mathematical formula that predicted the frequency-specific reverberation times as a function of a room's volume and the absorptive power of its specific materials. Sabine's success in Symphony Hall launched other acoustic-architectural collaborations, including the development of new sound-absorbing building materials such as Akoustolith and Celotex.

Thompson shows how these new materials were increasingly employed, not simply to control sound, but to eliminate noise. Against the cacophony of modernism, psychologists, physicists, and urban progressives identified noise abatement as a means to create better workers by improving mental health, concentration, and productivity. Buildings like the New York Life Insurance tower exemplified this interior modernism, a modernism, Thompson argues, that worked like a chrysalis, moving outward toward the exterior. And, instead of architects evoking or applying "science" to their designs to render them modern (à la Meyer), she shows architects and scientists working intimately together to create both new architecture and new science. The union of acoustic science and architecture reshaped what it meant to be an architect and a scientist at their intersection, at the same time that it altered the "modern experience" of the inhabitants of these rationalized concert halls, apartments, and offices.

As Allan Brandt and David Sloane demonstrate, hospitals in their various forms were also quasi-public institutions; like symphony halls, hospitals carried their own disciplining message to patients and doctors, inscribing idealized patterns of comportment on both. In Brandt and Sloane's words, the hospital stood in a mediative position between science and public culture, and thus situated, offered a unique window into the architecture of science. Understanding the manifold designs of the American hospital as it developed from the 1800s to the present offers insight not only into that institution, but into the authority structure of scientific medicine itself. For though the current prevalence of hospitalization in our lives makes the institution a fixture from birth to death, it was not always so. In the nineteenth century most American health care was received in the home, the hospital being a site only for the most dire circumstances: for the poor, and for those without family or friends. Disease within the hospital made it a feared place, and the architects of the early hospital struggled, not always successfully, to alter that stigma. During the nineteenth century, neither medical research nor science had a prominent place in the hospital; the institution was above all a charitably funded refuge for the dispossessed. Its architecture reflected and reinforced the patient's supplicant identity, and simultaneously its pavilions represented a bulwark against the "bad air" theory of disease.

Charles Rosenberg has shown how the move to construct hospitals vertically built on a shift from the bad air theory to the germ theory of disease; pavilions isolated patients from wafting bad air, but such isolation was not necessary if germs were the danger and they could be killed through sterilization.[16] And within these new towers of medicine, so Brandt and Sloane contend, antiseptic, diagnostic, and monitoring technologies altered the practice of medicine. The vertical, "functional," research-teaching hospital of the early through mid–twentieth century became the standard site of biomedicine.

With the coming of the 1960s and 1970s, these modern towers of medicine were viewed with an increasingly critical eye, no longer as sleek exemplars of efficiency and science in the Bauhaus-internationalist spirit, but rather as oppressive, machinelike monsters. Architects, doctors, and community boards set about taking down the monoliths, rescaling them to human proportions, recasting the patient as consumer, the doctor as "health provider," and medicine as commodity. The modern tower has become a mall. When patients enter this "postmodern hospital," they come neither to be disciplined into a morally improved state, nor to be "operated upon" by the medicine machine; they come to comparison shop. In this world of consumer medicine, the very categories of disease and health alter, along with the shifting, ever reconstructed identity of patients who purchase and doctors who provide.[17]

Part IV: Is Architecture Science?

In many of the essays presented here, we see an immediate concern with the modalities by which architecture functions in the identity formation of science and scientists. But there is a reciprocal and equally important question alluded to in the essays by Forgan, Thompson, and Hays: How does science function in the self-understanding of architecture and the architect? In this section, we bring this question to a head: Is architecture a science? Was it once or ought it aspire to be? Antoine Picon, Alberto Pérez-Gómez, Kenneth Frampton, and Denise Scott Brown critically interrogate the scientistic aspirations of architects and, more generally, the architectural endeavor.

Picon suspects part of the contemporary concern for science-architecture links emerges from a certain nostalgia for a cultural unity that cannot hold, if it ever did. Instead of searching for a simple unity or an architecture that merely "reflects" the scientific, Picon locates both architecture and science in a broader cultural frame encompassing both. In one example, he identifies a characteristic way of thinking in the early modern era: Scientific and technological practice all came to invoke a notion of "analysis" of wholes assembled from primitive parts, where these parts were taken as human products, not natural atoms. D'Alembert considered Newton's inverse square law, for example, to be an "element" of physics, and the engineers of the Ecole des Ponts et Chaussées counted flat decks and bridge piers as "elements" of construction.

In a more contemporary case, Picon indicates how differently architecture functions in a world in which the very meaning of a "technological artifact" is uncertain. An automobile, for example, is no longer defined as a discrete object of steel, rubber, and glass, but instead, a "computer on wheels," a node in a vast electronic network. For Picon, such amorphously extended quasi-objects cannot be counterposed against the human, and instead technology becomes an environment, a landscape, a background in which we live. Architecture responds to this new techno-environment of things through a variety of strategies that blur the traditional dichotomies of order-disorder, rationalism-lyricism. Rem Koolhaas plays with intermediate scales between buildings and urban development; Jean Nouvel crystallizes technological landscapes by playing with moving surfaces of light and texture as if they were computer images. Whether in the eighteenth or the late twentieth centuries, Picon concludes, the identity of the architect is neither derived from technology nor divorced from it, but coexists with it within a larger culture that defines both.

Alberto Pérez-Gómez is less sanguine than Picon about the harmonious relation between the technological and architectural worlds. For Pérez-Gómez the story begins in the territory of antiquity explored by Pamela Long, where architecture and science constituted a single endeavor. Following Edmund Husserl, Pérez-Gómez locates the fundamental epistemic break of modernity in the seventeenth century, with Galileo and his successors. On the far side of that break is the life-world (according to Husserl), the "only real world, the one that is actually given through perception";[18] on our side of the break is the quantification of nature, the creation of a "formula-world" in which idealized magnitudes and their functional relations come to count as all there is. In Husserl's view, once this transformation from life-world to formula-world has taken place, the further elaboration of science may be productive, but at root, philosophically, it is more of the same: Galileo writ large and wide.[19]

Pérez-Gómez builds on this Husserlian metaphysical periodization. Before the Galilean break, architecture and science (he states) "were linked at the very inception of our Western tradition." Plato's *Timaeus* was the prototype for both architecture and science; architecture and science revealed truth by embodying the proportional relations of the cosmos. Only with the Galilean break does architecture become an instrument for accomplishing goals; only then, in the hands of late seventeenth-century architect Claude Perrault, does architecture become a search for "most probable" solutions and mathematically exact deductions. To Pérez-Gómez, who is frankly normative, all subsequent attempts to identify truth with science and science with application are doomed to superficiality. As an alternative, Pérez-Gómez proposes, architecture must be grounded through language in "history (stories)," a history that could be humanity's and architectural theory's "true normative discipline."[20]

Like Pérez-Gómez, architect-historian Kenneth Frampton vividly summons up scientific instrumentalism appropriated by twentieth-century architects, but for Frampton that instrumentalism emerges more sociopolitically and more recently than for Pérez-Gómez. From Alexander Klein's ergonomic housing of 1923–31, with its resonant Taylorism, to Richard Neutra's and Alvar Aalto's biotechnics of the 1940s and Sven Hesselgren's late 1950s (proto) semiotic exploration of the language of modern architecture, these nominally scientific attempts to conduct architecture by means of computer-aided studies, topological analyses, or bioclimatic calculations leave Frampton cold.

At root, the problem for Frampton is the underlying acceptance of the scientific program of an analysis of subsets and parts and the concomitant lack of attention to the larger "big picture" of the negative effects of science/technology/capitalism on the global environment that all these projects represent. To Frampton, these problems cannot be solved by any science or scientific architecture, since any such enterprise would, by definition, break the problem up into pieces and miss it in its entirety. He proposes, instead, that a solution must be found in the development of a critical theory of architecture that would redefine architecture's identity; this new repoliticized identity must, like architecture itself, be constructed from the ground up, piece by piece.

Although she approaches the scientism of architecture as a practitioner-theorist, not as a historian, in her antireductionism Denise Scott Brown concurs with Picon, Pérez-Gómez, and Frampton. Recalling her own education amid and experience with the assault on architecture by the scientism of the 1960s, she writes that "in architecture we are dogged by the model of the sciences." Social planners of the 1960s as well as the computational designers of the 1990s are all, she charges, bent on dismantling the architect's craft and artistic methods. Frampton argues similarly: A computer program can minimax adjacency problems, but it can do so only by ignoring the myriad other factors an architect must bear in mind. Scott Brown, intriguingly, represents the antipode to Hays's depiction of Hannes Meyer. Where Meyer desperately wanted architecture per se to vanish in favor of the engineer's rational calculus, Scott Brown sees just this impulse as the most destructive attitude possible for designers of the built world.

Instead of scientism, Scott Brown calls explicitly for architects to draw upon the far more romantic image of science sketched in the rhapsodic reflections of humanist-biologist Lewis Thomas. Scott Brown cites Thomas's *Lives of a Cell* approvingly: "Solutions cannot be arrived at for problems . . . until the science has been lifted through a preliminary, turbulent zone of outright astonishment. Therefore, what must be planned for, in the laboratories engaged in the work, is the totally unforeseeable." In this sense of the scientific, Scott Brown endorses the merging of architecture with science: the chaos of the charrette as laboratory discovery, not architecture as applied protocols of computer rationality. It is particularly appropriate, then, that Scott Brown's architectural firm, Venturi, Scott Brown and Associates, was commissioned to build the Lewis Thomas Laboratory, the "case study" of this volume. It is to that site that we now turn.

Part V: Princeton after Modernism: The Lewis Thomas Laboratory

What then of the architecture that surrounds the contemporary site of scientific production? To explore the relation between laboratory builders and users, we turned to a single building, the Lewis Thomas Laboratory for molecular biology at Princeton University.

We assembled commentaries by an exterior architect (Robert Venturi of Venturi, Scott Brown and Associates), an interior architect (James Collins, Jr., of Payette Associates Inc.), a former director of the laboratory (Arnold Levine), and a sociologist of science who has focused on the uses of space in science (Thomas Gieryn).

A laboratory, Venturi argues, is no place for heroic architectural gestures, no proper site for a masterpiece. Diminishing distraction, GENERIC ARCHITECTURE (as Venturi calls it, capitals in original) ought be flexible in its design, able to accommodate the shifting allegiances of research groups one with the other.[21] In planning for the Lewis Thomas Laboratory, he believed that there was no place for the "expression of function" as the Bauhaus would have had it. Now the building had to *accommodate* function*s*. Whereas the modern age of industrial architecture could use the machine as a literal model of efficiency and rationality, in this age after modernism, Venturi asserts, "industrial and engineering/structural imagery of space is incidental for now, while an ornamental/symbolic imagery of applique is valid for now." Symbolism is explicit where modernist dogma would deny it or make it implicit; decoration is celebrated where the modernist claimed to be stripping it away; and function, instead of determining form, is collapsed into a flexible pragmatism. Neomodernism in the architecture of the 1990s did involve a revival of engineering, but it was as "engineering expressionism," "in the end, an ironical architectural vocabulary based on industrial imagery as industrial *rocaille*."

Equally instrumental in designing the Lewis Thomas Laboratory was James Collins, Jr., of Payette, since the building's interior would determine most immediately how the scientists would use and move through their laboratories. Early in the project, Tom Payette had walked Levine and Venturi through the Fairchild Biochemistry building at Harvard, a project recently completed by his firm. Spaces in the Fairchild building were linked visually by glass and, unlike the "traditional modernist" laboratory, natural wood was used to "humanize" the environment and to give the building a luxurious quality previously quite foreign in laboratory spaces. Brandt and Sloane emphasize how the architecture of medicine has shifted to keep pace with the corporate ethos of the 1990s; "pure" science with its myriad links to biotechnology would not be left behind with its identity mired in the drab, functionalist metal and concrete factory labs of the 1950s.

Aside from the texture of surfaces, the new laboratory design also reflected the changed realities of funding. As Collins describes it, the shift from foundational to government funding entailed the need to support (architecturally) spaces that would allow the kind of interdisciplinary collaboration that the government wanted to support. Spending constraints meant a focus on ease of maintenance: access to charcoal filters, ductwork, and data network cabling trays. Cooling requirements and the acidity of organic and chemical waste presented other constraints. Payette grouped the service technologies around a central shaft (which aided maintenance) and kept the researchers near the windows, in more "liveable" space.

Arnold Levine, recruited by Princeton to spearhead the revival of its department of molecular biology, did not have fond memories of working with architects. As a young researcher at the University of Pennsylvania, he recalled toiling in a laboratory built by Louis Kahn.[22] Fried by the sunlight, Levine and his graduate student colleagues pinned newspapers on the windows to cool themselves and save their experiments. Kahn was furious at this desecration of the building and ordered Levine to cease and desist. Now that Levine was directing a laboratory to be designed ab initio, he would not repeat that experience. Venturi's "accommodation" suited Levine's search for a mixture of flexibility and humanity. (Venturi's own complex formal and ideological relationship with Kahn began when Kahn was on Venturi's thesis review board, and Venturi went on to work in Kahn's office for nearly a year.) "Molecular biology is a blue-collar science," Levine writes, a science that involves working long hours at repetitive labor seven days a week. Levine wanted spaces where these "blue-collar" scientists could enjoy their fourteen-hour days, meet in small and large groups to talk about results, and combine teaching and research.

The structure of the building as a whole was driven by the technology of research, the instruments that bound together activities more than any analytic schema ever could. So on the third floor, one would find microbial genetics groups using microorganisms; on the second floor, cell division would find its home, with a focus on viruses linked to cancer; and on the first floor would be developmental biology and mouse-based genetics. Recounting the dramatic tale of the discovery and analysis of the tumor-suppressing protein (p53), Levine concludes that the physical space made it possible to increase collaboration, finding "the same genes in bacteria, yeast, worms, flies, frogs, mice, and humans. That brings the first, second, and third floors together for joint meetings, shared graduate students and common research themes." It became possible for researchers, architecturally associated, to define an identity for genetic research that cut across sectors of biology too often splintered one from the other.

When Thomas Gieryn agreed to study the Lewis Thomas Laboratory, he decided to throw its features into relief by contrasting it with another molecular biology laboratory—also designed by Payette Associates Inc.—the Center for Advanced Biotechnology and Medicine (CABM), a facility shared by Rutgers University and the University of Medicine and Dentistry of New Jersey (UMDNJ).[23] Like many other authors in this volume, Gieryn explicitly navigates between an architectural determinism on one side (architecture determinative of the science conducted inside) and an architectural indifference on the other (architecture irrelevant to the science contained within its walls). In the place of these reductive binaries, Gieryn argues that the architecture should be seen as constructing a cultural and institutional identity for the new biotechnology it housed. For example, when a tunnel joined CABM to UMDNJ, Rutgers insisted on an aboveground passageway known colloquially as "the leash." No one might use it but, as the UMDNJ Dean commented to Gieryn, the passageway was "[a] [half] million dollars' worth of symbolism." CABM was, as Gieryn puts it, an architectural meiosis, pulled in two directions by the conflicting force centers of UMDNJ and Rutgers.

As Gieryn relates, metaphors multiplied: To boosters of the Lewis Thomas Laboratory, it was an Elizabethan manor house, an old schoolhouse, a New England factory, whereas CABM supporters saw their lab as a temple, a showplace, and a Lord & Taylor among Kmarts. In this age of postmodernism, the symbolic dimension may be more self-announcing and playful than in the age of Bauhaus functionalist edginess, but symbolic allusions are clearly always activated, from the isolated Renaissance study to the sprawling postmodern mall-hospital.[24]

As the architects and their clients worked through the design of these biotechnological laboratories, struggles over their referential structure were anything but incidental. Venturi himself hoped the laboratories would abandon modernism's celebratory functionalism and become in the process more than a scientific site: "Perhaps the conventional scientific laboratory within this definition is the prototype for a valid and vivid architecture for now as a whole." That "whole" becomes ever less metaphorical when one contemplates the massive laboratory-cities of the high energy physics centers that have become the exemplars of twentieth-century big science.

Part VI: Centers, Cities, and Colliders

Physicist and sculptor Robert R. Wilson's Fermilab (Fermi National Laboratory, built in the late 1960s in Batavia, Illinois) for the study of particle physics is about as different as it could be from Arnold Levine's Lewis Thomas Laboratory (LTL): Wilson, acting in a dual role of architectural supervisor and scientific director, celebrated modernist architecture, whereas LTL defied it. Where the LTL is essentially a shoe box in footprint with three floors of laboratory space, the Fermilab campus revolves around a massive inwardly sloping tower, and architectural issues arose in reference to this building and the various supporting structures that dotted the campus. Wilson presents his image of the professional architect, however, in much the same suspicious light as Levine. Both were terribly afraid that the architect would ignore the real working conditions of the experimental scientist. But where Levine was reacting against Louis Kahn, Wilson dreaded the intervention of the Atomic Energy Commission (AEC) builder-architects, a team he viewed as producing nuclear physics structures "looking cheap and being expensive."

For Wilson, the preferred direction was up. A high-rise avoided the clutter of many small buildings, centralized all physics and engineering functions, and offered a magisterial view over the Illinois plains. His template, clearly, was that of urban modernism: His favorite "model" building was the New York office tower of the Ford Foundation with its open, green atrium and extensive use of glass. Referring back to Forty's gloss on circulation, one can see in Wilson's Fermilab tower a circulation design par excellence: Self-enclosed, it created a world in which the engineers, theorists, and experimentalists could move back and forth without undo reference to the outside.

Moshe Safdie, the architect commissioned to design the (now unfunded) Superconducting Super Collider in Waxahachie, Texas, positioned his own design against Wilson's Fermilab. Wilson wanted to build up and strip away the miscellany of many low buildings. Safdie aimed to build out to fill an even more enormous space; he relished the assemblage of different styles and sites. To Wilson, verticality meant clean lines and the vigorous circulation of modernity; to Safdie, horizontality meant organic unity and a gently chaotic, urban sense of place.[25]

Safdie saw the Wilson tower drained of some of its energy as it had to support industrial buildings alongside the proton tunnel. Instead, Safdie strove to build horizontal interactions along a "street" that would mimic the random but concentrated foot traffic of an urban center. If Wilson learned from the massive, climate-controlled, atrium-centered Ford Foundation building, Safdie much preferred the European Center for Nuclear Research (CERN) located outside Geneva, Switzerland, with its horizontal disposition of buildings each of which grew up organically around the well-defined central facility in which library, administration, and main auditorium were located. But maintaining a center in the much larger Waxahachie facility was easier said than done. With a circular ring 54 miles in circumference, the experimental sites where the interactions of interest would occur were many miles from any central facility. But if the center couldn't hold "naturally," then it was up to the architect to create it. By using the cooling pond to create a sense of place, a central set of buildings would offer a public face to the east (education center, hotels, and villas) while the west side of the pond would host offices and laboratories. The private sector would grow organically, in layers, like the expansion of a mill town; this time, though, the spontaneity was planned.

Caroline Jones and I conclude these chapters with a comparative exploration of the specific architectural spaces in which art and science were restructured during and after the Second World War. In many ways prewar scientists and artists both saw themselves as essentially solitary in their confrontation with nature. By the mid-1940s, that isolated self-understanding, previously reinforced by small-scale labs and studios, began to break down as both scientists and artists encountered patterns of work and places of production altered by the vast factory production quotas of firms like Albert Kahn Associates and the Austin Company. Physical plants, social ordering, and a new subject position entered together in the government-sponsored factory-laboratory exemplified by Oak Ridge, Hanford, and Los Alamos. Neither laboratories "modeled" on industry nor industries "modeled" on laboratories, these city-sized laboratories fused as a hybrid of the two.

In our reading, despite these changes, "big science" laboratories and factory-studios of the 1950s and 1960s were still tied, if in an unstable way, to a single focal point; there remained a director, a constructed creative apogee from which the full deployment of machines and personnel would originate. But this organization altered as particle physics entered an age of distributed data during the mid-1970s and as artistic efforts turned to site-specific work around that same time. With elected and rotating executive councils replacing a single physics leader, with an ever-shifting team of 2,000 PhDs distributed over hundreds of net-linked laboratories, it was no longer even possible to think of a particle physics experiment as organized around a single piece of equipment, much less a single individual. As earthwork artists shifted attention away from any studio toward the "back and forth thing" between earthwork sites and photographs, films, plans, and sketches of those sites, the art object, the singular artist, and the studio all diffused. As Jones and I argue, authorship and authorial identity in these spatially dispersed worlds of physics and art are unstable, contested, negotiated, and in the end never fully specifiable.

It is our collective hope that this volume will encourage a great deal more inquiry into the encounters between architecture and science. For example, science studies, in its enthusiasm for the study of space and the spatial production and distribution of knowledge, clearly has much to learn from architectural studies. Architects along with architectural historians have set a standard for study of the formalism, facture, and historical analysis of the built world that science studies has by and large missed. Architects, too, are always grappling both implicitly and explicitly with sociological concepts of how their structures channel human action; we still know relatively little about how the sociology of scientific knowledge fits into or clashes with architectural plans and intentions. Conversely, science studies has by now assembled a vast literature on the history and diversity of research practices. Yet the results of this work have barely been incorporated into the literature of architecture practice; to date, architects, historians, and theorists of architecture have largely worked with highly schematized images of scientific work in which a few kinds of finished products (e.g., Galilean, Newtonian, or chaos theory) have stood in for all of scientific practice. Nonetheless, it may be useful to end this introduction with the words of the scientific philosopher Rudolf Carnap, who in 1926 addressed the Dessau Bauhaus, where he had come in large measure to celebrate its new architecture: "I work in science and you in visible forms; the two are only different sides of a single life." If we can understand the limits to such a statement, if we can understand where and when it could and could not be uttered, we will have understood a great deal more about the subjects of both science and architecture, in all the many senses of "subject."

Notes

1. The literature on the history of the sites of science, including the laboratory, is now vast. The following are useful entry points into the social, epistemic, and architectural functions of space: Gooding et al., eds., *Uses of Experiment;* Lenoir and Elkana, eds., "Theory and Experiment"; James, *The Development of the Laboratory;* Galison, *Image and Logic;* Galison and Hevly, *Big Science;* Schaffer, "A Manufactory of Ohms"; Gooding, *Experiment and the Making of Meaning;* Olesko, *Physics as a Calling;* Cahan, *Institute for an Empire;* Hermann et al.,

CERN I and *CERN II;* Heilbron and Seidel, *Lawrence and His Laboratory;* Kohler, *Lords of the Fly;* Smith and Wise, *Energy and Empire;* Lynch, "Laboratory Space"; Ophir, Shapin, and Schaffer, eds., *The Place of Knowledge*; Heilbron, "Churches as Scientific Instruments"; Heilbron, "Science in the Church"; De Chadarevian, "Country House"; Hughes, *American Genesis;* Murphy, *Sick Buildings;* and other references cited in the References.

2. As an entry into architectural history that has probed the relation between architecture and the techno-scientific domain, see Picon, *Architectes et ingénieurs;* Pérez-Gómez, *Architecture and the Crisis of Modern Science* and *Polyphilo;* Frampton, *Studies in Tectonic Culture* and *Modern Architecture;* Hays, *Modernism and the Posthumanist Subject;* Buddensieg, *Industriekultur;* Banham, *Theory and Design in the First Machine Age;* Giedion, *Mechanization Takes Command* and *Space, Time, and Architecture;* Forgan, "Context, Image, and Function" and "Architecture of Science and University"; Mark, ed., *Architectural Technology;* and Jones, *Machine in the Studio.* Discussions of intriguing new relations inaugurated by the shift from mechanical to information systems may be found in Kwinter, "La citta nuova"; and De Landa, *War in the Age of Intelligent Machines.* "Smart materials" may also precipitate new architectural-scientific discussions.

3. Dal Co, *Figures of Architecture and Thought,* p. 38. For Heidegger, see "Building, Dwelling, Thinking" and "Art and Space."

4. The historicization of the subject and subjectivity in the face of changing scientific practices is, of course, one of Foucault's main themes. One thinks not only of his explicitly architectural work associated with the Panopticon (*Discipline and Punish*) but also of "Les techniques de soi," among many other texts.

5. The multiple meanings of "subject" are explored by Hubert Damisch in his important book *The Origin of Perspective,* and his own argument is that the Renaissance "perspective paradigm" determines a new subject (in the sense of viewer, in the sense of the structure of the painting, and in the sense of the artist's own position). "[W]hen man comes to terms with the symbolic order, his being is, from the very start, entirely absorbed in it, and produced by it, not as 'man,' but as *subject*" (p. 20; see also pp. 425ff).

6. Georg Simmel, "Bridge and Door"; a fascinating alternative sociology of the door can be found in Latour, "Mixing Humans and Nonhumans."

7. Shapin, "House of Experiment"; Shapin and Schaffer, *Leviathan;* Ophir, Shapin, and Schaffer, "Place of Knowledge."

8. Findlen's work on gender, space, and Renaissance knowledge can be pursued in her *Possessing Nature* and her forthcoming *Fragmented Past.* Literature on gender and architecture includes Friedman, "Architecture, Authority and the Female Gaze"; Spain, *Gendered Spaces;* and Colomina, ed., *Sexuality and Space.*

9. Hannaway, "Laboratory Design and the Aim of Science." On Tycho's laboratory design, see also Shackelford, "Tycho Brahe." See too Newman's careful study of George Starkey, an American alchemist, *Gehennical Fire.*

10. See Shapin, *A Social History of Truth.*

11. For more on the architectural dimension of the rise of the new science at the end of the Renaissance, see Long, "Architectural to 'Scientific' Outlook."

12. On self-fashioning, see Greenblatt, *Renaissance Self-Fashioning,* and Biagioli, *Galileo, Courtier.*

13. On Victorian scientific and museum architecture, see, e.g., Forgan, "'But indifferently lodged. . . .'" and "Architecture & Display"; and Stocking, *Objects and Others.*

14. Without doubt, the Bauhaus under Hannes Meyer was the single most influential and sustained attempt to integrate architecture and technical fields. See Galison, "Aufbau/Bauhaus" and "Constructing Modernism." Architectural histories of the Dessau Bauhaus can be found in Wingler, *The Bauhaus: Weimar, Dessau, Berlin, Chicago;* Whitford, *Bauhaus;* Naylor, *The Bauhaus Reassessed;* Meyer, *Hannes Meyer 1889–1954, Architekt Urbanist Lehrer* and *Bauen und Gesellschaft;* and Hays, *Modernism and the Posthumanist Subject.* For more on the scientific-engineering approach to town planning, and the "systems" thinking about lighting, ventilation, and kitchens, cf. Blau, *Red Vienna;* Rowe, *Modernity and Housing;* and Cohen, *World to Come.*
15. See also Thompson, "Dead Rooms and Live Wires"; and Thompson, "Mysteries of the Acoustic."
16. Rosenberg, *Care of Strangers.* For more on hospital and clinic architecture, see Thompson and Goldin, *The Hospital;* Foucault, *Birth of the Clinic;* Forty, "Modern Hospital in England and France"; Prior, "Architecture of the Hospital"; Sloane, "Hospital Mall"; Sloane, *Mall Medicine;* Kramer, "Psychiatry in a Biedermeier Asylum"; and Fiset, "Architecture and Healing."
17. Anthony Vidler uses Donna Haraway's concept of the cyborg to introduce a new form of the architectural imaginary. As he puts it, "If, for the first machine age, the preferred metaphor for the house was industrial, a 'machine for living in,' the second machine age would perhaps privilege the medical: the house as at once prosthesis and prophylactic" (Vidler, *Architectural Uncanny,* p. 147). Vidler explores the contradictory impulses already embodied in domestic space as the home blurs boundaries between organic and inorganic, private and public, body and mind, human and technological; the hospital pushes that blurring even further and the mall-hospital Brandt and Sloane explore might well serve as a quintessential illustration of the Vidler-Haraway intersection.
18. Husserl, *The Crisis of European Science,* p. 49.
19. As Husserl states, "In principle nothing is changed by the supposedly philosophically revolutionary critique of the 'classical law of causality' made by recent atomic physics. For in spite of all that is new, what is essential in principle, it seems to me, remains: namely, nature, which is in itself mathematical; it is given in formulae, and it can be interpreted only in terms of the formulae" (*Crisis of European Sciences,* p. 53).
20. Pérez-Gómez, *Polyphilo.*
21. On these themes, see also Robert Venturi, *Iconography and Electronics;* Venturi, Scott Brown, and Izenour, *Learning From Las Vegas;* and Venturi, *Complexity and Contradiction in Architecture.*
22. On Kahn's laboratory work, see Anderson, "Louis I. Kahn."
23. Gieryn develops further the views he presents here in his forthcoming *Cultural Boundaries* and in "Biotechnology's Private Parts."
24. On early twentieth-century linkages between technical culture and architecture, see Buddensieg, *Industriekultur.*
25. For more on Safdie's views about city design, see his *City after the Automobile.*

Bibliography

Anderson, Stan. "Louis I. Kahn in the 1960s." *Boston Society of Architects Journal* 1 (June 1967): 21–30. Also printed in *Louis I. Kahn,* Tokyo: *A&U: Architecture and Urbanism* (1975): 300–308.

Banham, Reyner. *Theory and Design in the First Machine Age.* London: Architectural Press, 1960.

Biagioli, Mario. *Galileo, Courtier: The Practice of Science in the Culture of Absolutism.* Chicago: University of Chicago Press, 1993.

Blau, Eve. *The Architecture of Red Vienna, 1919–34.* Cambridge: MIT Press, forthcoming 1998.

Buddensieg, Tilmann, in collaboration with Henning Rogge. *Industriekultur: Peter Behrens and the AEG.* Translated by Iain Boyd Whyte. Cambridge: MIT Press, 1984.

Cahan, David. *An Institute for an Empire: The Physikalisch-Technische Reichsanstalt, 1871–1918.* Cambridge: Cambridge University Press, 1989.

Cohen, Jean-Louis. *Scenes of the World to Come: European Architecture and the American Challenge, 1893–1960.* Paris: Flammarion, 1990.

Colomina, Beatriz, ed. *Sexuality and Space.* Vol. 1 of *Princeton Papers on Architecture.* New York: Princeton Architectural Press, 1992.

Dal Co, Francesco. *Figures of Architecture and Thought.* New York: Rizzoli, 1990.

Damisch, Hubert. *The Origin of Perspective.* Translated by John Goodman. Cambridge: MIT Press, 1994.

De Chadarevian, Soroya. "Laboratory Science Versus Country-House Experiments: The Controversy Between Julius Sachs and Charles Darwin." *British Journal for the History of Science* 29 (1996): 17–41.

De Landa, Manuel. *War in the Age of Intelligent Machines.* New York: Zone Books, 1991.

Findlen, Paula. *Possessing Nature: Museums, Collecting, and Scientific Culture in Early Modern Italy.* Berkeley: University of California Press, 1994.

Findlen, Paula. *Fragmentary Past: Museums and the Renaissance.* Forthcoming.

Fiset, Martin. "Architecture and the Art of Healing." *Canadian Architect* 35 (March 1990): 25–26.

Forgan, Sophie. "The Architecture of Display: Museums, Universities, and Objects in Nineteenth-Century Britain." *History of Science* 32 (June 1994): 139–162.

Forgan, Sophie. "The Architecture of Science and the Idea of a University." *Studies in History and Philosophy of Science* 20 (1989): 405–434.

Forgan, Sophie. "'But Indifferently Lodged . . .': Perception and Place in Building for Science in Victorian London," In Smith and Agar, eds., *Making Space,* 195–214.

Forgan, Sophie. "Context, Image, and Function: A Preliminary Enquiry into the Architecture of Scientific Societies." *British Journal for the History of Science* 19 (1986): 89–113.

Forty, Adrian. "The Modern Hospital in England and France: The Social and Medical Uses of Architecture." In Anthony D. King, ed., *Buildings and Society: Essays on the Social Development of the Built Environment.* London: Routledge and Kegan Paul, 1980.

Foucault, Michel. *The Birth of the Clinic: An Archaeology of Medical Perception.* Translated by A. M. Sheridan. London, Tavistock, 1973.

Foucault, Michel. *Discipline and Punish.* Translated by Alan Sheridan. London: Penguin, 1977.

Foucault, Michel. "Les techniques de soi." In *Dits et écrits: 1954–1988,* 783–813. Paris: Gallimard, 1994.

Frampton, Kenneth. *Modern Architecture: A Critical History.* 3rd ed. New York: Thames and Hudson, 1995.

Frampton, Kenneth. *Studies in Tectonic Culture: The Poetics of Construction in Nineteenth and Twentieth Century Architecture.* Cambridge: MIT Press, 1995.

Friedman, Alice T. "Architecture, Authority and the Female Gaze: Planning and Representation in the Early Modern House." *Assemblage* 19 (1992): 40–61.

Galison, Peter. "Aufbau/Bauhaus: Logical Positivism and Architectural Modernism." *Critical Inquiry* 16 (1990): 709–752.

Galison, Peter. "Constructing Modernism: The Cultural Location of *Aufbau*." In Ronald N. Giere and Alan W. Richardson, eds., *Origins of Logical Empiricism.* Minneapolis: University of Minnesota Press, 1996.

Galison, Peter. *Image and Logic: A Material Culture of Microphysics.* Chicago: University of Chicago Press, 1997.

Galison, Peter, and Bruce Hevly. *Big Science: The Growth of Large-Scale Research.* Stanford, CA: Stanford University Press, 1992.

Giedion, Siegfried. *Mechanization Takes Command.* New York and London: W. W. Norton & Company, 1969.

Giedion, Siegfried. *Space, Time, and Architecture: The Growth of a New Tradition.* Cambridge: Harvard University Press; London: H. Milford, Oxford University Press, 1941.

Gieryn, Thomas. "Biotechnology's Private Parts (and Some Public Ones)." In Smith and Agar, eds., *Making Space,* 281–312.

Gieryn, Thomas. *Cultural Boundaries of Science: Credibility on the Line.* Chicago: University of Chicago Press, forthcoming.

Gooding, David. *Experiment and the Making of Meaning: Human Agency in Scientific Observation and Experiment.* Dordrecht, the Netherlands: Kluwer, 1990.

Gooding, David, Trevor Pinch, and Simon Schaffer, eds. *The Uses of Experiment: Studies in the Natural Sciences.* Cambridge: Cambridge University Press, 1989.

Greenblatt, Stephen. *Renaissance Self-Fashioning: From More to Shakespeare.* Chicago: University of Chicago Press, 1980.

Hannaway, Owen. "Laboratory Design and the Aim of Science: Andreas Libavius versus Tycho Brahe." *Isis* 77 (1986): 585–610.

Hays, K. Michael. *Modernism and the Posthumanist Subject: The Architecture of Hannes Meyer and Ludwig Hilberseimer.* Cambridge: MIT Press, 1992.

Heidegger, Martin. "Art and Space." Translated by Charles H. Seibert. *Man and World* 6 (1973): 3–8.

Heidegger, Martin. "Building, Dwelling, Thinking." In *Poetry, Language, Thought.* New York: Harper and Row, 1971.

Heilbron, John L. "Churches as Scientific Instruments." *Bulletin of the Scientific Instrument Society* 48 (1996): 4–9.

Heilbron, John L. "Science in the Church." *Science in Context* 3 (1989): 9–28.

Heilbron, John L. and R. W. Seidel. *Lawrence and His Laboratory: A History of the Lawrence Berkeley Laboratory.* Berkeley: University of California Press, 1989.

Hermann, Armin et al. *History of CERN.* 2 vols. Amsterdam and New York: North-Holland Physics Pub., 1987 and 1990.

Hughes, Thomas. *American Genesis: A Century of Invention and Technological Enthusiasm, 1870–1970.* New York: Viking, 1989.

Husserl, Edmund. *The Crisis of European Science and Transcendental Phenomenology: An Introduction to Phenomenological Philosophy.* Translated with an introduction by David Carr. Studies in Phenomenology and Existential Philosophy. Evanston, IL: Northwestern University, 1970.

James, Frank A. J. L. *The Development of the Laboratory: Essays on the Place of Experiment in Industrial Civilization.* Basingstoke, England: Macmillan Press Scientific and Medical, 1989.

Jones, Caroline A. *Machine in the Studio.* Chicago: University of Chicago Press, 1997.

Kohler, Robert. *Lords of the Fly: Drosophila Genetics and the Experimental Life.* Chicago: University of Chicago Press, 1994.

Kramer, Cheryce. "A Fool's Paradise: The Psychiatry of Gemueth in a Biedermeier Asylum." Ph.D. diss., University of Chicago, 1997.

Kwinter, Sanford. "La citta nuova." In Michel Feher and Sanford Kwinter, eds. *Zone 1\2: The Contemporary City.* New York: Urzone, 1986.

Latour, Bruno (Jim Johnson). "Mixing Humans and Nonhumans Together: The Sociology of a Door-Closer." *Social Problems* 35 (1988): 298–310.

Lenoir, Timothy, and Yehuda Elkana, eds. *Practice, Context, and the Dialogue Between Theory and Experiment. Science in Context* 2 (1988).

Long, Pamela. "The Contribution of Architectural Writers to a 'Scientific' Outlook in the Fifteenth and Sixteenth Centuries." *Journal of Medieval and Renaissance Studies* 15 (Fall 1985): 265–298.

Lynch, Michael. "Laboratory Space and the Technological Complex: An Investigation of Topical Contextures." *Science in Context* 4 (1991): 51–78.

Mark, Robert, ed. *Architectural Technology up to the Scientific Revolution: The Art and Structure of Large-Scale Buildings.* Cambridge: MIT Press, 1993.

Meyer, Hannes. *Bauen und Gesellschaft: Schriften, Briefe, Projekte.* Dresden, Germany: Verlag der Kunst, 1980.

Meyer, Hannes. *Hannes Meyer 1889–1954, Architekt Urbanist Lehrer.* Berlin: Wilhelm Ernst & Sohn, 1989.

Murphy, Claudette. "Sick Buildings." Ph.D. diss., Harvard University, expected 1998.

Naylor, Gillian. *The Bauhaus Reassessed: Sources and Design Theory.* New York: E.P. Dutton, 1985.

Newman, William R. *Gehennical Fire: The Lives of George Starkey, an American Alchemist in the Scientific Revolution.* Cambridge: Harvard University Press, 1994.

Olesko, Kathryn. *Physics as a Calling: Discipline and Practice in the Konigsberg Seminar for Physics.* Ithaca, NY: Cornell University Press, 1991.

Ophir, Adi, Steven Shapin, and Simon Schaffer, eds. *The Place of Knowledge: The Spatial Setting and its Relation to the Production of Knowledge. Science in Context* 4 (1991).

Pérez-Gómez, Alberto. *Architecture and the Crisis of Modern Science.* Cambridge: MIT Press, 1983.

Pérez-Gómez, Alberto. *Polyphilo, or the Dark Forest Revisited.* Cambridge: MIT Press, 1994.

Picon, Antoine. *Architectes et ingénieures au siècle des lumières.* Paris: Editions Parenthèses, 1988.

Prior, Lindsay. "The Architecture of the Hospital: A Study of Spatial Organization and Medical Knowledge." *British Journal of Sociology* 1 (March 1988): 86–113.

Rosenberg, Charles. *Care of Strangers: The Rise of America's Hospital System.* New York: Basic Books, 1987.

Rowe, Peter G. *Modernity and Housing.* Cambridge: MIT Press, 1993.

Safdie, Moshe with Wendy Kohn. *The City After the Automobile*. New York: Basic Books, 1997.

Schaffer, Simon. "A Manufactory of Ohms." In Robert Bud and Susan E. Cozzens, eds., *Invisible Connections: Instruments, Institutions, and Science,* 23–49. Bellingham, WA: SPIE Optical Engineering Press, 1992.

Shackelford, Jole. "Tycho Brahe, Laboratory Design and the Aim of Science: Reading Plans in Context." *Isis* 84 (1993): 211–230.

Shapin, Steven. "The House of Experiment in Seventeenth-Century England." *Isis* 79 (1988): 373–404.

Shapin, Steven. *A Social History of Truth: Civility and Science in Seventeenth-Century England.* Chicago: University of Chicago Press, 1994.

Shapin, Steven, and Simon Schaffer. *Leviathan and the Air-Pump: Hobbes, Boyle and the Experimental Life.* Princeton, NJ: Princeton University Press, 1985.

Simmel, Georg. "Bridge and Door." Translated by Mark Ritter. *Theory, Culture and Society* 11 (1994): 5–10. Reprinted in Neil Leach, *Rethinking Architecture: A Reader in Cultural Theory.* New York and London: Routledge, 1997.

Sloane, David C. *Mall Medicine: The Evolving Landscape of American Health Care.* Baltimore: Johns Hopkins Press, forthcoming.

Sloane, David C. "Scientific Paragon to Hospital Mall: The Evolving Design of the Hospital, 1900–1990." *Journal of Architectural Education* 48 (November 1994): 82–98.

Smith, Crosbie, and Jon Agar, eds. *Making Space: Territorial Themes in the Shaping of Knowledge.* Basingstoke, England: Macmillan, 1997.

Smith, Crosbie, and M. Norton Wise. *Energy and Empire: A Biographical Study of Lord Kelvin.* Cambridge: Cambridge University Press, 1989.

Spain, Daphne. *Gendered Spaces.* Chapel Hill: University of North Carolina Press, 1992.

Stocking, George. *Objects and Others: Essays on Museums and Material Culture.* History of Anthropology, vol. 3. Madison: University of Wisconsin Press, 1985.

Thompson, Emily. "'Mysteries of the Acoustic': Architectural Acoustics in America, 1800–1932." Ph.D. diss., Princeton University, 1992.

Thompson, Emily. "Dead Rooms and Live Wires: Harvard, Hollywood and the Deconstruction of Architectural Acoustics, 1900–1930." *Isis* 88 (December 1997): 597–626.

Thompson, John D., and Grace Goldin. *The Hospital: A Social and Architectural History.* New Haven, CT: Yale University Press, 1975.

Venturi, Robert. *Complexity and Contradiction in Architecture.* Chicago: Museum of Modern Art and the Graham Foundation for Advanced Studies in the Fine Arts, 1977.

Venturi, Robert. *Iconography and Electronics Upon a Generic Architecture: A View from the Drafting Room.* Cambridge and London: MIT Press, 1996.

Venturi, Robert, Denise Scott Brown, and Steven Izenour. *Learning From Las Vegas.* Cambridge and London: MIT Press, 1972, 1977.

Vidler, Anthony. *The Architectural Uncanny: Essays in the Modern Unhomely.* Cambridge: MIT Press, 1992.

Whitford, Frank. *Bauhaus.* London: Thames and Hudson, 1984.

Wingler, Hans M. *The Bauhaus: Weimar, Dessau, Berlin, Chicago.* Translated by Wolfgang Jabs and Basil Gilbert; edited by Joseph Stein. Cambridge: MIT Press, 1978.

I

Of Secrecy and Openness: Science and Architecture in Early Modern Europe

2

MASCULINE PREROGATIVES: GENDER, SPACE, AND KNOWLEDGE IN THE EARLY MODERN MUSEUM

Paula Findlen

In 1576 the Bolognese naturalist Ulisse Aldrovandi received Countess Catarina Sforza, a descendant of the famous ruling family of Milan, in his museum. The countess and her son, the Marchese di Vassi, had contacted Aldrovandi at his villa outside of Bologna to request a special tour. Ever mindful of the opportunities that important patrons might bring him, Aldrovandi rushed back to the family palace to prepare for their arrival. So noteworthy was the appearance of the Sforza in Bologna that virtually all the city's nobility decided to join them in the visit to Aldrovandi's famous "theater of nature." One day late in May, as Aldrovandi subsequently told his brother, "they came in fourteen or fifteen coaches and carriages that contained fifty Gentlewomen who were the flower of the first [families] of this City, accompanied by more than 150 Gentlemen."[1] Normally Aldrovandi did not entertain hundreds of visitors in his museum, so how and where to put them for the two hours that they surveyed his attempts to rival Aristotle as the naturalist in command of the greatest expanse of nature became crucial to the outcome of the affair. Unlike the *Studio Aldrovandi,* the public institution made out of his 1603 bequest to the Senate of Bologna, the museum in the family palace overlooking Piazza Santo Stefano was fairly small. Guests were scattered throughout the palace while Aldrovandi took small groups into his natural history museum, where he showed them "a few pictures, a few dried animals and other selected things."[2] He even gave the Marchese di Vassi a piece of gold as a souvenir.

Apparently the visit was a great success. Aldrovandi reported with great pride, "They were stupefied and full of amazement at such infinite things observed by me, which the Countess said she would remember forever and for which she would favor me on every occasion." Since it was not every day that Europe's leading naturalist entertained large quantities of the Italian nobility, Aldrovandi encouraged his brother to inform their cousin, Pope Gregory XIII, of "this visit to our Theater by these Most Illustrious [Ones]."[3] Perhaps it would be the catalyst in bringing him the favors and recognition that he felt he deserved.

Precisely because Catarina Sforza's visit to the Aldrovandi museum was so unheralded, it gives us a unique view of the activities that surrounded the practice of collecting nature, allowing us to identify more precisely what Aldrovandi thought the museum was (as well as what it was not). To accommodate the large number of visitors, Aldrovandi expanded his range of activities beyond the setting that he normally identified as the "museum" proper and into the wider domestic sphere of the household. "I tell you that even the courtyard, loggias, rooms, Museum and study were not enough to hold such honored company," he explained to his brother.[4] Normally Aldrovandi and his assistants were the only people to greet visitors to the museum. On this occasion, however, Aldrovandi included his wife, Francesca Fontana, in the activities. She and her sister met the countess and the Bolognese noblewomen at the door to the museum and attended them throughout the visit. The unusual mention of Aldrovandi's wife in the description of the visit reflected where it had taken place as well as who it had included. Fontana and her sister were crucial to the event not only because the presence of noblewomen required their attendance but also because the activities had occurred in rooms other than the museum proper. Spilling out into the wider terrain, this event demanded the presence of both genders in ways that the ordinary visit to a museum did not.

As the above episode suggests, women were not a standard feature of early modern museum life or more generally the life of scholarship. Although Aldrovandi celebrated the arrival of important female patrons in letters, significantly he did not ask them to sign the visitors' books that recorded the presence of noteworthy individuals in his home. These catalogues recorded a public culture of science in which women did not participate.[5] Certainly Aldrovandi's attitude in this regard was not unique; it reflected a common view of the diverse roles for the genders that envisioned scholarship as a pursuit for ideal citizens, who were implicitly male. Such attitudes defined the social relationships that shaped the humanist republic of letters and accordingly the scientific community. Women, wrote Montaigne, were incapable of true friendship, the cornerstone of the humanist definition of scholarly exchange.[6] Primarily for this reason, they appeared infrequently in the Renaissance "books of friends" that inspired visitors' books such as the one in Aldrovandi's museum.[7] These traditions also help explain the excessive ceremonial accompanying the visits of noblewomen to museums. Unlike male visitors, women could not simply enter in their own right. Instead they demanded the constant attentions of the women in Aldrovandi's household, whose presence not only honored their social rank but maintained the decorum of the encounter.

To my knowledge, only two women were ever included in Aldrovandi's official lists of visitors, and both enjoyed the sort of exceptional identity that called their gender into question. Ippolita Paleotti was a kinswomen of Aldrovandi's most important patron, the Archbishop of Bologna, and renowned for her learning in moral philosophy; in fact, she was probably Bologna's most learned woman. Similarly Lavinia Fontana was also a local prodigy: one of the most celebrated female artists in late Renaissance Italy and possibly a relation of Aldrovandi's second wife.[8] The only other woman to play an official role in Aldrovandi's world was the "hairy girl" from Spain whose portrait featured prominently in Aldrovandi's villa and in his publications. However, she—prodigious in her appearance rather than her actions, unlike Paleotti and Fontana—was not asked to sign the book.[9] (See figure 2.1.) The presence of these two signatures among some 1,600 entries serves only to highlight the absence of other female visitors from the museum's official record, particularly since other forms of documentation reveal the occasional presence of women in these rooms. How should we understand the disappearance and reappearance of women from Aldrovandi's museum? Exploring this question sheds light on the paradoxical nature of the scientific enterprise in early modern Europe.

2.1

Lavinia Fontana, *Daughter of Pedro Gonzales* (ca. 1583). The "hairy woman" was one of the many celebrities who visited the museum of Ulisse Aldrovandi in Bologna. But she did not sign the visitors' book, which was an important record of the public culture of museum going. The virtual absence of women's signatures from these books emphasizes the presence of male scholars and patrons and highlights the invisibility of women within this culture. Source: Pierpont Morgan Library, New York.

Although natural philosophers found innumerable ways to restrict and diminish the role of women, they nonetheless conducted their affairs in locations that made it difficult for them to live up to the ideal of a gender-segregated universe. The "private residences of gentlemen" highlighted in Steven Shapin's recent work on experimental culture in seventeenth-century England quite often included the presence of women.[10] As long as scientific activities were conducted in largely domestic settings, natural philosophers could never fully divest themselves from the female culture of the household; at best, they succeeded in establishing fragile yet permeable boundaries, separating the world of scholarship from the life of the household only as long as the natural philosopher maintained absolute control over the location in which he worked.[11] Such irresolvable tensions made ideal expressions of these divisions, in architecture and in prescriptive writings about male and female conduct, even more important. Not surprisingly, as scientific activities became more complex and demanded more elaborate uses of space, most notably the creation of museums that expanded beyond the restricted space of study from which they emerged, the need to define the site in which one worked became more acute. That this occurred during a period in which domestic architecture in general paid greater attention to the specialized uses of rooms suggests the necessity of viewing the physical reconceptualization of scientific activities as part of a broader cultural reformulation of public and private life in the evolution of the household.[12]

Aldrovandi's primary image of the museum conformed to a classically inspired vision of the life of scholarship in which mentors and disciples, friends and associates, congregated in settings designed to facilitate their interaction. He constantly celebrated the "many honorable gentlemen and learned scholars" whom he encountered and subsequently invited to his museum.[13] In doing so, Aldrovandi reflected a common tendency to envision the museum as the culmination of the humanist sodalities that formed the core of intellectual life in sixteenth- and seventeenth-century Europe. By the mid-sixteenth century the museum had emerged as a privileged site of knowledge. In this setting scholars examined texts and objects, exchanged ideas, and reformulated their vision of nature. Museums formed the core of the early modern scientific community.[14] They gave shape to the republic of letters that otherwise existed only on paper. In doing so, they (more often than not) reinforced the image of the scientific community as a predominantly male preserve.

Whereas recent scholarship, most notably David Noble's *A World Without Women,* has argued that this exclusion of women occurred as a result of the clerical and scholastic roots of Western scientific culture, I would like to suggest that other motivations informed the apparent segregation of women from knowledge.[15] A better understanding of the household politics of early modern patrician life and its relationship to civic culture sheds light on the seeming contradictions that emerge from a close inspection of museum activities. More than anything else, the simultaneous presence and absence of women in this particular scientific setting was a product of domestic norms, embedded in Renaissance architectural ideals as well as the social understanding and use of space.

Albertian Fantasies

Our most common image of the museum focuses on its interior. In classic instances, such as the image of Machiavelli's famous study where he shed the garments of everyday life to commune with the ancients, we are presented with an image of solitude. (See figure 2.2.) Like Petrarch, Machiavelli represented the *studium*—the prototype for the early modern museum—as a setting devoid of adornment and filled only with the company of the dead.[16] Elaborating upon a key Christian topos, both humanists implied that speaking to God and speaking to the ancients must be done in solitude. Accordingly, images of the study prior to the sixteenth century lacked the embellishments that we associate with the museum: a few books, certainly, but virtually no artifacts beyond those precious volumes. Equally important, they lacked the visitors so crucial to Aldrovandi's conception of his museum. Undoubtedly Petrarch and Machiavelli permitted others to enter their studies. But it is very difficult to know this with any certainty, precisely because both chose to represent themselves as solitary scholars inhabiting "hermit's cells" in which civil conversation need not intrude.[17]

2.2

Machiavelli's study. Sant' Andrea in Percussina, Florence. This famous, bare room typifies our image of the male humanist study out of which the earliest museums were created. It is a private space for contemplating affairs of state, a true "world in the home." Source: Allinari/Art Resource.

By the end of the sixteenth century, museums were increasingly described and depicted as sociable settings that highlighted the conversational and demonstrative culture of science. Artists such as Lorenzo Lotto had introduced the image of the collector into the visual repertoire of Renaissance elites, in this particular instance a prelate whose study overflowed with his possessions.[18] (See figure 2.3.) Yet who would be the audience for this embarrassment of riches? The 1599 illustration of the apothecary Ferrante Imperato's museum in Naples exemplifies the transformation of the study into the museum. (See figure 2.4.) In contrast to the traditional image of the study as a place of isolation, the depiction of Imperato's museum highlights the presence of visitors. It is a different space altogether, a microcosm of Italian patrician society rather than a refuge from it. In the text of his *Natural History* (1599), Imperato further underscored this effect by enumerating the virtuosi, leading citizens of Naples and Spanish viceroys who all witnessed his collecting of nature.[19] His museum was a product of a community formed by the constant appearance of visitors whose presence, like that of the Sforza countess in Aldrovandi's museum, activated the knowledge imbedded in its objects. In their visual representation, as in their textual depiction in Aldrovandi's catalogue, such visitors were implicitly the male patricians who brought honor and fame to the study of nature.

2.3

Lorenzo Lotto, *Study* (16th century). In one of the earliest images of a private study, Lotto captured the informal domesticity of Renaissance collections as well as their association with the activities of a learned elite. In Rome this group was overwhelmingly composed of clerics like the one depicted here. Source: British Library.

2.4

Museum of Ferrante Imperato (1599). In the first known illustration of a science museum, in late sixteenth-century Naples, the engraver emphasized the importance of viewing natural curiosities as an act of male sociability. Here Imperato's son offers two well-dressed patricians a tour. The dogs in the foreground are the only traces of the domestic context of this museum, which was located in the family palace. Source: Ferrante Imperato, *Dell'historia naturale* (Naples, 1599).

The dramatically changing images of museums reveal a great deal about emerging scholarly conventions and the spatial understanding of knowledge. Yet they are equally interesting for what they do not reveal. None of the images mentioned above indicates one of the most crucial features of the museum: its location within the home. It is all too tempting to discuss settings for scientific activities such as the museum, the academy, or the laboratory without any reference to the fact that they emerged from the domestic activities of the early modern elite. Since in many instances these sites ultimately became autonomous from the home itself, this oversight is all too understandable.[20] Yet we need to look beyond representations of museums, which typically focused on the accumulation of objects rather than the museum's place within the home, to understand precisely what the museum meant for Italian patricians such as Aldrovandi who made it a central feature of their scholarly and domestic lives.

Expanding on the idea of the study as a site in which men prepared to take their place in civil society and stored their most important possessions, the museum defined the juncture between the private life of the humanist, surrounded by objects and secrets, and the theatrical world of courtiers who privileged visibility. In his 1649 museum catalogue, the French collector Pierre Borel defined the museum as a "world in the home."[21] This phrase not only alluded to the image of the museum as a microcosm of all nature but also captured well the idea that nature could be domesticated. But where would nature be placed in the home? For patricians who routinely advised each other on the evils of letting strangers enter one's portals, the transportation of the world into the home involved an important reorientation of their image of the study. There civil conversation would intrude, for the world was defined not only by the possession of objects but also by their display to visitors. Collectors were left to determine how such transgressions of the study should occur and who should participate in them.

Studies in early modern households were invested with many charged meanings that help us understand the difficulties of letting women into the museum. The fifteenth-century Florentine humanist and architect Leon Battista Alberti offered the normative ideal of the study in Renaissance Italy when he proclaimed: "The husband and wife must have separate bedrooms. . . . Off the wife's bedroom should be a dressing room, and off the husband's a library (*libraria cella*)."[22] (See figure 2.5.) Spouses would meet through a secret door linking their bedrooms for conjugal visits, but Alberti otherwise envisioned their leading separate but parallel lives in which the study and the dressing room served gender-specific functions. Such a division not only preserved the chastity and modesty of a household's female members but also allowed men to carve out a space reserved exclusively for their own activities. For this reason, Mark Wigley aptly characterizes Alberti's study as the "first truly private space . . . an intellectual space beyond that of sexuality."[23] Although a handful of learned women in Alberti's time, such as Veronese humanist Isotta Nogarola, also enjoyed the pleasures of the study, Alberti and his contemporaries viewed it primarily as a setting for masculine pursuits.[24] Subsequently the museum, as an

2.5

Alberti's division of the home. This reconstruction of Alberti's description of an ideal study places it squarely in the male space of an Italian Renaissance palace. Alberti's conception became the prototype for our image of the space of scholarship, both in and out of the home. Source: Visual Arts Services, Stanford University.

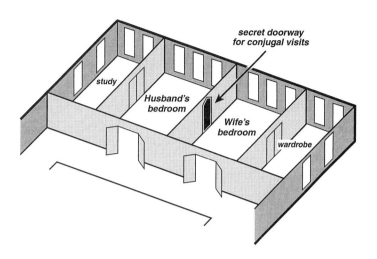

extension of the study, was invested with similar meaning. The locked chests that Alberti filled with papers became the numbered cabinets Aldrovandi stuffed with collectibles.

Controlling space became a primary means of asserting patriarchal authority and upholding virtue. In his *Books on the Family* (ca. 1434–37) Alberti further distinguished the study as a masculine world within the home into which women should not enter. Book three, for example, takes the reader on an imaginary tour of an ideal patrician household with one of Alberti's protagonists, Giannozzo, and his newly acquired wife. Giannozzo willingly shows his spouse everything in the palace, including his "household treasures," but halts his tour at the door of the study. "I never gave my wife permission to enter that place, with me or alone," he declared.

> I also ordered her, if she ever came across any writing of mine, to give it over to my keeping at once. To take away any taste she might have for looking at my notes or prying into my private affairs, I often used to express my disapproval of bold and forward females who try too hard to know about things outside the house and about the concerns of their husband and of men in general.[25]

Such statements, as Stephanie Jed has perceptively noted, reflect the overwhelming concerns about privacy among a merchant elite desirous of maintaining distinctions between civic and domestic affairs.[26] The study, containing materials reflective of "things outside the home," could not pertain to women, whom Alberti cloistered in his ideal palace. It was the final refuge not from the cares of public life but from female company.

Perhaps the most important thing to note about this description is not the vigilance with which Alberti guarded the study but how small a portion of the household it really occupied. Giannozzo's wife is shown the house as a prelude to being invested with the keys, a tangible sign of her assumption of the position of female head of household. Once in possession of those keys, she commanded the household in the sense that no facet of it, save for the study, would remain closed to her.²⁷ For these reasons, Alberti's seemingly authoritarian statements about the power of men in the home appear reactive rather than proscriptive. The study is invested with heightened meaning as if to deny its potential inadequacies as a center from which to regulate domestic life. Instead it is a perfectly appropriate setting in which to engage in other activities, among them the contemplation of nature. Yet these two occupations cannot be fully separated, and the secrecy of the Albertian study only increased their significance. The humanist in the study concerned himself with domestic conduct because it formed the bedrock of any public presentation of virtuous action. Yet he also applied these lessons to his intellectual life, since knowledge, let alone wisdom, could not be achieved without a proper ethic. Nature, placed in a centrally charged location within the home, needed a virtuous master. The peculiar physical confinement of scholarly life served to heighten its attractions since its power was clearly derived from a source other than the home itself.

The influence of Alberti's utopian architectural plans in the portrayal of scholarly life is borne out by the repetition of his advice in later guidebooks and by its realization in the placement of actual museums. Because sixteenth-century architects drew upon many of the same classical sources—Vitruvius's *Ten Books on Architecture* and Xenophon's *Oeconomicus*—that inspired Alberti, they shared a common framework of understanding. The growing number of architectural treatises, for example, Andrea Palladio's *Four Books on Architecture* (1570), that established canonical images of domestic architecture further elaborated on this sense of shared cultural values. These works continued to reinforce the norms Alberti articulated. As late as the 1570s, Benedetto Cotrugli advised his readers to place the study "in the most remote corner of the house" and "near the bedroom."²⁸

At approximately the same time, the prelate Ludovico Beccadelli bequeathed his collection and the room that contained it to his faithful secretary Antonio Giganti. Giganti subsequently recorded its contents and drew a plan of the *studio* in the Beccadelli family palace in Bologna, across the piazza from Aldrovandi's own museum. From this plan, we know that the only entrance to Giganti's *studio* was the "door that enters into the bedroom."²⁹ (See figure 2.6.) In this tiny room, Giganti sat beneath portraits of famous men to contemplate the works of nature. Like Aldrovandi's museum nearby, Giganti's *studio* was probably on the second floor of the Beccadelli palace at the top of a stairway that ascended directly from the ground floor into his quarters, conforming to architectural convention about the proper setting for the inspection of nature.

As museums expanded beyond the confines of a solitary room, Alberti's image of the study adjoining the bedroom had to be modified to accommodate a wider array of objects and to reflect new concerns about the differentiation of space. Collectors became increasingly aware of the gradations of secrecy and openness that their arrangements could

2.6

Studio of Antonio Giganti (1586). This image of a late-sixteenth-century museum in Bologna suggests the strong connections between architectural theory and practice. Alberti's recommendation that the study be attached to the male bedroom is realized in this plan, in which the door to the far right is labeled "door that enters into the bedroom." Source: Biblioteca Ambrosiana, Milan, ms.S.85 sup., f.235r.

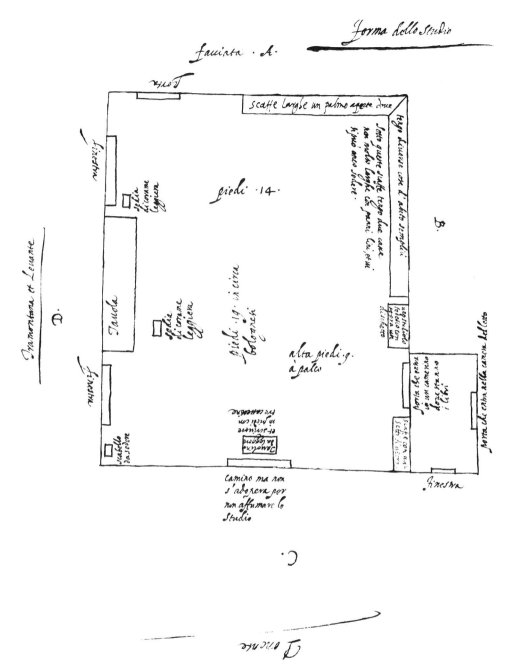

reflect, signifying the different levels of privilege accorded to a visitor. They were also sensitive to the different operations of the museum, where reading, observing, and recording nature all took place. Aldrovandi's museum occupied more than one room by the time of his death. A 1610 inventory described a vestibule, framed by "some whale's bones and other marine monsters," and an antechamber, overlooking the garden, through which one entered a "dark room" (to the right) and the "Museum" (to the left); the museum opened out into the two rooms containing the library, all of which surrounded a courtyard.[30] The complex description of the chests scattered throughout these rooms, containing Aldrovandi's most precious patrimony, precludes any doubt that they were an offshoot of Alberti's locked boxes, filled with curiously public "secrets" rather than the family treasures and histories hoarded in the more traditional study.

This multitiered image of the museum found its parallel in literary descriptions by architects of the period such as Palladio's student, Vincenzo Scamozzi. In his *Idea of Universal Architecture* (1615), Scamozzi placed the museum at the top of a set of "secret stairs" that led to the "consulting rooms" on the second floor. He divided them into three types: waiting rooms, reception rooms, and studies. The last were placed farthest away from the stairs, "and these rooms must be impenetrable because there within one deals with the substance, faculties, honor and life of men." Certainly this was how Aldrovandi envisioned his museum and his visitor's books, which recorded only honorable encounters. Elsewhere in the same chapter, Scamozzi reaffirmed the gender divisions of the household that placed the study firmly within the male domain when he observed: "Men's apartments are very separate from women's."[31] More than a century after Alberti sketched out his conception of the place of scholarship in domestic life, humanist collectors generally adhered to the rules that he laid out in his *On the Art of Building*.

BETWEEN THE MUSEUM AND THE VILLA

Aldrovandi was perhaps the most self-conscious disciple of Alberti, or more generally Albertian precepts, within the community of collectors. He typified the amateur architect, as described by Richard Goldthwaite, who read Alberti's *On the Art of Building* and subsequently applied Alberti's ideals to his practical work.[32] Early in his career as Italy's foremost naturalist he exhibited a highly refined understanding of the uses and divisions of space that were common currency among Renaissance architects. In fact, Aldrovandi viewed himself as a sort of architect through his choice of profession, medicine. The Roman physician Galen frequently compared the authority of physicians over all other medical practitioners to that of "architects over bricklayers," a phrase that Aldrovandi often invoked in his quest for authority in the study of nature.[33] The metaphorical image of the physician-naturalist as architect harmonized well with Renaissance definitions of this profession. As Alberti wrote, the architect was someone with a "well-informed and judicious mind" whose high social standing and faculty of reason made it possible for him to oversee complex projects.³⁴ Like the naturalist, he was capable of discernment. Like the physician, he enjoyed a high moral authority.

Aldrovandi's library catalogues revealed a well-developed interest in architecture. His immense library of more than 3,500 volumes included an entire section on *architectura* in which the works of Alberti, Vitruvius, Palladio, Sebastiano Serlio, and Antonio Vignola figured prominently.³⁵ Aldrovandi's interaction with these books was reflected in the careful distinctions that he drew among different sectors of his family palace in Bologna on occasions such as the Sforza visit and in the evolution of his supreme architectural fantasy: the villa he constructed in the late 1560s outside of Porta San Vitale, approximately one mile beyond the city walls of Bologna.³⁶ It was also in keeping with the image of the learned gentleman who took an amateur interest in the art of building as part of the construction of his country house and his general pursuit of humanistic subjects.³⁷ In this Aldrovandi followed the lead of Italian princes, such as the Medici, who patronized his natural history. As Richard Goldthwaite observes, "It is one of the notable features of Italian Renaissance culture that architecture began to arouse the interest of the wealthy and the powerful—popes, princes, and, eventually . . . local patriciates everywhere."³⁸

Alberti had recommended that "leading citizens" would benefit from the construction of houses away from the turmoils of the city, designed to put them in closer touch with nature. By the sixteenth century, the *villeggiatura* had become a common feature of Italian patrician life.³⁹ Aldrovandi envisioned his villa as a complement to and logical extension of his museum; he invited close associates to commune with nature in this setting, frequently transported books and artifacts between these two locations, and wrote many portions of his natural history there.⁴⁰ In time, the villa became virtually indistinguishable from the museum save for the important fact that Aldrovandi severely restricted the circle of associates who enjoyed the pleasure of discussing nature in that location, favoring friends such as the archbishop Paleotti over courtiers and traveling scholars. The villa, in essence, created an inner zone of privacy in ways that neither his study nor his museum in the family palace could accommodate as his fame and the size of his collection grew; just as it was the ideal city, as Reinhard Bentmann and Michael Müller have observed, it was also the ideal museum.⁴¹ There Aldrovandi truly studied nature, gathered not in the villa but in the countryside that surrounded it, and even months before his death worked feverishly on the revisions to his unpublished writings.⁴²

Once again reading provided a source of inspiration for the development of Aldrovandi's image as a villa owner. Books such as Agostino Gallo's *Ten Days of True Agriculture and the Pleasures of the Villa* and Alberto Lollio's *Letter . . . in which . . . He Celebrates the Villa*—the quintessential villa reading of the period—filled his library shelves.⁴³ As a good humanist, Aldrovandi never engaged in any activity based on experience alone; in his role as an amateur architect as in his studies of nature, reading provided an essential point of departure. How the humanists conceived of the villa and the museum mattered very much to Aldrovandi. Through his interactions with such texts as Alberti's *On the*

Art of Building he transformed early Renaissance architectural principles to fit the cultural aspirations of a late sixteenth-century naturalist.

In his description of an ideal villa, Alberti had recommended that the owners decorate it with pleasurable images; contrasting it to the sobriety of a city dwelling, he remarked, "in a villa the allures of license and delight are allowed."[44] Following Alberti's prescriptions, in the 1580s Aldrovandi embarked on a massive decoration of his villa. He commissioned several artists, most of whom were already employed as natural history illustrators in his museum, to complete a cycle celebrating the life of Ulysses in the great hall and to fill three smaller rooms with emblems that transformed the best and most dramatic objects in his museum into moral allegories about the life of a humanist encyclopedist.[45] This again conformed almost exactly to Alberti's suggestions that villas be decorated with heroic deeds of famous men and scenes from "the tales that poets make for moral instruction."[46] It also fulfilled more general recommendations by sixteenth-century connoisseurs of villa life who recommended that villa inhabitants exercise their minds in the sort of pleasurable recreations that the contemplation of allegories and emblems afforded.

As part of the embellishment of his villa, Aldrovandi had a series of *sententiae* placed above the doors to various rooms in the house. The content of these inscriptions allows us to understand more fully how Aldrovandi envisioned the different uses of a house's rooms. As Lina Bolzoni writes, "The inscriptions exalted family life and the joys of conjugal love as a central theme."[47] They provide an imaginative guide to Aldrovandi's own ideas regarding humanist conversation and domestic ideology. Many of the sayings on the walls of the great hall, for example, exhorted visitors to embrace the concept of humanist sodality that Aldrovandi practiced in his museum and in his villa. They also testified to his absorption of principles laid out not only in Alberti's ethical writings but also in conduct books such as Baldassare Castiglione's *Book of the Courtier* and Stefano Guazzo's *Civil Conversation,* both of which he owned. "It is fitting that worthy men collect the common fruit of the prosperity of civil consort," announced the inscription above the first door. Another praised the "company and frequency of like-minded [men]" as a source of good, and one door opening out to a walkway highlighted the close and lasting associations that came from the "harmony of souls and the search for honest pleasure."[48] Clearly visitors were expected to perceive the great hall as a public space within the house in which conversations about nature, among many other things, might occur.

Aldrovandi spent time at the villa in the company of his wife, Francesca Fontana, as well as in communion with scholars and patrons. We know that Fontana accompanied him to the villa because, in the one letter that survives in her hand in his manuscripts, she apologizes for not responding immediately to the wife of another naturalist because "I have been at the Villa for many days, far from the City."[49] Fontana's dowry had financed the villa's construction after her marriage to Aldrovandi in October 1565; she as much as Aldrovandi played a part in its evolution and in essence was the primary owner of this site of knowledge. In decorating the villa, Aldrovandi not only filled its walls and entryways with messages celebrating the virtues of civic life and humanist scholarship; he also embellished one room and the attached dressing room (*salvaroba*), undoubtedly Fontana's quarters, with inscriptions addressed to his wife. All the inscriptions evoked classical virtues for women reminiscent of Alberti's own prescriptions. "For me the dowry is not that which is so-called but modesty and decency."[50] Surely this was a particularly revealing image given Aldrovandi's calculated use of his wife's own dowry to finance the villa? Yet another celebrated the pleasures of virtuous love. All in various ways underscored the ideal gender divisions of villa life, in which men and women led largely separate lives.[51]

Two inscriptions are particularly noteworthy in light of Alberti's strong statements about the need to design a villa that segregated women from men in their daily affairs. "It is not fitting that virgins mix with the crowd" (*virgines in turbam progredi non est honestum*), exhorted one inscription, as if to remind Fontana that the "civil consort" of the great hall was not her domain.[52] Over the entrance to the dressing room, the setting Alberti had singled out as belonging specifically to the wife just as the study belonged to the husband, Aldrovandi placed his most decisive pronouncement on the different duties of men and women: "It is proper that women be clever not in civic but in domestic [affairs]" (*mulierem non in civilibus sed in domesticis oportet ingeniosam esse*).[53]

Such definitive moral advice recalls many of the pronouncements made by Alberti on the restrictions he hoped to place upon women within the home. "[A]ny place reserved for women ought to be treated as though dedicated to religion and chastity," he declared in his *On the Art of Building*.[54] Surely this advice must have gone over well with a naturalist who declared in his autobiography that his future spouse ought to be a chaste virgin?[55] More strikingly the inscriptions echoed Alberti's comments in his *Books on the Family* regarding those "bold and forward females" who inquired into knowledge of their husband's affairs. After praising his own wife's modesty, Alberti's protagonist Giannozzo told his listeners that he refused to speak with her about subjects outside her purview: "Furthermore, I made it a rule never to speak with her of anything but household matters or questions of conduct, or of the children."[56] Like Alberti, Aldrovandi attempted to inscribe limits on the presence of women in public affairs—in which natural history, as he conceived it, played an important role—by literally writing his code of conduct on the walls of his villa.

In light of the inscriptions, the relative absence of women in Aldrovandi's catalogue of visitors becomes further comprehensible. They belonged to such settings as the loggia, the courtyard, and the rooms of his palace, in which the female members of his household figured prominently; their temporary presence in the museum during the 1576 visit had done nothing to alter this, and it was therefore unthinkable that they should be asked to leave a record of their visit behind. Aldrovandi's precision about this matter, both in the construction of the visitors' books and in the decoration of his villa, indicates how deeply early modern patricians felt these social conventions. As Bartolomeo Taegio advised in his *Villa* (1559), "The woman who reads puts herself in too great danger. . . . I conclude that the *otio* of letters is for men, not women, the duty of whom is to learn to manage the family well and not to read."[57] Literacy, a basic precondition for any participation in the humanist study of nature, was placed in a specific location in the house, where women were loath to be present.

In its official image, the museum was a setting in which women did not enter; accordingly they did not publicly participate in any of its activities, including the production of scientific knowledge. Yet as descriptions of early modern domestic and scientific life suggest, this demarcation was ambiguous if not precarious.[58] Whereas architectural convention served to clarify the household's gender divisions, social practice did not. Even Alberti alluded to the difficulties of segregating men and women when he remarked that the female head of household "should be accommodated most effectively where she could monitor what everyone in the house was doing."[59] Women played important and powerful roles in the early modern domestic economy; idealized distinctions between civic and domestic life notwithstanding, it was genuinely hard to separate the two in any decisive way. If a scholar conducted a great deal of his "public" life in the household, then how were such boundaries to be effectively maintained? Instead they quite often broke down in practice.

Francesca Fontana provided more than just the resources to finance her husband's pastoral enclave. She was a literate and well-educated woman whose talents contributed significantly to her husband's own research. *Uxor dilectissima,* Aldrovandi called his wife in the 1595 will in which he gave her use of their palace and villa during her lifetime and possession of all the furniture in both homes.[60] Possibly Fontana's learning had led him to acquire such works as Luigi Dardano's *Excellent and Learned Defense of Women,* which argued the virtues of this gender.[61] On several occasions, Aldrovandi recalled with pleasure his wife's willingness to help him complete the natural history that would best any work of the ancients. At the end of a five-volume *Lexicon of Inanimate Things,* for example, he thanked his wife for putting it together: "Francesca Fontana, wife of the Most Excellent Ulisse Aldrovandi, bound in these books the observations of inanimate things. I undertake nothing vile and abject out of which profit is not reaped at some time."[62] From this entry we learn that his wife participated in some of the daily tasks of accruing knowledge that defined his work as a naturalist.

Fontana apparently played a similar role in the formation of Aldrovandi's greatest work of reference: the *Pandechion Epistemonicon,* which offered its readers passages from ancient and modern authorities on every imaginable subject pertaining to natural history. Participation in this project required an even higher level of expertise, because it involved reading and selecting passages from his voluminous library on an endless variety of topics. In his autobiography, Aldrovandi depicted his wife as the very model of a learned female humanist: "she was young with beauty and incomparable wisdom, skilled with her talent at every discipline and art, of which he availed himself sometimes in the additions to his *Epistemonicon* and other literary matters (*pratiche delle lettere*)."[63] Thus despite his Albertian proclivities, Aldrovandi violated one of the key precepts of this philosophy when he acknowledged that he had allowed his wife to participate in those "literary matters" that theoretically belonged to the world of men. As part of her domestic activities, she had entered the museum.

How far exactly did these responsibilities extend? Upon Aldrovandi's death in 1605, Fontana emerged briefly as an important figure in the editing and publication of his manuscripts and in the transferral of his museum to the *Palazzo Pubblico*, following his 1603 bequest to the Senate of Bologna. She and not Johann Cornelius Uterwer, Aldrovandi's successor to his professorship in natural history, wrote the Latin preface to his first posthumous publication, *On the Remains of Bloodless Animals* (1606). Dedicating the work to the Senate of Bologna, she exhorted them to fulfill the promises made to "your citizen, my husband" (*civem vestrum, meum coniugem*) who had brought so much fame to the city.[64] Beyond this, Fontana delayed the release of her husband's museum to the Senate for over a decade; not until the printer who bought the rights to Aldrovandi's publications had paid her an annual stipend did she cede the museum to the Senate.[65] Thus for the ten years after her husband's death, Fontana controlled his famous theater of nature as part of a widow's prerogative to settle the affairs of her husband in accordance with his wishes. In doing so, she fulfilled some of the worst fears of male humanists such as Francesco Barbaro who, in his *On Wifely Duties* (1416), expressed concerns about the actions of rich women and widows whose autonomy might lead them to disregard men's wishes. Ironically this was yet another book that found its place in the Aldrovandi library.[66]

Few if any visitors came to the museum during that period. In Aldrovandi's absence and prior to its installment in the Palazzo Pubblico in 1617, it was no longer a space of knowledge. Implicitly it had been reabsorbed, albeit temporarily, into the domestic space of the household, a world ruled by women rather than men. Only in that context did Fontana feel it appropriate to add her voice to her husband's publication (which was, in essence, their shared work) through the composition of a Latin preface. For what Alberti did not say but implied regarded the status of those secrets and papers locked in the study after the death of their keeper: At that moment, those possessions might potentially fall into the hands of a woman, as they did in the case of Aldrovandi, who left behind no direct male heirs.[67]

Worlds of Women

As the case of Aldrovandi suggests, early modern naturalists who created new forms of knowledge through the transformation of preexisting space were often highly self-conscious about the uses to which their museums could be put. Nurtured in a humanist culture that made the investigation of nature a supremely civic act, they drew upon preexisting conventions about the study as a domain for the more private contemplation of public affairs to integrate the museum into civic consciousness.[68] At the same time, they also created new forms of privacy to remove themselves occasionally from the public culture of museum going that they had invented. As Stephanie Jed notes in her study of fifteenth-century humanists, the "anxiety about maintaining a conceptual distance between public and private space" was quite real for these individuals.[69] Architecture and other official representations of their activities provided a means by which such distinctions could be refined and maintained.

Prescriptive discussions of the uses of space connected new activities such as collecting to preexisting discourses about domestic ideology in order to integrate them fully into the activities of the urban elite. In this context, gender played an important role. As Renaissance moralists observed frequently, the inappropriate mixing of the genders, either in household arrangements or in conversation, threatened to disrupt the conception of order essential to the harmonious functioning of society. In his *Civil Conversation* (1574), for example, Aldrovandi's contemporary Stefano Guazzo reminded his readers that conversation with women was often "vain and unprofitable." To bolster his claims, he cited examples of women—surely he must have had the courtesans in mind?—whose conversation would lead male scholars out of the study in pursuit of pleasure.[70] Only a few decades later, Federico Cesi, founder of the Lincean Academy, which numbered Galileo among its members, disparaged the institution of marriage precisely because it distracted true philosophers of nature from their calling. "In fact, those who busy themselves with a wife, family and house of their own, are not able to dedicate themselves regularly to the reading and composition of books," he affirmed.[71] Implicit in both statements was the image of the study (Cesi dubbed his the *musaeum*) as an Albertian fortress, filled with closed boxes containing objects and papers, that divided the civic world of men from the domestic world of women. Men who ventured out of their own realm would cease to use the skills—reading, writing, and more generally thinking—that defined their capacity to reason.[72]

Despite the numerous efforts to create clear and discernible divisions between these two realms, the early modern scientific world was nonetheless a world filled with women. Even in the wake of studies such as Londa Schiebinger's *The Mind Has No Sex* and Erica Harth's *Cartesian Women,* we know very little about these women precisely because the printed documents from this period recorded such a small portion of their activities.[73] Yet manuscript sources are rich with portraits like the one I have sketched for Aldrovandi's wife, precisely because the written record was both a civic and a domestic entity (and as such recorded both facets of people's lives). Controlling the household frequently entailed managing the study or entering the museum as well. Since scientific activities did not yet have any precise place in the public consciousness—for despite Aldrovandi's best efforts to convince his patrons otherwise, it was not yet clear what the utility of knowing nature might be—they resided somewhere between the domain of politics and the realm of personal knowledge. This ambiguity provided a certain amount of room for educated women such as Francesca Fontana to play an important role.

The architectural fantasies of late Renaissance naturalists offered a means of articulating the shape that the new scientific culture would take. Indeed it is noteworthy that not a single depiction of a science museum prior to the eighteenth century included women, despite the continued visits of noblewomen such as Catarina Sforza (or later Christina of Sweden) to such settings. The decision to place the museum in a particular location within the house did indeed have ramifications for its evolution as a place of knowledge; as naturalists such as Aldrovandi understood well, what really counted was not what one did but how it was publicly represented. That such decisions confined, quite physically, female participants in scientific culture is undeniable. We see this sense of confinement acutely in moments when women deliberately crossed these implicit thresholds in their words and their actions.

I would like to close with two episodes in particular. The first concerns the noblewoman Margherita Sarrocchi who from the 1580s until her death in 1617 conducted a lively scientific salon in Rome attended by noteworthy figures such as the Duke of Mantua, Torquato Tasso, Galileo, and other members of the Accademia dei Lincei. As a ward of the Vatican librarian Guglielmo Sirleto, Sarrocchi participated in a powerful intellectual network that spread word of her aptitude for learning well beyond Rome.[74] By 1609 Sarrocchi enjoyed the tutelage and close companionship of the mathematician and Lincean Luca Valerio and increasingly turned her interests toward mathematics and natural philosophy; admiring friends described her home as an "Academy of the leading virtuosi in Rome."[75] Yet as the increasingly vitriolic criticisms of her activities by members of the Accademia dei Lincei suggest, the scope of Sarrocchi's intellectual activities was fairly restricted. Her attempts to gain intellectual autonomy threatened deeply rooted conventions about the place of learned women in Renaissance culture.

The death of her husband Carlo Birago in 1613 offered Sarrocchi new possibilities. As her friend Valerio noted, "Signora Margherita Sarrocchi, as before, will have more free room to philosophize, having become a widow."[76] Perhaps the *libertas philosophandi* that Valerio observed in Sarrocchi, not unlike that of Francesca Fontana in the wake of Aldrovandi's death, included possession of the space of study that had been tacitly reserved for men alone, even in a home that boasted a thriving scientific salon. Noticeably Sarrocchi's reputation plummeted after this event. The "freedoms" that came with philosophizing—among them, her association with Valerio as a widow—cast doubt upon her virtue. Such criticisms stand in marked contrast to the honors accorded Fontana in Bologna who, never having claimed the title of philosopher herself, presented an appropriate demeanor in all of her interactions with the Senate of Bologna. Fontana managed but did not claim the study, while Sarrocchi hoped to be a philosopher as well as manage an academy.

The image of Sarrocchi clearing out "room to philosophize" in her Roman palace provides an interesting point of comparison for yet another image of women attempting to claim some portion of scientific knowledge for themselves. In 1600 a work entitled *Women's Worth* appeared in Venice under the pseudonym of Moderata Fonte. It had been written by the Venetian noblewoman Modesta Pozzo in the early 1590s. Fonte recorded the imaginary exchange of seven noblewomen who came together for two days in a garden to enjoy friendship and "domestic conversation."[77] There, "impeded in no way by the presence of men, they conversed among themselves about those things that most intrigued them."[78] In contrast to Sarrocchi's bold attempt to claim the space for philosophizing primarily reserved for men, the imaginary women of Fonte's dialogue do not encroach upon masculine space within the household. Instead they disparage "virile conversation," preferring the company of women.[79] By not crossing the threshold of the museum, they successfully claim much of the knowledge produced by men for use in their conversations, including a very lengthy discussion of contemporary questions in natural history. Men provide knowledge, they remark wittily, but women make use of it.

Fonte's decision to place her female academy in the garden rather than in the museum was a deliberate act that subverted the close association between knowledge and place that was a distinguishing feature of early modern scientific culture. Natural history could just as well be discussed by women in the garden as by men in the museum, she argued; the location, after all, did not determine the discourse.[80] Despite the strength of her arguments, I find it rather telling that Fonte could not imagine a group of literate noblewomen discussing natural history in the museum. The subject had been moved, but the site of its production was largely left intact. Women could ultimately converse about nature but they, even in Fonte's passionate defense of their right to speak, could not produce knowledge.

"A world without women did not simply emerge," writes David Noble, "it was constructed."[81] The museum culture of the sixteenth and seventeenth centuries contributed in very interesting ways to this process. Even after the transferal of Aldrovandi's museum from his family palace to the Palazzo Pubblico in Bologna, where it resided until 1742, it continued to be a setting into which women rarely entered, reserved instead for the male virtuosi ("doctors and other foreigners") who had appropriate letters of introduction to the custodian or members of the local political and scientific elite who had access to the *Studio Aldrovandi*. (See figure 2.7.) Unlike their predecessors, they now had the pleasure of viewing the famous portrait of the "hairy woman" (figure 2.1), because it had been moved from the villa and united with the other objects in Aldrovandi's museum.[82] In its domestic setting, the Aldrovandi museum occupied a specifically male sector of the home; its removal from the home in the seventeenth century served to further clarify its gendered connotations by placing it in the most public and political building in Bologna. In 1732 the noblewomen who attended the ceremonies surrounding Laura Bassi's degree, the first awarded to a woman at the University of Bologna, may have attended

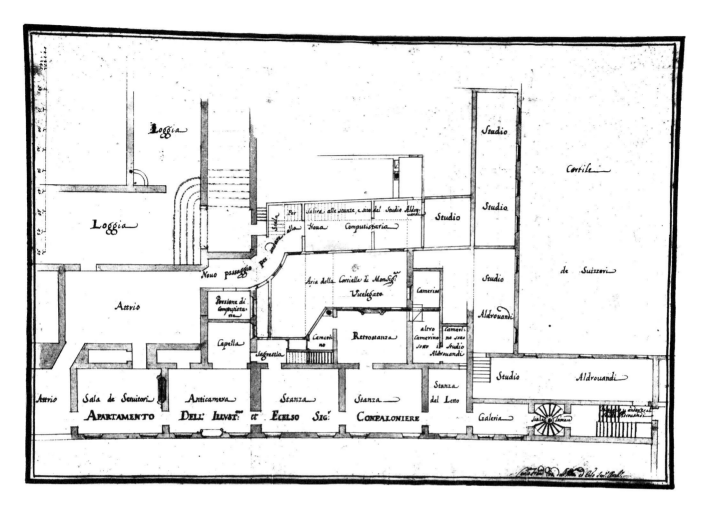

2.7

Studio Aldrovandi (early 18th century). This early plan of a public museum helps us visualize the museum's transition from its domestic origins to the sort of civic buildings that we now associate with museum culture. Located in the *Palazzo Pubblico* in Bologna, the *Studio Aldrovandi* was contiguous to the quarters of the *gonfaloniere* and the papal legate, the two heads of government in the second city of the papal state. Even in its nascent public form, the early modern museum continued to realize the close connection between male authority and intellectual life. Source: ASB, *Assunteria di Munizione, Recapiti, 32, Palazzo Pubblico ed annessi,* vol. I, b.2, 1698–1700, n.8.

49 | MASCULINE PREROGATIVES

her in the rooms reserved for Aldrovandi's museum.[83] By then, however, it was no longer a place of knowledge, as it had been throughout the sixteenth and seventeenth centuries, but a historical curiosity.

I would like to suggest that we take the metaphor of construction, at least in these episodes, quite literally. Throughout the Renaissance and well into the Baroque period, naturalists created museums that in their location as well as in their design realized the humanist dreams of sodality that linked scientific pursuits with civic culture. Their configuration of sites of knowledge in domestic life established important preconditions for the public understanding of scientific space, as museum and laboratories emerged from the homes of aristocrats and gentlemen to enjoy a new autonomy. Such institutions, even when divested of their former location, continued to incorporate a host of assumptions about the appropriateness of women in sites of knowledge. Certainly they reflected emerging conventions about work within this new scientific culture, but they were also indebted to the long tradition of domestic ideology and social behavior outlined above. Perhaps this was one of the reasons why, when Alberti's contemporary Christine de Pisan envisioned a "city of ladies," she had her protagonists build it, brick by brick. She was certainly rebelling against a culture of the study that allowed little scope for her own activities, creating a counteredifice in which women occupied the central place.

Notes

Thanks to Cristelle Baskins, Lina Bolzoni, Alice Friedman, Deborah Harkness, Pamela Long, Giuseppe Olmi, Nancy Siraisi, and audiences at Harvard, CUNY, Cornell, and Stanford for help with this essay.

1. Biblioteca Universitaria, Bologna (hereafter BUB), *Aldrovandi,* ms. 35, c. 204r (Aldrovandi to Teseo Aldrovandi, Bologna, 29 May 1576).
2. Ibid.
3. Ibid.
4. Ibid.
5. BUB, *Aldrovandi,* mss. 41 and 110. For a more detailed discussion of these catalogues, see Findlen, *Possessing Nature,* pp. 136–146.
6. Aymard, "Friends and Neighbors," p. 454.
7. The "books of friends" are a relatively unexplored subject, but see Klose, *Corpus Alborum Amicorum.*
8. BUB, *Aldrovandi,* ms. 110 (Ippolita Paleotti appears as a studiosa); ms. 136, XXIV, cc.21v–35v (Lavinia Fontana along with her father Prospero appears in this list of additions to Aldrovandi's catalogue). The latter is discussed briefly in Olmi and Prodi, "Gabriele Paleotti, Ulisse Aldrovandi e la cultura a Bologna," p. 224; for general background on Fontana, see Cantaro, *Lavinia Fontana, bolognese.* On Ippolita Paleotti, see Masetti Zannini, *Motivi storici della educazione femminile,* p. 38.

9. For more on the *puella hirsuta,* see Fanti, "La villeggiatura di Ulisse Aldrovandi," p. 36; and Bolzoni, "Parole e immagini," pp. 346–347. Around 1583, Lavinia Fontana also sketched a portrait of this same girl that she entitled *Portrait of the Daughter of Pedro Gonzales;* it is reproduced in Kenseth, *The Age of the Marvelous,* p. 333. Possibly Fontana made this sketch in Aldrovandi's museum as well as the portrait for the villa? The connection seems like more than a coincidence.
10. Shapin, "The House of Experiment," p. 378; idem, *A Social History of Truth.* Shapin's work is an important point of departure for this sort of work, though he does not focus on the gendered dimensions of space per se. Since the initial writing of this essay, Deborah Harkness has completed a very interesting study of John Dee's household; see her "Managing an Experimental Household."
11. Owen Hannaway also comments on the permeability of the barriers erected to demarcate public and private life in early modern household laboratories; see his "Laboratory Design," esp. pp. 600–601.
12. On this subject, see particularly Girouard, *Life in the English Country House,* esp. pp. 164–180; and Friedman, *House and Household,* pp. 146–147. The most accessible English language works on domestic architecture in Italy focus on slightly different periods but are nonetheless useful for general background: Goldthwaite, *The Building of Renaissance Florence;* and Waddy, *Seventeenth-Century Roman Palaces.*
13. Raimondi, "Le lettere di P. A. Mattioli ad Ulisse Aldrovandi," p. 38 (Prague, 29 January 1558).
14. For a more detailed discussion of the role of museums in early modern scientific culture, see Findlen, *Possessing Nature.*
15. Noble, *A World Without Women.* As Daphne Spain writes, "Women and men are spatially segregated in ways that reduce women's access to knowledge and thereby reinforce women's lower status relative to men's." (*Gendered Spaces,* p. 3.)
16. On Machiavelli's *scrittoio,* see de Grazia, *Machiavelli in Hell,* pp. 24–27, 370–372.
17. Della Casa, *Galateo,* p. 4.
18. Wolfgang Liebenwein argues that this is the first image of a collector in Western Europe; see his *Studiolo,* p. 108.
19. Imperato, *Dell'historia naturale,* preface.
20. Shapin comments on this transformation in "The House of Experiment," p. 404. Similarly Owen Hannaway also urges historians of science to consider early modern laboratories as specialized domestic sites: "Laboratory Design," p. 586.
21. Borel, *Catalogue des choses rares de Maistre Pierre Borel,* p. 132.
22. Alberti, *On the Art of Building in Ten Books,* p. 149. Alberti's study is also discussed in Liebenwein, *Studiolo,* p. 52; Ranum, "The Refuges of Intimacy," pp. 217–218; and most comprehensively in Wigley, "Untitled: The Housing of Gender," pp. 327–389.
23. Wigley, "Untitled: The Housing of Gender," p. 347. For an interesting discussion of Alberti, particularly his less well known writings, see Jarzombek, *On Leon Battista Alberti;* on the uses of architecture during this period, see Smith, *Architecture in the Culture of Early Humanism.*
24. Isotta Nogarola's study was described in the exact same terms as Alberti's ideal study: a *libraria cella.* Yet hers was private for different reasons than his own. The female scholar did not retire to the study to keep secrets from the household, as Alberti's ideal husband did, but

to remove herself from the threat of impropriety through isolation and spiritual contemplation. On this subject, see King, "Book-Lined Cells," p. 74; and idem, *Women of the Renaissance,* p. 198.

25. Alberti, *I libri della famiglia,* translated by Renée Neu Watkins as *The Family in Renaissance Florence,* p. 209.
26. Jed, *Chaste Thinking,* p. 86. Jed also discusses this same passage from Alberti on p. 80.
27. Mark Wigley also underscores this point in his "Untitled: The Housing of Gender," p. 348.
28. Benedetto Cotrugli, *Della mercatura e del mercato perfetto* (1573), p. 86, in Franzoni, "'Rimembranze d'infinite cose,'" p. 307.
29. Biblioteca Ambrosiana, Milan, ms.S.85 sup., f. 235r. Giganti's activities are well described in Fragnito, *In museo e in villa,* pp. 159–214.
30. Archivio di Stato, Bologna (hereafter ASB), *Senato, Instr.* C, b.33, n.48; published as Appendix I in Scappini and Torricelli, *Lo Studio Aldrovandi in Palazzo Pubblico,* pp. 93–94.
31. Scamozzi, *L'idea della architettura universale,* pp. 254, 256.
32. Goldthwaite, *The Building of Renaissance Florence,* pp. 97–98. As Goldthwaite points out, Alberti's architectural writings were intended more for a general public than for architects, making Aldrovandi a sixteenth-century version of his ideal reader.
33. Ulisse Aldrovandi, *Discorso naturale* (ca.1572–73), in Tugnoli Pattaro, *Metodo e sistema,* p. 203.
34. Alberti, *On the Art of Building,* p. 37.
35. BUB, *Aldrovandi,* ms.147 (*Bibliotheca secundum nomina authorum qui penes se habentur, in alphabeticum ordinem non exiguo labore ac studio digesta*), cc.309r, 409r, 441r; ms.148 (*Ulyssis Aldrovandi. Bibliothecarum thesaurus secundum titulos librorum variasque materias ordine alphabetico in duodecim tomos distinctus. Opus sane perutile ac necessarium omnibus cuiuscumque professionis hominibus et praesertim iis qui in conscribendis diversarum rerum historiis operam suam impendunt*), I, cc.430v–432r. The catalogue was composed between October 1582 and February 1583. Thanks to Giuseppe Olmi for supplying the specific pages.
36. Of course this accords well with Alberti's advice that the villa "must be located at no great distance from the city"; Alberti, *On the Art of Building,* p. 141.
37. For similar behavior on the part of English aristocrats, see Friedman, *House and Household,* p. 33.
38. Goldthwaite, *The Building of Renaissance Florence,* p. 96.
39. Alberti, *On the Art of Building,* p. 126; Coffin, *The Villa in the Life of Renaissance Rome,* pp. 14–15. On the cultural role of the villa, see Ackerman, *The Villa,* esp. pp. 108–123; Bentmann and Müller, *The Villa as Hegemonic Architecture;* and Fragnito, *In museo e in villa,* pp. 65–108.
40. This accords well with Palladio's recommendation that one use the villa for contemplation and "scholarly studies," in imitation of the ancients who were "visited by virtuous friends and relatives" at their country estates. Palladio, *I quattro libri dell'architettura,* p. 142.
41. Bentmann and Müller, *The Villa as Hegemonic Architecture,* p. 59.
42. On 18 July 1604 Aldrovandi transported all three volumes of his *Ornithology* (1599), one volume of his study of insects, a two-volume work of Calepinus (?), an edition of Aristotle, Aldus Manutius's *Adages,* Avicenna's commentary on monsters and portents, and Franciscus Valesius's *Commentary on Galen* to the villa. Aldrovandi also brought the manuscript catalogue of his library, a "Book of the pictures of the Palace," seven sacks of "observations," and

reams of writing paper. During two trips in October he transported eleven more books, particularly several editions of Petrarch—the quintessential "villa reading" of the period; BUB, *Aldrovandi,* ms.136, XXXII, cc.292r–296r. Some of this material is reproduced in Fanti, "La villeggiatura," pp. 39–40. Many contemporaries of Aldrovandi remarked on the advantages of the villa for observing nature. See Scamozzi, *L'idea della architettura universale,* p. 266; more generally, Ackerman, *The Villa,* pp. 110–111.

43. BUB, *Aldrovandi,* ms.147, cc.13r, 62r.
44. Alberti, *On the Art of Building,* p. 294.
45. The *invenzioni* in Aldrovandi's villa are discussed in Lina Bolzoni, "Parole e immagini," pp. 317–348.
46. Alberti, *On the Art of Building,* p. 299.
47. Bolzoni, "Parole e immagini," p. 345.
48. BUB, *Aldrovandi,* ms.99; reproduced in Fanti, "La villeggiatura," p. 32. On Castiglione and Guazzo, see BUB, *Aldrovandi,* ms.147, cc.69v, 554v.
49. BUB, *Aldrovandi,* ms.21, IV, c.317r (Francesca Fontana to the wife of Giuseppe Casabona, Bologna, 13 October 1586).
50. BUB, *Aldrovandi,* ms.99, reproduced in Fanti, "La villeggiatura," p. 33.
51. Aldrovandi's views parallel many of the comments made by contemporary celebrants of villa life such as Agostino Gallo and Bartolomeo Taegio. As James Ackerman notes, women were a new subject in Renaissance treatises on the villa, because they had not appeared in ancient writings. Yet the discussions of women underscored their segregation from men by excluding them from the pleasures of hunting, reading, and civil conversation. Perhaps these divisions were so strong on paper because they were so weak in practice and because women increasingly chose to participate in such activities? See Ackerman, *The Villa,* pp. 120–123, 131.
52. BUB, *Aldrovandi,* ms.99, reproduced in Fanti, "La villeggiatura," p. 33.
53. Ibid., p. 34.
54. Alberti, *On the Art of Building,* p. 149.
55. Frati, "La vita di Ulisse Aldrovandi," p. 15.
56. Alberti, *The Family in Renaissance Florence,* p. 210. On the limits placed upon women in early modern households, see Friedman, *House and Household,* pp. 47–51; and idem, "Architecture, Authority, and the Female Gaze," pp. 44–46.
57. Bartolomeo Taegio, *La villa* (Milan, 1559), in Ackerman, *The Villa,* p. 121.
58. Friedman, "Architecture, Authority, and the Female Gaze," esp. p. 46; see also Harkness, "Managing an Experimental Household."
59. Alberti, *On the Art of Building,* p. 149.
60. ASB, *Fondo Notarile, Rogiti di Ser Achille Canonici,* Book III, c.63, in Fanti, "La villeggiatura," p. 21.
61. BUB, *Aldrovandi,* ms.147, c.394v.
62. BUB, *Aldrovandi,* ms.96, V.
63. Aldrovandi, *Vita,* in Frati, ed., *Intorno alla vita,* p. 15. On women humanists, King's "Book-Lined Cells" is a good point of departure. Lisa Jardine's "'O Decus Italae Virgo,'" pp. 799–819, emphasizes the paradoxes of female learning in Renaissance Italy.
64. Francesca Fontana, "Illustrissimis ac Prudentissimis Senatoribus Bonon[iensis?]," in Aldrovandi, *De reliquis animalibus exanguibus libri quatuor,* n.p.

65. Scappini and Torricelli, *Lo Studio Aldrovandi,* pp. 24, 55 (n.13), 95–98.
66. BUB, *Aldrovandi,* ms.147, c.292v. Barbaro is discussed in Jordan, *Renaissance Feminism,* p. 46; see generally her section on "Household Government," pp. 40–54. On the power of widows, see Friedman, "Architecture, Authority, and the Female Gaze," pp. 47, 50; and Klapisch-Zuber, *Women, Family and Ritual,* pp. 117–131.
67. Aldrovandi made his nephews, Alessandro and Giuseppe Griffoni, his provisional heirs on the condition that they change their last name.
68. For more on this subject, see Hannaway, "Laboratory Design."
69. Jed, *Chaste Thinking,* p. 86. Mark Wigley also emphasizes this point, from the perspective of controlling sexuality; see his "Untitled: The Housing of Gender," p. 336.
70. Guazzo, *The Civile Conversation,* vol. 1, pp. 232, 243. I have modernized the English spelling.
71. Federico Cesi, *Linceografo* (written 1605, published 1624), in Lombardo, "'With the Eyes of a Lynx,'" p. 42. Mario Biagioli's interesting discussion of Lincean homosociability appeared too late to be fully incorporated in this essay; see his "Knowledge, Freedom and Brotherly Love."
72. Mark Wigley also discusses the dangers of being in the wrong place in his "Untitled: The Housing of Gender," p. 335.
73. Schiebinger, *The Mind Has No Sex;* Harth, *Cartesian Women.*
74. Gabrieli, "Luca Valerio Linceo," p. 843.
75. Biblioteca Nazionale, Florence, *MS. Galileiani,* Part VI, Tome XIV, c.31 (Guido Bettoli to Margherita Sarrocchi, Perugia, 4 June 1611), in Favoro, "Amici e corrispondenti di Galileo Galilei, I, Margherita Sarrocchi," p. 576.
76. Ibid., p. 570 (Valerio to Galileo, n.d.).
77. Fonte, *Il merito delle donne,* p. 14.
78. Ibid. I follow Margaret King's translation here in *Women of the Renaissance,* p. 228. She discusses Fonte on pp. 228–232.
79. Fonte, *Il merito delle donne,* p. 54.
80. Of course the garden, like the museum, also had a long tradition as a setting for philosophizing. In Renaissance Italy, women and men often entered gardens together as well as separately. Patricia Waddy provides an interesting description of the garden and *casino* in the women's quarters of the Palazzo Barberini as the complement of the famous library of Cardinal Francesco Barberini in the men's quarters; Waddy, *Seventeenth-Century Roman Palaces,* p. 29.
81. Noble, *A World Without Women,* p. 43.
82. Scappini and Torricelli, *Lo Studio Aldrovandi,* pp. 29, 110. The only woman whose visit was anticipated with great public ceremony during the seventeenth century was Christina of Sweden, who arrived in Bologna in 1655. Unfortunately her travel plans changed, making it impossible for her to see the museum.
83. Findlen, "Science as Career in Enlightenment Italy," p. 449.

BIBLIOGRAPHY

Ackerman, James. *The Villa: Form and Ideology of Country Houses.* London: Thames and Hudson, 1990.

Alberti, Leon Battista. *On the Art of Building in Ten Books.* Translated by Joseph Rykwert, Neil Leach and Robert Tavernor. Cambridge: MIT Press, 1988.

Alberti, Leon Battista. *The Family in Renaissance Florence.* Translated by Renée Neu Watkins. Columbia: University of South Carolina Press, 1969.

Aldrovandi, Ulisse. *De reliquis animalibus exanguibus libri quatuor, post mortem eius editi nempe de mollibus, crustaceis, testaceis, et zoophytis.* Bologna, 1606.

Aymard, Maurice. "Friends and Neighbors." In Roger Chartier, ed., *Passions of the Renaissance,* 447–492. Vol. 3 of *A History of Private Life,* ed. Philippe Ariès and Georges Duby. Cambridge: Harvard University Press, Belknap Press, 1989.

Bentmann, Reinhard, and Michael Müller. *The Villa as Hegemonic Architecture.* Translated by Tim Spence and David Craven. Atlantic Highlands, NJ: Humanities Press, 1992.

Biagioli, Mario. "Knowledge, Freedom, and Brotherly Love: Homosociality and the Accademia dei Lincei." *Configurations 3* (1995): 139–166.

Bolzoni, Lina. "Parole e immagini per il ritratto di un nuovo Ulisse: l' 'invenzione' dell'Aldrovandi per la sua villa di campagna." In Elizabeth Cropper, Giovanna Perini, and Francesco Solinas, eds., *Documentary Culture: Florence and Rome from Grand-Duke Ferdinand I to Pope Alexander VII,* 317–348. Bologna, Italy: Nuova Alfa Editoriale, 1992.

Borel, Pierre. *Catalogue des choses rares de Maistre Pierre Borel.* In *Les antiquitez, raretez, plantes, mineraux, & autres choses considerables de la Ville, & Comte de Castres d'Albigeois.* Castres, 1649.

Cantaro, Maria Teresa. *Lavinia Fontana, bolognese: "pittore singolare" 1552–1614.* Milan, Italy: Jandi Sepi, 1989.

Coffin, David R. *The Villa in the Life of Renaissance Rome.* Princeton, NJ: Princeton University Press, 1979.

de Grazia, Sebastian. *Machiavelli in Hell.* Princeton, NJ: Princeton University Press, 1989.

della Casa, Giovanni. *Galateo.* Translated by Konrad Eisenbichker and Kenneth R. Bartlett. Toronto: Dovehouse Editions, 1990.

Fanti, Mario. "La villeggiatura di Ulisse Aldrovandi." *Strenna storica bolognese* 8 (1958): 17–43.

Favoro, Antonio. "Amici e *corrispondenti* di Galileo Galilei, I, Margherita Sarrocchi." *Atti del R. Istituto Veneto di scienze, lettere ed arti,* ser. 7, 5 (1893–94): 552–580.

Findlen, Paula. *Possessing Nature: Museums, Collecting and Scientific Culture in Early Modern Italy.* Berkeley: University of California Press, 1994.

Findlen, Paula. "Science as Career in Enlightenment Italy: The Strategies of Laura Bassi." *Isis* 84 (1993): 441–469.

Fonte, Moderata. *Il merito delle donne,* edited by Adriano Chemello. Mirano, Italy: Eidos, 1988.

Fragnito, Gigliola. *In museo e in villa: Saggi sul Rinascimento perduto.* Venice: Arsenale, 1988.

Franzoni, Claudio. "'Rimembranze d' infinite cose': Le collezioni rinascimentali di antichità." In Salvatore Settis, ed., *Memoria dell'antico nell'arte italiana,* vol. 1, 299–360. Turin, Italy: Einaudi, 1984.

Frati, Ludovico, ed. "La vita di Ulisse Aldrovandi scritta da lui medesimo." In *Intorno alla vita e alle opere di Ulisse Aldrovandi.* Bologna: L. Beltrami, 1907.

Friedman, Alice T. "Architecture, Authority, and the Female Gaze: Planning and Representation in the Early Modern Country House." *Assemblage* 19 (1992): 40–61.

Friedman, Alice T. *House and Household in Elizabethan England: Wollaton Hall and the Willoughby Family.* Chicago: University of Chicago Press, 1989.

Gabrieli, Giuseppe. "Luca Valerio Linceo: Un episodio memorabile della *vecchia* accademia." In *Contributi per la storia dell'Accademia de' Lincei,* vol. 1, 835–864. Rome: Accademia Nazionale dei Lincei, 1989.

Girouard, Mark. *Life in the English Country House: A Social and Architectural History.* New Haven, CT: Yale University Press, 1978.

Goldthwaite, Richard A. *The Building of Renaissance Florence: An Economic and Social History.* Baltimore: Johns Hopkins University Press, 1980.

Guazzo, Stefano. *The Civile Conversation of M. Steeven Guazzo.* Translated by George Pettier. New York: Knopf, 1925.

Hannaway, Owen. "Laboratory Design and the Aim of Science." *Isis* 77 (1986): 585 610.

Harkness, Deborah. "Managing an Experimental Household: The Dees of Mortlake and the Practice of Natural Philosophy." *Isis* 88 (1997): 247–262.

Harth, Erica. *Cartesian Women: Versions and Subversions of Rational Order in the Old Regime.* Ithaca, NY: Cornell University Press, 1992.

Imperato, Ferrante. *Dell'historia naturale.* Naples, 1599.

Jardine, Lisa. "'O Decus Italae Virgo,' or the Myth of the Learned Lady in the Renaissance." *The Historical Journal* 28 (1985): 799–819.

Jarzombek, Mark. *On Leon Battista Alberti.* Cambridge: MIT Press, 1989.

Jed, Stephanie H. *Chaste Thinking: The Rape of Lucretia and the Birth of Humanism.* Bloomington: Indiana University Press, 1989.

Jordan, Constance. *Renaissance Feminism: Literary Texts and Political Models.* Ithaca, NY: Cornell University Press, 1990.

Kenseth, Joy, ed. *The Age of the Marvelous.* Hanover, NH: Dartmouth College, Hood Museum of Art, 1991.

King, Margaret L. "Book-Lined Cells: Women and Humanism in the Early Italian Renaissance." In Patricia A. Labalme, ed., *Beyond Their Sex: Learned Women of the European Past,* 66–90. New York: New York University Press, 1980.

King, Margaret L. *Women of the Renaissance.* Chicago: University of Chicago Press, 1991.

Klapisch-Zuber, Christiane. *Women, Family and Ritual in Renaissance Italy.* Translated by Lydia G. Cochrane. Chicago: University of Chicago Press, 1985.

Klose, Wolfgang. *Corpus Alborum Amicorum.* Stuttgart, Germany: A. Hierseman, 1988.

Liebenwein, Wolfgang. *Studiolo: Storia e tipologia di uno spazio culturale.* Edited by Claudia Cieri Via. Ferrara, Italy: Istituto per gli Studi Rinascimentali, 1988.

Lombardo, Richard. *"With the Eyes of a Lynx": Honor and Prestige in the Accademia dei Lincei.* Master's thesis, University of Florida, Gainesville, 1990.

Masetti Zannini, Gian Ludovico. *Motivi storici della educazione femminile,* Vol. 2, *Scienza, lavoro, giuochi.* Naples, Italy: M. D'Auria, 1982.

Noble, David. *A World Without Women: The Christian Clerical Culture of Western Science.* New York: Knopf, 1992.

Olmi, Giuseppe, and Paolo Prodi. "Gabriele Paleotti, Ulisse Aldrovandi e la cultura a Bologna nel secondo Cinquecento." in *Nell'età di Correggio e dei Carracci: pittura in Emilia dei secoli XVI e XVII.* Bologna, Italy: Nuova Alfa Editoriale, 1986.

Palladio, Andrea. *I quattro libri dell'architettura.* Edited by Licisco Magaguato and Paola Marini. Milan, Italy: Edizioni il Polifilo, 1980.

Raimondi, C. "Le lettere di P. A. Mattioli ad Ulisse Aldrovandi." *Bullettino senese di storia patria* 13, f.1–2 (1906): 3–67.

Ranum, Orest. "The Refuges of Intimacy." In Roger Chartier, ed., *Passions of the Renaissance,* 207–264. Vol. 3 of *A History of Private Life,* ed. Philippe Ariès and Georges Duby. Cambridge: Harvard University Press, Belknap Press, 1989.

Scamozzi, Vincenzo. *L'idea della architettura universale.* Venice, 1615.

Scappini, Cristiana, and Maria Pia Torricelli. *Lo Studio Aldrovandi in Palazzo Pubblico (1617–1742).* Edited by Sandra Tugnoli Pattaro. Bologna, Italy: CLUEB, 1993.

Schiebinger, Londa. *The Mind Has No Sex? Women in the Origins of Modern Science.* Cambridge: Harvard University Press, 1989.

Shapin, Steven. "The House of Experiment in Seventeenth-Century England." *Isis* 79 (1988): 373–404.

Shapin, Steven. *A Social History of Truth: Civility and Science in Seventeenth-Century England.* Chicago: University of Chicago Press, 1994.

Smith, Christine. *Architecture in the Culture of Early Humanism: Ethics, Aesthetics, and Eloquence 1400–1470.* Oxford: Oxford University Press, 1992.

Spain, Daphne. *Gendered Spaces.* Chapel Hill: University of North Carolina Press, 1992.

Tugnoli Pattaro, Sandra. *Metodo e sistema delle scienze nel pensiero di Ulisse Aldrovandi.* Bologna, Italy: CLUEB, 1981.

Waddy, Patricia. *Seventeenth-Century Roman Palaces: Use and the Art of the Plan.* New York: Architectural History Foundation; Cambridge: MIT Press, 1990.

Wigley, Mark. "Untitled: The Housing of Gender." In Beatriz Colomina, ed., *Sexuality and Space,* 327–389. Vol. 1 of *Princeton Papers on Architecture.* Princeton, NJ: Princeton Architectural Press, 1992.

3

Alchemical Symbolism and Concealment: The Chemical House of Libavius

William R. Newman

State of the Question

The sixteenth and seventeenth centuries are often viewed as forming a critical juncture in the history of science, when older practices of secrecy came to be abandoned in favor of a new scientific openness.[1] Stock examples of this new attitude are often found in the assertion that the proprietary character of trade secrecy gave way to the free exchange of information in the early scientific societies, while the jealous retention of alchemical recipes yielded to the development of a new genre of chemical textbooks.[2] In contrast to this still prevalent view, historians of alchemy and early chemistry have recently begun to question the dichotomy of an "open" chemistry versus a "closed" alchemy, with a growing consensus that the evidence is not so transparent.[3] The present essay adds to the revisionist evidence by showing that the Saxon chemist Andreas Libavius (c. 1560–1615), one of the very founders of the early modern chemical textbook tradition, was by no means an advocate of the unrestricted promulgation of chemical knowledge. Indeed, he designed a "house of chemistry" (*domus chemiae*) that segregated the casual visitor from the laboratory, and more importantly, sequestered the master alchemist's transmutational work space from that of his less trustworthy technicians. This division of space, moreover, was not the mere product of casual expedience: It was a conscious adaptation of principles used in alchemical texts themselves to delude the unwary and prevent the overly easy acquisition of knowledge. Nor did Libavius restrict himself to the structural embodiment of textual principles of alchemical concealment: He openly designed his laboratory around one of the most famous graphic symbols of the alchemists, the *monas hieroglyphica* invented by the Elizabethan magus John Dee. As we shall see, a close analysis of Libavius' *domus chemiae* throws serious doubt on the much vaunted march of openness, a prominent feature of the "grand narrative" of the scientific revolution, while also revealing the densely emblematic character of early modern alchemy.[4]

Libavius must be approached first through the pioneering work of Owen Hannaway, who has made his laboratory the object of a particular study.[5] In 1606 Libavius published a commentary on his encyclopedic *Alchemia* of 1597; in the commentary Libavius included a detailed description and plan of an ideal chemical laboratory also meant to serve as the artificer's residence. (See figure 3.1.) As Hannaway points out, Libavius fashioned this *domus chemiae* in response to a far more grandiose scheme published by the famous astronomer Tycho Brahe in 1598. Tycho's laboratory formed a small part of his sumptuous Uraniborg, an observatory complex located on the island of Hven in the Danish sound. According to Hannaway, Tycho had placed his lavish laboratory underground, "condemn[ing] it into the hands of the god of the underworld" to keep his secrets from the public gaze.[6] Alchemical secrets were reserved for an aristocratic elite. A subterranean setting for alchemy also seemed to fit Tycho's printed description of alchemy as "terrestrial astronomy." Engaging in the old conceit of the Emerald Tablet of Hermes, Tycho claimed that when he looked up, he saw down and when he looked downward, he observed what was on high: As Hermes had said, "That which is above is also below"—alchemy and astronomy were tellurian and cosmic twins.[7]

Libavius consciously placed his own laboratory above ground and eschewed, in Hannaway's view, the mannered symbolism and aristocratic ethos of his lordly predecessor. Hannaway goes so far as to view Libavius's chemical house as an embodiment of the ideal of civic humanism, which the historian reduces to three main elements. First, civic humanism stressed the importance of an active life over that of "scholarship and contemplation." Second, it valued the family as the basis of society, and third, it rejected "the notion that the perfect life is that of the sage and instead celebrates the citizen" intent on serving his fellow man.[8] Thus where Tycho advised secrecy in alchemical affairs, Libavius promoted the promulgation of alchemically produced medicines. Where Tycho insisted that alchemical knowledge was fit only for the aristocratic few, Libavius brought it to the *res publica* for the service of the common good. In this fashion Hannaway depicted a schism between the old, Hermetically sealed way of doing alchemy and the new, permeable chemistry of Libavius, the latter a fully systematized, open discipline, free to the access of all.

Hannaway's brilliantly argued dichotomy of Libavius versus Tycho has come under recent attack by Jole Shackelford, who debunks at length the notion that Tycho was re-

3.1

Plan and elevation of the "chemical house." From Andreas Libavius, *Alchymia* (1606).

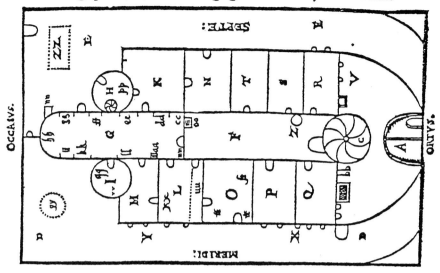

61 | ALCHEMICAL SYMBOLISM AND CONCEALMENT

ally so enraptured by cosmic harmonies as Hannaway would have it.⁹ In this essay I do not discuss Tycho further, however, but focus on the laboratory of Libavius. Of the fact that Libavius considered himself a humanist there can be no doubt. He wrote in a tortuous Horation Latin, replete with references to Vitruvius and Cicero, and the latter's ideology of civic responsibility Libavius consciously emulates. In Hannaway's view, Libavius was so committed to the model of civic humanism that he viewed alchemical symbolism with nothing but scorn. It was the province of would-be elitists and Paracelsian obscurantists, charlatans who had no real knowledge to offer. Indeed, Libavius held the followers of the sixteenth century magus Paracelsus in the greatest contempt: In addition to their nugatory bragging, they were impious enthusiasts, plebeian boors, and espousers of illicit magic. Libavius, in Hannaway's analysis, was the representative of a new, natural alchemy divorced from obscure terminology, superstition, and the ideology of concealment: In other words, he was one of the first chemists.

THE HIEROGLYPHIC MONAD

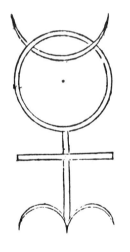

3.2

The hieroglyphic monad. From John Dee, *Monas Hieroglyphica* (1564).

Let us now pass from this summary to the words of Libavius himself. In the pages preceding his description of the chemical house, Libavius delivers a short dissertation on alchemical symbolism.¹⁰ Among the symbols discussed, Libavius includes the *monas hieroglyphica*, the hieroglyphic monad of the famous Elizabethan polymath John Dee. The monad is a traditional symbol for mercury, to which Dee added a point symbolizing gold and a foundation representing fire. (See figure 3.2.) Dee thought that the ancient wise men created alchemical symbols by reference to the monad, as elements of a sort of artificial language. Just as the Romans supposedly rotated a "V" for 5 to arrive at an "L" for 50, so the primordial alchemists created the various astronomical and alchemical symbols by rotating elements of the monad's horned circle around its cruciform base.¹¹ Thus according to Dee and Libavius, the monad comprises the symbols of all seven metals, for the alchemists traditionally represented the metals by means of the planetary symbols.¹² The astronomical symbol of the sun is merely a circle with a central dot and that of the moon a crescent, and the symbols of the other planets can be formed by moving the circle and crescent around the cruciform base below. By removing the horned circle from the base, one can obtain the cross by itself, which then represents the four elements. In Libavius's view, Dee intended the monad as a symbol for the "philosophers' mercury," the first matter out of which the philosophers' stone, the agent of metallic transmutation, was supposed to be made.

According to Hannaway, Libavius's apparent approbation of the *monas* is nothing but sarcasm.¹³ Committed to his humanist ethos of openness, Libavius could hardly subscribe to the obfuscation of John Dee. But if we listen to Libavius himself, a different story emerges. He says of the monad:

One should know that artificers have endeavored to make the symbols of the materials and operations of their art agree with [the old] sources, instead of being fabricated irrationally, arbitrarily, and monstrously, as the Paracelsians are accustomed madly to do.[14]

The mad practice of the Paracelsians to which Libavius alludes refers to their penchant for neologisms such as *spagyria* for alchemy or *xenexton* for a medical amulet. To his mind, this neologizing is a reflection of the Paracelsian hatred of authority, be it civil or scientific. Libavius thinks that Dee's monad is a salutary remedy to this impious, willful degeneration of language and wisdom, for it is composed of venerable old astronomical and alchemical symbols. (See figure 3.3.)

Libavius is not, however, an unequivocal acolyte of the monad. He points out that there is virtually no symbol that cannot be reduced into straight and curvilinear elements if one so chooses. The whimsicality of such analysis is a minor peccadillo compared to the fact that John Dee attributed a deep theosophical significance to the monad, however, linking its components to the Neoplatonic and Cabalistic emanations connecting the physical world to God. Libavius is outraged at this incursion of theosophy into the domain of alchemy.[15] Nonetheless, he retains the monad, saying that it does not displease him if someone should combine the various signs for chemical essences in one "hieroglyphic symbol," for then it will be possible for a skilled interpreter to deduce directly from the symbol that something has been artificially produced from the initial chemicals.[16] As examples of how this might work, Libavius refers to the common symbols for sublimed and precipitated mercury. Here we are back in the realm of Dee's attempt to use the monad as the basis for an artificial language.

3.3

Libavius's version of the hieroglyphic monad, showing its relationship to planetary symbols. From Libavius, *Alchymia* (1606).

None of this sounds like sarcasm on Libavius's part, nor does he seem sarcastic when he reintroduces the monad in his description of the chemical house. First, however, he describes the elaborate symbolism of Tycho's Uraniborg, pointing out that Tycho incorporated globes, circles, triangles, and rectangles into his structure, because philosophers and alchemists had used these figures to represent the structure of the cosmos as well as the philosophers' stone. Libavius adds that he too admires Platonic globes and triangles, but wishes to pass beyond such cosmic symbolism to the world of the archetypes itself. Thus he asserts:

> as [Plato] says in the *Symposium,* in the beginning man was made round, but he displeased the gods, and therefore was split into a cylindrical length with the roundness retained in his nobler parts: Indeed, the Stoics assigned the quadrate and sphere to the most perfect sage. We therefore desire this [combination of shapes] to be observed in our plan, which chemical vessels also display most prominently.[17]

Relying on the authority of Plato, Libavius asserts that both the figure of man and that of a flask share the elements of a rectangle surmounted by a sphere. Since he is building a chemical house for men to inhabit, it is only natural that these should form its elements as well. The meaning of this does not become entirely clear until Libavius arrives at his description of the laboratory proper. Contrasting his work space with Tycho's underground laboratory, Libavius says:

> We dignify this [laboratory] with sunlight and erect it in the middle of a field with a round head, but with a body drawn out like the figure of the microcosm, and like the hieroglyphic monad.[18]

The laboratory, therefore, preserves the rectangle and circle of the Stoics. But here Libavius reintroduces Dee's hieroglyphic monad, saying that it too preserves this shape, as does man. Clearly Libavius has derived his laboratory plan, with the odd apsidal termination of the rectangle in a complete hemicycle, from Dee's monad.[19] (See figure 3.4.) Libavius has employed the monad as a design concept by taking the rectangular hall of the laboratory from the monad's elemental cross or trunk, and by using its round solar head to arrive at the proportions of his apse.

Why, after supposedly ridiculing alchemical symbolism, does Libavius here reintroduce the alchemical monad and claim that it serves as the model for his laboratory? One answer to this question lies in the explicit connection that Libavius draws between Dee's monad and the human body, an association already made by Dee himself.[20] Libavius was probably thinking of the architectural tradition stemming from the famous *De architectura* of the Roman imperial architect Vitruvius, whom he cites. Vitruvius begins Book III with a discussion of temples, which the architect says should be symmetrical. This principle of symmetry is to be derived from the human body: If a compass is used to draw a circle with its center at the navel, the hands and feet can both be placed on the cir-

3.4

The hieroglyphic monad superimposed over Libavius's laboratory. This montage shows how the monad served Libavius as a design concept.

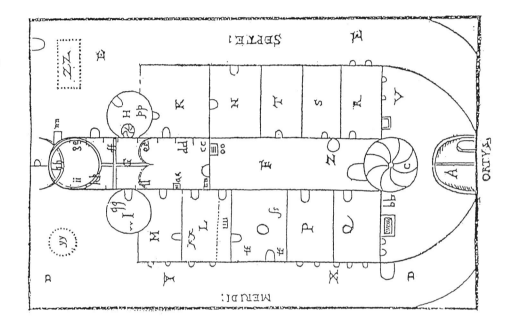

65 | ALCHEMICAL SYMBOLISM AND CONCEALMENT

cumference. But if we measure the length of the body and the arm span, they will be found to equal one another, giving the basis for a square. As Rudolph Wittkower showed many years ago, the popularity of the "Vitruvian Figure" in the Renaissance led to a widespread acceptance that centrally planned churches were better than the traditional longitudinal basilica. Nonetheless, the Vitruvian tradition of using the human body as a design concept came to be extended even to longitudinal structures. Thus the famous quattrocento architect Francesco di Giorgio employed a supine figure as the basis for an ideal church (see figure 3.5), and the lesser-known Juan Bautista Villalpando used it as a means of understanding the Temple of Solomon.[21] Libavius is merely drawing on a well-established architectural tradition when he refers to the human body as a design concept.

One might then reasonably ask why Libavius uses the monad at all, instead of simply employing the body, as do Giorgio and Villalpando. Is it not perhaps because the traditional goals of alchemy, the transmutation of metals and the secrecy that surrounded its processes, are dearer to Libavius's heart than Hannaway would admit? The monad, let us recall, was based on the traditional symbol for mercury, and the "philosophers' mercury" was the initial ingredient of the philosophers' stone.

3.5

Francesco di Giorgio's design for a longitudinal church based on the human body. From Rudolph Wittkower, *Architectural Principles in the Age of Humanism* (London, The Warburg Institute, 1949).

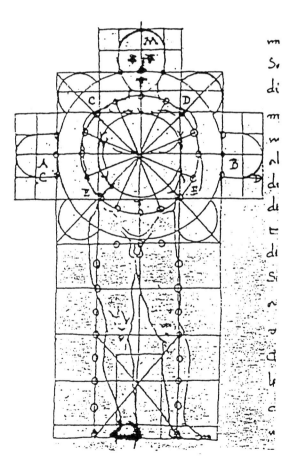

In fact, the original printing of Libavius' *Alchemia* was accompanied by four *Commentationes,* of which three directly concern transmutatory alchemy: the *De mercurio philosophorum, De azotho,* and *De lapide philosophorum.* In these works Libavius expressly defends alchemists' traditional claims, arguing that alchemy can indeed transmute base metals into noble ones. Moreover, he defends alchemists' secrecy, saying alchemists have hidden their processes in exotic imagery to avoid the prostitution of their knowledge.[22] Only he who is most knowledgeable about nature in general is admitted to the secrets of the art. Such a man must be close to perfection, not a plebeian operator. This is why the alchemists originally adopted the use of symbolic writing: so that the art might not fall into the hands of fools and frauds, thereby being prostituted.[23]

Libavius gives his fullest depiction of the ideal alchemist for whom this secret is reserved in his treatise on the philosophers' stone. Relying primarily on the late thirteenth-century *Summa perfectionis* of Geber, Libavius says that the sages recommend that only a pious man frequent in prayer undertake the art. He should not care for money or mundane pleasure, but revel in the study of nature's marvels.[24] Moreover, the alchemist must have at heart the goal "of aiding the health and necessities of his needy neighbor," rather than his own interest.[25] Without these attributes, Libavius says, "let no one dream of success."[26] What Libavius has done here is to recapitulate the medieval motif of the *donum Dei,* the notion that God will only bestow alchemical success upon his chosen few, the sons of doctrine. But he has also described the character of an alchemist that fits very well the model of civic humanism as described by Hannaway. Like the civic humanist, Libavius's alchemist is an active participant in nature, by virtue of his experimentation. Moreover, he is virtuous and imbued with the civic ideal of helping his fellow man. If Libavius has rejected the model of the perfect sage, as Hannaway claims, it is clearly not the alchemical sage of Geber and other medieval authors whom he has expelled. Libavius ends this passage by contrasting his ideal alchemist with the nefarious Paracelsians, who are known to conjure spirits. Is he not invoking the authority of medieval alchemy in the hope of expelling the neoteric plebeians? This interpretation fits quite well with Libavius's earlier defense of the hieroglyphic monad. There he said that part of the monad's value lay in its justification of traditional alchemical symbolism: It was therefore a useful scourge for the upstart Paracelsians.

It seems therefore that Libavius is operating well within the ethos of a medieval alchemist like Geber, the author of the *Summa perfectionis.* Like Geber, Libavius feels that the highest secrets of transmutatory alchemy should be reserved for an elect. Indeed, they will do only harm to those of the plebs who acquire them without God's willing assent. Given this appeal to the traditional motifs of medieval alchemy, it seems less surprising than before that Libavius would model his laboratory on the hieroglyphic monad.

Hannaway's assertion that this is all mere sarcasm seems increasingly untenable. Indeed, the hieroglyphic monad itself surfaces at a number of points in Libavius's defenses of alchemy. In describing the philosophers' mercury, Libavius says that it is the "one thing, circle, and hieroglyphic unity" of which all philosophers have spoken.[27] It is one thing, containing in itself one, two, three, four, or a circle that can be squared.[28] Here Libavius overtly links the hieroglyphic unity or monad to an old alchemical trope found in one of the late medieval florilegia that go by the title *Rosarium philosophorum*. It is the *figura* of squaring the circle, which he quotes directly at another point: "make a round circle from the male and the female, from this extract a square and from that a triangle, and again a round circle, and thus you will have the philosophers' stone."[29] (See figure 3.6.) By way of commentary, Libavius states that the author of the *Rosarium* intends "the hieroglyphic unity." This must be turned into a square and triangle by means of cooking alone, or with sublimation, "which is also an exaltation."[30] It sounds then as though Libavius viewed the hieroglyphic monad as a sort of shorthand symbolic representation of the process for the philosophers' stone, as the following quotation makes clear (see figure 3.7):

> If you draw out lines from [the central point of the monad's circle] to the arms of the cross under the circle, you will form a triangle. And [if you draw out lines] from the points of intersection of the sun and moon to the same [arms of the cross], you will describe a rectangle, so that you may understand the chemical squaring of the circle to be explained by this figure, that the *Rosarium philosophorum* mentions together with the triangle; both [the triangle and the rectangle] are erased at the same time and yield to the circle, and therefore these lines are not expressed in the figure.[31]

In this fashion Libavius gave a graphic representation of the process by which the initial ingredient of the philosophers' stone, the philosophical mercury (represented as a circle), was to be cooked or sublimed until it passed through the successive stages (the triangle and rectangle) that would lead to the philosophers' stone (again represented by a circle).[32] Since Dee's text contains no such attempt to derive a rectangle and triangle from the horned circle, we must conclude that Libavius's own sincere appreciation of the monad led him to this elaboration.

Let me now summarize the points made up to this juncture. There can be no doubt that Libavius believed in transmutatory alchemy and was a firm defender thereof, despite his hatred of the Paracelsians: Even Hannaway admits this. It should be clear from Libavius's references to the *Rosarium philosophorum* that the alchemical trope of squaring the circle was one of his favorite metaphors for the manufacture of the philosophers' stone. It is also evident that Libavius equated Dee's hieroglyphic monad with the squared circle of the *Rosarium,* for he uses the two terms interchangeably. There can be no ground then for dismissing Libavius's appreciation of the monad as simple sarcasm, and by extension we cannot reduce his use of the monad as a design concept to the status of jocularity. Let

3.6

The alchemical trope of the squaring of the circle. From Michael Maier, *Atalanta Fugiens* (1618).

3.7

Libavius's application of the hieroglyphic monad to the alchemical riddle of squaring the circle. Adapted from the discussion in his *Alchymia* (1606).

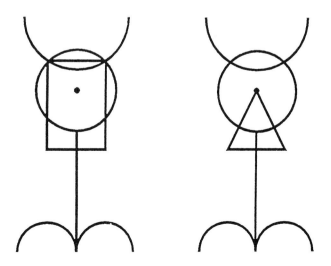

69 | ALCHEMICAL SYMBOLISM AND CONCEALMENT

us therefore return to Libavius's chemical house to see what a realization of his serious affection for the monad may yield us.

A few lines after contrasting his rectangular and circular "monadic" laboratory to the rotund Tychonian one, Libavius launches into a description of their furnaces:

> [Our laboratory] is equipped with ten furnaces and chemical hearths along the circuit of its head and sides. But Tycho filled and packed his laboratory with sixteen furnaces, which he was able to do because of its spacious roundness. We have separated some furnaces from the crowd [*turba*]. For all [the works] cannot be traversed in one place. And there are certain works in secret [places] to which no one is admitted except one quite close to perfection, who has already demonstrated to his master [that he has] industry, faith, and labor, with rather full knowledge of inferior things.[33]

Hannaway interprets this as another sarcastic thrust aimed at Tycho's use of architectural symbolism and its accompanying appeal to secrecy. Despite the vaunted rotundity of Tycho's laboratory, he has stuffed it to overcapacity with furnaces. He has been forced to this extreme out of a desire to bury his alchemical secrets, thereby protecting them from the prying eyes of the vulgar. Libavius, to the contrary, has avoided such overcrowding by placing some of his furnaces in other rooms, thus producing an ordered, workable laboratory. Hannaway's interpretation hinges on the word *turba*, "crowd," which he takes to refer to the crowded mass of furnaces in Tycho's laboratory. Let us however consider what we have learned by examining Libavius's use of the hieroglyphic monad throughout his works. We know that he disdained the vulgar, grasping sort of alchemist as much as any medieval sage could do. Indeed, to Libavius the Paracelsians were the living embodiment of such a shiftless plebs. Is it not more likely that the crowd from which certain furnaces must be sequestered refers to the mass of the inquisitive vulgar, intent on acquiring the alchemist's transmutational secrets? In fact, only three pages after the cited passage, Libavius specifically advises the alchemist to escape the vulgar crowd (*turba*) by performing his secret operations in a private space removed from the general laboratory.[34] It would seem, therefore that Libavius is not criticizing Tycho for excessive occultation when the latter buries his laboratory: Instead, Libavius is recommending a more commodious means of attaining the very secrecy that Tycho, by placing all his furnaces in a single room, forgoes.

The Secret Room

Let us therefore consider the topic of a secret space more carefully. When Libavius asserts in the above passage that in his laboratory all the works cannot be traversed in one place, he means that by dispersing his furnaces in different locales, he has avoided exposing them all to the view of a casual observer. This meaning is borne out by his next comment, which expressly states that some of his processes are to be carried out in secret rooms, where only one "quite close to perfection" is allowed to enter. Surely this faithful, industrious, laborious, and knowledgeable attendant is the student alchemist described at length in Libavius's *Commentationes,* where Libavius outlined the qualities necessary for success in transmutation, derived from such medieval sources as the *Summa perfectionis*. It is interesting to note that Libavius also knew the medieval technique the alchemists called "the dispersion of knowledge." As he says, the alchemists "do not treat the parts of the art in order, but disperse them, and through their method of teaching, vary [the words]."[35] This tradition of *dispersa intentio* derives from the tradition of Geberian alchemy. As Geber says at the end of the *Summa:*

> we have not passed on our science in a continuity of discourse, but . . . we have strewn it about in diverse chapters. This is because both the tested and untested would have been able to take it up undeservedly, if the transmission were continuous.[36]

It is highly likely that Libavius is incorporating the dispersion of knowledge directly into his laboratory. By removing some of his furnaces from the laboratory's main space, Libavius meant to insure that an outside observer could not easily follow the chain of processes leading from furnace to furnace. Thus in addition to the laboratory proper, he describes a preparatory room for ores (see figure 3.1) (L), a "coagulatorium" (O), and an assaying workroom (I). More important by far, however, is Libavius's *adytum,* or "sanctuary," which the diagram tells us is supposed to contain a "philosophical furnace." This *adytum* was not meant merely to serve as "the inner retreat for the chemist," as Hannaway suggests.[37] It was rather the locus for producing the philosophers' stone. This was to be the secret room *par excellence,* which no one who was not "quite near to perfection" could enter. As Libavius says, "we have established the furnace of the philosophers here as in a secret recess away from the other parts of the workshop."[38] He identifies this furnace further, designating it as "the athannor," whose strategic location in the sanctuary allows the alchemist "to cheat the crowd" in the more public spaces by performing his more delicate operations in secrecy.[39] But how do we know that Libavius meant these operations to be those of transmutatory alchemy, rather than the mere production of pharmaceuticals or other mundane products? An easy answer is given by the fact that Libavius's *Alchemia* delivers a detailed portrayal of the athannor, here considered a type of furnace specifically adapted to the incubation of the philosophical mercury in an egg-shaped vessel so that it might mature and become the philosophers' stone.[40] (See figure 3.8.)

There can be no doubt, then, that Libavius meant to sequester the more arcane processes of his transmutatory alchemy from the eyes of the vulgar by performing them in his sanctuary. His close adherence to the propaedeutic of the *Summa perfectionis* makes it likely that he viewed such division and sequestering of his laboratory as an architectural

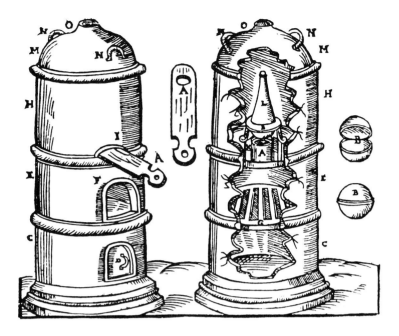

3.8

Exterior and cutaway views of an athannor. From Libavius, *Alchymia* (1606).

realization of Geber's principle of the dispersion of knowledge. Given that Libavius openly designed his workshop around the graphic symbol of the hieroglyphic monad, it should arouse little surprise that he would have incorporated such textual elements into the structure of his workshop.

In this paper I have presented a picture of Libavius that is almost the inverted image of Hannaway's portrayal. Where Hannaway sees Libavius's civic humanism as consciously inimical to the alchemical ideal of the "perfect sage," I have shown that Libavius engaged in no such dichotomy. Indeed, it is arguable that some of the elements of Libavius's civic ideal stem from the texts of medieval alchemy themselves, and it is not a matter of argument but of certainty that Libavius managed to fuse the two traditions. As for Libavius's supposed irony in using John Dee's hieroglyphic monad as a design concept, this interpretation collapses as soon as one realizes that Libavius adopted the monad as a symbol for the philosophical mercury, a substance that the German iatrochemist defended with all his rhetorical might.

But what can we generalize from this new interpretation of Libavius's laboratory? The fact that one of the major proponents of scientific openness has been shown to build the textual principles of alchemical secrecy into his very laboratory raises serious questions about the restraint and revelation of knowledge in early modern science. It was not the ancients or medievals but the neoteric Paracelsians whom Libavius hoped to expose and supplant when he wrote his *Alchemia*. His primary weapons for launching this attack were drawn from the very arsenal of medieval alchemy. To put it simply, did

Libavius really see his chemistry as diverging radically from the secrecy of its medieval counterpart, or have we been seduced by historiographical commonplaces into mistaking the terms of the argument?

NOTES

I thank James Ackerman for his help in placing Libavius's *domus chemiae* within an architectural context, and Monika Asztalos for her gracious evaluation of my translations of Libavius's Latin.

1. The classic statement on the importance of organized public knowledge to the birth of modern science is Ornstein, *Role of Scientific Societies.* A useful critique of Ornstein's influence may be found in Lux, "Societies, Circles, Academies, and Organizations," pp. 23–43.
2. Eamon, "From the Secrets of Nature," pp. 333–334, where Eamon argues forcefully for the radical shift in viewpoint taking place in the sixteenth and seventeenth centuries. In the corresponding chapter of Eamon's recent book, he seems to have modified his stress on the open character of early modern science: Eamon, *Science and the Secrets of Nature,* pp. 319–350. The traditional view that early modern science was inimical to a secrecy epitomized by alchemy is also stressed by Dobbs, "From the Secrecy of Alchemy," pp. 75–94; and to a lesser degree by Long, "Openness of Knowledge," pp. 318–355. In a more recent work, Long has come to the revisionist conclusion that military secrecy was little evident before the early modern period, but that it is largely the product of our own era. See Long and Roland, "Military Secrecy," pp. 259–290.
3. Here I bow to the reigning convention of distinguishing "alchemy" from "early chemistry," though it is by no means clear that they were distinguished in a meaningful fashion in the seventeenth century. In the remainder of the essay, I use the terms "alchemy" and "chemistry," as well as "alchemical" and "chemical," interchangeably, in accordance with early modern usage. For a discussion of this issue, see Newman, *Gehennical Fire,* pp. xi–xii. For the reassessment of secrecy in early modern chemical literature, see Golinski, "Chemistry in the Scientific Revolution," pp. 367–396; Principe, "Robert Boyle's Alchemical Secrecy," pp. 63–74; Clucas, "Correspondence of a XVII-Century 'Chymicall Gentleman,'" pp. 147–170; and Newman, *Gehennical Fire,* pp. 54–78.
4. I take the term "grand narrative" from Lindberg and Westman, *Reappraisals,* pp. xvii–xxv. Like them, I have in mind the "canonical version" of the scientific revolution, inherited from Ornstein and Alexandre Koyré, among others.
5. Hannaway, *Chemists and the Word;* Hannaway, "Laboratory Design," pp. 585–610.
6. Hannaway, "Laboratory Design," p. 604.
7. The classic study of the Emerald Tablet of Hermes remains Ruska, *Tabula Smaragdina.*
8. Hannaway, "Laboratory Design," pp. 606–607.
9. Shackelford, "Tycho Brahe," pp. 211–230.
10. Andreas Libavius, *Alchymia . . . recognita, Commentariorum,* Part 1, pp. 84–88.
11. Josten, "A Translation," pp. 161–163, 169–171.
12. For a medieval example of this conventional alchemical usage, see Newman, *Summa perfectionis.*

13. Hannaway, "Laboratory Design," pp. 588, 608.
14. "Id scire convenit artifices studuisse; ut signa materiarum & operationum artis suae cum istis fontibus convenirent, neque sine ratione monstrose & temere confictae apparerent quomodo insanire paracelsici solent." Libavius, *Alchymia, Commentarorium,* Part 1, p. 86.
15. Josten, "A Translation," p. 96.
16. "In praesens Chymican monadem cur deseramus, nulla nobis apparet caussa. Itaque enim non displicet nobis eorum industria, qui characteres essentiarum ex integris aliqua nota hieroglyphica insignitis, ita effingunt, ut pateat liquido, modo quis gnarus sit, scriptionis primae, esse quid artificialiter ex illis productum, sicuti indices mercurii sublimati & praecipitati docent." Libavius, *Alchymia, Commentariorum,* Part 1, p. 88.
17. "Non contemta Platonica rotunditate & triangulis Homericam mundi catenam, & ad coelum usque extensos Atlantas minime praetereamus, & quia, uti est in symposio, initio rotundus homo factus displicuit superis, & ideo sectus postea est in longum teretemque, orbe tamen in nobiliorubis partibus retento, quin perfectissimo sapienti Stoici rotundum & quadratum assignarunt: hoc in nostra ichnographia observatum cupimus, quod etiam Chymica vasa potissimum prae se ferunt." Libavius, *Alchymia, Commentariorum,* Part 1, p. 93. Hannaway mistakenly translates "quod etiam Chymica vasa potissimum prae se ferunt" intransitively, paraphrasing the expression as "the vessels (instruments) of chemistry stand out and reveal themselves most prominently." In reality "quod" is the object of "prae se ferunt," and refers to the sphere and quadrate.
18. "Nos id Solis dignamur luce, & extruimus in media planicie rotundo quidem capite, sed corpore inde producto in longum ad figuram microcosmi, & hieroglyphicam monadem." Libavius, *Alchymia, Commentariorum,* Part 1, p. 94.
19. As James Ackerman has pointed out to me, the laboratory as depicted in Libavius's diagram would be structurally unsound because it would lack supporting walls extending out from the base of the hemicycle. This provides further evidence, if there be need of corroboration, for the fact that Libavius is really using Dee's monad as a "blueprint" for his structure, rather than deriving it from built structures.
20. Josten, "A Translation," p. 165.
21. For Francesco di Giorgio, see Wittkower, *Architectural Principles,* pp. 10–13, et passim. For Villalpando, see Taylor, "Architecture and Magic," pp. 81–109.
22. Libavius, *Alchymia, Commentariorum,* Part 2, pp. 29, 32. Most of Libavius's *De mercurio philosophorum,* here cited, consists of a dialogue between two iatrochemists named Euthymus and Philiatrus. From the agreement between Philiatrus's positions and those expressed by Libavius elsewhere, it is clear that Philiatrus is his mouthpiece.
23. "Ne itaque cum in manus non initiatorum & fraudibus inhiantium incidissent, prostitueretur artificium, commenti sunt characteres, eosque in illa disciplina ita familiares, ut hieroglyphiae Aegyptiae instar sacerdotem suum statim intelligeret aedituus, & magistrum discipulus saltem prolato signo." Libavius, *Alchymia, Commentariorum,* pp. 1, 85.
24. "Sed venio ad operationes ipsas, ad quas iubent artifices accedere initio hominem pium, & ad preces frequentem, quique tantum artificium non lucri & mundanarum voluptatum caussa, sed studio mirabilium Dei in natura investigandorum, unde gloria existat creatoris: deinde etiam iuvandae sanitatis & commodorum proximi egentis proposito, aggrediatur & perficiat. Ubi enim absque oratione labor est, & propositum malum, nemo sibi ne somniet

quidem successum." Libavius, *Alchymia, Commentariorum,* Part 2, p. 61. Libavius does not name his main source here, but a parallel passage at *Commentariorum,* Part 1, p. 84, reveals it to have been the *Summa perfectionis* of Geber: "Geberus in discipulo requirit naturae vim magnam, & mentem subtiliter scrutantem principia naturalia, artificiaque, quae imitari naturam possunt actionis suae proprietatibus. . . . Longe gravissimum, quod omnes uno ore pronunciant, ingeminantque impossibilem esse artis consecutionem nisi quis singulari Dei gratia illustretur, aut eam ab artifice impetret."

25. This appeal to charity is lacking in the propaedeutic of the *Summa* but may be found in other alchemical texts, such as the *De secretis naturae* of pseudo-Ramon Lull. This work begins with a preface in which a Benedictine monk begs Ramon to write down his alchemical wisdom in the following words: "I therefore ask out of love of Christ that you bestow the great mercy upon us of composing a most secret compendium about the remedies of illnesses so that we might have the fruit of your art or science, since you know us to be variously oppressed and tortured by diverse sorts of diseases." Needless to say, Ramon complies. For the text, consult Pereira, "Filosofia naturale Lulliana," pp. 762–763 (my translation).

26. Libavius, *Alchymia, Commentariorum,* Part 2, p. 61.

27. "Hoc tibi certum esto, mercurium philosophorum, qui propriissime ita appellatur, esse illam rem unam & circulum, & unitatem hieroglyphicam, de qua docent philosophi omnes." (The speaker is Philiatrus.) Libavius, *Alchymia, Commentariorum,* Part 2, p. 32.

28. "Illa res una est, quae in se continet, unum, duo, tria, quatuor, seu circulus, qui quadrari possit." (The speaker is Philiatrus.) Libavius, *Alchymia, Commentariorum,* Part 2, p. 32. See also Part 2, p. 30.

29. "Ros. philos: Fac de masculo & foem. circulum rotundum, ex hoc extrahe [symbol of a square], ex hoc [symbol of a triangle], & iterum circulum rotundum: Et habebis lap. phil." Libavius, *Alchymia, Commentariorum,* Part 1, p. 72.

30. "Omnino circulus & hieroglyphica unitas proponitur. Ea ad quadratum triangulumque redigenda, reducendaque sola COCTIONE, vel sublimatione, quae & exaltatio." Libavius, *Alchymia, Commentariorum,* Part 1, p. 72.

31. "Nec temerarius est punctus in medio circuli; sed signat ab unitate, quae est principium numerorum rerumque numeratarum progressum, & ad eandem recessum. Si ab hoc puncto lineas duxeris ad brachia crucis sub circulo, triangulum efformabis: Et a punctis intersectionis solis & lunae ad eadem, quadratum describes, ut intelligas ex hac ipsa figura explicari posse etiam circuli quadraturam Chymicam, de qua est apud Rosarium philosophicum, una cum triangulo, quae duo tandem obliterantur vicissim, inque circulum concedunt: Et ideo hae delineationes non exprimuntur in figura." Libavius, *Alchymia, Commentariorum,* Part 1, p. 86.

32. Libavius's explanation of the alchemical squaring of the circle is remarkably vague for one who has been lauded as a promoter of openness in chemistry. Possibly he intends the triangle and rectangle to indicate ingredients as well as stages in the *magnum opus.* For discussion of these conventional regimens or stages, see Newman, *Gehennical Fire,* pp. 44, 47, 119, 125, 126, 140, figure 3F.

33. "In ambitu vero capitis & laterum decem focis, furnisque Chymicis est instructum, quanquam Tycho suum laboratorium sedecim fornacibus frequentaverit stipaveritque, id, quod ei facere in rotunditate capaci licuit. Nos quasdam separavimus a turba. Neque enim in uno loco peragi possunt omnia; Et sunt quaedam opera in arcanis, ad quae non admittitur, nisi

perfectioni vicinior, quique industriam, fidem & laborem cum pleniore cognitione inferiorum iam probavit magistro." Libavius, *Alchymia, Commentariorum,* Part 1, p. 94.

34. "Adytum, in quo Athannor subjectum ei est posticum, ut magister transire obireque aedes queat, & si necessitas postulaverit, atria servantem postica fallere turbam, ut Horatius monebat," Libavius, *Alchymia, Commentariorum,* Part 1, p. 97. The reference is to Horace's *Epistles* I. 5. 31: in the translation of William Coutts, "there is room for several uninvited guests. But banquets if too crowded are spoiled by offensive (goatish) odours. Write me back word among how many you would like to be, and laying business aside, elude by the back-door your client, who is waiting in the hall." See Coutts, *Works of Horace,* p. 169.

35. "Dictum esse aliquid, si quis praesciverit artem, cuius partes non tractant ordine, sed dispergunt, & modo docendi ita variant, ita mutant vocabula, ut dissentire omnes videantur, neque quicquam certi percipere," Libavius, *Alchymia, Commentariorum,* Part 1, p. 84. This passage comes only a few lines after an overt reference to Geber and probably reflects Libavius' use of the *Summa perfectionis.*

36. Newman, *Summa perfectionis,* pp. 784–785.

37. Hannaway, "Laboratory Design," p. 605.

38. "Furnum Philosophorum hic collocamus, utpote in recessu secreto ab aliis officinae partibus," Libavius, *Alchymia, Commentariorum,* Part 1, p. 95.

39. For "fallere turbam," see note 34.

40. Libavius, *Alchymia, Commentariorum,* Part 1, pp. 165–168.

Bibliography

Clucas, Stephen. "The Correspondence of a XVII-Century 'Chymicall Gentleman': Sir Cheney Culpeper and the Chemical Interests of the Hartlib Circle." *Ambix* 40 (1993): 147–170.

Coutts, William. *The Works of Horace.* London: Longman, Green, 1898.

Dobbs, Betty Jo Teeter. "From the Secrecy of Alchemy to the Openness of Chemistry." In Tore Frängsmyr, ed., *Solomon's House Revisited: The Organization and Institutionalization of Science,* 75–94. Canton, OH: Science History Publications, 1990.

Eamon, William. "From the Secrets of Nature to Public Knowledge." In Lindberg and Westman, eds., *Reappraisals of the Scientific Revolution,* 333–365.

Eamon, William. *Science and the Secrets of Nature.* Princeton, NJ: Princeton University Press, 1994.

Golinski, Jan. "Chemistry in the Scientific Revolution: Problems of Language and Communication." In Lindberg and Westman, eds., *Reappraisals of the Scientific Revolution,* 367–396.

Hannaway, Owen. "Laboratory Design and the Aim of Science: Andreas Libavius versus Tycho Brahe." *Isis* 77 (1986): 585–610.

Hannaway, Owen. *The Chemists and the Word: The Didactic Origins of Chemistry.* Baltimore: Johns Hopkins Press, 1975.

Josten, C. H. "A Translation of John Dee's 'Monas Hieroglyphica' (Antwerp, 1564), with an Introduction and Annotations." *Ambix* 12 (1964): 84–221.

Libavius, Andreas. *Alchymia . . . recognita, emendata, et aucta.* Frankfurt: Joannes Saurius, 1606.

Lindberg, David C., and Robert S. Westman, eds. *Reappraisals of the Scientific Revolution.* Cambridge: Cambridge University Press, 1990.

Long, Pamela O. "The Openness of Knowledge: An Ideal and Its Context in 16th-Century Writings on Mining and Metallurgy." *Technology and Culture* 32 (1991): 318–355.

Long, Pamela O., and Alex Roland. "Military Secrecy in Antiquity and Early Medieval Europe: A Critical Reassessment." *History and Technology* 11 (1994): 259–290.

Lux, David S. "Societies, Circles, Academies, and Organizations: A Historiographic Essay on Seventeenth-Century Science." In Peter Barker and Roger Ariew, eds., *Revolution and Continuity: Essays in the History and Philosophy of Early Modern Science,* 23–43. Washington, DC: Catholic University of America Press, 1991.

Newman, William R. *Gehennical Fire: The Lives of George Starkey, An American Alchemist in the Scientific Revolution.* Cambridge: Harvard University Press, 1994.

Newman, William R. *The Summa perfectionis of pseudo-Geber.* Leiden, the Netherlands: Brill, 1991.

Ornstein, Martha. *The Role of Scientific Societies in the Seventeenth Century.* Chicago: University of Chicago Press, 1928.

Pereira, Michela. "Filosofia naturale Lulliana e Alchimia." *Rivista di Storia della Filosofia* 41 (1986): 747–780.

Principe, Lawrence. "Robert Boyle's Alchemical Secrecy: Codes, Ciphers and Concealments." *Ambix* 39 (1992): 63–74.

Ruska, Julius. *Tabula Smaragdina: Ein Beitrag zur Geschichte der Hermetischen Literatur.* Heidelberg, Germany: Carl Winter's Universitaetsbuchhandlung, 1926.

Shackelford, Jole. "Tycho Brahe, Laboratory Design, and the Aim of Science: Reading Plans in Context." *Isis* 84 (1993): 211–230.

Taylor, René. "Architecture and Magic: Considerations on the Idea of the Escorial." In Douglas Fraser et al., eds., *Essays in the History of Architecture Presented to Rudolf Wittkower,* 81–109. London: Phaidon, 1967.

Wittkower, Rudolf. *Architectural Principles in the Age of Humanism.* London: The Warburg Institute, 1949.

4

Openness and Empiricism: Values and Meaning in Early Architectural Writings and in Seventeenth-Century Experimental Philosophy

Pamela O. Long

Openness and empiricism were two of the most conspicuous values of the new experimental philosophy in the seventeenth century. The scientific societies that developed in the 1650s and 1660s such as the Accademia del Cimento in Florence, the Royal Society in London, and the Parisian Royal Academy of Sciences made explicit their method of experimentation and their belief in openness. In letters and other writings, members remarked that their organizations were created to facilitate the communication of experimental results to all interested or appropriate parties. They suggested that this openness would lead to progress in knowledge, and some noted that it would allow experiments to be repeated to validate results.[1]

The image of science as a disinterested, truth-yielding process of open inquiry based on experiment began to be fashioned in the seventeenth century and still holds wide currency today. At the same time, criticisms of such a view on the basis of its descriptive accuracy constitute a complex twentieth-century historiography, represented most recently by sociological and contextual investigations of both early modern and modern science.[2] Recent studies concerning modern science have focused on openness and experimentation, bringing into question the view that they function as self-evident values in science.

For example, David Hull's *Science as a Process* has shown twentieth-century biology to be characterized as much by secrecy and rivalry as by openness, whereas Marcel LaFollette's *Stealing into Print* gives concrete meaning to the notion of scientific theft. Concerning the role of experiment, Peter Galison's study of twentieth-century particle physics, *How Experiments End,* paints a complex picture of the ways theoretical and experimental processes interrelate and the various points at which theory is taken to have been proven by experimental results.[3]

Likewise, historians have developed pictures of the workings of early modern science that are far more complex than that of open, empirical inquiry interested only in the search for truth. Studies by Paula Findlen and Mario Biagioli, among others, have shown that openness became associated not so much with expeditious publication for the furtherance of knowledge as with proper timing consonant with the establishment of priority, with the felicitous advancement of patronage relationships, and with display for the purpose of advancing the courts' legitimacy and power.[4] For Germany, Bruce Moran and Pamela H. Smith have pursued contextual investigations of early modern alchemy in which the interests of princely courts play a central role in both experimentation itself and in the dissemination of "results."[5] For the Royal Society in England, Steven Shapin has pointed to the difference between public demonstration for the display of already accepted truths and private experiment for validation, the latter process being dependent on the character of witnesses whose perceived impartiality in turn depends on their social status as gentlemen.[6]

This scholarship has focused on the ways experimentation and the dissemination of empirical knowledge are embedded within particular societies and cultures. There has been far less interest in the question of how and why the values of empiricism (including observation, physical measurement, and experimentation) and of openness became central to seventeenth-century experimental philosophy in the first place. Recent contextual studies have not addressed sufficiently an extensive earlier tradition of scholarship represented by the work of Robert Merton, Edgar Zilsel, and Paolo Rossi, among others, on the role of technical and artisanal culture in the development of early modern science.[7] Such studies have established, however, that both "empiricism" and "openness" carry various meanings depending on context. Empiricism usually implies positive valuation of some sort of practice, observation, experience, measurement, and/or manipulation in the material world.[8] Openness becomes meaningful with a view to readership or audience.

This paper explores the linked values of empiricism and openness in four texts located within traditions of architectural authorship before the seventeenth century. The first two come from the ancient Mediterranean world: one a Greek treatise authored by Philo of Byzantium (3rd century B.C.E.), a military engineer working in the Library and Museum of Ptolemaic Alexandria; the second, the Latin *De architectura* by the Roman architect Vitruvius (fl. 20s B.C.E.). The second pair of treatises are from fifteenth-century Italy; first the Latin *De re aedificatoria* by Leon Battista Alberti (1404–1472), the famous humanist scholar whose patrons included the Pope Nicholas V; second, the Italian treatise *Trattato di architettura,* which describes the design and construction of a new city called Sforzinda, by the goldsmith and architect, Antonio Averlino, known as Filarete (ca. 1400–ca. 1469). Each of these four authors explicates in very different ways the values of openness and empiricism. Their treatises, taken together, constitute a significant (although by no means exclusive) aspect of what Hans-Georg Gadamer called the traditionary material[9] appropriated by experimental philosophy in the seventeenth century for its own particular ends.

At the outset, it should be recalled that architecture both in antiquity and in the fifteenth century encompassed a larger subject matter than it does today and included, among other things, what we would subsume under engineering. Philo's discussion of the origins of architecture was therefore quite appropriate within a treatise on military machines. Vitruvius posits that architecture has three parts: building, dialing (i.e., clock making), and *machinatio* (the construction and operation of machines). He also insists that architecture is made up of a rational aspect (*ratiocinatio*) and construction (*fabrica*).[10] Fifteenth-century architectural treatises also describe a discipline broader than its modern counterpart, one that included the construction of machines for lifting heavy weights and civil and military engineering.[11]

Openness and Empiricism in the Writings of Philo of Byzantium and Vitruvius

Historians of science and technology have often taken for granted that ancient cultural elites had contempt for handwork, which they associated with slavery. However, recently some scholars have challenged the view that such contempt was characteristic of all of antiquity, arguing instead that there were also positive ancient traditions of handwork and of the mechanical and technical arts.[12] Philo of Byzantium and Vitruvius are representatives of this positive ancient tradition.

Philo of Byzantium was a military engineer who worked in Alexandria, presumably at the Ptolemaic Museum and Library. This institution, which housed an enormous collection of books, was a center for teaching and scholarship. The Ptolemaic kings created and aggressively supported the institution to further their cultural power and hegemony in the Mediterranean world. Philo, who was clearly an experienced military engineer, wrote a partly extant treatise on the mechanical arts.[13] In one of these books, the *Belopoeica,* which mostly concerns siege machines, Philo describes the discovery of architectural proportions. He first discusses the improvement of catapults in terms of ongoing experimentation. Turning to architecture, he asserts that "everything cannot be accomplished by the theoretical methods of pure mechanics, but that much is to be found by experiment." He illustrates this point by recounting the step-by-step, experimental discovery of building proportions:

the correct proportions of buildings could not possibly have been determined right from the start and without the benefit of previous experience, as is clear from the fact that the old builders were extremely unskilful, not only in general building, but also in shaping the individual parts. The progress to proper building was not the result of one chance experiment. Some of the individual parts, which were equally thick and straight, seemed not to be so, because the sight is deceived in such objects, taking no account of perspective. By experimentally adding to the bulk here and subtracting there, by tapering, and by conducting every possible test, they made them appear regular to the sight and quite symmetrical, for this was the aim in that craft.[14]

Philo saw the development of "proper" building proportions as a result of gradual experimentation. His description reveals no interest in secrecy or competition and indeed suggests an implicitly open and explicitly cooperative process of testing and adjusting.

Philo describes a similar process of gradual, apparently open experimentation in the development of catapult machinery. He himself appears to have been involved in catapult manufacture; he writes, he says, to offer improvements in design.[15] Although we lack concrete evidence concerning how subsequent generations used his texts, there is no evidence either of a view that they should be concealed, or of actual concealment. In her detailed study of Hellenistic mechanical books, including those written by Philo, Astrid Schürmann emphasizes that such writings functioned as part of the enlarged compass of learning the Ptolemies promoted to further their own cultural hegemony in the Mediterranean world.[16] Such a project, which demands the open display of learning, was accomplished in part by writings such as those of Philo of Byzantium.

Whereas openness seems to be implicit in the technological processes Philo described, in contrast the Roman architect Vitruvius advocated openness explicitly and at the same time promulgated the value of hands-on empirical practice. Vitruvius, who wrote in the 20s B.C.E., was heir to Hellenistic traditions represented by Philo. He worked as a military engineer and later enjoyed the patronage of Octavia, sister to the emperor Augustus. Imperial patronage allowed him to write (or to complete) *De architectura,* the only fully extant architectural treatise that has survived from the ancient world.[17] Vitruvius suggests that the progress of architecture emerged out of gradual experimentation, and he links that progress to the ascent of human civilization itself. Modifying Lucretius, he writes that after humans had discovered fire and learned to speak with one another, they came together in one place. They stood upright and looked at the magnificence of the stars and the world and also "easily handled with their hands and fingers whatever they wished." They began to make shelters, some out of leaves, some dug from caves in hills, others constructed from mud and wattle in imitation of the nests of swallows. Then, "observing the houses of others and adding to their ideas new things from day to day, they produced better kinds of huts." Humans were of "an imitative and teachable nature" and boasted of their achievements. "Exercising their talents in rivalry," they acquired better judgment daily. Nature had equipped humans with perception like other animals, but also "armed their minds with ideas and purposes," so that "from the construction of buildings they progressed by degrees to other crafts and disciplines, and they led the way from a savage and rustic life to a peaceful civilisation."[18] Vitruvius thus describes architecture as a discipline that has progressed by various experiments, observations, even rivalries, to more complex structures, to a better knowledge of symmetry, and to other branches of knowledge as well.

Vitruvius's account of the progressive development of architecture and of civilization suggests an open process. Elsewhere, the Roman architect explicitly advocates openness as it concerns knowledge about the crafts and other kinds of learning as well. He elucidates the value of openness with a story about Socrates. As we know from Plato's *Apology*, the Delphic Apollo declared Socrates to be the wisest of all men. However, Vitruvius's story is rather different from the one told by Plato, in which Socrates's wisdom is understood to be based on the recognition of his own ignorance. In the Vitruvian version, the wisdom of Socrates rests on his belief that the hearts of men should have open windows so that they "might not keep their notions hidden, but open for inspection." Vitruvius wished that nature had followed Socrates's opinion, making human ideas "explicit and manifest." If humans had been constructed in this way, "not only would the merits or defects of human minds be seen at once, but the knowledge of disciplines also, lying under the view of the eyes, would be tested by no uncertain judgments," thereby adding authority both to those who were learned and to the skilled.[19]

Stipulating that many kinds of knowledge would benefit from openness, Vitruvius focuses on the crafts. He laments that some skilled craftsmen have gained reputation, but that others, equally skilled, have not. Indeed, his own authorship seems to have come out of his failure to obtain architectural commissions. He complains that though "the ignorant excel in influence rather than the learned, I judge that we must not rival the ignorant in their intrigues; but I will rather display the excellence of our knowledge by the publication of these rules."[20]

Vitruvius advocates the open, written transmission of knowledge as crucial to the progress of all disciplines. He also defends the integrity of his own authorship and insists that past authors be honored and accorded proper credit. He writes:

> Truly Caesar, I do not bring this compendium before the public with the titles of other people's books changed and my own name inserted, nor did I set out to win approval for myself by finding fault with anyone's ideas. Instead I offer infinite gratitude to all writers, because they have furnished abundant supplies of knowledge, each in his own field, by assembling over the ages outstanding achievements of human talent. As if drawing water from springs and moving it to serve our own objectives, we have in these supplies more eloquent and readily available resources for [our own] writing, and placing our trust in such authors, we dare to compose new teachings.[21]

The Roman architect expresses gratitude for particular authors who have written down their precepts, including philosophers and those who had written about the natural world. He names, among others, Thales, Democritus, Anaxagoras, Plato, Aristotle, Zeno, and Epicurus.[22]

Yet Vitruvius's notion of openness and credit to past authorship is limited in a particular way. Whereas he condemns literary theft, he also denounces any criticism of dead authors. His description of the fate of Zoilus, a critic of Homer, is remarkably unsympathetic. Some say Zoilus was crucified, some that he was stoned, some that he was burned alive. Whichever version of his death is the correct one, the penalty was deserved—the just desert of someone who "calls into court those who can no longer reply publicly when asked what they meant by their writings."[23]

Vitruvius was indebted to Hellenistic traditions of authorship as well as to the Hellenistic belief that experimentation leads gradually to the principles of various disciplines and thus to the progress of knowledge. Yet his explicit advocacy of open, written transmission, his concern to credit past writers, and his harsh condemnation of those who criticize dead authors all arise from his particular Roman context. Vitruvius's views concerning openness, authorship, and credit, I would suggest, represent a response to the Augustan revival of traditional Roman religious values.

In the 20s B.C.E., the decade in which Vitruvius was probably writing the relevant passages, Augustus was attempting to reinstate traditional Roman religious values as one of many ways to secure Rome's stability and the legitimacy of his own power. The emperor inaugurated a great Roman building program that included the repair and construction of more than 185 temples. He also appointed numerous priests, filling offices that had been left vacant for decades. Traditional Roman religion involved a form of ancestor worship. Roman households included altars for the *Lares,* who represented, among other things, the living spirit of ancestors. Augustus revived the cult of the *Lares* and associated it with his own family, as well as with the newly revived state religion. Vitruvius's exhortation to honor past authors and his condemnation of their critics should be understood in the context of this reinstatement of traditional Roman religion. The ancestors of authors are the writers of the past. They are crucial both to the progress of knowledge and to the edifice of accumulated knowledge. Vitruvius uses his terminology very precisely. Those who steal the writings of others are *impius:* guilty of impiety.[24]

Both Philo and Vitruvius wrote under the aegis of royal or imperial patronage. Both belonged to ancient empirical traditions that valued experimentation. Philo's implicit openness, I suggest, was related to his part in displaying Ptolemaic intellectual and technical superiority. Vitruvius's ideals of openness and credit to authorship were tied to the Augustan revival of traditional Roman religion that inculcated pious respect for ancestors. Philo and Vitruvius both characterized the development of architecture as a gradual, experimental process that relied on openness. Yet both were oriented more to technical design and production than to any ongoing system for the creation of new knowledge. Both men had been practitioners, military engineers who clearly, from their writings, had some hands-on experience. Their positive views concerning empirical practice were not readily transferable to a methodology pertaining to the production of knowledge in a more general sense. Neither can it be assumed that the social and cultural elites of their times widely shared their values.

Techne and *Praxis* in the Exoteric Writings of Alberti and Filarete

In contrast to that in the ancient Mediterranean world, the legitimacy of rulership in fifteenth century Italy came to be closely tied to aspects of technical production, including architecture and construction, painting, and other decorative arts. Jürgen Habermas, among others, has observed that in antiquity the ability to govern was perceived to rest on character rather than on technology, whereas early modern political and military power became closely allied with technique.[25] From a different conceptual framework, Christine Smith has noted the quattrocento "use of the built environment as evidence for the authority of the state," whereas Richard Goldthwaite has characterized the Renaissance economy as one based on the construction of magnificent palaces and other great buildings, both civic and secular, and on the consumption of luxury goods by the ruling elites. Randolph Starn and Loren Partridge have analyzed three Italian political regimes in terms of the great decorated halls that represented them.[26] Princes and oligarchs transformed the urban spaces they ruled. They redesigned city streets and built great public buildings, churches, and magnificent palaces. They commissioned painting and sculpture, as they also supported the luxury trades including silks, fine ceramics, and jewelry. As bankers and merchants became princes and rulers during the fifteenth and sixteenth centuries, they used construction and ornamentation as modalities for legitimation. *Praxis* (action, including political action) and *techne* (the production and manipulation of material objects) came to be far more closely connected than they had been before.[27]

One result of this closer connection was that the linkage between the value of openness and the positive valuation of empiricism, handwork, and experimentation, already present in a minority tradition of antiquity, developed much greater cultural significance during the fifteenth century. Skilled architects, engineers, painters, sculptors, and builders could practice their trades within the context of the guilds, which often fostered secrecy and exclusion. Yet such skilled practitioners could sometimes find patrons instead and move away from the requirements of guild membership to the very different social and professional context of a patronage relationship. The shift to a patronage milieu encouraged practitioners to explicate various constructive arts in treatises that they dedicated to patrons. Such written explication tended toward openness as it also continued to affirm the value of empirical practice.[28]

From a social position very different from that of practitioners, university-educated humanists such as Alberti, also motivated by patronage, wrote learned treatises on subjects such as architecture, painting, and sculpture that formerly had been confined for the most part to craft practice.[29] Humanist authorship such as Alberti's facilitated the complex process whereby some artisan crafts, especially architecture, became learned disciplines. As princes and oligarchs transformed both private and public space as a way of legitimizing their political power, the empirical practices that lay at the foundation of such transformations were, perhaps not surprisingly, explicated openly in writing and illustration for the greater glory of the ruler and for the benefit of his client, the author, as well.

It is not to turn away from the necessity of investigating particular cases and local situations to suggest also that the very fundamental processes alluded to above ultimately contributed not only to the reordering of urban space, to new ways of representing political power, and to a great plethora of books and writings on the practical and mechanical arts, but ultimately to a transformation of knowledge concerning the natural world. I say "contributed to," not "caused." I have no desire to simplify the immensely complex and interesting group of diverse developments and transformations that many of us no longer want to call the "scientific revolution." I merely point out that the linkage between the values of empiricism and of openness was highly familiar, not to say commonplace, well before they became important tenets of experimental philosophy in the seventeenth century, and to point out that architecture, and specifically writings on architecture, were important sites for that linkage.

Here it is only possible to elucidate this claim with two examples, albeit important ones, from the fifteenth century: the architectural treatises of Battista Alberti and Antonio Averlino, known as Filarete. Alberti, one of the most important humanists of the fifteenth century, wrote a learned Latin treatise on architecture. Filarete was a practitioner trained as a goldsmith. His Italian treatise was clearly inspired by Alberti's. Both works elucidate in very different ways the joined values of openness and empiricism expressed within a context of civic and courtly praxis.

In *De re aedificatoria,* Alberti underscores the close relationships of the constructive and mechanical arts, civic and moral life, and political power, just as he also emphasizes the importance of clear terminology and openness. Alberti's interest in (and criticism of) the text of Vitruvius, his insistence on clarity and lucid terminology, and his vehement construction of the discipline of architecture as an art central to ethics and to civic life are all consistent with his humanist orientation. Alberti wrote treatises on painting, sculpture, and architecture in which he separates these arts from craft practice and suggests that they are founded upon mathematics, harmony, proportion, and perspective. By means of his own authorship, he places them among the learned disciplines. Yet John Oppel has cogently observed that although it is usually said that Alberti raised architecture to the status of one of the liberal arts, "it would be at least as true and historically more appropriate to argue that he did just the reverse, that he brought the liberal arts down to the level of the mechanical ones."[30]

Alberti situates the visual arts among the learned disciplines, but he does not thereby dislodge them from the constructed world. He dedicated the Italian translation of his treatise on painting to his friend Brunelleschi, who had invented a version of painter's perspective, had designed and built the dome of the Florentine Cathedral, and had invented numerous mechanisms such as lifting machines to aid in its construction. Alberti had himself invented a form of perspective, undoubtedly influenced by Brunelleschi, and had tried his own hand at sculpture, drawing, and painting. Although his exact role as architect is often tenuously documented, Alberti was involved in the design of several major buildings, for example, in Rimini, Florence, and Mantua. He also contributed to Nicholas V's renewal of Rome in the mid–fifteenth century and perhaps was involved in the planning of Pienza, the town redesigned (1459–1464) under Pius II.[31]

Separating architecture from craft practice by means of mathematics and design, Alberti also grounds it in engineering for the improvement of civic life. In the *De re aedificatoria,* completed around 1452, he defines the architect not as a carpenter, but rather as one "who by sure and wonderful reason and method, knows both how to devise through his own mind and energy, and to realize by construction, whatever can be most beautifully fitted out for the noble needs of man, by the movement of weights and the joining and massing of bodies." He expounds upon architecture as a discipline essential to human life and health. Not only houses, but walks, swimming pools, and baths have helped to keep men healthy. Vehicles, mills, timepieces, and other smaller inventions play a vital role in everyday life. Architecture devised methods of drawing up vast quantities of water for many purposes, and it produced buildings for divine worship, such as "memorials, shrines, sanctuaries, temples, and the like." Further, cutting through rock, tunneling through mountains, filling in valleys, restraining the waters of the sea and lakes, draining marshes, building ships, altering the course of rivers, dredging the mouths of rivers, constructing harbors and bridges—through doing all these things the architect has not only "met the temporary needs of man, but also opened up new gateways to all the provinces of the world," allowing nations to exchange food and goods as well as experience and knowledge. Architecture moreover has provided ballistic engines and machines of war, fortresses and other things that protect a country's liberty. Alberti posits the architect's skill and ability as the central item for victory in most wars, more important than the command and foresight of any general.[32]

Alberti's own interest in machines and mechanical problems is notable. In 1447 he had supervised the raising of an ancient ship from Lake Nemi and had written a small book about it, *Navis,* which is lost.[33] In his architectural treatise, he points to several of his own inventions that have to do with rigging gangplanks of ships to prevent successful boarding by enemies and other inventions for sinking and burning enemy ships. He also mentions that he will deal with war machines at greater length elsewhere, implying, perhaps, that he was planning to write a treatise on the subject.[34]

The solution of engineering problems represents just one aspect of the intrinsic relationship between the work of the architect and civic life as a whole. Alberti writes that "when you erect a wall or portico of great elegance and adorn it with a door, columns, or roof, good citizens approve and express joy for their own sake, as well as for yours, because they realize that you have used your wealth to increase greatly not only your own honor and glory, but also that of your family, your descendants, and the whole city." He emphasizes that "the security, dignity, and honor of the republic depend greatly on the architect."[35]

Alberti insists upon an extremely close relationship between civic life, the patron's social and political position, and architecture. He admonishes the patron that "it is your duty [among other things] . . . to take into account your own social standing as the one who commissions the building: it is the sign of a well-informed and judicious mind to plan the whole undertaking in accordance with one's position in society and the requirements of use." To accomplish perfection in architecture, he advises the patron to follow the knowledge and advice of "honest and impartial" experts, rather than "personal whim and feeling."[36] Elsewhere, he warns those who would commission a building "to take careful account of the current situation, both in terms of public affairs and in terms of your own personal circumstances and those of your family, to insure that in unsettled times your undertaking does not incite envy if you continue construction nor cause needless expense if you abandon it."[37] Alberti insists finally that the type of rulership should determine the ruler's architectural choices: Tyrants must make different architectural decisions than magistrates.[38]

If architecture is almost organically grounded in moral and political life, it is also a discipline that must be learned and that can achieve progress. To acquire knowledge concerning architectural design, Alberti advocates empirical study and observation as well as collecting advice from the past: "[W]e shall collect, compare, and extract into our own work all the soundest and most useful advice that our learned ancestors have handed down to us in writing, and whatever principles we ourselves have noted in the very execution of their works. We shall go on to report things contrived through our own invention, by careful, painstaking investigation, things we consider to be of some future use."[39] Elsewhere, Alberti stresses inventiveness as a desirable approach to the architectural orders: "although other famous architects seem to recommend by their work either the Doric, or the Ionic, or the Corinthian, or the Tuscan division as being the most convenient, there is no reason why we should follow their design in our own work, as though legally obliged; but rather, inspired by their example, we should strive to produce our own inventions. . . ."[40]

In laying out the precepts of architecture, Alberti emphasizes openness, by which he means clarity of technical language:

> But since it is our desire to be as limpid, clear, and expeditious as possible in dealing with a subject otherwise knotty, awkward, and for the most part thoroughly obscure, we shall explain, as is our custom, the precise nature of our undertaking. For the very springs of our argument should be laid open, so that the discussion that follows may flow more easily.[41]

Elsewhere, Alberti laments the obscurity and the poor Latinity of Vitruvius's text. He stresses the difficulty of his task of writing on architecture. He would rather his speech "seemed lucid than appeared eloquent," and believes that what he has written is "in proper Latin, and in comprehensible form."[42]

Unlike the university-educated Alberti, Filarete ("lover of virtue") was trained as an artisan—in a Florentine workshop. Subsequently, he worked as a goldsmith in Rome and then in the early 1450s acquired the patronage of Francesco Sforza in Milan, who placed him in charge of the Fabbrica of the Duomo and other major construction projects over the protests of local craftsmen. Filarete completed his architectural treatise on the ideal city, Sforzinda, in the early 1460s and dedicated the first copy to Francesco Sforza and a later copy to Piero di Medici of Florence. His career was similar to that of other fifteenth-century artisan-practitioners (such as Piero della Francesca and Francesco di Giorgio) who wrote treatises on practical and technical subjects. Like many of them, Filarete was geographically mobile and used patronage to circumvent the restrictions of local guilds. He wrote his treatise on architecture in the context of patronage and includes within it an elaborate portrait of Sforzinda and of an ideal architect-patron relationship, in which he portrays himself as the architect and Francesco Sforza as the patron.[43] Filarete counsels his patron to listen to his advice intently, just as if he were hearing military reports concerning his troops' defense or reconquest of one of his valued possessions,[44] thereby underscoring the intrinsic relationship between princely power and architecture.

Filarete's lengthy treatise involves extravagant display: of Sforza power, glory, and magnanimity, of the princely city as sign of princely power, and of the architect as preeminent and key facilitator of princely glory. Openly displayed as well is the discipline of architecture itself. Having created the treatise as a showcase for princely glory and architectural virtue, Filarete also vigorously advocates the value of empirical, hands-on practice and experience in building design and construction. He advises his readers that he will discuss all parts of building—design, measurement, materials, construction—and also the requisite knowledge of architects; the means and construction of building a city, its site, and the location of its buildings, squares, and streets; and finally, ancient buildings and practice.[45]

The architect, he suggests, must exercise close day-to-day oversight over the construction process: He should be directly responsible to the patron and should also have enough authority to enforce his will in matters that affect the building's ultimate quality.

In return, the patron should honor and love the architect, insofar as he wants his building to go well, nor should that love and devotion be different from what is given to "that one without which man cannot generate [i.e., the patron's wife]." The architect should be esteemed for his knowledge. The patron has chosen the architect as organizer and executor and has put his soul into the architect's hands, just as the architect has put his own soul into the patron's hands and will satisfy the patron's desires with love, managing the building project that the patron loves and on which he spends so much of his treasure "solely for the desire that he has to see it completed."[46]

The architect is paid well and is treated as a near equal to the prince. He hunts with the nobles from the area surrounding the new city and converses and dines regularly with the prince himself. He is honored in every way.[47] What is significant here is not so much Filarete's fantasied transformation of himself, the artisan, into an intimate companion of the prince, but his resolute retention of the values of handwork and empirical knowledge in the process. The architect himself goes "to lay out the cords," in the first step of laying out the city. The architect should understand mathematics but should also be able "to understand more practices and also to demonstrate them with the work of his hand, with rules of measure, and of proportion and of quality, and of suitability." Reiterating the importance of handwork, Filarete insists that if the architect does not have the skill of making things "with his own hand, he will never know how to show or to explain a thing which may be good. It is necessary that he be resourceful and that he conceive of making various things, and [that] he demonstrate with his own hand." Note that the architect's skill at making things is tied directly here to his capacity for show, explanation, and demonstration.[48]

Such a capacity to demonstrate appears to be crucial for transmitting architectural knowledge, not so much, it seems, to other artisans, but to patrons. In one episode, the architect enjoys a dinner with the patron and other nobles. During the dinner he explains the many good omens that have appeared during the building of the city, after which "so great was the love that the [l]ord and the others had for me that they gave me enough to live honorably." Moreover, the lord's son is so taken with love for the architect and so pleased with the reasoning of architecture that he begs his father to let him become the architect's student. The father grants his son's request. Thereafter, the two are always together. The architect continually teaches his young pupil, who in turn increasingly gains his own skill and always delights in his instruction.[49]

The close association of the architect-practitioner and the noble patron is paralleled in the new city by institutions in which learned and craft disciplines are taught side by side. Filarete proposes a school for impoverished youth that should provide instruction in letters and in good habits but should also include instruction in "every branch of knowledge and every skill" that he prescribes. Instructors should include a doctor of laws, of medicine, of canon law, of rhetoric and poetry. The school would be unusual, however, because "although they may not have as much dignity," he intends that "manual skills" will also be taught there by craftsmen, including a painter, a silversmith, sculptors of marble and of wood, a turner, an iron smith, a master of embroidery, a tailor, a pharmacist, a glassmaker and a master of clay.[50] All children could thereby be trained under the same roof in the discipline to which their soul and intelligence is most suited.

The proposed salaries of the various instructors make it clear that some have much higher status than others (as was of course the case in his own society). Nevertheless, Filarete himself recognizes the significance of the arrangement whereby both the learned and the skilled are taught in the same school:

> This will be a thing that will always endure and a thing that has never been done, although there are in these lands universities where students are boarded in this way. In many places they pay a certain amount, but this convenience does not exist if one is not a student of letters. Other skills are also necessary and noble, and there are good masters of them, and indeed all intellects are not equal. Thus it will be possible for every intellect to be trained.[51]

In addition to the school, Filarete's city contains a house of vice and house of virtue. In the latter, instruction is offered in the seven liberal arts, areas are designated for military exercises, and in addition, all the crafts and trades are both practiced and taught. When the students are "judged by good masters" and educated there, they are "given the degree like the doctors." A foreigner who arrives and is a good master can also show an example of his work, be examined, and if he passes muster, be honored similarly with a doctorate. The governors of the temple of virtue are three: the first, one of the doctors; the second, one who had received honors for his feats of arms; and the third, also "from those artisans who are learned."[52]

Filarete gives the architect who builds the house of virtue a name that is an anagram of Antonio Averlino, that is, himself. The architect's dwelling is situated close to the house of virtue. To enter the house, one is required to pass an examination that tests skills relevant to architecture, for example, in "drawing, measure, and knowing how to do many things with his own hands as this one did." When the person entering has been found to be satisfactory, he is greatly honored and is permitted to stay in the architect's beautiful and ornately decorated room.[53]

Thus learned and craft subjects are pursued within one institution in Filarete's ideal city. Similarly, traditional class alliances are mixed when Filarete portrays the young prince as a student avidly interested in technological and craft processes. The prince wants to know how metal is melted and how furnaces are made for melting bronze, and he asks for information about glass furnaces. Filarete's treatise contains one of the earliest descriptions of a modern blast furnace as well as a detailed description of how to make a plaster that does not stain, and it includes sections on drawing, perspective, and painting. Filarete's interest in technical subjects went beyond the treatise that we have in hand.

He mentions books that he is in the process of writing (none of which, if they were completed, are extant) on agriculture, on technical matters, and on engines.[54]

Filarete was initially trained as an artisan. Yet his open, written exposition of architectural principles stands in contrast to the secrecy associated with late medieval craft guilds. He could dispense with such guilds because he successfully opted for patronage as an alternative form of protection. His exoteric treatise, which he presented in elegant manuscript copies, concerned a subject of essential interest to Renaissance princes and could only assist his fortunes as a client. Yet he carefully preserved in his treatise the values of handwork, intrinsic to artisanal practice, as well.

In terms of authorship and of architectural commissions, both Filarete and Alberti worked within patronage systems. Nevertheless, Alberti's university education, his family status, and his outstanding abilities as a Latinist allowed him to transcend patronage relationships more readily than could Filarete. Both authors undoubtedly intended their architectural treatises as comprehensive presentations of general principles. Alberti was far more successful, in part because he wrote in Latin, in part because he wrote with less local specificity than did Filarete (whose patron appears numerous times within the treatise itself). Ultimately, Alberti's social status and learning ensured that his treatise was printed, an event that removed it from the vagaries of a local patronage system. Filarete's, in contrast, remained in manuscript form. Yet it is significant that both combined the values of open explication and empirical practice.

Conclusion

In what ways did the positive valuation of empirical practice and of openness within traditions of architectural authorship influence the development of experimental philosophy? To summarize, the treatises of Philo and of Vitruvius were among those that provided exemplary material from the ancient world. The contexts respectively of Ptolemaic Egypt and Augustan Rome that constituted the initial conditions of their authorship, both practical and conceptual, were not of course either apparent or relevant to readers in the fifteenth century. The Renaissance interest in and appropriation of these texts occurred at a time when political praxis and technological production were far more closely allied than they had been in antiquity. The increasing importance in the fifteenth century of patronage systems based on both the construction of material objects, including buildings, and the display of both knowledge and things, created a favorable climate for the ideal of openness with reference to the principles of technological and architectural practice.

In the sixteenth century, the broad purview of fifteenth-century architecture broke apart into more specialized categories, such as civic architecture, fortification, artillery, and other aspects of engineering. Yet practitioners within these more narrowly defined disciplines continued to produce not only buildings and machines, but treatises and books. Printing greatly encouraged in practice the openness that had been advocated both in the ancient and more recent past.[55] The authors of practical and technical books often either implicitly or explicitly advocated openness. They also discussed empirical practices and advocated the positive value of those practices.

Alliances among architecture, engineering, power, and knowledge remain clearly in evidence for much of the sixteenth century. How then, can such alliances be reconciled with the picture being developed by some historians of late sixteenth and seventeenth-century mathematics and experimental philosophy that suggests that experimental philosophers separated themselves from practice and practitioners, devaluing them in the process (even when technicians were needed to carry out experiments). Some aspects of this picture of practitioner devaluation are the subject of ongoing scholarly debate; in some cases, the picture is valid for some contexts, but not for others. Yet practitioner devaluation is unquestionably in evidence in at least some arenas of experimental philosophy.[56]

If experimental philosophy devalued practitioners and separated itself from practice, how did it come to be experimental and/or empirical in the first place, and why was that empiricism tied to openness? My answer to this question would include the observation that ancient architectural and engineering traditions provided important models for empirical or experimental values linked to openness. In the fifteenth century and later, these models were taken up and developed in very different ways as architecture-engineering became self-consciously elaborated as intrinsic to princely power and civic life and was explicated in treatises as a rational body of knowledge. Architecture played a central role in the legitimization of political power as it also became validated as a rational discipline in written treatises. Experimental philosophy appropriated values (including empiricism and openness) from these broadly based traditions.

Practitioners of natural philosophy in the sixteenth and seventeenth centuries frequently adopted the express values of openness and empiricism, while they also in actual practice, as Paula Findlen and William R. Newman show in their essays in this volume, established hierarchies of openness and access that were regulated in a variety of ways. The actual practices of access and indeed of empiricism were not identical to the expressed ideals of openness and empiricism or the linkage of such ideals to each other. Both practices and ideals were crucial to the project of legitimation carried forward by the new experimental philosophy. In this essay I have argued that the linkage of openness and empiricism had ancient roots within written architectural traditions that were strenuously reinforced in the fifteenth century and beyond in architectural treatises produced in a context of close interaction between political legitimization and the constructive arts. I have been able to illustrate the thesis with only two examples; there are numerous others, as I have demonstrated at least in part elsewhere.[57] The twin values of openness and empiricism were repeatedly enunciated in writings on the constructive arts and engineering in the fifteenth and sixteenth centuries. As linked values that came out of a broadly based cultural heritage, they were available to anyone who wanted to appropriate them in the following century.

Seventeenth-century experimental philosophers often distanced themselves from specific architectural, practical, and technical traditions as well as from direct association with monarchical or other kinds of political power. Such philosophers enhanced the status of experimental knowledge by placing it in a special preserve, cordoned off from the practical concerns of architecture, technology, and practice as well as from the arena of political power. Whereas experimental philosophy became legitimized as valid "disinterested" knowledge of the world, architecture, technology, and other practices came to be seen as separate and often derivative—important, undoubtedly, for life itself and for political and military power—but distinct from the new edifice of knowledge about the natural world. Yet this separation of architecture, political power, and knowledge about the world should not obscure a complex prior history in which the linked values of openness and empiricism flourished within architectural traditions and then were appropriated as central values of the new experimental philosophy.

NOTES

I would like to thank Bob Korn, Priscilla Long, Graham Hammill, and Julie R. Solomon for discussion and critiques, which improved this paper greatly.

1. One example can be found in the report of the experiments, or *Saggi,* of the Florentine Accademia del Cimento published in 1667. The preface of the *Saggi* notes that the patron of the academy, Prince Leopold, had decided to "verify the value" of the assertions of great authors "by wiser and more exact experiments" and to present the results as a precious and desirable "gift" to whoever was "anxious to discover Truth." The academy's conclusions were intended to encourage other people "to repeat the same experiments with the greatest rigor." The academy wished "for nothing else but a free communication from the various Societies" throughout Europe. It insisted that it had "always cited the authors, as far as they were known" to them. Finally, the "clearest proof of the open sincerity" of their procedures was that "all may see the liberality with which we have always shared them with anyone passing through these parts who showed a desire to enjoy some account of them, whether as an act of courtesy, or because he esteemed learning, or from the incentive of noble curiosity." Middleton, *Experimenters,* p. 91 (citations from the preface of the *Saggi di naturali esperienze fatte nell'Accademia del Cimento,* first published in 1667 and here translated by Middleton). For the Royal Society, Henry Oldenburg noted in *Philosophical Transactions* 1 (6 March 1665): 1–2, that there was "nothing more necessary for promoting the improvement of Philosophical Matters, than the communicating to such, as apply their Studies and Endeavours that way, such things as are discovered or put in practise by others. . . . To the end, that such Productions being clearly and truly communicated, desires after solid and usefull knowledge may be further entertained, ingenious Endeavors and Undertakings cherished, and those, addicted to and conversant in such matters, may be invited and encouraged to search, try, and find out new things, impart their knowledge to one another, and contribute what they can do to the

Grand design of improving Natural knowledge, and perfecting all *Philosophical Arts,* and *Sciences.*" For the values of openness and cooperation in the Parisian Royal Academy, see Hahn, *Anatomy of a Scientific Institution,* pp. 3 and 25–26.

2. Earlier twentieth-century social construction of science might schematically be divided into Marxian and Mertonian strands. For an introduction, see especially Freudenthal et al., "Controversy," pp. 105–191; Freudenthal, "Towards a Social History," pp. 193–212; Hessen, "Social and Economic Roots," pp. 149–212; Merton, *Science, Technology and Society,* esp. pp. 137–261; and Shapin, "Understanding the Merton Thesis," pp. 594–604. More recent scholarship in the sociology of science was initially influenced by Kuhn, *Structure of Scientific Revolutions,* who proposed that revolutions in scientific thought occurred by means of paradigm shifts among communities of scientists. The paradigm shift that followed among communities of historians is exemplified by Shapin's *Social History of Truth,* among many others.
3. Hull, *Science as a Process;* LaFollette, *Stealing into Print;* Galison, *How Experiments End.*
4. Findlen, "Controlling the Experiment"; Biagioli, "Social Status of Italian Mathematicians"; and Biagioli, *Galileo, Courtier.*
5. Moran, "German Prince-Practitioners"; Moran, ed., *Patronage and Institutions;* and Smith, *Business of Alchemy.*
6. Shapin, "House of Experiment"; and Shapin and Schaffer, *Leviathan and the Air Pump,* esp. pp. 110–224.
7. Merton, *Science, Technology, and Society,* esp. pp. 137–198; Zilsel, "Sociological Roots of Science"; and Rossi, *Philosophy, Technology, and the Arts.* For further bibliography, see Long, "Contribution of Architectural Writers to a 'Scientific' Outlook," pp. 265–266; and see Bennett, "Mechanics' Philosophy."
8. Recent discussions of "empiricism," "experiment," "experience," "mechanics," "mechanical philosophy," and the influence of practical traditions on these entities include Bennett, "Mechanics' Philosophy"; Gabbey, "Between *ars* and *philosophia naturalis*"; Gooding, Pinch, and Schaffer, eds., *Uses of Experiment;* Hahn, "Mechanistic Age"; Holmes, "Do We Understand Historically"; Long, "Contribution of Architectural Writers to a 'Scientific' Outlook"; Schmitt, "Experience and Experiment"; Tiles, "Experimental Evidence vs. Experimental Practice"; and Voss, "Between the Cannon and the Book."
9. Gadamer, *Truth and Method,* esp. pp. 265–379.
10. Vitruvius, *De architectura,* 1.3.1 for the parts of architecture (*aedificatio, gnomonice, machinatio*), and 1.1.1 for the *ratiocinatio* and *fabrica* on which architecture depends. Throughout this paper, I have used Granger's edition but have in some cases made changes in the translation.
11. Alberti, *Art of Building,* pp. 3–4.
12. For the view that ancient culture was characterized by contempt for handwork, see esp. Rossi, *Philosophy, Technology, and the Arts,* pp. 13–15; and for further bibliography, Long, "Contribution of Architectural Writers to a 'Scientific' Outlook," p. 271, n. 14. For the revised view, see Ferrari and Vegetti, "Science, Technology and Medicine," pp. 200–202; Long, "Invention, Authorship, 'Intellectual Property,' and the Origin of Patents," pp. 848–858; and Whitney, *Paradise Restored,* esp. pp. 23–55.
13. For Philo, see *Dictionary of Scientific Biography,* s. v. "Philo of Byzantium," by Drachmann; Fraser, *Ptolemaic Alexandria,* 1: 428–434 and 2: 619–620; and Schürmann, *Griechische*

Mechanik, esp. 7–8 and 61–92. For the Ptolemaic Museum and Library, see especially Delia, "Romance to Rhetoric," pp. 1449–1467; Fraser, *Ptolemaic Alexandria,* 1: 305–335 and 2: 462–494; and Wendel and Göber, "Griechisch-Römische Altertum," pp. 62–88.

14. Marsden, *Artillery: Treatises,* pp. 108–109 (Philon, *Belopoeica,* 50.24–51.7) for the text and English translation. Shortly before this passage, Philo mentions Polyclitus, a fifth-century B.C.E. sculptor who wrote a treatise on the canon of proportions. It seems possible that Philo's view of the development of architectural proportions is derived from this treatise. For a discussion of Polyclitus, see *Oxford Classical Dictionary,* 2d ed., s. v. "Policlitus (2)," by Richter; and Stewart, "Canon of Polykleitos."

15. Marsden, *Artillery: Treatises,* pp. 106–109 (Philon, *Belopoeica,* 49–50), "In the old days, some engineers were on the way to discovering that the fundamental basis and unit of measure for the construction of engines was the diameter of the hole. This had to be obtained not by chance or at random, but by a standard method. . . . It was impossible to obtain it except by experimentally increasing and diminishing the perimeter of the hole. The old engineers, of course, did not reach a conclusion . . . but they did decide what to look for. Later engineers drew conclusions from former mistakes, looked exclusively for a standard factor with subsequent experiments as a guide, and introduced the basic principle of construction, namely the diameter of the circle that holds the spring." Philo specifies that his own interest in writing has to do with the issues of the machine's proportions and the hole's diameter.

16. Schürmann, *Griechische Mechanik,* pp. 13–32; and Long and Roland, "Military Secrecy," pp. 271–272, who point to the lack of evidence for secrecy with regard to weapons technology in the ancient world.

17. See Baldwin, "Date, Identity, and Career of Vitruvius," pp. 425–434, for a discussion of the many uncertainties and conflicting viewpoints concerning Vitruvius's life and identity.

18. Vitruvius, *De architectura,* 2.1.1–7. Vitruvius elaborated that within the discipline of architecture itself, humans moved from huts to houses on foundations, to brick walls, and to wooden and tile roofs. Finally, "by the observations made in their studies they were led on from wandering and uncertain judgments to the assured method of symmetry." For the primitive hut in the history of thought after Vitruvius, see Rykwert, *Adam's House in Paradise.*

19. Vitruvius, *De architectura,* 3.pref.1. Silvio Ferri (Vitruvius, *Architettura,* p. 90, n. 1) suggests that the probable source for Vitruvius's story is the myth (reported by Lucian, *Hermotimus,* 20) in which Athena, Poseidon, and Hephaestus quarrel about who is the best artist. To resolve the quarrel, they have a contest in which Poseidon makes a bull, Athena designs a house, and Hephaestus constructs a man. The judge, Momus, criticizes Hephaestus because he did not make windows in the man's chest so that everyone could see his desires and thoughts and whether he was lying or telling the truth.

20. Vitruvius, *De architectura,* 3.pref.3.

21. Vitruvius, *De architectura,* 7.pref.10. The translation of this passage is my own.

22. Vitruvius, *De architectura,* 7.pref.2.

23. Vitruvius, *De architectura,* 7.pref.8–9. See also Fraser, "Aristophanes," pp. 121–122.

24. For brief discussions of Augustus's revival of traditional religion and of his building program with further bibliography, see Stockton, "Founding of Empire," pp. 133–135 and 149; and Wilson, "Roman Art and Architecture," pp. 364–374 and pp. 399–400; and see Ogilvie, *Romans and Their Gods in the Age of Augustus,* esp. pp. 112–123. On the *Lares,* see *Oxford Classical Dictionary,* 2d ed., s. v. "Lares," by Rose. Vitruvius, *De architectura,* 7.pref.3, writes that

those stealing [*furantes*] the works of past authors, who "appropriate them [the writings] to themselves; writers who do not depend upon their own ideas, but in their envy boast of other men's goods whom they have robbed with violence, should not only receive censure but punishment for their impious manner of life."

25. Habermas, *Theory and Practice,* pp. 41–81; and see also Dunne, *Back to the Rough Ground,* pp. 168–226. Aristotle drew the distinctions between *praxis* (action, including political and military action), *techne* (material production), and *episteme* (knowledge): see particularly *Nichomachean Ethics,* 6.4–6.5 (1140a–1141a).

26. Smith, *Architecture in the Culture of Early Humanism,* p. 50; Goldthwaite, *Building of Renaissance Florence* and *Wealth and the Demand for Art,* esp. pp. 212–255; and Starn and Partridge, *Arts of Power.*

27. See Long, "Power, Patronage, and the Authorship of *Ars*" for the influence of the fifteenth-century alliance between *praxis* and *techne* on the development of authorship in the mechanical arts.

28. For craft secrecy in the late medieval guilds, see Long, "Invention, Authorship, 'Intellectual Property,' and the Origin of Patents," esp. pp. 870–881; and see Long, "Power, Patronage and the Authorship of *Ars*" for the fifteenth-century expansion of authorship in the mechanical arts under the influence of patronage.

29. Long, "Power, Patronage, and the Authorship of *Ars,*" esp. pp. 21–31.

30. Oppel, "The Priority of the Architect," pp. 251–267, citation on p. 251. See also Jarzombek, *On Leon Baptista Alberti,* who argues for an integrative approach to Alberti's writings that would consider together his moral, technical, and aesthetic works; and Smith, *Architecture in the Culture of Early Humanism,* who emphasizes the eclectic nature of Alberti's architectural theory as well as its indebtedness to particular rhetorical traditions.

31. See especially Alberti, *On painting,* p. 120, n. 21 (on Alberti's painting and sculpting activities); Borsi, *Leon Battista Alberti* (for his architectural works); *Dizionario Biografico degli Italiani,* s. v. "Alberti, Leon Battista," by Grayson and Argan; Smith, *Architecture in the Culture of Early Humanism,* pp. 19–39 (who emphasizes that Alberti admired Brunelleschi for his engineering accomplishments) and 98–129 (for Pienza); Westfall, *In This Most Perfect Paradise;* and for a critical appraisal of Westfall's work, Tafuri, "'Cives esse non licere': The Rome of Nicholas V" (for Nicholas V and urban planning in Rome).

32. Alberti, *On the Art of Building,* pp. 3–4. In all citations, I have used the translation by Rykwert, Leach, and Tavernor and have also consulted, for the Latin text and valuable notes and commentary, Alberti, *L'architettura* [*De Re Aedificatoria*], edited and translated by Orlandi, with an introduction by Portoghesi.

33. Alberti, *Art of Building,* p. 136 and n. 43.

34. Alberti, *Art of Building,* pp. 135 and 137.

35. Alberti, *Art of Building,* pp. 4–5 and p. 367, n. 12. For an illuminating discussion of Alberti's civic and moral views as well as his debt to Cicero, see Onians, *Bearers of Meaning: The Classical Orders,* pp. 147–157.

36. Alberti, *Art of Building,* p. 37.

37. Alberti, *Art of Building,* p. 58.

38. Alberti, *Art of Building,* pp. 117–125.

39. Alberti, *Art of Building,* p. 7.

40. Alberti, *Art of Building,* p. 24.
41. Alberti, *Art of Building,* p. 7.
42. Alberti, *Art of Building,* p. 155. See also Krautheimer, "Alberti and Vitruvius."
43. I have used Filarete, *Treatise on Architecture,* translated by Spencer, throughout; see pp. xviii–xix for Filarete's biography. I have also consulted the Italian edition: Filarete, *Trattato di architettura,* edited and translated by Finoli and Grassi, with introduction and notes by Grassi; see pp. lxxxviii–xci for his biography. See also *Dizionario Biografico degli Italiani,* s. v. "Averlino (Averulino), Antonio, detto Filarete," by Romanini; and Lang, "Sforzinda, Filarete and Filelfo"; Onians, "Alberti and Filarete"; Saalman, "Early Renaissance Architectural Theory and Practice"; and Tigler, *Die Architekturtheorie des Filarete.* For other practitioners whose careers were similar to Filarete's, see Long, "Power, Patronage, and the Authorship of *Ars,*" pp. 31–39.
44. Filarete, *Trattato,* 1: 11–12; and Filarete, *Treatise on Architecture,* 1: 5.
45. Filarete, *Trattato,* 1: 12; and Filarete, *Treatise on Architecture,* 1: 5.
46. Filarete, *Trattato,* 1: 44–45, "quella sanza la quale generare uomo non si può"; "solo per lo desiderio che hai di vederlo fornito"; and Filarete, *Treatise on Architecture,* 1: 17–18.
47. Filarete, *Trattato,* 1: 46–47, "Dico che l'architetto dee essere onorato e premiato di degno salario conveniente di tale scienza." ["I say the architect ought to be honored and rewarded with a worthy salary, suitable to such knowledge."]; 1: 81–89 (hunting with nobility); and 1: 112–147 (dining with patron); and Filarete, *Treatise on Architecture,* 1: 19 (Filarete also mentions an "architect he knows," probably himself, whose salary had been reduced by a sixth); 1: 35–38 (hunting); 1: 47–54 (dining with patron).
48. Filarete, *Trattato,* 1: 158, "a tendere tutte le corde"; 2: 428, "di più esercizii intendere, e anche coll'opera della mano dimostrarle, con ragioni di misure e di proporzioni e di qualità, e conveniente"; "di sua mano, non saprà mai mostrare, né dare a 'ntendere cosa che stia bene. Bisogna che sia ingegnoso e che immagini di fare varie cose e di sua mano dimostri"; and Filarete, *Treatise on Architecture,* 1: 70–71 and 198.
49. Filarete, *Trattato,* 1: 177, "tanto fu l'amore che mi prese il Signore e tutti, che mi donò tanto che io potevo vivere onoratamente"; and Filarete, *Treatise on Architecture,* 1: 79.
50. Filarete, *Trattato,* 2: 494–495, "tutte queste generazioni di scienze ed esercizii"; "Benché non abbino tanta dignità"; "esercizii di mano"; and Filarete, *Treatise on Architecture,* 1: 228.
51. Filarete, *Trattato,* 2: 500–501, "Questa sarà una cosa che sempre durerà e una cosa che mai non fu fatta, benché sia in queste terre di studio i luoghi dove stieno scolari in dozzina a quel modo. In molti luoghi pagano un tanto, poi non è se none in lettere questa comodità; e anche gli altri esercizii sono di necessità e degni, chi ne viene buono maestro, e ancora gl'ingegni non sono tutti a una cosa iguali. Si che e'si vuole che ogni ingegno si possa esercitare"; and Filarete, *Treatise on Architecture,* 1: 231.
52. Filarete, *Trattato,* 2: 552–553, "quando erano giudicati buoni maestri"; "come i dottori s'adottoravano"; "di quegli artigiani i quali sieno dotti"; and Filarete, *Treatise on Architecture,* 1: 254–255.
53. Filarete, *Trattato,* 2: 557–558, "in disegno, in misure, in sapere fare di mano più cose"; and Filarete, *Treatise on Architecture,* 1: 257.
54. Filarete, *Trattato,* 2: 503–505 (for furnaces), 2: 561 (plaster); 2: 601 (reference to a treatise on engines), 2: 682 (where he notes that two books of his treatise on agriculture are complete, and see books 21, 22, and 23 for his discussion of drawing, perspective, and painting).

In the English translation, see Filarete, *Treatise on Architecture*, 1: 232–233, 1: 258–259, 1: 277, and 1: 317. See also John R. Spencer, "Filarete's Description of a Fifteenth Century Italian Iron Smelter at Ferriere."

55. For the influence of printing on openness, see especially, Eisenstein, *Printing Press as an Agent of Change;* and Eamon, *Science and the Secrets of Nature.*
56. For discussions of the practitioner's low status, see Biagioli, "Social Status of Italian Mathematicians"; and Shapin, *Social History of Truth,* pp. 355–407 ("Invisible Technicians").
57. See especially Long, "Power, Patronage, and the Authorship of *Ars*" and "Contribution of Architectural Writers to a 'Scientific' Outlook."

BIBLIOGRAPHY

Alberti, Leon Battista. *On the Art of Building in Ten Books.* Translated by Joseph Rykwert, Neil Leach, and Robert Tavernor. Cambridge: MIT Press, 1988.

Alberti, Leon Battista. *L'Architettura* [*De Re Aedificatoria*]. 2 vols. Edited and translated by Giovanni Orlandi. Introduction and Notes by Paolo Portoghesi. Milan, Italy: Edizioni Il Polifilo, 1966.

Alberti, Leon Battista. *On Painting.* Translated by John R. Spencer. New Haven, CT: Yale University Press, 1956.

Aristotle. *Nichomachean Ethics.* Translated by W. D. Ross, revised by J. O. Urmson. In vol. 2 of Jonathan Barnes, ed., *The Complete Works of Aristotle,* pp. 1799–1800. Princeton, NJ: Princeton University Press, 1984.

Baldwin, Barry. "The Date, Identity, and Career of Vitruvius." *Latomus* 49 (April–June 1990): 425–434.

Bennett, J. A. "The Mechanics' Philosophy and the Mechanical Philosophy." *History of Science* 24 (1986): 1–28.

Biagioli, Mario. "The Social Status of Italian Mathematicians, 1450–1600." *History of Science* 27 (1989): 41–95.

Biagioli, Mario. *Galileo, Courtier: The Practice of Science in the Culture of Absolutism.* Chicago: University of Chicago Press, 1993.

Borsi, Franco. *Leon Battista Alberti.* Milan, Italy: Electa Editrice, 1975.

Delia, Diana. "From Romance to Rhetoric: The Alexandrian Library in Classical and Islamic Traditions." *American Historical Review* 97 (December 1992): 1449–1467.

Dictionary of Scientific Biography. S. v. "Philo of Byzantium" by A. G. Drachmann, New York: Charles Scribner's Sons, 1974. Vol. 10, pp. 586–589.

Dizionario Biografico degli Italiani. S. v. "Alberti, Leon Battista" by Cecil Grayson and Giulio Carlo Argan. Rome: Istituto della Enciclopedia Italiana, 1960, vol. 1, pp. 703–713.

Dizionario Biografico degli Italiani. S. v. "Averlino (Averulino), Antonio, detto Filarete," by Angiola Maria Romanini. Rome: Instituto della Enciclopedia Italiana, 1962, vol. 4, pp. 662–667.

Dunne, Joseph. *Back to the Rough Ground: 'Phronesis' and 'Techne' in Modern Philosophy and in Aristotle.* Notre Dame, IN: University of Notre Dame Press, 1993.

Eamon, William. *Science and the Secrets of Nature: Books of Secrets in Medieval and Early Modern Culture.* Princeton, NJ: Princeton University Press, 1994.

Eisenstein, Elizabeth L. *The Printing Press as an Agent of Change: Communications and Cultural Transformations in Early-Modern Europe.* 2 vols. Cambridge: Cambridge University Press, 1979.

Ferrari, Gian Arturo, and Mario Vegetti. "Science, Technology and Medicine in the Classical Tradition." In *Information Sources in the History of Science and Medicine,* 197–200. Butterworth's Guides to Information Services. London: Butterworth and Co., 1983.

Filarete [Antonio Averlino]. *Treatise on Architecture.* 2 vols. Translated and edited by John R. Spencer. New Haven, CT: Yale University Press, 1965.

Filarete [Antonio Averlino]. *Trattato di Architettura.* 2 vols. Edited by Anna Maria Finoli and Liliana Grassi. Introduction and notes by Liliana Grassi. Milan, Italy: Edizioni Il Polifilo, 1972.

Findlen, Paula. "Controlling the Experiment: Rhetoric, Court Patronage and the Experimental Method of Francesco Redi." *History of Science* 31 (1993): 35–64.

Fraser, Peter M. "Aristophanes of Byzantion and Zoilus Homeromastix in Vitruvius: A note on Vitruvius VII, Praef. 4–9." *Eranos* 68 (1970): 115–122.

Fraser, Peter M. *Ptolemaic Alexandria.* 3 vols. Oxford: Clarendon Press, 1972.

Freudenthal, Gideon. "Towards a Social History of Newtonian Mechanics. Boris Hessen and Henryk Grossmann Revisited." In Irme Hronszky, Márta Fehér and Balázs Dajka, eds., *Scientific Knowledge Socialized: Selected Proceedings of the Fifth Joint International Conference on the History and Philosophy of Science, Organized by the IUHPS, Veszprém, 1984.* Dordrecht, the Netherlands: Kluwer Academic Publishers, 1988.

Freudenthal, Gideon, et al. "Controversy: The Emergence of Modern Science out of the Production Process." *Science in Context* 1 (March 1987): 105–191.

Gabbey, Alan. "Between *ars* and *philosophia naturalis:* Reflections on the Historiography of Early Modern Mechanics." In J. V. Field and Frank A. J. L. James, eds., *Renaissance and Revolution: Humanists, Scholars, Craftsmen and Natural Philosophers in Early Modern Europe,* 133–145. Cambridge: Cambridge University Press, 1993.

Gadamer, Hans-Georg. *Truth and Method.* 2d rev. ed. Translated by Joel Weinsheimer and Donald G. Marshall. New York: Continuum, 1993.

Galison, Peter. *How Experiments End.* Chicago: University of Chicago Press, 1987.

Goldthwaite, Richard A. *The Building of Renaissance Florence: An Economic and Social History.* Baltimore: Johns Hopkins University Press, 1980.

Goldthwaite, Richard A. *Wealth and the Demand for Art in Italy, 1300–1600.* Baltimore: Johns Hopkins University Press, 1993.

Gooding, David, Trevor Pinch, and Simon Schaffer, eds. *The Uses of Experiment: Studies in the Natural Sciences.* Cambridge: Cambridge University Press, 1989.

Habermas, Jürgen. *Theory and Practice.* Translated by John Viertel. Boston: Beacon Press, 1973.

Hahn, Roger. *The Anatomy of a Scientific Institution: The Paris Academy of Sciences, 1666–1803.* Berkeley and Los Angeles: University of California Press, 1971.

Hahn, Roger. "The Meaning of the Mechanistic Age." In James J. Sheehan and Morton Sosna, eds., *The Boundaries of Humanity: Humans, Animals, Machines.* Berkeley and Los Angeles: University of California Press, 1991.

Hessen, Boris. "The Social and Economic Roots of Newton's 'Principia.'" In Joseph Needham, ed., *Science at the Cross Roads,* 2d ed. Introduction by P. G. Werskey, 149–212. London: Frank Cass & Co., 1971.

Holmes, Frederic L. "Do We Understand Historically How Experimental Knowledge is Acquired?" *History of Science* 30 (1992): 119–136.

Hull, David L. *Science as a Process: An Evolutionary Account of the Social and Conceptual Development of Science.* Chicago: University of Chicago Press, 1988.

Jarzombek, Mark. *On Leon Baptista Alberti: His Literary and Aesthetic Theories.* Cambridge: The MIT Press, 1989.

Krautheimer, Richard. "Alberti and Vitruvius." In *The Renaissance and Mannerism,* 42–52. Vol. 2 of *Acts of the Twentieth International Congress of the History of Art.* Princeton, NJ: Princeton University Press, 1963.

Kuhn, Thomas S. *The Structure of Scientific Revolutions.* 3rd ed. Chicago: University of Chicago, 1996.

LaFollette, Marcel C. *Stealing into Print: Fraud, Plagiarism, and Misconduct in Scientific Publishing.* Berkeley and Los Angeles: University of California Press, 1992.

Lang, S. "Sforzinda, Filarete and Filelfo." *Journal of the Warburg and Courtauld Institutes* 35 (1972): 391–397.

Long, Pamela O. "The Contribution of Architectural Writers to a 'Scientific' Outlook in the Fifteenth and Sixteenth Centuries." *Journal of Medieval and Renaissance Studies* 15 (Fall 1985): 265–298.

Long, Pamela O. "Invention, Authorship, 'Intellectual Property,' and the Origin of Patents: Notes toward a Conceptual History." *Technology and Culture* 32 (October 1991): 846–884.

Long, Pamela O. "Power, Patronage, and the Authorship of *Ars:* From Mechanical Know-How to Mechanical Knowledge in the Last Scribal Age." *Isis* 88 (March 1997): 1–41.

Long, Pamela O., and Alex Roland. "Military Secrecy in Antiquity and Early Medieval Europe: A Critical Reassessment." *History and Technology* 11 (1994): 259–290.

Lucian. "Hermotimus or Concerning the Sects." Translated by K. Kilburn. Loeb Classical Library. Vol. 6, pp. 259–415. London: William Heinemann, 1959.

Marsden, Eric W. *Greek and Roman Artillery: Technical Treatises.* Oxford: Clarendon Press, 1971.

Merton, Robert K. *Science, Technology and Society in Seventeenth-Century England.* 1938. Reprint, New York: Harper & Row, 1970.

Middleton, W. E. Knowles. *The Experimenters: A Study of the Accademia del Cimento.* Baltimore: Johns Hopkins Press, 1971.

Moran, Bruce T. "German Prince-Practitioners: Aspects in the Development of Courtly Science, Technology, and Procedures in the Renaissance." *Technology and Culture* 22 (April 1981): 253–274.

Moran, Bruce T., ed. *Patronage and Institutions: Science, Technology, and Medicine at the European Court, 1500–1750.* Rochester, NY: Boydell Press, 1991.

Oldenburg, Henry. *Philosophical Transactions* 1 (March 6, 1665): 1–2.

Ogilvie, R. M. *The Romans and Their Gods in the Age of Augustus.* London: W. W. Norton, 1969.

Onians, John. "Alberti and ΦΙΛΑΡΕΤΗ." *Journal of the Warburg and Courtauld Institutes* 34 (1971): 96–114.

Onians, John. *Bearers of Meaning: The Classical Orders in Antiquity, the Middle Ages, and the Renaissance.* Princeton, NJ: Princeton University Press, 1988.

Oppel, John. "The Priority of the Architect: Alberti on Architects and Patrons." In F. W. Kent and Patricia Simons, eds., with J. C. Eade, *Patronage, Art, and Society in Renaissance Italy,* 251–267. Canberra, Australia: Humanities Research Centre, 1987.

Oxford Classical Dictionary. 2d ed. S. v. "Lares," by Herbert Jennings Rose. Oxford: Clarendon Press, 1970, pp. 578–579.

Oxford Classical Dictionary. 2d ed. S. v. "Polyclitus (2)" by Gisela M. A. Richter. Oxford: Clarendon Press, 1970, pp. 854–855.

Rossi, Paolo. *Philosophy, Technology and the Arts in the Early Modern Era.* Translated by Salvator Attanasio. Edited by Benjamin Nelson. New York: Harper and Row, Harper Torchbooks, 1970.

Rykwert, Joseph. *On Adam's House in Paradise: The Idea of the Primitive Hut in Architectural History.* New York: Museum of Modern Art, 1972.

Saalman, Howard. "Early Renaissance Architectural Theory and Practice in Antonio Filarete's *Trattato di Architectura.*" *Art Bulletin* 41 (March 1959): 89–106.

Schmitt, Charles B. "Experience and Experiment: A Comparison of Zabarella's View with Galileo's in *De Motu.*" *Studies in the Renaissance* 16 (1969): 80–138.

Schürmann, Astrid. *Griechische Mechanik und Antike Gesellschaft: Studien zur Staatlichen Förderung einer technischen Wissenschaft.* Stuttgart, Germany: Franz Steiner Verlag, 1991.

Shapin, Steven. "Understanding the Merton Thesis." *Isis* 79 (December 1988): 594–604.

Shapin, Steven. "The House of Experiment in Seventeenth Century England." *Isis* 79 (September 1988): 373–404.

Shapin, Steven. *A Social History of Truth: Civility and Science in Seventeenth-Century England.* Chicago: University of Chicago Press, 1994.

Shapin, Steven, and Simon Schaffer. *Leviathan and the Air Pump: Hobbes, Boyle, and the Experimental Life.* Princeton, NJ: Princeton University Press, 1985.

Smith, Christine. *Architecture in the Culture of Early Humanism: Ethics, Aesthetics, and Eloquence, 1400–1470.* New York: Oxford University Press, 1992.

Smith, Pamela H. *The Business of Alchemy: Science and Culture in the Holy Roman Empire.* Princeton, NJ: Princeton University Press, 1994.

Spencer, John R. "Filarete's Description of a Fifteenth Century Italian Iron Smelter at Ferriere." *Technology and Culture* 4 (Spring 1963): 201–206.

Starn, Randolph, and Loren Partridge. *Arts of Power: Three Halls of State in Italy, 1300–1600.* Berkeley and Los Angeles: University of California Press, 1992.

Stewart, Andrew. "The Canon of Polykleitos: A Question of Evidence." *Journal of Hellenic Studies* 98 (1978): 122–131.

Stockton, David. "The Founding of Empire." In *Oxford History of the Classical World.* Vol. 2: *The Roman World,* pp. 121–149. Edited by John Boardman, Jasper Griffin, and Oswyn Murray. Oxford: Oxford University Press, 1988.

Tafuri, Manfredo. "'Cives esse non licere': The Rome of Nicholas V and Leon Battista Alberti: Elements toward a Historical Revision." *Harvard Architectural Review* 6 (1987): 60–75.

Tigler, Peter. *Die Architekturtheorie des Filarete.* Berlin: Walter de Gruyter & Co., 1963.

Tiles, J. E. "Review Article: Experimental Evidence vs. Experimental Practice?" *British Journal of the Philosophy of Science* 43 (1992): 99–109.

Vitruvius. *Architettura (Dai Libri I–VII).* Edited by Silvio Ferri. Rome: Fratelli Palombi Editori, 1960.

Vitruvius. *On Architecture* [*De Architectura*]. 2 vols. Edited and translated by Frank Granger. Loeb Classical Library. Cambridge, MA: Harvard University Press, 1931.

Voss, Mary J. "Between the Cannon and the Book: Mathematicians and Military Culture in Sixteenth-Century Italy." Ph.D. diss., Johns Hopkins University, 1994.

Wendel, Carl, and Willi Göber. "Das Griechisch-Römische Altertum." In Fritz Milkau and Georg Leyh, eds., *Handbuch der Bibliothekswissenschaft,* 51–145. Vol. 3.1 of *Geschichte der Bibliotheken.* Wiesbaden, Germany: Otto Harrasowitz, 1955.

Westfall, Carroll William. *In This Most Perfect Paradise: Alberti, Nicholas V. and the Invention of Conscious Urban Planning in Rome.* University Park: Pennsylvania State University Press, 1974.

Whitney, Elspeth. *Paradise Restored: The Mechanical Arts from Antiquity through the Thirteenth Century.* Transactions of the American Philosophical Society, vol. 80, Part 1, 1990.

Wilson, R. J. A. "Roman Art and Architecture." In *Oxford History of the Classical World.* Vol. 2: *The Roman World,* pp. 361–400. Edited by John Boardman, Jasper Griffin, and Oswyn Murray. Oxford: Oxford University Press, 1988.

Zilsel, Edgar. "The Sociological Roots of Science." *American Journal of Sociology* 47 (January 1942): 544–562.

II

Displaying and Concealing Technics in the Nineteenth Century

5

Architectures for Steam

M. Norton Wise

With the defeat of Napoleon sealed in Paris in 1814, Prussia began to represent its victory to itself in a variety of cultural forms, stressing always its distance from "French" values and its intention to join Britain in the pursuit of industrial power. Landscape gardens provided fertile ground for such representations. Between 1815 and 1850 numerous gardens were constructed or reconstructed in the naturalistic "English" style in Berlin and Potsdam. The largest belonged to the Hohenzollern royal family. They were built by two famous architects in the departments of gardens and of buildings in the royal administration, the landscape architect Peter Josef Lenné and the building architect Karl Friedrich Schinkel. Designed for display on a massive scale, these immense works of art exist today as public parks, somewhat like Kensington Park or Regency Park in London but even more like the great estates at Windsor Castle or Stowe.

English gardens in Berlin, however, contain a striking anomaly in comparison with those in England: They feature buildings that once housed steam engines. Focusing largely on these engine houses, the following account considers the gardens' role in transferring steam technology from England to Prussia. It shows how the gardens served aesthetically to integrate steam engines into Prussian culture by representing the engine in the vocabulary of past and present Prussian ideals and by identifying engine power with Prussian power. At the same time, it illustrates, using Hermann Helmholtz as exemplar, how the famous generation of scientists that emerged in Berlin at midcentury participated in this cultural projection.

5.1–5.4

Panorama of *Glienicke,* August C. Hahn, after Wilhelm Schirmer, ca. 1845, Staatliche Schlösser und Gärten Berlins, Charlottenburg. The enlarged portions show, from left to right, (5.2) the *Teufelsbrücke,* (5.3) the *Glienicke* engine house, and (5.4) the steamboat *Alexandra* with the *Babelsberg* engine house behind the bridge. The immense size and naturalistic beauty of this artificial landscape celebrate the power of steam to remake the world in an aesthetically pleasing image. See also color plates 1–4.

NATURALIZING STEAM

Looking east over the river Havel near Potsdam, a present-day visitor encounters a scene much like that painted in 1845 by August Haun. (See figure 5.1.) This natural-looking landscape of forested hills, however, is almost entirely artificial. It consists of two immense gardens. In the center of the panorama stands the engine house at *Glienicke* (figure 5.3). To its right is the tower of the actual palace with its associated "casino" prominent on the riverbank. Both were designed by Schinkel for Prince Carl, the anglophile third son of King Friedrich Wilhelm III (who died in 1840). On the far right, occupying the hill behind the *Glienicke* bridge (of spy-trading fame during the Cold War), is *Babelsberg,* where the new castle Schinkel designed for the second son, Prince Wilhelm, stands on the hillside. Its engine house appears just to the right of the small paddle steamer on the river. This striking black presence (figure 5.4) is the *Alexandra,* used by the royal family, under King Friedrich Wilhelm IV from 1840, to travel from the palaces and gardens at *Sanssouci* in Potsdam (off to the right) to a third splendid garden on the *Pfaueninsel* (barely visible on the left), where yet another engine house marked the shoreline.[1] On the way they passed a curious waterfall beneath a ruined bridge (the *Teufelsbrücke,* figure 5.2) and the Church of Saint Peter and Paul, at Nikolskoe, whose tower is visible on the ridge. The tranquil vista is a product of steam, and it celebrates its origins. The painting reiterates that celebration, as do numerous others from the same period.

This interpretive theme has long ago been expressed for *Glienicke* by a remarkable nineteenth-century witness, Helmut von Moltke. The later field marshal and count visited Prince Carl in 1841 and wrote the following account to his wife. "I wish I could escort you around in this exquisite park. For as far as the eye can see, the grass is of the freshest green; the hills are crowned with beautiful deciduous trees; and the stream and the lakes weave their blue ribbon through a landscape in which castles and villas, gardens and vineyards lie scattered. Certainly the *Glienicke* park is one of the most beautiful in Germany." Here the garden presents its face of peaceful beauty. But power lies beneath the romantic production and surges forth in the Count's exposition:

> It is unbelievable what art has known how to make out of this barren earth. A steam engine works from morning til night to lift the water out of the river Havel up to the sandy heights and to create lush meadows where without the engine only weeds of the heath [heather] would survive. A powerful cascade roars over cliffs under the arch of a bridge [*Teufelsbrücke;* see figure 5.5], half washed away, seemingly from its violence, and abruptly rages fifty feet down to the Havel onto a terrain where prudent Mother Nature would not have thought to let a pail of water flow, because the parched sand would have immediately swallowed it. Trees forty feet high are planted where they would have had to stand forty years to achieve this mightiness. Scattered around lie enormous boulders, which one day geologists will give up guessing about if notice has not reached them that they wandered here from Westphalia through Bremen and Hamburg.[2]

5.5

"Teufelsbrücke," A. Lompeck, 1852, Staatliche Schlösser und Gärten Berlins, Charlottenburg. The entire assembly of the "bridge of the devil"—ruin, rocks, waterfall and vegetation—is constructed. Its gothic imagery projects modern steam power into the mythological past of uncontrollable forces of nature, thus also acknowledging its fearsome character, but in a domesticated pageant.

Evoking fearsome power and arresting beauty simultaneously, von Moltke calls up an image of the "technological sublime." The steam engine brings forth "Nature." But its creation is unmistakably a work of technical art. It is just the technological character of this masterwork, in its extravagance and perfection, that von Moltke appreciates. The engine makes beauty where the original of Mother Nature would not have allowed it to exist, any more than she would have brought boulders through several hundred miles of equally artificial canal-linked waterways or would have installed mighty trees on a sandhill. If this world-creating engine is not directly viewed, its presence is nevertheless everywhere felt through the irony of naturalistic artificiality that it creates.

It enriches our sense of how budding young scientists like Hermann Helmholtz participated in the cultural life of Potsdam and Berlin to place him directly into this remade landscape of steam that so impressed von Moltke. Lenné and Schinkel had been busy constructing the landscape while Helmholtz, born in 1821, was growing up nearby in Potsdam. As a boy he took long walks through the local countryside and gardens with his father, a professor at the Gymnasium, no doubt remarking on the progress of artificial nature as they went.[3] From 1838, while he pursued medical studies at the *Friedrich-Wilhelms-Institut* for military doctors in Berlin, Helmholtz traveled often across this landscape. Usually he went *par vapeur* on a railroad line (see figure 5.6) which opened in 1838, the first in Prussia and a model for the promotion of railroad building by Friedrich Wilhelm IV. Alternately, he went by coach or occasionally even by foot on a five-hour "tour" along the so-called Royal Road from *Sanssouci* in Potsdam to the palace in the center of Berlin, thus crossing the bridge with the engine houses at *Glienicke* and *Babelsberg* to the left and right. From 1843 to 1847, while he served as an army doctor in Potsdam, his service included attending patients at *Glienicke* with some regularity. Occasionally he continued his walk across the garden to the heights of Nikolskoe, which would have taken him past the *Teufelsbrücke*. At the church he visited a friend, the pastor C. J. T. Fintelman, who belonged to a well-known family of court gardeners. From the front entrance they looked out on the *Pfaueninsel,* where G. A. Fintelman, a protégé of Lenné, directed the King's garden. Another of the family, C. J. Fintelman, was celebrated for the numerous varieties of grapes that he developed in Potsdam, some of which Helmholtz used in 1845 for experiments on fermentation.[4] Thus the gardens were a part of Helmholtz's everyday life, and it is of interest to observe how he built his career as a scientist within the same culture that built this landscape of steam.

Prince Carl had purchased *Glienicke* in 1824 following the death of its former owner, Prince Hardenberg, Prussian chancellor from 1810 to 1822 and the architect of a famous program of economic reform in that period. Hardenberg himself had rented *Glienicke* from 1811 but was able to purchase it only after his triumphal entry into Paris with Friedrich Wilhelm III in 1814, when he was raised to princely status. It was Hardenberg who commissioned the young Lenné in 1816 to help him turn the estate into a showpiece of naturalistic freedom in the English style and of simultaneous economic lib-

5.6

"Die Berlin-Potsdamer Bahn," Adolph Menzel, 1847, Staatliche Museen zu Berlin, Preußischer Kulturbesitz, Nationalgalerie, NG 780. The city of Berlin is in the background. The locomotive for the railroad, which opened in 1838, came from England, before firms like August Borsig's began to produce "native" locomotives in the 1840s. Hermann Helmholtz often traveled on this line while in training as a military doctor in Berlin and afterward. The painting was acquired originally by Helmholtz's immediate superior in the army and Menzel's close friend, Dr. Wilhelm Puhlmann.

eralism. Into the landscape he integrated a brick and tile manufactory employing more than sixty workers. It was located not far from his *Lustgarten* (pleasure garden, including flower beds), which he equipped with a horse-powered irrigation and fountain system featuring the latest in experimental technology, including several hundred feet of cast zinc pipe (which corroded within three years and had to be replaced).[5] This combination of aesthetics, economics, and technological modernization characterized Hardenberg's policy for all of Prussia. It continued to develop under administrators who got their start in his government, albeit with substantial opposition.[6] Thirty years later the policy had produced the showpiece landscape depicted in the panorama of figure 5.1. In it we should recognize some of the cultural foundations for Prussia's industrial revolution, which entered its "takeoff" phase by the mid-forties. By then, Berlin's entrepreneurs and their ministerial promoters could see themselves as challengers to the British industrial monopoly.

No doubt Hardenberg would have preferred a steam engine to power his garden, but not only were there no engine builders in Berlin in 1816, there were no engines either, or rather there was precisely one, at the Royal China Factory (KPM). English engines were prohibitively expensive to import, when such importation was not explicitly prohibited by English law, and impossible to maintain, unless one brought over an English mechanic as well. The first Prussian steamboat, for example, was built in Berlin in 1816 by an imported Englishman named John Barnett Humphrey and fitted with an imported English engine. For exorbitant fees, Humphrey carried passengers between Potsdam and Berlin, until 1824, when his Steamboat Travel Co. went bankrupt. The situation for manufacturing engines was equally grim. Hardenberg therefore devoted considerable energy to overcoming this economically debilitating deficiency in the technologies of iron and steam. His ministry invented numerous schemes for promoting engine building, engine use, and engine espionage. For example, they sent the promising young mechanic, F. A. J. Egells, to England on a study tour of machine works with the expectation that he would then establish his own works in Berlin, which he did by 1825.[7] Simultaneously, however, they were sending Lenné and Schinkel on tours of English gardens and buildings with similar modernizing intentions.

Much of this activity was organized through the Ministry of Trade and Industry under Hans von Bülow (until 1825). Its direct agent, however, was an ambitious young liberal named Peter Beuth, who had risen rapidly in Hardenberg's administration and in 1818 had subjected the existing agencies for industrial development to a scathing critique. The king and his ministers responded by appointing Beuth to head the so-called Technical Deputation, where he was to carry out himself the proposals he had made for overhauling the entire apparatus. From among four old and four new members of his reorganized deputation, Beuth chose one who became his closest friend and ally, the architect Schinkel. Beuth made his own first study tour of British machinery and manufactures in 1823, visiting London, Manchester, Leeds, Edinburgh, Glasgow, and other towns. "The wonders of the new age," he reported to Schinkel, "are to me the machines and their buildings, called factories." In Manchester he marveled at the forest of tall chimneys marking the engines that powered a hundred factories, each eight stories high and "forty windows" long, constructed of iron columns and beams. It was a scene of wonder, especially at night, when the thousands of windows announced unceasing work within under bright gas lighting, as yet nonexistent in Berlin. Equally impressive was the diversity of machinery for every purpose, some lacking even a name in German. To read Beuth's letters is to understand immediately the fascination of the factory tour. In 1826 he returned with Schinkel for an even more extended study of what the "new age" seemed to require both architecturally and mechanically.[8] But his own organization of Prussian industrialization had already taken shape.

As a political and economic liberal who placed a premium on youth and education, P. C. W. Beuth founded in 1821 a new school for raising future leaders of private industry, the *Gewerbeschule* or Technical Institute, which he personally administered. It formed the centerpiece of his project. Its carefully selected students were to acquire a basic knowledge of physics, chemistry, mathematics, and construction, as well as practical knowledge of metalworking and manufacturing techniques. And they were to spend part of their time apprenticed to newly established manufacturers and machine builders like Egells before they went on to establish their own firms.[9] Thus the *Gewerbeschule* soon required a wide range of experts to train its students, experts who were in short supply. To alleviate the problem, Beuth integrated the *Gewerbeschule* into a network of recently established technical schools in Berlin that shared their teachers. These schools included the *Vereinigte Artillerie- und Ingenieurschule* (United Artillery and Engineering School, 1816), the *Kriegschule* (Military Academy, 1810), the *Gärtnerschule* (School of Gardens, *Königliche Gärtner-Lehranstalt zu Schöneberg und Potsdam,* 1823) and the *Bauschule* (School of Architecture, 1799; *Bauakademie* after 1848). Schinkel, for example, taught in the *Gewerbeschule,* although his main duties were in the *Bauschule.* Crucially for the science-industry linkage that Beuth was promoting, the teaching network extended to more traditionally academic institutions, the *Friedrich-Wilhelms-Institut* (for military doctors like Helmholtz) and the university. Many professors of natural science, in fact, taught simultaneously at the university and at one or more of the technical schools during the 1830s and 1840s. They included the chemists Mitscherlich and Magnus, the physicists Turte and Dove, and the botanist Link. (I return to these scientists below as the teachers of Hermann von Helmholtz.) Sharing the same building with the *Gewerbeschule* and with his Technical Deputation was Beuth's second major institution for promoting industrial interests, a kind of social and technical support group called the *Verein zur Beförderung des Gewerbefleisses* (Union for the Advancement of Industry). Its facilities included a library for technical publications and drawings; a collection of models, machines, and engines (including many acquired from Britain either entire or as parts or drawings, in violation of the export prohibition); laboratories for experiments on industrial processes; and workshops for model building and teaching.[10] Operating something like a scientific society, the *Verein* held regular meetings for the presentation of the latest technical advances and inventions, which were then published in the society's *Verhandlungen,* a high-quality publication displaying numerous copper engravings. The membership of the *Verein* indicates the breadth and depth of Beuth's project for integrating industry into Prussian culture. By 1840 it included all of the royal princes except Wilhelm, a considerable number of liberal ministers and advisors, and a wide variety of trading and manufacturing people, including those in the new iron and machine industries like Egells and his former apprentice and student of the *Gewerbeschule,* August Borsig, soon of locomotive fame. It has not usually been remarked about the *Verein,* however, that it included as well a considerable number of natural scientists, typically those, like Helmholtz's professors, who taught at two or more schools for technical education and who regarded their expertise as a critical support base for industrial development. By far the most prominent of the scientists, however, and their prime exponent and patron was Alexander von Humboldt, who was also closely associated with Beuth and Schinkel. Humboldt played a prominent role at court (*Wirkliche Gehiemerrath und Kammerherr*) and became a kind of live-in guest of Friedrich Wilhelm IV at his new house and gardens, *Charlottenhof* (1835). Thus there existed a modernizing network of institutions, associations, and friendships extending from the heights of royalty and ministers of state to low but rising mechanics and across a broad middle level including Beuth, Schinkel, Lenné, and a variety of professors.

This network built the gardens in which Helmholtz walked so casually on his daily rounds in Potsdam. But before considering the gardens in more detail, a first qualifying word on "modernization" is important. Although Beuth, Schinkel, and Lenné all sought to reproduce aspects of the "civilization" of Britain—its material wealth, personal freedom, and national power—they sought to avoid reproducing the blight of British industrial towns, their poverty, choking air, and disease. They aimed rather at producing an aestheticized version of industrial society in architecture and design as well as the conditions of life. Aesthetics played a prominent role in both the *Verein* and the *Gewerbeschule* and figured in the prizes awarded at the technological exhibitions they mounted in 1822, 1827, and especially 1844. At this last exhibition Borsig won a gold medal for his new locomotive, named "Beuth" (see figure 5.7), the show's centerpiece.[11] Schinkel may have set the artistic standard, but Beuth led the movement. He pursued inexhaustibly the vision of an industrial culture integrated organically into the existing landscape of rural towns, thus avoiding the horrors of the British "Coketowns" while gaining a materially comfortable and spiritually refined citizenry. One of Beuth's assistants described their goal as "to integrate art with industry," and Schinkel painted Beuth in 1837 as a Greek god riding Pegasus over a factory town (see figure 5.8).[12] It would be a mistake to see in its smoking chimneys a sign of the British social problem. For the town is small and well ordered, and it sits nestled against the hills with a refreshing breeze dissipating the smoke into the surrounding countryside, at least as Beuth envisaged it.

5.7

"Beuth," August Borsig's prize-winning locomotive in 1844, from Borsig, "Beschreibung der Lokomotive 'Beuth'," plate V. A source of immense pride at the Berlin Exhibition, the locomotive symbolized Prussia's entry into industrial competition with Britain under Beuth's leadership.

5.8

"The Aesthetic View from Pegasus," K. F. Schinkel, in Matschoss, *Preussens Gewerbeförderung*, p. 72. Schinkel here captures the humanistic ideals that informed the modernizing vision of his intimate friend Beuth: Steam-powered industry integrated into a landscape of rural towns.

Lenné also sought to realize this principle of the complementarity of utility and beauty in his steam-powered gardens and in the institutions for gardening culture that he helped to establish in parallel with Beuth's institutions for industrial culture: the *Verein zur Beförderung des Gartenbaues* (1822) and the *Gärtnerschule,* which he directed.[13] In a grand design with Schinkel for the city of Berlin, Lenné stressed the need to intersperse green public parks amid neighborhoods of workers' dwellings and nearby factories. However idealistic the vision, the alien beasts of steam engines and factories were to meld harmoniously into the garden landscape. The following depiction of the engine houses in the royal gardens of Potsdam and Berlin attempts to show how this naturalization was handled symbolically by integrating engines into architectural forms that appealed to contemporary tastes, whether looking to the past or the present for aesthetic validation.

The Gardens of the Princes

All of the institutions for the progress of industry and gardens were already in place when Prince Carl obtained *Glienicke* in 1824, after Hardenberg's death. The efforts of Hardenberg's government to modernize Prussian industry were just beginning to pay off. Carl immediately set out to display his support for the newly emerging economic and aesthetic order: by commissioning Schinkel to rebuild his palace in a neoclassical style and Lenné to extend his English garden; by attempting to acquire, through Humphrey's bankruptcy, an English steam engine to power the garden; and by sprinkling English phrases in his conversation and letters. Successful with palace and garden, he failed initially with the engine, for unknown reasons.[14] By the mid-thirties, however, Egells had his Berlin engine works in full operation, ready to supply the princes' gardens as well as local industry. The process of representing the power of Prussia to remake itself through the power of steam to remake nature accelerated rapidly.

At *Glienicke* in 1836–38, Schinkel's favored student and heir apparent, Ludwig Persius, housed the new eighteen-horsepower engine and its full-time mechanic in an Italianesque villa on the bank descending to the Havel (see figure 5.9). The villa style, in both country and city versions, promoted by Crown Prince Friedrich Wilhelm IV as well as the royal architects, was becoming popular among newly wealthy industrialists around Potsdam. Persius constructed several examples, including his own *Villa Persius* (1836–37) and two buildings that graced the view from *Glienicke,* the *Villa Schöningen* (1843) immediately across the bridge and the *Villa Jacobs* (1835) on a direct line of sight from the engine house, with a mirroring tower.[15]

The tower suggests implicitly something that should be noted more often about stationary steam engines at the time. The working engine included not only its machinery proper, but also its house and its keeper. They formed an integral system. An appropriate analogy might be a locomotive in motion, with engineer and stoker at their work. Thus building a house for a pumping engine was nothing like building a garage for a car. The

5.9

The engine house at *Glienicke*. Photograph by the author. Schinkel's heir apparent, Ludwig Persius, housed the engine and its supporting system—coal, water reservoir, chimney, and mechanic and family—in the popular villa style being promoted by the Crown Prince and royal architects for Potsdam's newly emerging upper middle class. Thus a housing style for the agents of progress contains the engine of progress.

structure of the house was also the support structure for the engine, pumps, smokestack, piping, and a reservoir, as well as an apartment for the mechanic and his family, to keep him in constant attention. In figure 5.9 the machinery is at lower left. The tower contains the smokestack, and on successive floors: (1) coal storage, (2 and 3) the mechanic's apartment, (4) a belvedere looking out over the Havel, and (5) a water reservoir of 1,800 cubic feet. On the far right is the garden house, with an apartment for the full-time gardener.[16] To unite this system's technical requirements with the estate's aesthetic requirements demanded considerable architectural ingenuity.

From the pumping installation ran a network of cast iron pipes to feed the garden's fountains, lakes, and streams as well as to irrigate its plants. In addition, one pipe filled a special reservoir on the hill to supply a torrent of water for the display at the *Teufelsbrücke* that so impressed von Moltke. Prince Carl loved pageantry. On special occasions his "fleet," consisting of a three-masted schooner kept permanently anchored in front of the engine house and a variety of smaller boats, would parade in front of the estate bedecked with colorful flags. He enjoyed also assimilating his engine to the mythical powers of the German *Volk* and its warrior heroes, in what would look now like a Wagnerian *Gesamtkunstwerk*. One should imagine a group of favored guests arriving via the steamboat *Alexandra* on the Havel. As they approached, masses of water penned up in the special reservoir would be let loose down a gorge and over the cliff to crash fifty feet below on the shore directly in front of them, while the prince's cannons boomed out a ten-gun salute.[17]

Such elaborate effects were no doubt exceptional. Carl's brothers, however, went to considerable lengths also to incorporate engines into their visions of themselves and Prussia. The most politically conservative of these projections is *Babelsberg,* where Schinkel and Persius, after numerous frustrations with their employers, produced a version of a medieval castle for Prince Wilhelm (later Kaiser Wilhelm I) and Princess Augusta. Lenné initiated the large garden, although again not to the satisfaction of Wilhelm and Augusta, who preferred the gothic designs of their friend Prince Hermann Pückler-Muskau. These politically reactionary aristocrats looked to the mythical German past for their symbols of power and directed their architects accordingly. Schinkel and Lenné, as liberals both economically and politically, failed to appreciate their taste.

After Schinkel's death in 1841, Persius received the commission to build an engine house for *Babelsberg* in a compatible medieval style (see figure 5.10). Completed in 1845, it supported a large pumping engine of forty horsepower from Egells. Among its functions, the engine filled a large reservoir high on the hill, from which the water descended with great pressure to drive a towering fountain, rising 130 feet from the River Havel in front of the castle.[18] Wilhelm thereby bettered his brother King Friedrich Wilhelm IV in possessing the most powerful geyser in the family, for the new king's fountain at *Sanssouci* (below) reached only 126 feet. Whatever the significance of this particular accomplishment, it may stand here, along with the directly symbolic engine house itself,

5.10

The engine house at Babelsberg, Carl Graeb, 1853, Staatliche Schlösser und Gärten Berlins, Charlottenburg. If the engine house at *Glienicke* symbolizes modernization in its familiar progressive-liberal guise, Babelsberg projects the same technological modernization into a romanticized, authoritarian regime.

for the fact that Wilhelm vigorously opposed his brother's impulses toward liberal political reform even while promoting technological and industrial modernization himself. Thus the engine house at *Babelsberg* serves as a reminder of a second qualification concerning technological modernization: It could be politically reactionary.[19] As someone like Hermann Helmholtz crossed the bridge between the engine houses and palaces at *Glienicke* and *Babelsberg* he was walking between the architectural symbols and the political forces of moderate liberalism (Hardenberg and Prince Carl) and absolutism (Prince Wilhelm), both promoting not only technology-based industry but science as well.

In a style more typical for the liberal vision of Prussian modernization, Schinkel remodeled a palace in the city for the youngest, Prince Albrecht. Lenné designed the garden, for which Schinkel added a neoclassical engine house directly adjacent to the palace (see figure 5.11). Its elegant chimney stands sentinel over the water that the machinery moves and that fills the foreground.[20] In the background rises *Kreuzberg,* the hill south of the city on which stands a remarkable memorial to the war of independence against Napoleon.

Schinkel designed the memorial in 1818. It is actually the remnant of a grand cathedral that he had originally proposed to honor Prussia and its people in their great victory. The symbolic remnant preserves only one tower, and that one is cast in iron rather than carved in stone (see figure 5.12). On top stands the iron cross, recalling the famous medal that Schinkel had designed for Friedrich Wilhelm III in 1813. The iron is important. It celebrated first of all the loyalty of the king's subjects in turning in their jewelry to support the war in return for cast iron rings and patriotic medals. But it celebrated also the coming economy of iron and steam, for it pushed the technological capacity of the Royal Foundry to its limits to cast the complex figures that stand in the niches around the monument in memory of twelve crucial battles. These heroic figures have faces modeled on those of the king, his family, and his generals.[21] Thinking of all this from the perspective of Prince Albrecht's palace in the city, one can imagine him looking out over his engine house and garden toward *Kreuzberg,* where he could almost see his family immortalized in iron beneath the iron cross, a picture simultaneously of Prussian independence, economic development, and military strength.

These symbols merit one more look, in Schinkel's (and Lenné's) most exquisite production for the Hohenzollern princes and the earliest with an engine, this one for Crown Prince Friedrich Wilhelm IV, completed in 1827 with interior modifications through the 1830s. Named *Charlottenhof,* it is a small house and garden near the enormous *Neue Palais* of Friedrich Wilhelm III at *Sanssouci.* One interpreter has given an elegant reading of the house and grounds as a realization of the ideals of nature promoted by Alexander von Humboldt, a confidant of the crown prince who spent extended periods at *Charlottenhof* in a guest room reserved for him.[22] Schinkel lavished his attention on every detail of this intimate retreat for his friends the crown prince and princess, not only on the building but on its interior: chairs, tables, desks, chandelier, fabrics, even personal items, perhaps

5.11

Prince Albrecht Palais, looking toward Kreuzberg, unknown artist, ca. 1835, Staatliche Schlösser und Gärten Berlins, Charlottenburg, Nachlass Sievers. Schinkel's neoclassical engine house brings humanistic rationality into relation with Lenné's naturalistic garden.

5.12

The Kreuzberg monument, memorializing Prussia's victory in the Napoleonic wars. Photograph by the author. Schinkel here cast the spire of a gothic cathedral in iron. This transformative construction is multiply symbolic of the past, the future, the spirit of the people, and the military might of Prussia. It is topped by the iron cross and depicts the royal family as the heroes of independence.

in part because Friedrich Wilhelm IV was himself an architect of some talent. Deeply religious in their lives and their tastes, the royal couple hung on the walls numerous reproductions of biblical scenes from the Italian Renaissance, especially renderings by Raphael and Michelangelo. I want to call attention, however, to a particular image, also full of religious symbolism, that figures throughout the design: candelabra. Most dramatically, as the visual focus looking out from the house and terrace toward a large pond, Schinkel installed a stately candelabrum atop a broad pedestal (see figure 5.13). It reproduces a classical Roman architectural form that became popular again in the Renaissance. Used in public buildings, these monumental constructions in marble or bronze "had for the base a pedestal resembling a little altar which carried a heavy shaft, frequently decorated with row on row of acanthus leaves,"[23] which describes rather well Schinkel's re-creation in cast iron. But here the shaft and altar constitute the smokestack and low-lying house for a three-horsepower steam engine.

This remarkable presentation established the eastern pole on the east-west axis of the architectural design, the pole of the rising sun and of daylight, a theme reinforced in the light blue walls and ceiling of the portico on the eastern side of the house. Sitting in the portico, one looked out across the formal terrace with its fountain rising from a stemmed

5.13

Charlottenhof, with engine house as candelabrum, from Schinkel, *Architektonischer Entwürfe,* plate 110. Schinkel apparently intended the candelabrum, taken from Roman and Renaissance forms and recalling a candle on an altar, as a focal point in his creation for the Crown Prince. It serves, along with interior candelabra, to juxtapose the powers of steam with those of Christianity and the Prussian state.

basin to the candelabrum of the engine, complete with decorative wisps of smoke trailing in the wind, at least in Schinkel's imagination. At the opposite pole is the entrance to the house, where he repeated these themes a bit later and on a smaller scale. In the entry hall one faces another basin and fountain (designed originally for the *Gewerbeschule*) that nestle between dual stairways on either side. On the corner posts of the railings at the top, two tall, gilded candelabra (1835) stand as the symbolic lights for this space of the setting sun and of nighttime, represented in a dark blue window over the doorway carrying a field of stars. Ascending the staircase and moving left, one comes to the crown prince's bedchamber, with figures of Christ, Elija, and Moses (after Raphael) above the headboard (see figure 5.14). Here Schinkel redid the candelabra as bedposts (1828), and with a new meaning. They now protected the crown prince's sleep and supported the Prussian eagle, symbol of monarchical power.[24]

Considered as a whole, then, the candelabra at *Charlottenhof* unite religious devotion with the powers of engines and the Prussian state in one of the most aesthetically pleasing of all Schinkel's designs. If the point needed reiteration that engine houses did not merely camouflage their somewhat threatening interior machinery, but instead naturalized and even celebrated its presence among cherished images of Western culture and Prussian tradition, *Charlottenhof* would supply convincing evidence.

5.14

Bedroom of Friedrich Wilhelm IV, from Bergdoll, *Schinkel*, p. 147. Photograph by Erich Lessing. Here Schinkel's candelabra carry the eagle, symbol of glory in the military state.

123 | Architectures for Steam

5.15

"Ansicht des Königlichen Schlosses Sanssouci mit den Terrassen," Carl Daniel Freydanck, 1843, cat. no. 24, KPM-Archiv (Land Berlin), Schloss Charlottenburg, Berlin. Rising 126 feet to the top of Frederick the Great's eighteenth-century palace, this central fountain evoked Prussia's return to greatness among European states along with its mastery of the new technologies of steam.

But the point goes further. Engine architecture for the gardens of the Hohenzollern princes not only legitimized the present in terms of past glories and present fashions, it also projected a vision of the future in which iron and engines would support a new Renaissance, the opening to Beuth's aesthetically pleasing industrial culture. When the crown prince and his brothers Carl and Albrecht lent their status to the Union for the Advancement of Industry, they aligned themselves symbolically with the most liberal faction of the state administration. But when they integrated engines into their gardens between 1827 and 1837, they went beyond symbolism to put their personal financial resources behind the entrepreneurial spirit that Beuth was promoting. Schinkel and Lenné applied their creative energy to designing that spirit into the material reality of the landscape, to render the vision visible in interior and exterior space.

New Gods of Steam

The most dramatic example of the reified symbols of Prussia's modernization came only in the 1840s, not on the personal estates of the princes, but in the most lavish garden of the state itself, *Sanssouci,* built originally by Frederick the Great in an attempt to rival *Versailles.* Lenné had begun its transformation soon after the war of independence from the formal French style, with broad straight avenues and geometrical symmetry, to the natural curves and textures of the English style. And already in 1819 he and Humphrey had tried to sell Friedrich Wilhelm III on an English engine to supply irrigation and a fountain rising forty-five feet. They argued for the unity of utility and aesthetics in such an installation, as well as for the economy of the engine. But Friedrich thought Humphrey's price of 11,500 taler a poor showing of economy. Besides, he noted, the engine would destroy the livelihood of the peasants who watered the garden with buckets.[25] *Sanssouci* had to wait for the crown prince to ascend the throne in 1840.

When Friedrich Wilhelm IV took up Lenné's plan for a complex system of fountains and waterworks for the buildings and gardens of *Sanssouci,* a master German machine builder of the second generation had come onto the market. Educated partly in Beuth's *Gewerbeschule* in the early twenties and for ten years in Egells' shop, August Borsig set up independently in 1837 with two horses to power his iron works. But he was soon building heavy machines and engines with an eye toward locomotive construction. His first locomotive emerged in 1841. A year later he had constructed a stunning eighty-horsepower engine for *Sanssouci,* one of the largest and certainly the most elegant in Prussia. It drove fourteen pumps simultaneously to send water through two ten-inch-diameter pipes at more than two atmospheres pressure, part of it going directly to the gardens and fountains and part to a large reservoir high on a hill behind the palace (the *Ruinenberg,* which provided a vista of "ancient" ruins). Returning from this reservoir, the water could be distributed throughout the gardens and fountains, but most spectacularly it drove a great spout 126 feet over the heads of strolling visitors, reaching higher than the roof of the summer palace on the hill behind it (see figure 5.15).[26]

5.16

"Ansicht der neuen Dampfmaschine bei Sanssouci," Carl Daniel Freydanck, 1843, cat. no. 50, KPM-Archiv (Land Berlin), Schloss Charlottenburg, Berlin. Drawing on the exotic images of the Orient, Islam, and Venetian canals, Persius incorporated the steam power of Sanssouci into a stunning mosque.

5.17

Borsig's engine at Sanssouci. Photograph by Manfred Hamm. Designed to inspire awe, Borsig's great engine rises like an altar before the visitor entering through wide front doors.

The symbol bespoke its origins. Behind this awesome fountain stood an engine for a new political economy. Its power had been doubled on the advice of Beuth, who never missed an opportunity to promote industry with art. Persius (with advice from Friedrich Wilhelm IV) housed it in a comparably imposing building, an exotic mosque with a minaret for a chimney (see figure 5.16). He intended the engine to be seen and honored inside its temple, where it sat beneath the dome in what would normally have been the central place of worship.[27]

Visitors entered through wide front doors to marvel at its grandeur, an experience still available today (see figure 5.17). The ornate columns and arches of this space are also the main support members of the machine. They exhibit its dynamic action against a background of Moorish decoration extending to the top of the dome. The elaborately transposed religious display inevitably suggests the presence of a new spiritual power. And indeed, the crowds of visitors who came to view its wonders were taught to see it in that light. A tourist guide of 1843 likened the idea of building the fountains and engine to "the spark of heavenly fire which Prometheus, son of a Titan, stole from the gods in order to animate his human model of clay and earth."[28]

126 | M. Norton Wise

To fully appreciate this godlike quality, visitors were advised to go and stand before the engine and gaze up at its ominously quiet operation, ensured by hardwood gear teeth meshing with iron. "The hushed quiet and uniform motion of the imposing masses and forces of this hydraulic steam engine arouse a stirring of amazement at the vast power of the human spirit, which itself knows how to put the elements to use, to accomplish truly titanic work without strain."[29] The human spirit is here specifically the German spirit, symbolically present in the form of the Prussian eagle poised above the governor that regulated the giant machine (see figure 5.18).

Domestic industry had been able to construct the entire engine with only two colossal cranks imported from England. "We are justifiably delighted at this advance of German national industry, as one of the proofs of how Germany more and more emancipates itself from the British industrial monopoly that has prevailed until now." Even more specifically, German national industry appeared here in the shape of Borsig's machine works, which was expanding as fast as the railroads. In 1844 he produced his twenty-fourth locomotive, the prize-winning "Beuth" at the Berlin Industrial Exhibition. By 1854, when he died of a stroke, his firm had built 500 locomotives.[30] Entrepreneurial capitalism was flourishing in Berlin. With it came a change in the steam-powered garden.

5.18

The governor of the engine at Sanssouci. Photograph by the author. Perched high in the mosque's dome, the Prussian eagle atop the twirling brass balls of the governor maintains symbolic order in the machinery below.

5.19

"Das Borsigsche Etablissement zu Moabit," etching by Josef Michael Kolb after a photograph by Johannes Rabe, 1854, from Rellstab, *Berlin und seine nächsten Umgebungen,* p. 238. Photograph from the Berlin Museum. In a telling reversal of the sites of power, rather than introduce the engine discreetly into a princely garden, Borsig brought the entire garden into the midst of the steam, smoke, and clamor of his expanding machine works.

In 1847, just prior to the revolutions that briefly swept away the old order, Borsig moved much of his operation to Moabit, north of the city, where Schinkel and Lenné were laying out a new multiuse suburb and where some of the newly wealthy were establishing themselves. There, in the midst of his works, Borsig built his own villa on the banks of the Spree (see figure 5.19), with the architectural assistance of another Schinkel student, Johann Heinrich Strack. Spreading behind the house and leading down to the river was a garden Lenné himself designed. Fairly small to be sure, it nevertheless contained an extensive *Wasserkunst,* including an impressive fountain driven by an engine from the plant. In greenhouses constructed of glass and cast iron, also from his works, and warmed by its drainage water, Borsig pursued his avocation of exotic plant culture, sharing his accomplishments with Lenné, Beuth, and other members of the Union for the Advancement of Horticulture, which he had joined in 1835. In 1852 he brought into bloom the first *Victoria Regia* rose in Berlin.[31] Perhaps no single achievement could

129 | Architectures for Steam

better epitomize the now familiar ideal of realizing aesthetic goals within industrial and scientific culture, remaking the English garden in a landscape of domesticated machinery.

But the English garden in Borsig's hands represented something quite new. If earlier it had projected the symbols of British achievement into German culture through the medium of princely wealth and status, it announced here the attained economic might of entrepreneurs and engines in a transformed German state. No longer did the engine simply power the garden and project a new culture; it paid for the garden and powered the culture. The middle class had begun to compete with princes for possession of the symbols that had figured so importantly in their own rise to economic prominence, if not yet to political authority.

The Culture of Steam and the Science of Work

If August Borsig represented the new Titans of steam technology, he soon had a counterpart in science, Hermann Helmholtz, whom we have already noted frequenting the Potsdam gardens. When Borsig's great engine first charged the fountains of *Sanssouci* in 1842, Helmholtz was just completing his doctorate with the famous Johannes Müller. From 1843 to 1847 he served nearby in the prestigious *Königliches Garde-Husaren-Regiment*. In fact, Helmholtz served as assistant to the regimental physician, Wilhelm Puhlmann, a neighbor of the Helmholtz family with whom he became close friends and whose niece Olga von Velten he would marry in 1849. Puhlmann provided a critically important point of contact for the ambitious young Helmholtz with the cultural life of the aristocracy and higher civil servants. Puhlmann, with personal ties to the royal family, was a leading promoter of the arts in Potsdam, had founded the *Potsdamer Kunstverein* in 1833, and was an important art collector, especially of works by his intimate friend, the well-known painter Adolf Menzel, with whom Helmholtz would also become well acquainted. (Figure 5.6 is Menzel's painting of the Berlin-Potsdam railroad).[32]

In brief, Helmholtz came of age as a striving scion of the middle class who moved just on the margins of the elite society that he sought to join through a dedicated pursuit of medicine and science coupled with a developed taste and talent for art and music. Another important point of contact with this higher social level came through the Leithold family, relatives of his mother with whom he sometimes spent a Sunday evening while a student in Berlin. One of the family, H. R. C. von Leithold, was a young lieutenant in the engineer corps of the *Vereinigte Artillerie- und Ingenieurschule* whose architectural drawings earned him space in the official exhibitions of the Berlin Academy of Art in 1838, 1839, and 1840. His apparent combination of technological and artistic sensibilities mirror Helmholtz's own ideals for himself, as well as those of Borsig and of Helmholtz's lifelong friend from the late 1840s, Werner Siemens. As a lieutenant in the artillery brigade of the same school, Siemens too exhibited a drawing at the Academy, of a cannon emplacement in 1838, when Leithold exhibited a fortification. In a rather

5.20

"Das Neue Museum in Berlin," Carl Daniel Freydanck, 1838, cat. no. 3, KPM-Archiv (Land Berlin), Schloss Charlottenburg, Berlin. The utilitarian engine house to the right of Schinkel's famous museum was a remnant of his and Lenné's plans to establish an extraordinary *Lustgarten* in the square in front of the king's palace, including a grand fountain and engine tower. The Berlin Physical Society originated within sight of the *Lustgarten* and palace, in the home of Gustav Magnus, just down the *Kupfergraben* canal on the left.

different mode, Leithold contributed to the engine-and-garden genre when he presented in 1840 *Der Dampfmaschine zum Springbrunnen im Lustgarten*.[33] It probably consisted of a structural drawing and elevation for the house of a small version of Egell's engine, with chimney just visible in the painting of figure 5.20, which supplied a fountain in the formal pleasure garden across from the royal palace in Berlin. This *Lustgarten* provided a public square in front of Schinkel's grand monument to state-supported art, the *Kunstmuseum* (soon the *Altes Museum*), completed in 1830. Schinkel had planned a monumental eighty-eight-foot engine tower and much larger fountain to accompany a more friendly and colorful public garden, a *Volksgarten* designed by Lenné, with orange trees and flower beds, but Frederick III did not appreciate either the glory of engines and gardens or their public-spirited use to the degree that his sons and his liberal administrators did.[34]

131 | ARCHITECTURES FOR STEAM

In any case, Helmholtz's contact with the Schinkel-Lenné landscapes and the cultural ideals that they represented was a matter not only of intellectual engagement but of daily life, of where he walked and to whom he talked. Among the gardens he frequented were those of *Sanssouci* and *Charlottenhof,* where he and Olga von Velten were friends of the court gardener Hermann Selo and his wife, who happened to be Persius's sister. Selo was a favored protégé of Lenné, who introduced him at *Charlottenhof.* Soon Friedrich Wilhelm IV was speaking with Selo in the familiar and had Schinkel and Persius build for him at *Charlottenhof* the structure that quickly developed into the first example of villa architecture in Potsdam, the Court Gardener's House (1829–30). Here Alexander von Humboldt, who would intercede for Helmholtz in obtaining his release from military service in 1848, had his private room. As Selo's position advanced, Persius built a new Court Gardener's Villa (1841–42) for him and his family at *Sanssouci,* again as an integral part of the garden landscape.[35]

Hermann Helmholtz and Olga von Velten moved in this privileged domain with friendly familiarity. For example, on the public announcement of Hermann's engagement to Olga, Selo was the first to congratulate him in person. The two men stood on the terraced hillside in front of the summer palace at *Sanssouci* (see figure 5.15), "where it presently is wonderfully beautiful," no doubt looking out on the great fountain rising directly before them, powered by Borsig's engine in Persius's mosque.[36]

The scene, so everyday and yet so symbolic of the titans of steam, has a certain appropriateness for Helmholtz's scientific image and his work, for in the same letter in which he recounted the meeting to Olga he told of having sent off the corrected proofs of the pamphlet that would soon identify him as another sort of titan of steam. *On the Conservation of Force (as Work or Energy)* served as the manifesto of the group of young reformers who in 1845 had constituted themselves the *Berlin Physikalische Gesellschaft* (Berlin Physical Society) and who set out to remake all the natural sciences in the image of modern physics and modern technology, meaning especially to remake physiology as the physics of work, of work done by muscles and engines, remembering always that this pursuit was not to be divorced from aesthetic judgment. Of particular importance for Helmholtz were Emil Du Bois-Reymond, Ernst Brücke, and Carl Ludwig, among the physiologists, and Werner Siemens, who numbered among at least twelve engineers and mechanics out of a total membership of fifty-four. Identifying themselves as the party of progress, the *Gesellschaft* reported on the latest advances in what they regarded as the physical foundations of all natural science in their annual review, *Die Fortschritte der Physik* (Progress in Physics).[37]

The *Gesellschaft* and its *Fortschritte* might well be compared with Beuth's *Verein* (Union for the Advancement of Industry) and its *Verhandlungen,* for Helmholtz and his "progressive" friends were direct heirs of the institutions and values that Beuth and his associates in "advancement" had established in the 1820s and 1830s. As observed previously, Helmholtz's professors in the natural sciences commonly taught students from both the *Friedrich-Wilhelms-Institut* for military doctors and the university, and most participated as well in the network of technical schools and unions for progress that connected academic science at the university to the more practical goals of engineering and industry. Eilhard Mitscherlich, Helmholtz's professor of both theoretical and organic chemistry and one of the leading chemists in Europe, epitomizes the Berlin ethos and practice. Maintaining a lifelong interest in practical chemistry, he studied mines and metallurgical works in Sweden in the early twenties with Berzelius, made the factory tour in England in 1824, and later took his students to visit industrial sites in Berlin. His famous *Lehrbuch der Chemie* (1st ed. 1829, 4th ed. 1847) was appropriately translated into English as *Practical Experimental Chemistry Adapted to Arts and Manufactures* (1838). The renowned Mitscherlich taught chemistry and physics also at the *Kriegschule* and was a member of Beuth's *Verein*.[38]

In similar fashion, H. F. Link, professor of botany and natural history and director of the botanical garden, both joined Beuth's *Verein* and served as the long-term director of the *Verein zur Beförderung des Gartenbaues,* where Lenné was the leading light and where Borsig displayed his roses. K. D. Turte, professor of physics, taught at the *Artillerie- und Ingenieurschule* and held the rank of major, and H. W. Dove, another professor of physics (from 1845), lectured both at the *Gewerbeschule* and the *Artillerie- und Ingenieurschule* and lived and taught at the *Kriegschule*.[39]

Certainly these university professors taught at the technical schools in part simply to increase their incomes. But their commitment to technical education and industrial progress was real. Indeed, they carried that commitment into their university teaching and research as well. Another of their number, of great importance to Helmholtz's scientific development but whom he apparently came to know only in 1845, was Gustav Magnus, who in that year became full professor of technology at the university, having been steered in the technological direction since the early thirties by the minister of culture, Karl Freiherr von Altenstein. Magnus helped administer the section for chemistry and physics of Beuth's *Verein,* taught both technology and physics at the university, and at various times in the 1830s had taught physics at the *Artillerie- und Ingenieurschule* and chemistry at the *Gewerbeschule*. In 1830 he had given his trial lecture as *Privatdozent* on the power obtainable from steam, "Die Lehre von den Dämpfen," and he regularly incorporated in his courses an excursion to Borsig's and other works. Traveling himself to what he still recognized in 1835 as "the country of industry" (England), he visited numerous industrial plants much as Beuth, Schinkel, and Mitscherlich had done a decade earlier and rode the first passenger railway in England from Liverpool to Manchester.[40] Magnus, in short, wove his own activities thoroughly into the fabric of technical and scientific institutions that Beuth and his allies were so energetically promoting. Their ideal for a *Gewerbeschule* mirrors the network whose resources it required. It should "concern itself with a thorough education in physics, chemistry, the mathematical sciences, principles of construction for craftsmen and mechanics, line drawing, and free-hand drawing of architectural and other illustrations." In thus uniting theory and practice, "the greater difficulty lay in the lack of capable teachers for this purpose."[41] Magnus fit himself to the requirement. His close association also with the artistic component of the Beuth vision, especially through his brother Eduard, one of the foremost realist portrait painters of the day and an intimate of Menzel, can only have enhanced his position.

As the son of a well-to-do banking family who married a wealthy woman in 1840, Magnus set himself up in a rather splendid house within sight of Schinkel's *Altes Museum,* Lenné's *Lustgarten,* and the Royal Palace. Its small rear garden adjoined a larger one next door, designed by Lenné in 1831 for Finanzrat A. L. Kerll, a member of the *Gartenbauverein* and friend of Schinkel.[42] In his new home, evidently in the west wing looking out on the garden, Magnus installed a lecture room and the best-equipped laboratory for chemistry and physics in Berlin, housing instruments and apparatus that he had been assembling since 1833 with funds promised by von Altenstein,[43] which suggests another analogy with Beuth's facilities for the *Gewerbeschule.* Here Helmholtz carried out his fermentation experiments on Fintelmann's grapes in 1845. Here too the *Berlin Physikalische Gesellschaft* organized itself in the same year.

Their program rested on the thesis that all natural processes—whether physical, chemical, or physiological—involved nothing other than conversions of physical "force" from one form to another, precisely like a steam engine converting heat to work. Helmholtz's conservation paper generalized this principle from the analysis of steam engines by French engineers Sadi Carnot and Emile Clapeyron, with the all-important change that the work done by the engines derived not from a "fall" of heat from high to low temperature but from the consumption of heat, which was itself nothing but the motion of atoms. With respect to physiology, muscles produced work and heat from "force" released in chemical reactions. Helmholtz and Du Bois-Reymond developed this perspective to great effect in their research on muscle action and nerve transmission in frogs. Perhaps the single most important tool for this work was Helmholtz's adaptation of the graphical recording instruments used on steam engines—the indicators and indicator diagrams originally employed by James Watt—to record the temporal course of the production of work in muscles.[44] He thereby assimilated the tiny quantities of work produced by frog muscles to the immense productive power of a Borsig engine.

The symbolic impact of this relation was not lost on his fellow reformers. "It is a spectacle for the gods," Du Bois-Reymond wrote to Helmholtz in 1852, "to see the muscle working like the cylinder of a steam engine." Nor was the power of their achievement lost on more traditional physiologists, as Helmholtz reported to Du Bois-Reymond in the same year with respect to an attack by Rudolph Wagner: "He blasts the physical-physiological school as a young race of Titans who are growing the wings of Icarus."[45] Thus not everyone in the sciences approved of the young gods' assimilation of physiology to engineering technology, anymore than everyone in the Prussian administration approved of Beuth's program for industrial progress through technical education and cultural projection.⁴⁶ But Helmholtz and his friends had aligned themselves definitively with the modernizing wing of the social elite. Their scientific achievements were incubated in the climate promoted by that elite and were deeply grounded in the resources it provided.

Conclusion

I have attempted to show how the Schinkel-Lenné landscapes coupled with the new educational institutions and societies for progress in architecture, gardens, and industry helped to shape the expectations and accomplishments of a new technological and scientific elite in Prussia, projecting a new cultural landscape in which English advances in steam technology could be naturalized and transformed into independent Prussian achievements. If the program failed in that the reality of urban blight soon overshadowed Beuth's ideal of small industrial towns organically integrated into the landscape, it nevertheless succeeded in nurturing industrial giants like Borsig to compete with the British. Perhaps unexpectedly, it also nurtured scientific giants like Helmholtz, who were able to turn the tools and ideas of steam technology into the physics and physiology of work.

In 1853, in a late version of the English tours of Beuth, Schinkel, Lenné, Egells, and his own teachers, Helmholtz went to see for himself the ever present yardstick of Prussia's accomplishment. He wanted to meet his British scientific peers, to attend the annual meeting of the British Association for the Advancement of Science, and to advertise his work. At the meeting he found that 600 people, including 175 women, had paid at least one pound each to attend the opening ceremony, where they heard him introduced as a man who had made one of the most important advances of continental science. "My paper 'On the Conservation of Force,'" he wrote to his wife, "is better known here than in Germany." As attendance at the meeting grew to 850, including 236 women who seemed well instructed in the natural sciences, Helmholtz was staggered by the public character of science and technology in England and by what he perceived as the support it received from the state. On a visit to Westminster Abbey he had already seen this cultural phenomenon carved in stone. "Professors of physics and chemistry there lie between the kings. . . . Newton, James Watt, Humphrey Davy, Thomas Young. . . . To have had such men and to see them so honored, is something grand."⁴⁷ No doubt the image of prestigious scientists in the service of a wealthy and powerful state captured Helmholtz's vision of himself.

On visiting London, he wrote to Carl Ludwig: "England is a great country, and one feels here what a grand and glorious thing civilization is when it penetrates all the most minute relations of life. Relative to London, Berlin and Vienna are only villages." And in writing to Olga about the civilization of this great "Babylon," Helmholtz focused on this paper's central image, English gardens, but with the most un-Prussian spectacle of ordinary citizens walking at will on the grass. "Imagine to yourself immense lawns with very short beautiful clean grass, with here and there large old trees or tree groups, traversed by a few well-defined paths which are actually only intended for rainy weather, while in good weather everyone walks freely through the grass; then you have the ideal that you would like to make out of our garden in the *Domstraße*."[48] Apparently the young Helmholtzes shared the vision of rising out of the middle classes to imitate princes that Borsig had so recently expressed in his garden. While Borsig built real, work-producing engines to establish his social position and his freedom, Helmholtz was busy building a general theory of the work they accomplished to establish his. The English garden still symbolized their goals.

NOTES

An abbreviated version of this paper has appeared previously: See Wise, "English Gardens in Berlin," in the Bibliography for this essay. For detailed commentary, I thank Mario Biagioli, Robert Brain, John Carson, Peter Galison, Mary Morgan, Simon Schaffer, Otto Sibum, Deborah Silverman, Mary Terrall, and especially Elaine Wise.

1. Michael Seiler describes Hahn's painting in "Entstehungsgeschichte," p. 425, cat. no. 312; full reproduction in color at p. 168.
2. Von Moltke, *Gesammelte Schriften*, 6: p. 3; quoted in von Krosigk and Wiegand, *Glienicke*, p. 47. Painting described by Michael Seiler in Julier, et al., eds., *Schloss Glienicke*, p. 413, cat. no. 281.
3. Hermann Helmholtz to Ferdinand Helmholtz, 1 December 1838, in Cahan, ed. *Letters*, p. 54.
4. Helmholtz to Olga von Velten, 30 July 1847, in Kremer, ed., *Letters*, pp. 22–23. Hermann Helmholtz to Ferdinand Helmholtz, 19 December 1845, in Cahan, ed., *Letters*, p. 110.
5. Seiler, "Entstehungsgeschichte," p. 114.
6. Brose, in *Politics of Technological Change*, focuses on the contested character of the movement for industrial modernization within the ministries, the Mining Corps, and the military. His deeply researched study informs my discussion of Beuth, below.
7. Matschoss, *Entwicklung der Dampfmaschine* 1: 172f, 176f.
8. Matschoss, *Preussens Gewerbeförderung*, pp. 49–52; with two letters from Beuth to Schinkel (1823), pp. 127–132 and a selection of Schinkel's 1826 letters and diary, pp. 132–145, from von Wolzogen, *Aus Schinkels Nachlaß*.
9. Matschoss, *Preussens Gewerbeförderung*, pp. 37–40, 147–149.
10. Matschoss, *Preussens Gewerbeförderung*, pp. 41–45.
11. Borsig, "Beschreibung der Lokomotive 'Beuth'," which includes detailed drawings.
12. Brose, *Politics of Technological Change*, ch. 3, "The Aesthetic View from Pegasus," gives a most illuminating account of Beuth's activities; quotation, p. 115; Schinkel's painting, p. 99, from Matschoss, *Preussens Gewerbeförderung*, p. 72.
13. Gröning, "Lenné und der 'Verein zur Beförderung des Gartenbaues'," pp. 82–92; and Wefeld, "Lenné und die erste Gärtnerschule," pp. 91–97.
14. Seiler, "Entstehungsgeschichte," p. 138.
15. Bohle-Heintzenberg and Hamm, *Persius*, pp. 23–37, 126.

16. Eggeling, "Persius als Architekt in Glienicke," pp. 66–71. Bohle-Heintzenberg and Hamm, *Persius,* pp. 17–19, 112–114.
17. Seiler, "Entstehungsgeschichte," pp. 141, 149, citing *Journal über Glienicke,* entry of 1862.
18. Bohle-Heintzenberg and Hamm, *Persius,* pp. 21, 79–80, 162.
19. Brose, in *Politics of Technological Change,* pp. 188–189, suggests using the term applied to the Weimar and Nazi periods by Herf in the title of his *Reactionary Modernism* to describe the views of Prince Wilhelm and others, including von Moltke.
20. The picture, made originally with a camera obscura, is reversed; see von Buttlar, *Lenné,* p. 208, cat. no. 93.
21. Nungesser, *Kreuzberg,* pp. 22–30.
22. Hoffmann, *Charlottenhof,* interior furnishings, passim; engine, p. 7. Bergdoll, *Schinkel,* pp. 141–167, gives the Humboldt story. On Friedrich Wilhelm IV as architect, see Bohle-Heintzenberg and Mann, *Persius,* pp. 9–11, 51–65.
23. "Candelabrum," in *Encyclopedia Britannica* (1967).
24. I thank Elaine Wise for pointing out Schinkel's continuation of the candelabrum theme throughout his design.
25. Hinz, *Lenné und seine Schöpfungen,* pp. 50–51.
26. Hinz, *Lenné und seine Schöpfungen,* p. 67. Matschoss, *Die Entwicklung,* 1: 185, 801. Wagenbreth and Wächtler, eds., *Dampfmaschinen,* p. 266; pictures of engine on pp. 217, 272, 279, 304–5.
27. Belani, *Der Fontainenanlagen in Sanssouci,* p. 41. Bohle-Heintzenberg and Mann, *Persius,* pp. 74–77.
28. Belani, *Der Fontainenanlagen in Sanssouci,* p. 40. The popular Prometheus metaphor for the creative power of the age of iron and steam aptly reappears in Landes's classic account, *The Unbound Prometheus.*
29. Belani, *Der Fontainenanlagen in Sanssouci,* p. 67.
30. Belani, *Der Fontainenanlagen in Sanssouci,* p. 68. Matschoss, *Entwicklung,* 1: 185–186, 801; see pictures on p. 802.
31. Galm, *Borsig,* pp. 87–90.
32. Kremer, ed., *Letters,* pp. xiii, xv, 4n; Cahan, ed., *Letters,* p. 104; Lammel, *Menzel,* pp. 46–49.
33. Leithold's drawings are listed in Börsch-Supan, ed., *Die Kataloge,* 2: 1838, No. 917; 1839, No. 966; 1840, No. 892; Siemens appears in 1838, No. 908. From 1820 to 1846 the catalogs regularly include a section of "Zeichnungen von der Königlichen Artillerie- und Ingenieurschule" with ten to fifteen drawings per year.
34. Schinkel's and Lenné's aspirations are described in Rave, *Schinkel: Stadtbaupläne,* pp. 116–118; and Hinz, *Lenné,* pp. 139–140.
35. von Buttlar, *Lenné,* p. 204; Bohle-Heintzenberg and Hamm, *Persius,* pp. 14–15, 37; Kremer, ed., *Letters,* p. 27n.
36. Helmholtz to Olga von Velten, 8 September 1847, in Kremer, ed., *Letters,* p. 27.
37. Lenoir, "Laboratories, Medicine, and Public Life."
38. Szabadváry, "Mitscherlich," pp. 423–426; and Ladenburg, "Mitscherlich, Eilhard M.," in *Allgemeine Deutsche Biographie,* 22: 15–22.
39. Cahan, ed., *Letters,* pp. 50n (Turte) and 58n (Link); Dove, "Dove," pp. 51–69; and Neumann, *Dove.*

40. Kant, "Die Entwicklung der Physik," pp. 187–191; and Wolf, "Magnus," pp. 20–26.
41. From Beuth's report, "Votum des Ministers des Innern für Handel und Gewerbe betreffend das Gewerbe-Schulwesen in der Provinz Schlesien" (1833) in Matschoss, *Preussens Gewerbeförderung,* pp. 148–149. On technical education see Lundgreen, *Techniker in Preussen;* and Schubring, "Mathematics and Teacher Training."
42. Becker and Jacob, eds., *Das Magnus-Haus,* pp. 51–52; and Becker, "Geschichte des Magnus-Hauses," p. 108.
43. Wolf, "Magnus," p. 20.
44. This thesis is developed in Brain and Wise, "Muscles and Engines." Lenoir, "Farbensehen," develops a similar thesis for Helmholtz's sensory physiology.
45. Du Bois-Reymond to Helmholtz, 9 February 1852, and Helmholtz to Du Bois-Reymond, 20 June 1852, in Kirsten et al., eds., *Dokumente einer Freundschaft,* pp. 123, 132.
46. Brose, *Politics of Technological Change,* Introduction and passim, has called particular attention to the conflicts.
47. Helmholtz to Olga von Velten, 8–14 September, 1853, in Kremer, ed., *Letters,* pp. 129–132.
48. Helmholtz to Ludwig, August 1853, in Koenigsberger, *Helmholtz,* 1: 202. Helmholtz to Olga von Velten, 20 August 1853, in Kremer, ed., *Letters,* pp. 110–111.

BIBLIOGRAPHY

Becker, Christine. "Zur Geschichte des Magnus-Hauses." In Hoffman, ed. *Magnus und sein Haus,* 99–121.

Becker, Christine, and Brigitte Jacob, eds. *Das Magnus-Haus in Berlin-Mitte: Geschichte, Wandel und Wiederherstellung eines barocken Palais.* Berlin: Bruckmann, 1994.

Belani, H. G. R. [Karl Ludwig Häberlin]. *Geschichte und Beschreibung der Fontainenanlagen in Sanssouci unter Friedrich dem Großen und Sr. Majestät dem Könige Friedrich Wilhelm IV.* Potsdam: Otto Janke, 1843.

Bergdoll, Barry. *Karl Friedrich Schinkel: An Architecture for Prussia.* Photographs by Erich Lessing. New York: Rizzoli, 1994.

Beuth, August. "Beschreibung der in der Maschinenbau-Anstalt des herrn A. Borsig hierselbst erbauten Lokomotive 'Beuth'." With 4 plates. *Verhandlungen des Vereins zur Beförderung des Gewerbefleisses* 46 (1846): 75–82.

Bohle-Heintzenberg, Sabine, and Manfred Hamm. *Ludwig Persius, Architekt des Königs.* Berlin: Gebrüder Mann, 1993.

Börsch-Supan, Helmut, ed. *Die Kataloge der Berliner Akademie-Ausstellungen 1786–1850.* 2 vols. & Registerband. Berlin: Bruno Hessling, 1971.

Borsig, August. "Beschreibung der in der Maschinenbau-Anstalt des Herrn A. Borsig hierselbst erbauten Lokomotive 'Beuth'." *Verhandlungen des Vereins zur Beförderung des Gewerbfleisses in Preussen,* 1846: 75–82 and plates V–IX.

Brain, Robert M., and M. Norton Wise. "Muscles and Engines: Indicator Diagrams and Helmholtz's Graphical Methods." In Krüger, *Helmholtz,* 124–145.

Brose, Eric Dorn. *The Politics of Technological Change in Prussia: Out of the Shadow of Antiquity, 1809–1848.* Princeton, NJ: Princeton University Press, 1993.

Cahan, David, ed. *Letters of Hermann von Helmholtz to his Parents: The Medical Education of a German Scientist, 1837–1846.* Stuttgart: F. Steiner, 1993.

Cunningham, A., and P. Williams, eds. *The Laboratory Revolution in Medicine.* Cambridge: Cambridge University Press, 1992.

Dove, Alfred. "Heinrich Wilhelm Dove." In *Allgemeine deutsche Biographie.* Reprint, Berlin: Duncker & Humblot, 1971, vol. 48, pp. 51–69.

Eggeling, Tilo. "Ludwig Persius als Architekt in Glienicke." In Julier et al., eds., *Schloss Glienicke,* 63–79.

Galm, Ulla. *August Borsig.* Berlin: Stapp, 1987.

Gröning, Gert. "Peter Joseph Lenné und der 'Verein zur Beförderung des Gartenbaues in den Königlich Preussischen Staaten.'" In von Buttlar, ed., *Lenné: Volkspark und Arkadien,* 82–90.

Günther, Harri, and Sibylle Harksen. *Peter Joseph Lenné: Katalog der Zeichnungen.* Tübingen: Wassmuth, 1993.

Herf, Jeffrey. *Reactionary Modernism: Technology, Culture and Politics in Weimar and the Third Reich.* New York: Cambridge University Press, 1985.

Hinz, Gerhard. *Peter Joseph Lenné und seine bedeutendsten Schöpfungen in Berlin und Potsdam.* Berlin: Deutscher Kunstverlag, 1937.

Hoffman, Dieter, ed. *Gustav Magnus und sein Haus.* Stuttgart, Germany: 1995.

Hoffmann, Hans. *Schloss Charlottenhof und die Römische Bäder.* 3rd ed., rev. by Renate Möller. Potsdam-Sanssouci: Chemnitzer Verlag, 1991.

Julier, Jürgen, Susanne Leiste, and Margret Schütte, eds. *Schloss Glienicke: Bewohner, Künstler, Parklandschaft* (essays and catalog of the exhibition). Berlin: Staatliche Schlösser und Gärten Berlin, 1987.

Kant, Horst. "Entscheidende Impulse für die Entwicklung der Physik in Berlin." *Physik in der Schule* 15 (1977): 187–191.

Kirsten, Christa, et al., eds. *Dokumente einer Freundschaft: Briefwechsel zwischen Hermann von Helmholtz und Emil Du Bois-Reymond, 1846–1894.* Berlin: Akademie Verlag, 1986.

Koenigsberger, Leo. *Hermann von Helmholtz.* 3 vols. Braunschweig: Friedrich Vieweg, 1902–3.

Kremer, Richard L., ed. *Letters of Hermann von Helmholtz to His Wife, 1847–1859.* Stuttgart, Germany: F. Steiner, 1990.

Krüger, Lorenz, ed. *Universalgenie Helmholtz: Rückblick nach 100 Jahren.* Berlin: Akademie, 1994.

Ladenburg, "Mitscherlich, Eilhard M." In *Allgemeine deutsche Biographie.* Reprint, Berlin: Duncker & Humblot, 1970, vol. 22, pp. 15–22.

Lammel, Gisold. *Adolph Menzel und seine Kreise.* Dresden, Germany: Verlag der Kunst, 1993.

Landes, David. *The Unbound Prometheus: Technological Change and Industrial Development in Western Europe from 1750 to the Present.* Cambridge: Cambridge University Press, 1969.

Lavin, Irving, ed. *Meaning in the Visual Arts: Views from the Outside: A Centennial Commemoration of Irwin Panofsky (1892–1968).* Princeton, NJ: Princeton University Press, 1995.

Lenoir, Timothy. "Farbensehen, Tonempfindung und der Telegraph: Helmholtz und die Materialität der Kommunikation." In Rheinberger and Hagner, eds., *Experimentalisierung des Lebens,* 50–73.

Lenoir, Timothy. "Laboratories, Medicine, and Public Life in Germany 1830–1849." In Cunningham and Williams, eds., *The Laboratory Revolution in Medicine,* 14–71.

Lundgreen, Peter. *Techniker in Preussen während der frühen Industrialisierung.* Berlin: Colloquium Verlag, 1975.

Matschoss, Conrad. *Die Entwicklung der Dampfmaschine: Eine Geschichte der ortsfesten Dampfmaschine und der Lokomobile, der Schiffsmaschine und Lokomotive.* 2 vols. Berlin: Julius Springer, 1908.

Matschoss, Conrad. *Preussens Gewerbeförderung und ihre Grossen Männer.* Berlin: Verein der Deutschen Ingenieure, 1921.

Neumann, Hans. *Heinrich Wilhelm Dove: Eine Naturforscherbiographie.* Liegnitz: H. Krumbharr, 1925.

Nungesser, Michael. *Das Denkmal auf dem Kreuzberg von Karl Friedrich Schinkel.* Berlin: W. Arenhövel, 1987.

Rave, Paul Ortwin. *Karl Friedrich Schinkel: Lebenswerk.* Berlin: *Stadtbaupläne, Brücken, Straßen, Tore, Plätze.* 2nd. ed. Berlin: Deutscher Kunstverlag, 1981.

Rellstab, Ludwig. *Berlin und seine nächsten Umgebungen in malerischen Originalansichten: Historisch-topographisch beschrieben.* Darmstadt: 1854.

Rheinberger, Hans-Jörg and Michael Hagner, eds. *Die Experimentalisierung des Lebens: Experimentalsysteme in den biologischen Wissenschaften, 1850–1950.* Berlin: Akademie, 1993.

Schinkel, Karl Friedrich. *Sammlung Architektonischer Entwürfe.* Berlin: Duncker, 1866.

Schubring, Gert. "Mathematics and Teacher Training: Plans for a Polytechnic in Berlin." *Historical Studies in the Physical Sciences,* 12 (1981): 161–194.

Seiler, Michael. "Entstehungsgeschichte des Landschaftsgartens Klein-Glienicke." In Julier et al., eds., *Schloss Glienicke,* 109–156.

Szabadváry, F. "Mitscherlich, Eilhard." *Dictionary of Scientific Biography.* New York: Charles Scribner's Sons, 1973, vol. 9: 423–426.

von Buttlar, Florian, ed. *Peter Joseph Lenné: Volkspark und Arkadien* (catalog of the exhibition). Berlin: Nicolai, 1989.

von Krosigk, Klaus, and Heinz Wiegand. *Glienicke.* Berlin: Haude and Spener, 1984.

von Moltke, Helmut. *Gesammelte Schriften und Denkwürdigkeiten des General-Feldmarschalls Grafen Helmuth von Moltke.* Vol. 6, *Briefe, 3. Sammlung.* Berlin: E. S. Mittler, 1892.

von Wolzogen, Alfred. *Aus Schinkels Nachlaß: Reisetagebücher, Briefe und Aphorismen.* 4 vols. Berlin: R. Decker, 1862–1864.

Wagenbreth, O., and E. Wächtler, eds. *Dampfmaschinen: Die Kolbendampfmaschine als historische Erscheinung und technisches Denkmal.* Leipzig: Fachbuchverlag, 1986.

Wefeld, Hans Joachim. "Peter Joseph Lenné und die erste Gärtnerschule." In von Buttlar, *Lenné: Volkspark und Arkadien,* 91–97.

Wise, M. Norton. "English Gardens in Berlin: Aesthetics, Technology, Power." In Lavin, ed. *Meaning in the Visual Arts,* 237–254.

Wolf, Stefan. "Gustav Magnus—Sein Weg zum Ordinariat an der Berliner Universität." In Hoffman, ed., *Magnus und sein Haus,* 33–53.

6

Illuminating the Opacity of Achromatic Lens Production: Joseph von Fraunhofer's Use of Monastic Architecture and Space as a Laboratory

Myles W. Jackson

Most physicists and historians of science are familiar with Joseph von Fraunhofer and the so-called Fraunhofer lines that dissect the solar spectrum. His "discovery" of those dark lines and his subsequent work on optical glass manufacturing and diffraction gratings either gave rise to, or greatly advanced, a large spectrum of disciplinary research during the nineteenth century, including spectroscopy, X-ray technology, photochemistry, and of course stellar and planetary astronomy. He was a self-educated, working-class artisan who manufactured the best achromatic lenses[1] for telescopes the world had hitherto seen. He changed the way Gauss, Schumacher, Olbers, and Bessel viewed the cosmos. Fraunhofer's workmanship played a critical role in the temporary transfer of an optical monopoly from Britain to Bavaria during the early nineteenth century. Hence, numerous Germanic accounts romanticize Fraunhofer as epitomizing German *Handwerkerkultur* as well as praise the sagacity of German governments in recognizing and coordinating the labor of skilled artisans.

This essay argues that Fraunhofer drew on the preexisting architectural space and layout of a secularized Benedictine monastery to construct his lenses. A study of Fraunhofer can therefore offer an insight into the more general relationships between the scientific enterprise and architectural space by illustrating how a preexisting Benedictine architecture that instantiated three aspects so crucial to the Rule of Saint Benedict—labor, silence, and secrecy—played a major role in Fraunhofer's execution of optical glass manufacture.[2]

In so doing, this essay is not contextually deterministic. I do not claim that a Benedictine monastery was the only site where optical lenses could have been manufactured. However, of all the abundant resources which Joseph von Utzschneider, former Bavarian Privy Councillor, entrepreneur, and co-owner of the Optical Institute, had at his disposal, a Benedictine cloister best suited his needs for a glass house. First, Benediktbeuern was located in the midst of a large forest, where wood for fuel was in abundant supply. Second, a quarry of quartz, a key ingredient of glass, was only ten kilometers away. Third, Benedictine monks and artisans from the surrounding communities were well versed in optical theory and practice. Fourth, monasteries and cloisters during the late eighteenth and early nineteenth century possessed lavish collections of physical instruments and texts dealing with glass manufacture and optical theory. Fifth, the cloister's large space enabled Fraunhofer to devise an experiment such that the rays of light emitted from sodium lamps were parallel when they struck the face of a glass prism whose refractive and dispersive indices were to be determined. This proved crucial for his understanding of the two lines emitted by sodium lights, later to be called the sodium lines. Finally, Benedictine architecture reflected the public and private rituals associated with monastic life. Indeed, monasteries contained a historical legacy of secret spaces. Hence, Benediktbeuern could easily be transformed into an optical enterprise that required a demarcation between public and private access. All of these factors made Benediktbeuern the ideal location for the Optical Institute.

Fraunhofer and the Dark Lines of the Solar Spectrum

Ever since Isaac Newton published his *Opticks* in 1704, it has been generally well known that white light is composed of colors, each with a specific refrangibility; that is, each colored ray of light refracts through a specific angle after passing through a prism. Violet refracts through the greatest angle, red through the least. The differences in refrangibility lead to chromatic aberration, the appearance of colored rings in images under high magnification.

During the mid–eighteenth century, John Dolland and Chester Moor Hall in England developed a method that corrected chromatic aberration. Both flint and crown glass were used in the construction of achromatic lenses.[3] Combining a convex lens of one type of glass with a concave lens of the other corrected the aberration by focusing the red and violet rays into a single point. Fraunhofer adopted this technology for determining more precisely the refrangibility of glass production for a portion of the solar spectrum. Formerly, the refractive indices of glass blanks were determined by taking the arithmetic average of the extreme rays of the spectrum: red and violet.[4] From late 1813 to early 1815, Fraunhofer devised an experimental technique to improve his production of achromatic lenses. He published the culmination of his attempt in his essay on the Six Lamps Experiment, "*Bestimmung des Brechungs- und Farbenzerstreuungs-Vermögens verschiedener Glasarten in Bezug auf die Vervollkommung achromatischer Fernröhre*" ("Determination of the Refractive and Dispersive Indices for Differing Types of Glass in Relation to the Perfection of Achromatic Telescopes").[5]

Fraunhofer's Six Lamps and Solar Light Experiments

Fraunhofer's essay, written from 1814 to 1815, was the culmination of his experiments on perfecting the manufacture of achromatic lenses; however important the work later was taken to be for spectroscopy, Fraunhofer's essay was certainly not an attempt to explain theoretically the nature of the solar dark lines (later called absorption or Fraunhofer lines) or the lamp lines (later called emission lines). Its sole purpose was to publish his method for improving the construction of achromatic lenses for telescopes. Indeed, the essay commences by claiming that:

> The calculation of an achromatic object glass, and generally that of every achromatic telescope, necessitates a precise knowledge of the ratio of the sines of incidence and refraction [Snel's Law], and of the ratio of various types of glass which are used in the construction of telescopes. . . . Experiments repeated during many years have led me to discover new methods of obtaining these ratios, and I have therefore obeyed the wishes of several scholars [astronomers and experimental natural philosophers] in publishing these experiments, in the order I made them, with the necessary modifications that the experiments themselves forced me to introduce.[6]

He concludes his piece by admitting:

> In making the experiments of which I have spoken in this memoir, I have considered principally their relations to practical optics. My leisure did not permit me to make any others, or to extend them any farther. The path that I have taken in this memoir has furnished interesting results in physical optics, and it is therefore greatly hoped that skilful investigators of nature would condescend to give them some attention.[7]

Beginning with Newton's work on the spectrum during the late seventeenth and early eighteenth centuries, experimental natural philosophers and opticians had attempted, in vain, to determine the dispersion and refraction of each colored ray. But since colors of the spectrum seem to be continuous, they could detect no precise limits. Fraunhofer himself admitted this in his essay:

> It would be of great importance to determine for every species of glass the dispersion of each separately colored ray. But since the different colors of the spectrum do not present any precise limits, the spectrum cannot be used for such.[8]

Like many others before him, Fraunhofer attempted to circumvent this problem by focusing his attention on colored glasses and prisms filled with colored fluids with the hope of determining the refractive index of the glass sample for a specific color of light supplied by the monochromatic source. That is, Fraunhofer and his predecessors hoped that the glasses and prisms would permit only homogeneous light to pass. Despite various attempts to produce such a glass or fluid, the emergent light never proved to be truly monochromatic; a mixture of spectral colors always resulted. Fraunhofer also attempted to use colored flames produced by burning alcohol and sulfur. But these flames, too, produced a spectrum when viewed through a prism.[9] He did, however, notice during these investigations that a clearly defined line in the orange region marked the spectra produced by the alcohol and sulfur flames. This line, later discovered to be two lines very close to each other and called the D-sodium couplet, proved to be crucial to Fraunhofer's subsequent research. (Indeed, the sodium line later served as the basis for Fraunhofer's Six Lamps Experiment, which was designed to isolate homogeneous light.)

After giving up on colored glasses and liquid-filled prisms (but before turning to the sodium lines), Fraunhofer returned to using white light from a sodium lamp. He wished to view the spectrum produced by a sodium lamp, refracted by a prism and viewed through the telescope located on a modified theodolite, an ordinance-surveying instrument, built by Georg Reichenbach, originally designed to measure angles for the production of maps. The modified theodolite could measure the angle of emergence from the prism for each colored ray. Unfortunately, the rays of light falling onto the subject prism would not be parallel, so that the angle of incidence would not be the same for each one, rendering the modified theodolite's measurement useless. To ensure that the rays striking the prism would be parallel, Fraunhofer substantially increased the distance between the lamp and the prism. (See Figure 6.1.) But he noted that although the rays now all had measurably the same angle of incidence, the increased distance resulted in some of the refracted rays' missing the prism altogether. To ensure that the incident rays remained parallel, that an entire spectrum would be generated, and that the intensity of the light would be sufficient to be seen through minute slits at such a large distance, Fraunhofer used six sodium lamps.

Fraunhofer placed the six lamps behind a shutter 1.5 Bavarian inches (approximately two English inches) high and .007 inches thick (see figures 6.2 and 6.3). The shutter was pierced by six narrow slits each of which was slightly less than 1.5 inches high and .05 inch wide. The six lamps were placed .58 inch from each other directly be-

6.1

The importance of maintaining a great distance between the lamp and the prism in Fraunhofer's experiments. In the diagram, a sodium lamp (L) shines its light into a prism (H). If the lamp is close to the prism (I), the rays of light that strike the prism's face are not parallel. As the distance between the lamp and prism is increased (II), however, the rays of light striking the prism's face become parallel.

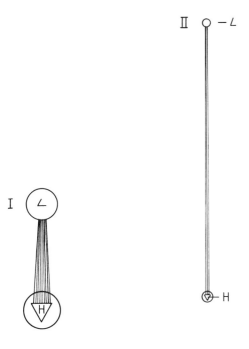

hind a shutter, putting each lamp centrally behind one of the slits. The shutter itself was placed 13 Bavarian feet (slightly more than four meters) from prism A, which was made of flint glass and an angle of approximately 40 degrees. The prism now refracted the light, having passed through the slits, and decomposed it into colors. The dispersed light then traveled through a second slit placed directly behind the prism, which accordingly blocked a portion of the emergent beam. Some of the rays were channeled to the site of a theodolite located in Fraunhofer's laboratory at the very great distance of 692 Bavarian feet from the six lamps.

The six-shutter mechanism controlled the angles at which light from each lamp struck the surface of prism A, thereby determining the locus of the corresponding spectrum. For example, from the lamp at C, red rays refracted to E and violet to D. From lamp B, the red rays traveled toward F and the violet rays toward G. On the theodolite, Fraunhofer placed a prism (H) whose index of refraction for the different colored rays was to be determined. He then adjusted the distances of the six-shutter mechanism from prism A, of A from the single shutter, and of the single shutter from prism H in such a manner that prism H received only red rays from lamp C and only violet rays from B. The intermittent lamps supplied the other colors of the spectrum.[10] The spectrum of rays passing through the small aperture below A and then through prism H appeared in the modified theodolite's telescope as depicted in figure 6.2, where I is the violet, K the blue, L the green, and so on; each spectral color appeared at a unique locus. Fraunhofer ground down the angle of prism H until all the rays from the six lamps sensibly emerged

145 | Illuminating the Opacity of Achromatic Lens Production

6.2

A page from Fraunhofer's manuscript, "Bestimmung des Brechungs- und Farbenzerstreuungsvermögens verschiedener Glasarten," detailing the layout of his Six Lamps Experiment. Reprinted with the kind permission of the Deutsches Museum, Munich, photograph 33018.

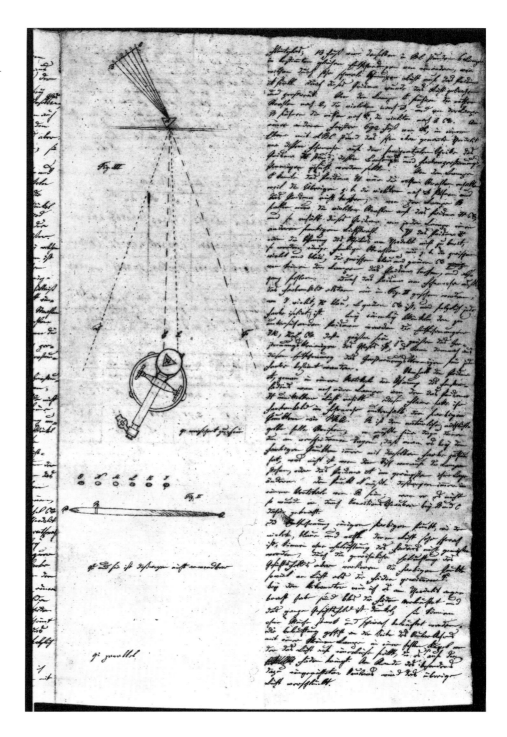

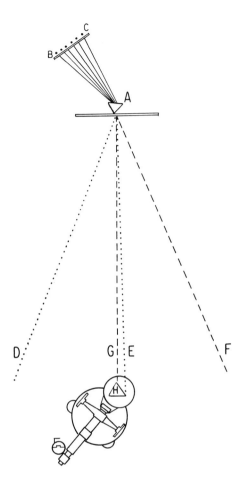

6.3

The design of Fraunhofer's Six Lamp Experiment. The light from sodium lamps placed between B and C travels through slits in a screen to a *Zuordnungsprism* (A) and is refracted through the slit of a second screen to the prism (H) whose refractive and dispersive indices are to be measured. Prism H is mounted on a modified theodolite.

from H at a single point (though, of course, each colored ray exits from that point at a unique angle with respect to the face normal there). The object lens of the modified theodolite's telescope was aimed at that point, thereby enabling Fraunhofer to see the entire spectrum and measure each color's dispersion. The distances ON, NM, and so forth increased with dispersive power under these conditions. Because the modified theodolite could measure these distances and the incident angle to the nearest arc-second, Fraunhofer could now determine the index of refraction for each colored ray for each type of refracting substance.

To see whether other sources of light produced the same sort of lines as the sodium lamps had, Fraunhofer decided to use the sun as his source. He placed his modified theodolite and prism in a darkened room with a window covered by a shutter. He cut a vertical slit in the window shutter, 15 arc-seconds wide and 36 arc-minutes high, with respect to the center of the theodolite, allowing the solar rays to fall on a flint glass prism with an angle of 60 degrees mounted on the theodolite 24 feet from the window. The prism was placed in front of the telescopic objective lens in such a manner as to ensure symmetric passage[11] (because, as Newton had argued a century earlier, that position minimizes the effect of errors in setting the incidence). Fraunhofer remarked,

> In looking at this spectrum for the bright [sodium] line, which I had discovered in a spectrum of artificial light, I discovered instead an infinite number of vertical lines of different thicknesses. These lines are darker than the rest of the spectrum, and some of them appear entirely black.[12]

Fiddling with the window shade aperture and varying the distance of the theodolite from the window did not obliterate the lines. Between B and H, Fraunhofer counted 574 dark lines. Because the lines persisted no matter how he rearranged the distances, Fraunhofer became convinced that these lines were not an experimental artifact but an inherent property of solar light.

This discovery leads us to Fraunhofer's second major contribution to experimental optics. He was clever enough to use those dark lines as a natural grid that demarcates minute portions of the spectrum. He could now obtain refractive indices for an extraordinary precise portion of the spectrum: "As the lines of the spectrum are seen with every refractive substance of uniform density, I have employed this circumstance for determining the index of refraction of any substance for each colored ray."[13] He then chose the most obvious (i.e., the thickest and clearest) lines to determine the refractive indices of a glass prism: B, C, D, E, F, G, and H (see figure 6.4). They could easily be aligned with the cross-hatchings on his theodolite. Fraunhofer would simply read off the angle from the instrument's vernier. He made five measurements for each line.

To compute the index, Fraunhofer used a standard equation found in several eighteenth-century optics textbooks for prisms:[14]

If

$\delta =$ the angle of the incident solar ray,[15]

$\rho =$ the angle of the emergent ray,

$\psi =$ the angle of the prism, and

$n =$ the index of refraction,

then

$$n = \frac{\sqrt{(\sin \rho + \cos \psi \sin \delta)^2 + (\sin \psi \sin \delta)^2}}{\sin \psi}.$$

As we have seen, Fraunhofer devised his experiments such that the angle of the incident ray (for the D line) equaled that of the emergent ray (i.e., symmetric passage). If μ (the angle of deviation) is the angle between the incident and the emergent rays, then, under these circumstances

6.4

Fraunhofer's 1814 depiction of the solar spectrum and the spectral lines. These dark lines provided a natural grid that Fraunhofer lined up with the cross-hatchings of a modified theodolite to obtain the angle measurements used in determining the refractive indices of his glass samples. Reprinted with the kind permission of the Deutsches Museum, Munich, photograph 43952. See also color plate 5.

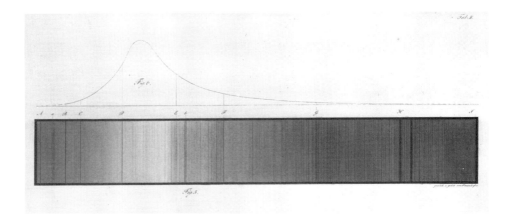

$$n = \frac{\sin\left(\frac{\mu + \psi}{2}\right)}{\sin\left(\frac{\psi}{2}\right)}.$$

The angle μ was measured by the modified theodolite, as were the arcs BC, CD, DE, EF, FG, and GH. If n_E is the refractive index for the ray E, then

$$n_E = \frac{\sin\left(\frac{\mu + \psi + DE}{2}\right)}{\sin\left(\frac{\psi}{2}\right)},$$

$$n_F = \frac{\sin\left(\frac{\mu + \psi + DE + EF}{2}\right)}{\sin\left(\frac{\psi}{2}\right)},$$

$$n_G = \frac{\sin\left(\frac{\mu + \psi + DE + EF + FG}{2}\right)}{\sin\left(\frac{\psi}{2}\right)},$$

and so on. Fraunhofer created scores of tables listing the refractive indices of the rays (each ray corresponding to a line) for different substances: flint glass, crown glass, oil of turpentine, and water, to name just a few. He then created tables of indices for combinations of refracting media to determine the combination that would correct chromatic aberration for the red and violet rays of the spectrum.

Fraunhofer had now provided opticians and experimental natural philosophers with a vastly more precise method for determining the refractive indices of glass samples than had ever before been available. Previously, Fraunhofer himself had determined the relative dispersive and refractive indices of two kinds of glass by cementing them together, forming a single prism. If the two spectra produced by this compound prism appeared at the same place, without any reciprocal displacement, he concluded that their dispersive and refractive powers were the same and equal to the arithmetic average of the two extreme rays: red and violet. After the discovery of the lines, however, he quickly realized that two pieces of glass, which appeared to have the same refrangibility when employing this early method of testing, could actually have slightly different powers, as revealed by the existence in the overlap region of two lines where there should be only one. Fraunhofer had now accomplished what Clairaut had attempted some thirty years earlier, namely to destroy the secondary spectrum resulting from the different partial dispersions of crown and flint glass.

This is the rather well known account of Fraunhofer's contribution to physical optics. Such an account, however, fails to uncover a fascinating sociocultural history of the relationship between public and private knowledge, the tension between scientific and entrepreneurial interests, and the importance of monastic secrecy to the development of optical theory and technology. This paper intends to offer precisely this sort of sociocultural history.

The Importance of Monastic Culture to the Inception of Munich's Optical Institute

The purpose of the Optical Institute was to provide optical lenses for astronomical and ordnance-surveying instruments for the joint Bavarian and French *Bureau topographique.* In 1801 Napoleon ordered this Bureau of Topography to produce accurate topographical maps of this new ally for his troops, and more accurate maps were also needed for the reforms initiated by Prince Maximilian Joseph IV's new Bavarian government.[16] A Land Registry, or *Bureau de cadastre,* was set up in 1808 to produce highly detailed and accurate maps used in determining property boundaries for tax assessment.[17] This institute, which was a part of the newly founded Mathematico-Mechanical Institute (MMI), produced the optical lenses for instruments manufactured by the MMI, such as achromatic telescopes, theodolites, and heliometers.

Utzschneider was a key member in these enterprises, responsible for the direction of both the Bureau of Topography and Land Registry. He was also the codirector of the MMI and the director and owner of the Optical Institute, which employed Fraunhofer.[18] Utzschneider drew on the Benedictine monastic community to hire the astronomers, mapmakers, surveyors, and opticians for the Optical Institute, Bureau of Topography, and Land Registry. For example, he hired Pater Ulrich Schiegg, a Benedictine monk and professor of mathematics and astronomy at the University of Salzburg. Schiegg had surveyed property around his monastery for taxation purposes some fifteen years earlier.[19] Monasteries had owned and ruled over vast areas of land. Indeed, the abbots had possessed an incredible amount of wealth and power in Bavaria until the secularization of monasteries and cloisters in 1803, at which time the Bavarian government usurped the riches of the Roman Catholic Church.[20] Schiegg also possessed extensive knowledge of astronomical instruments, because he was an astronomer. His knowledge of telescopes, theodolites, and sextants proved crucial for Fraunhofer and the Optical Institute. Utzschneider also hired Joseph Niggl, an optician by trade who was trained at the observatory of the Benedictine cloister at Rott, where he was taught optics by and produced reading glasses for Benedictine monks. He also manufactured lenses for astronomical telescopes and theodolites. The quality of his lenses, however, was disastrous. Utzschneider scoured monasteries for glassmakers, because Benedictine monks possessed a 1,000-year tradition in the manufacturing, cutting, and polishing of glass, particularly of stained glass for cathedrals.[21] Indeed, monastic texts claimed that glass was the manifestation of the divine light and the light of truth (*lux divinus* and *lux veritatis*) on earth. These texts even established the analogy between glass and the Holy Spirit.[22] Knowledge of glassmaking led to an understanding of God and his relationship with humankind. In addition to this theological reason for glassmaking was an economic factor: Monasteries prospered by making stained glass for town halls and the private residences of the aristocracy and wealthy merchants. Monasteries had owned a large number of glass houses throughout Bavaria, Bohemia, and the Swiss cantons until the secularization.[23] Glassmaking was normally restricted to common table glass, window glass, Benedictine brandy glasses, and reading lenses.

As a result of the blockade of 1806, which cut Great Britain off from continental Europe, Utzschneider needed to increase efforts in optical glass production, as Britain had been the Continent's chief supplier of optical lenses. He hired the nineteen-year-old Fraunhofer to assist Niggl and the Swiss glassmaker Pierre Louis Guinand. Peter Ulrich Schiegg instructed Fraunhofer in optics and mathematics, while Niggl taught him the art of lens production. Fraunhofer learned quickly, and by 1809, merely three years after his arrival, was appointed manager of the Optical Institute. Napoleon's blockade guaranteed a market for Fraunhofer's and Utzschneider's institute throughout continental Europe.

Monastic Labor, Architecture, and Space

In 1805 Utzschneider purchased from the government the secularized Benedictine cloister at Benediktbeuern, some fifty-five kilometers south of Munich, in the foothills of the Bavarian Alps.[24] This purchase was significant for two reasons. First, the inhabitants of monasteries and cloisters were uniformly well versed in optical theory. For example, Pater Udalricus Riesch, a coworker of Fraunhofer's at the Optical Institute, had been a mapmaker at Benediktbeuern and possessed a thorough knowledge of surveying instruments.[25] Pater Josef Maria Wagner taught optics to the young apprentices working with Fraunhofer in the cloister.[26] Riesch and Wagner were Fraunhofer's closest friends.[27] Recall that Pater Ulrich Schiegg was a Benedictine monk and that Joseph Niggl had been trained in optics by the Benedictines at Rott. Benedictine monasteries also possessed impressive libraries that included texts on physical optics, and many had lavish collections of physical instruments, including stellar telescopes, theodolites, and sextants.[28] The Benedictine cloister of Kremsmünster in Austria had been a renowned astronomical observatory and center for ordnance surveying calculations since the third quarter of the eighteenth century.[29] Second, cloisters and monasteries also possessed skilled artisans from the Benedictine order as well as from the surrounding communities, including instrument makers, brewers, distillers of brandy, and glassmakers.[30]

Utzschneider and Fraunhofer could therefore draw on the skills of a preexisting community. Skilled and unskilled artisans could provide assistance in glassmaking, glass cutting, and polishing lenses. Skilled laborers also included the forest scientists of Benediktbeuern, who calculated the amount of lumber available in the surrounding forests for its use as fuel for glass melting. But also needed, of course, were Guinand's and Fraunhofer's abilities to produce achromatic lenses, which are more difficult to produce than ordinary glass.

Labor had always played a major role in Benedictine culture. The Benedictine monastery consisted of the school of divine service and the workshop, where the monks labored at the *artes spirituales;* work was worship: *laborare est orare.*[31] *The Rule of Saint Benedict* ordered monks to practice manual labor in any art that would benefit the monastery, order, and humankind.[32] It clearly explicated the importance of labor to the order: "Idleness is an enemy of the soul. Because this is so the brethren ought to be occupied at specified times in manual labor. . . ."[33] Craftsmen from the surrounding areas were requested "to ply their trade in the monastery."[34]

The importance of labor to Benedictine culture has been the object of much attention. Herbert Workman suggests that

> Benedict's success in linking Monasticism with labor was the first step in a long evolution. . . . The change . . . had been accompanied by the glorification and systematization of toil. . . . We have the organized community . . . whose axes and spades cleared the densest of jungles, drained pestilent swamps, and by the alchemy of industry turned the sands into waving gold, and planted centres of culture in the hearts of forests.[35]

Both Max Weber and Lewis Mumford alluded to the importance of Benedictine notions of labor to the development of Western culture. Weber claimed that *The Rule of Saint Benedict* "had developed a systematic method of rational conduct with the purpose of overcoming the *status naturae,* to free man from the power of irrational impulses and his dependence on the world and on nature."[36] Mumford asserted that Benedictine monasteries were the first example of a fusion of technology with moral force. Monasteries abandoned the master-slave relationship in economic organization of labor in favor of "free man's moral commitment."[37] Monastic culture organized "a new ritual of daily activity" around manual and spiritual labor.[38] The Benedictine monastery "laid down a basis for order as strict as that which held together the early megamachines: the difference lay in its modern size, its voluntary constitution and the fact that its sternest discipline was self-imposed."[39] The Benedictine *Rule* was successful in fusing economic and spiritual concerns precisely because it viewed the two as being closely related.[40] This union formed the cornerstone upon which Benedictine culture was erected.[41]

Another important characteristic of Benedictine culture was its emphasis upon silence and secrecy. Chapter 4 of *The Rule of Saint Benedict* insisted that a Benedictine should not be a murmurer or be fond of much talking.[42] Monks were to "practice silence at all times. . . . If any one shall be found breaking this rule of silence he shall be punished severely. . . ."[43] Silence performed two crucial functions in the lives of the Benedictines. First, it guaranteed self-discipline. Second, it ensured that the monks' lifestyle was kept secret to outsiders. Guests were allowed to visit and stay only in the monastery's guest quarters. Strict measures were taken to assure that visitors were kept separated from the majority of the brethren. Workshops and sacred portions of the monastery were off limits unless the guests themselves were monks. *The Rule of Saint Benedict* forbade monks either to associate with or speak to the visitors.[44] The monastery's boundaries were rather difficult to negotiate; only with the permission of the abbot could the brethren breach the monastic enclosure.[45] Those who received permission to leave were forbidden to tell the outside world what went on inside the monastery's walls and were also not permitted to tell the brethren what went on in the outside world.[46]

The Rule of Saint Benedict, including its provisions regarding monastic labor practices and the enforcement of silence and secrecy, influenced the architecture of Benedictine monasteries and cloisters. Indeed, the architecture of Benedictine monasteries and cloisters was an instantiation of *The Rule of Saint Benedict*.[47] These monasteries and cloisters manifested the varying degrees of privacy required for different aspects of Benedictine life. Individual monks' cells in the cloisters guaranteed privacy for meditation as well as helped police silence.[48] Visitors were prohibited from entering the cells. Monks were permitted only occasionally to enter into each other's quarters. Artisanal workshops were usually located either immediately outside the monastery's compound (yet still on the monastery's property) or at its periphery.[49] Seminaries, where communication of thoughts and ideas was paramount, were designed in a classroom-like fashion.[50] The monastery's library was always located near the monks' cells and seminary for easy access. The basilica, the most sacred portion of the monastery or cloister, was off limits to the artisans from the surrounding communities who did not belong to the order. In short, *The Rule of Saint Benedict* required a sharp delineation, both temporally and spatially, between the different aspects of the monks' lives. It rigidly plotted the daily routine of reading, worship, meals, labor, and silence.[51] Similarly, there were proper places where the monks ate, meditated, worked, and worshiped. Benediktbeuern, like all Benedictine cloisters, was made up of spaces with varying degrees of public access. This extensive, highly controlled space proved crucial to Fraunhofer's experimental design described at the beginning of this paper.

The Practice of Secrecy: Public and Private Knowledge at Cloister Benediktbeuern

Because Fraunhofer was an integral member of a business, the Optical Institute, strict measures were taken to differentiate between public and private knowledge. On the one hand, Fraunhofer desired to increase the institute's visibility to increase sales. Hence, he needed to publicize aspects of his enterprise, which he accomplished by publishing lists of the Optical Institute's products as well as by entertaining visiting experimental natural philosophers and demonstrating to them his method for calibrating achromatic lenses. Fraunhofer also wished to obtain credit for his work from the scientific community. This he was able to do by publishing his essay on the experiments that enabled such calibration in the prestigious journal of the Royal Bavarian Academy of Sciences. On the other hand, certain forms of his artisanal knowledge needed to be kept secret. Hence, Fraunhofer policed both the written word and social space. He published neither the procedures nor recipes for producing achromatic lenses, nor did he ever permit the experimental natural philosophers to witness the skills involved in lens production.

Although there are numerous examples of distinction between public and private knowledge in Benediktbeuern, this section discusses only four. First, on 14 April 1810, co-owner of the Optical Institute Utzschneider wrote to Fraunhofer, "If the French physicists [interested in optics] come to Benediktbeuern, do not show them the flint glass production. Also, be sure that Guinand does not speak to them. M. Morelle de Serres must be shown nothing."[52]

A second instance also involved Guinand, the Swiss glassmaker at Benediktbeuern who taught Fraunhofer his trade. As a part of Guinand's contract of 20 February 1807, he received from Utzschneider an apartment for him and his wife and five cords of wood "on the condition that he never communicates the secret of the fabrication of flint and crown glass to anyone."[53] After numerous quarrels with Fraunhofer, Guinand left Benediktbeuern in December 1813. Utzschneider agreed to pay him an annual pension as long as he and his wife promised not to divulge the flint and crown glass manufacturing secrets or work for another optical institute that could rival Benediktbeuern.[54] Such deals were commonplace. Guinand agreed but broke his promise shortly thereafter. The third example of secrecy involved Johann Salomon Christoph Schweigger, a professor of physics and chemistry at the University of Erlangen (and later of Halle), who wished to visit

Fraunhofer at the Optical Institute. Fraunhofer wrote Schweigger expressing his delight to entertain such a prestigious guest but informed the experimental natural philosopher that although he would be glad to teach him how to use the dark lines of the solar spectrum to calibrate lenses, he would not be able to show the professor the glass hut or divulge any information on the manufacture of achromatic glass.[55]

Finally, the most informative example of secrecy involved Fraunhofer and John Herschel, Britain's leading experimental natural philosopher. On 19 September 1824, Herschel visited Benediktbeuern on behalf of the Joint Committee of the Royal Society and Board of Longitude for the Improvement of Glass for Optical Purposes. British lens makers and experimental natural philosophers had felt threatened by Fraunhofer's enterprise. Herschel hoped to take away information regarding Fraunhofer's glassmaking techniques. Imagine Herschel's surprise, as representative of reformed, regal, and industrial Britain, where glass houses were located only in major cities, when he was greeted in a former cloister situated in the foothills of the Alps. He wrote to Charles Babbage, "I saw Fraunhofer and all his works but the one most desirable to see: his glass-house, which he keeps enveloped in thick darkness."[56]

Although as we have just seen, visitors increase the visibility and fame of an enterprise, they can also be potentially dangerous, since they can witness any methods for glass manufacture. Fraunhofer and Utzschneider therefore needed to switch from controlling the written word to policing social space. They controlled this space by prohibiting access to certain portions of their institute. Such restrictions could be executed rather efficiently, because Fraunhofer and Utzschneider could draw upon the preexisting Benedictine architecture, which was a manifestation of the combination of public and private activities that constituted monastic life.

Secrecy and Architecture of Benediktbeuern

As we have seen, Fraunhofer and Utzschneider's practice of silence and secrecy mirrored the centuries-old labor practices of the Benedictine monks, though for utterly different reasons. For decades, the labor practice and recipes for glassmaking had been closely guarded secrets. To the monks, their silence fulfilled a vow of secrecy with God. Skilled laborers who joined the monks in their glassmaking were also secretive, though on guild protection rather than theological grounds. But whether for religious or craft reasons, glassmakers were a taciturn lot, going as far as establishing symbols, not unlike alchemical ones, to encode their recipe books, techniques, and practices.[57] By the late Middle Ages, as literacy spread, guild members had shifted from basing their teachings on oral instructions to writing down their technical secrets for future generations. This was particularly true for monastic craftsmen, among whom literacy rates had been much greater than among secular artisans.[58] These texts, however, were still not discernable to the public. As Eamon has argued, these recipe books were "the record of trial-and-error experimentation. They [we]re the accumulated experience of practitioners boiled down to a rule."[59] These recipe books and manuscripts were preserved in monasteries where the arts were practiced.[60] But it must be emphasized that the knowledge of artisans was most accurately transmitted by doing and by imitating and emulating the master, rather than by reading the printed word.[61]

Some glassmakers encoded their recipe books with numbers, each number corresponding to a particular letter. Those who could not break the code could not reproduce the glass.[62] One example of number coding was found in a recipe book from a monastic glass house in the forests of Bavaria. The key to the puzzle was to decode the word "crystal." The letters of the German word "Christalen" corresponded, from left to right, to the numbers from one to ten, so that "c" was 1, "h" was 2, "r" was 3, and so on.[63] Another extreme measure taken to ensure secrecy was the master glassmakers' decision rarely to disclose their last names for fear of being kidnapped by rulers of other regions.[64] Utzschneider and Fraunhofer's use of monks and local artisans for their sources of labor and optical knowledge was rather clever, because it ensured that all of their employees originated from cultures in which secrecy was paramount. They were able to exploit a preexisting monastic privacy and secrecy, even after the secularization of Bavarian monasteries in 1803, to create a scientific, technological, and private enterprise.

The monks' washroom at Benediktbeuern (A in figure 6.5) had been a public place where monks freely engaged in various discourses. It was transformed into the achromatic glass melting room. Common glassmakers, glass cutters, grinders, and polishers could all enter these premises to assist in the manufacture of achromatic lenses. The glass house, however, was off limits to visiting opticians and experimental natural philosophers, such as Herschel.

Entrance to Fraunhofer's laboratory (B in figure 6.5) was limited to those workers of Benediktbeuern who had optical expertise. His laboratory was built within the preexisting monks' cells, which were designed to reflect the importance of silence in *The Rule of Saint Benedict*. It was therefore private. However, visiting opticians and experimental natural philosophers were taken to Fraunhofer's laboratory where he could demonstrate to them his achromatic lens calibration technique. Hence, the notion of public and private was reversed, depending on the individuals present. By showing such visitors how he used the dark lines of the spectrum in producing achromatic lenses, Fraunhofer ensured his institute's optical hegemony without having to show them how the lenses were actually constructed.

Conclusion

Over the past decade or so, a small, yet significant, literature has described social life and its practices as being contingent upon their spatial organization.[65] Hence, public-private and openness-restriction can be fruitfully analyzed in two ways.[66] First is the concept of

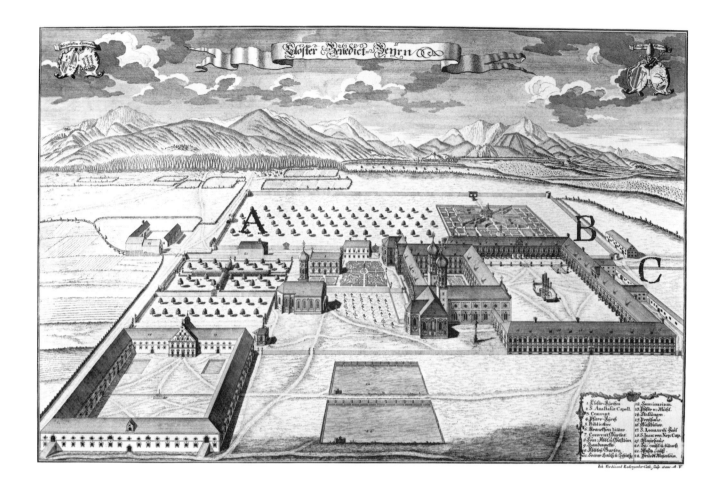

6.5

A copper engraving of the Benedictine cloister Benediktbeuern, circa 1800. The monks' washroom (A) was transformed into the glass hut where the glass for the achromatic lenses was produced. Fraunhofer's laboratory (B) had previously been the monks' living quarters. At (C) were located the cloister's artisanal workshops. Reprinted with the kind permission of the Deutsches Museum, Munich, photograph 30270.

social space. Social space assigns appropriate places for practices, ensures continuity and cohesion of those practices, incorporates practices, permits analysis of a particular community or culture, and establishes restricted areas.[67] Where one worked and what one did in Benediktbeuern certainly depended on who one was. The second related and relevant concept is the notion of boundary. Boundaries separating social spaces are informative to the historian since historians can probe the permeability of such borders.[68] Where secrecy is not an issue, such boundaries tend to be negotiable. Boundaries and their junction points, where boundaries are breached, are often sites of controversy and of friction. They therefore undoubtedly vary with the particular type of culture the historian is analyzing. Such an analysis is very similar to Steven Shapin's pioneering paper on Robert Boyle's laboratory.[69] In this work, Shapin draws a compelling connection between the empiricist processes of knowledge making and the participants' spatial distribution. By looking at who could go in certain social spaces and how the entry into such spaces was regulated, Shapin illustrates how trust is crucial to the evaluation of experimental knowledge. Similarly, in a more recent article included in this volume, William Newman has convincingly argued how critical a laboratory's social space was in establishing a gradient of privacy in Andreas Libavius's Chemical House.[70] As was the case in Benediktbeuern, certain spaces of Libavius's fictional Chemical House were off limits to those other than Libavius himself. The architecture of the Chemical House enabled Libavius to differentiate between public and private knowledge.

The notion of social space and its analysis greatly add to our attempt to understand the scientific enterprise as a cultural activity. By examining the account of Fraunhofer's production of achromatic lenses I have offered in this paper, it becomes clear that Fraunhofer and Utzschneider were rather skilled at manipulating a preexisting culture in which the architecture reflected a more encompassing epistemology. Benedictine architecture instantiated a gradient of privacy that was precisely what Utzschneider and Fraunhofer needed to ensure their optical monopoly.

NOTES

This research has been funded, in part, by the Herbert C. Pollack Award of the Dudley Observatory, Schenectady, New York.

1. Achromatic lenses do not produce colored fringes (the so-called secondary spectrum), which distort resolution under high magnification.
2. In so doing, I also employ Henri Lefebvre's notion of social space as a heuristic tool to offer a microhistorical account of how Fraunhofer manufactured his lenses; Lefebvre, *The Production of Space.*
3. Flint glass differs from crown glass in that it contains lead oxide, which increases the degree of refraction.
4. See Jackson, "Artisanal Knowledge and Experimental Natural Philosophers," pp. 552–557, and "Buying the Dark Lines."
5. Fraunhofer's *Gesammelte Schriften,* pp. 3–31. The article originally appeared in *Denkschriften der Königlichen Akademie der Wissenschaften zu München für die Jahre 1814 u. 1815,* vol. 5, which was not published until 1817. Fraunhofer sent his paper to his good friend, the mathematician, ordnance surveyor, and member of the Royal Bavarian Academy of Sciences Johann von Soldner. Inclusion in this periodical meant instant recognition as a *Naturwissenschaftler,* something that the working-class artisan desperately sought. Indeed, as a result of this essay's merits, Fraunhofer became a corresponding member of the Academy in 1817. All references to this work are cited as "Fraunhofer," followed by the page numbers corresponding to Lommel's edited collection.
6. Fraunhofer, p. 3.
7. Ibid., p. 27.
8. Ibid.
9. David Brewster and John Herschel also attempted, in vain, to isolate monochromatic light. See, for example, Brewster, "Description of a Monochromatic Lamp," pp. 433–444.

10. Note that this setup permits overlap of colors from the intermittent lamps; for example, the lamp adjacent to B might supply prism H with a small portion of violet rays as well. However, since prisms map angle onto position, and since the incident rays fall parallel (as a result of such a large distance between the two prisms) on prism H, all lamp light of the same wavelength is mapped onto some position as seen by the telescope of the theodolite.
11. The symmetric passage is with respect to the D line.
12. Fraunhofer, p. 10.
13. Ibid., pp. 13–14.
14. See, for example, Boscovich, *Dissertationes Quinque*, p. 142.
15. All of these angles are defined relative to the normal of the glass prism's surface.
16. Weiss, "Bayern und Frankreich," pp. 559–560.
17. Ziegler, *Vom Grenzstein zur Landkarte*, p. 15; and Messerschmidt, "Die wissenschaftlichen Grundlagen," p. 11.
18. von Utzschneider, *Kurzer Umriß*, pp. 5–6.
19. Veit, "P. Ulrich Schiegg von Ottobeuern," pp. 153–167.
20. Seitz, "Der Münchner Optiker Joseph Niggl," pp. 150–151.
21. For the most comprehensive account of the skilled labor of monks in glassmaking, see Lerner, *Geschichte des Deutschen Glaserhandwerks;* Vopelius, *Entwicklungsgeschichte der Glasindustrie Bayerns;* Haller, *Historische Glashüttensagen;* Friedl, *Glasmachergeschichten und Glashüttensagen;* Lobmeyr, *Die Glasindustrie;* and Blau, *Die Glasmacher in Böhmer- und Bayernwald.*
22. For a discussion of these texts, see Ganzenmüller, *Beiträge zur Geschichte der Technologie und der Alchemie*, pp. 46–51. Ganzenmüller lists Theophilus's *De diversis artibus* of 1123 (it should be noted that Theophilus was a Bavarian Benedictine monk); Paracelsus's reference to glass and the divine light; and Ostwald Crell's *Basilica Chymica* of 1609. Johannes Kunckel's *Ars Vitraria Experimentalis* discusses texts dealing with glass and the divine light written during the sixteenth and seventeenth centuries.
23. Utzschneider himself even conducted, in the Benedictine cloister, achromatic glass trials of Ettal's glass house in Grafenanschau, but failed to produce optical glass of sufficient quality: Kirmeier and Treml, eds., *Glanz und Elend der alten Klöster*, p. 342.
24. Ibid., p. 342.
25. Ibid., p. 232; and Staatsbibliothek Preußicher Kulturbesitz Berlin, Fraunhofer Nachlaß, Box 5, "Lohne der Arbeiter."
26. Schmitz, *Handbuch zur Geschichte der Optik,* vol. 2, p. 148.
27. "Aufzeichnungen des Lehrers Aloys Rockinger in Benediktbeuern," reprinted in Seitz, *Joseph Fraunhofer,* pp. 94–97, here, p. 95.
28. Kirmeier and Treml, eds., *Glanz und Elend der alten Klöster,* pp. 260–261.
29. Fellöcker, *Geschichte der Sternwarte der Benediktiner-Abtei Kremsmünster.*
30. Stutzer, *Klöster als Arbeitgeber um 1800.*
31. Cardinal Gasquet, translator, *The Rule of Saint Benedict,* p. xxiii.
32. Ibid., p. xxvi.
33. Ibid., p. 84.
34. Ibid., pp. 96–97.
35. Workman, *The Evolution of the Monastic Ideal*, pp. 219–220. See also Ovitt, *The Restoration of Perfection,* p. 88.

36. Weber, *The Protestant Ethic,* pp. 118–119; and Ovitt, *The Restoration of Perfection,* p. 89. I concur with Ovitt that Weber was misguided in claiming capitalist origins lay in Benedictine culture. Capitalism views labor and its products quantitatively, as a means of increasing productivity, whereas monasticism cherishes the labor process itself. See Ovitt, *The Restoration of Perfection,* p. 106.
37. Mumford, *The Myth of the Machine,* vol. 1, p. 264, and Ovitt, *The Restoration of Perfection,* p. 89.
38. Mumford, *The Myth of the Machine,* vol. 1, p. 264.
39. Ibid.
40. Ovitt, *The Restoration of Perfection,* p. 104.
41. For the importance of labor to the Benedictines, see White, "Cultural Climates and Technological Advances in the Middle Ages."
42. Cardinal Gasquet, translator, *Rule of Saint Benedict,* pp. 19–20.
43. Ibid., p. 77.
44. Ibid., p. 93.
45. Ibid., p. 119.
46. Ibid., pp. 118–119.
47. See Eschapasse, *L'Architecture Bénédictine en Europe,* pp. 11–21.
48. Ibid., p. 17.
49. These included, among other things, glass huts, sawmills, slaughterhouses, brandy distilleries, and breweries. See, for example, Kirmeier and Treml, eds., *Glanz und Elend der alten Klöster,* pp. 186–193.
50. Kirmeier and Treml, eds., *Glanz und Elend der alten Klöster,* p. 193.
51. Cardinal Gasquet, translator, *Rule of Saint Benedict,* pp. 84–87: "Idleness is an enemy of the soul. Because this is so the brethren ought to be occupied at specified times in manual labour, and at other fixed hours in holy reading. We therefore think that both these may be arranged for as follows: from Easter to the first of October, on coming out from Prime, let the brethren labour till about the fourth hour. From the fourth till close upon the sixth hour let them employ themselves in reading. On rising from the table after the sixth hour let them rest on their beds in strict silence. . . .

    Let None be said somewhat before the time, about the middle of the eighth hour, and after this all shall work at what they have to do till evening. . . ." (Pp. 84–85.)
52. Staatsbibliothek Preußischer Kulturbesitz Berlin I, Utzschneider Nachlaß, Box I, Utzschneider and Fraunhofer Correspondence, 14 October 1810.
53. Quoted in Seitz, *Joseph Fraunhofer,* p. 23.
54. Quoted in ibid., pp. 49–50.
55. Fraunhofer's letter to Schweigger, 2 August 1817. Deutsches Museum Archiv 7414.
56. Royal Society of London Archives, Herschel to Babbage, 3 October 1824, Herschel Letters, HS.2.199.
57. Blau, *Die Glasmacher in Böhmer- und Bayerwald,* vol. 1, pp. 51–56 and 175–178. It should also be noted that distilling and brewing processes, which were executed in Benediktbeuern from the sixteenth century until its secularization in 1803, were originally closely related to secret alchemical processes.
58. Eamon, *Science and the Secrets of Nature,* p. 81.

59. Ibid., p. 7.
60. Ibid., p. 30.
61. Smith, *The Business of Alchemy*, p. 37.
62. Lerner, *Geschichte des Deutschen Glaserhandwerks*, vol. 1, pp. 46–47; and Blau, *Die Glasmacher in Böhmer- und Bayerwald*, vol. 1, pp. 51–56 and 175–178.
63. Blau, *Die Glasmacher in Böhmer- und Bayerwald*, vol. 1, p. 51, and "Das geheime Rezeptenbuch," pp. 12–20.
64. Friedl, *Glasmachergeschichten und Glashüttensagen*, pp. 11–17. This particular work discusses the importance of secrecy in glassmaking as reported in the various stories and myths of Bavarian glassmakers.
65. Giddens, *The Constitution of Society*, particularly chap. 3; Hillier and Hauser, *The Social Logic of Space*, pp. ix–xi, 4–5, 8–9 and 19; Foucault, "Questions on Geography," pp. 63–77, and *Discipline and Punish: The Birth of the Prison;* and Ophir, "The City and the Space of Discourse."
66. Lefebvre suggests both ways in *The Production of Space*.
67. Ibid., pp. 32–36.
68. Ibid., pp. 86–87.
69. Shapin, "The House of Experiment." Shapin's article can be seen as an elaboration of Owen Hannaway's work on the laboratory as a site of knowledge production, "Laboratory Design and the Aim of Science." For other works on the laboratory as a social space of knowledge production, see Galison, "Bubble Chambers and the Experimental Workplace"; and Owens, "Pure and Social Government."
70. See Newman, this volume.

BIBLIOGRAPHY

Blau, Josef. "Das geheime Rezeptenbuch des Glasmeisters Joh. Bapt. Eisner in Klostermühle 1842–1862." *Glastechnische Berichten* 18 (1940): 12–20.

Blau, Josef. *Die Glasmacher in Böhmer- und Bayerwald in Volkskunde und Kulturgeschichte*. Vol. 1. Regensburg: Verlag Michael Lassleben, 1954.

Boscovich, Roger J. *Dissertationes Quinque ad Dioptricum*. Vindobonae: Johannis Thomae, 1767.

Brewster, David. "Description of a Monochromatic Lamp for Microscopial Purposes, &c., with the Remarks on the Absorption of the Prismatic Rays by Coloured Media." *Transactions of the Royal Society of Edinburgh* 9 (1823): 433–444.

Eamon, William. *Science and the Secrets of Nature: Books of Secrets in Medieval and Early Modern Culture*. Princeton, NJ, and London: Princeton University Press, 1994.

Eschapasse, Maurice. *L'Architecture Bénédictine en Europe*. Paris: Editions des Deux-Mondes, 1963.

Fellöcker, Pater Sigmund. *Geschichte der Sternwarte der Benediktiner-Abtei Kremsmünster*. Linz: J. Feichtinger, 1864.

Foucault, Michel. *Discipline and Punish: The Birth of the Prison*. Translated by Alan Sheridan. New York: Vintage Press, 1979.

Foucault, Michel. "Questions on Geography." In Collin Gordon, ed., *Power/Knowledge: Selected Interviews and Other Writings, 1972–1977*. Brighton: Harvester Press, 1980.

Fraunhofer, Joseph von. *Joseph von Fraunhofer's Gesammelte Schriften: Im Auftrage der Mathematisch-Physikalischen Classe der königlichen bayerischen Akademie der Wissenschaften.* Edited by E. Lommel. Munich: Verlag der königlichen Akademie in Commission bei G. Franz, 1888.

Friedl, Paul. *Glasmachergeschichten und Glashüttensagen aus dem Bayerischen Wald und dem Böhmerwald.* Grafenau: Verlag Mosak, 1973.

Galison, Peter. "Bubble Chambers and the Experimental Workplace." In Peter Achinstein and Owen Hannaway, eds. *Observations, Experiment, and Hypothesis in Modern Physical Science,* 307–373. Cambridge: MIT Press, 1985.

Ganzenmüller, W. *Beiträge zur Geschichte der Technologie und der Alchemie.* Weinheim: Verlag Chemie, 1956.

Cardinal Gasquet, trans. *The Rule of Saint Benedict.* New York: Cooper Square Publishers Inc., 1966.

Giddens, Anthony. *The Constitution of Society: Outline of the Theory of Structuration.* Cambridge, UK: Polity Press, 1984.

Haller, Reinhald. *Historische Glashüttensagen in den Bodenmaiser Wäldern. Ein Beitrag zur Geschichte des Glases im Bayersichen Wald.* Bodenmais: Joska-Glaskunstwerkstätten, 1975.

Hannaway, Owen. "Laboratory Design and the Aim of Science: Andreas Libavius versus Tycho Brahe." *Isis* 77 (1986): 585–610.

Hillier, Bill, and Julienne Hauser. *The Social Logic of Space.* Cambridge and New York: Cambridge University Press, 1984.

Jackson, Myles W. "Artisanal Knowledge and Experimental Natural Philosophers: Focusing on the British Response to Joseph von Fraunhofer's Optical Institute." *Studies in History and Philosophy of Science* 25 (1994): 549–575.

Jackson, Myles W. "Buying the Dark Lines of the Solar Spectrum: Joseph von Fraunhofer and His Standard for Optical Glass Production." *Archimedes* 1 (1996): 1–22.

Kirmeier, Josef, and Manfred Treml, eds. *Glanz und Elend der alten Klöster. Säkularisation im bayerischen Oberland 1803.* Munich: Haus der Bayerischen Geschichte, 1991.

Lefebvre, Henri. *The Production of Space.* Translated by Donald Nicholson-Smith. Oxford: Blackwell, 1984.

Lerner, Franz. *Geschichte des Deutschen Glaserhandwerks.* Schondorf: Hofmann-Verlag, 1981.

Lobmeyer, L. *Die Glasindustrie, ihre Geschichte, gegenwärtige Entwicklung und Statistik.* Stuttgart: W. Spemann, 1874.

Messerschmidt, E. "Die wissenschaftlichen Grundlagen der bayerischen Landesvermessung." *Das öffentliche Vermessungswesen in Bayern. Festschrift zum 175jährigen Bestehen der bayerischen Vermessungsverwaltung.* Munich: Bayerisches Staatsministerium der Finanzen- und Vermessungsverwaltung, 1976.

Mumford, Lewis. *The Myth of the Machine.* Vol. 1, *Technics and Human Development.* New York: Secker and Warburg, 1967.

Ophir, Aphir. "The City and the Space of Discourse: Plato's Republic—Textual Acts and Their Political Significance." Ph.D. diss., Boston University, 1984.

Ovitt, Jr., George. *The Restoration of Perfection: Labor and Technology in Medieval Culture.* New Brunswick, NJ, and London: Rutgers University Press, 1986.

Owens, Larry. "Pure and Social Government: Laboratories, Playing Fields, and Gymnasia in the Nineteenth-Century Search for Order." *Isis* 76 (1985): 182–194.

Schmitz, E.-H. *Handbuch zur Geschichte der Optik.* 2 vols. Bonn: Verlag J. P. Wayenborgh, 1982.

Seitz, Adolf. "Der Münchner Optiker Joseph Niggl." *Central-Zeitung für Optik und Mechanik* (1923): 150–155.

Seitz, Adolf. *Joseph Fraunhofer und sein Optisches Institut.* Berlin: Julius Springer, 1926.

Shapin, Steven. "The House of Experiment in Seventeenth-Century England." *Isis* 79 (1988): 373–404.

Smith, Pamela H. *The Business of Alchemy: Science and Culture in the Holy Roman Empire.* Princeton, NJ, and London: Princeton University Press, 1994.

Stutzer, Dietmar. *Klöster als Arbeitgeber um 1800. Die bayerischen Klöster als Unternehmenseinheiten und ihre Sozialsysteme zur Zeit der Säkularisation 1803.* Göttingen: Vanderhoeck & Rupprecht, 1986.

Utzschneider, Joseph von. *Kurzer Umriß der Lebens-Geschichte des Herrn Dr. Joseph von Fraunhofer, königlich bayerischen Professors und Akademikers, Ritters des königlichen bayerischen Civil-Verdienst, und des königlich dänischen Dannebrog-Ordens, Mitgliedes mehrerer gelehrten Gesellschaften, etc.* Munich: Rösl'schen Schriften, 1826.

Veit, H. "P. Ulrich Schiegg von Ottobeuern (1752–1810) und die bayerische Landesvermessung." *Beiträge zur Geschichte der Abtei Ottobeuern* 73 (1962): 153–167.

Vopelius, Eduard. *Entwicklungsgeschichte der Glasindustrie Bayerns bis 1806.* Stuttgart: J. G. Cotta'schen Buchhandlung, 1895.

Weber, Max. *The Protestant Ethic and the Spirit of Capitalism.* Translated by Talcott Parsons. New York: Charles Scribner's Sons, 1958.

Weiss, Eberhardt. "Bayern und Frankreich in der Zeit des Konsulats und des Ersten Empire (1799–1815)." *Historische Zeitschrift* 237 (1983): 559–595.

White, Jr., Lynn. "Cultural Climates and Technological Advances in the Middle Ages." *Viator* 2 (1971): 171–201. Reprinted in idem, *Medieval Religions and Technology: Collected Essays.* Berkeley, Los Angeles, and London: University of California Press, 1978.

Workman, Herbert. *The Evolution of the Monastic Ideal.* London: Kelly, 1913.

Ziegler, Theodor. *Vom Grundstein zur Landkarte.* Stuttgart: Konrad Wittwer, 1989.

7

THE SPACES OF CULTURAL REPRESENTATION, CIRCA 1887 AND 1969: REFLECTIONS ON MUSEUM ARRANGEMENT AND ANTHROPOLOGICAL THEORY IN THE BOASIAN AND EVOLUTIONARY TRADITIONS

George W. Stocking Jr.

Glossing my topic as "the architecture of anthropological knowledge," and glossing "architecture" loosely as having to do with the organization of the spaces in which knowledge is constructed, it seemed appropriate to preface—one might say to "situate"—a consideration of "the spaces of cultural representation" with some general comments on the types of spaces (both physical and metaphorical) in which anthropology as the study of human cultural variation has, over the last century or more, been conducted. I group these under three headings: the places or spaces of anthropological inquiry, the subdisciplinary architecture of anthropological knowledge, and the structure of its paradigmatic traditions.

The space of anthropological inquiry that comes most immediately to mind is "the field," the place in which the ethnographic raw material of anthropological knowledge is produced by a process called "fieldwork." Practiced *avant la lettre* before 1900, so denominated in 1903 by Alfred Haddon, given its mythical charter by Bronislaw Malinowski in

1922 and its metaphorical equivalence to "the laboratory" by Margaret Mead in 1928, fieldwork has since then been the dominant mode not only of anthropological knowledge gathering, but of the generation of the anthropologist as knower.[1] Recently, however, the very notion of the field and fieldwork as the privileged place and process of anthropology has begun to be called into question,[2] and in the context of this critical disciplinary self-analysis, we are reminded of other spaces, past, present, and future, in which the construction of anthropological knowledge may be carried on.

In archetypal terms, before the field there was "the armchair," a space at once more solitary and more interconnected. Although isolated physically in the scholar's study, it was linked textually to the various bodies of literature on which nineteenth-century comparativists based their syntheses and to the discourses into which the latter were incorporated.[3] Even today, when fieldwork is still a central feature of anthropological inquiry, anthropologists perform much of their knowledge-constructing activity in spaces loosely analogous to those of the armchair: on the one hand, the academic office, the library, the archive, the body of notes, the memory traces in or through which the remembered fieldwork experience is interpreted[4]; on the other, the lecture hall, the seminar room, the library, and all the spaces that intellectual discourse implies, which in the postmodern computational age of cyberspace may be very different from those of either the armchair or the field.[5]

Persisting in this metaphoric usage of the term "architecture," we may locate within this multiplex and historically evolving space of anthropological inquiry an architecture of anthropology as a discipline, which may perhaps be related in significant ways to the actual physical structures in which certain kinds of anthropological inquiry have been carried on. Taking anthropology in its widest "embracive" (i.e., North American, and to a lesser degree, British) sense, as "the science of man," it may be seen as encompassing several methodologically and epistemologically diverse forms of inquiry, often spoken of as "the four fields": physical (or biological) anthropology, archaeology, linguistic anthropology, and social or cultural anthropology (or, in its early–twentieth century incarnation, "ethnology"). The unity of those four fields, however, though often proclaimed, has been epistemologically and historically problematic. In contrast to some traditional models of scientific development that see the differentiation of fields in terms of fission from a common trunk, anthropology developed as a fusion of traditions of inquiry with quite diverse historical roots: antiquarianism, natural history, philology, and moral philosophy.[6] From this point of view, the architecture of anthropological inquiry is a rather ramshackle structure whose connecting beams were most tightly articulated a hundred years or more ago, when human differentiation could be seen as a single, multifaceted evolutionary process. In this context, anthropology, before it entered the modern university, was previously institutionalized in museums of several different types—natural historical, archeological, ethnographic, and cultural historical—that reflected (though with imperfect correlation) the diverse intellectual roots of anthropological inquiry.[7]

As a final situating reflection, I would comment on the structuring of inquiry within the ramshackle disciplinary architecture I have sketched. If, as Clifford Geertz once suggested,[8] the fundamental question anthropology addresses is how to reconcile the tremendous diversity of human cultural manifestations with the biological unity of humankind, then it may be argued that the answers to this question may be reduced to several different ideal types. These may be called "paradigmatic traditions," insofar as they represent structured models of inquiry that have constituted enduring explanatory alternatives over long periods of Western European intellectual history. For present purposes, two such paradigmatic traditions are relevant: one that may be called "ethnological" and one that may be called "developmental," the former rooted ultimately in the Bible, the latter in classical speculation, archetypically in Lucretius. The former tradition approaches the phenomenon of human diversity as a differentiation of peoples migrating through geographical space into different environments. It is concerned with characterizing and explaining their distinguishing features and with reconstructing historically their movements and relationships. Its focus is the component ethnic units (people, tribe, nation, race) of the human species. The latter of the two traditions treats the variety of cultural forms as manifestations of a single developmental or evolutionary process common to all humankind. There is of course more to be said about the relation of each of these traditions to nineteenth-century speculations about race or their relation to Darwinian evolutionary thought. But for present purposes, this polar paradigmatic opposition must suffice.[9]

Keeping in mind these remarks on the places or spaces of anthropological inquiry, on the architecture of anthropological subdisciplines, and on the structure of its paradigmatic traditions, let us consider an exchange that took place in the letter columns of the journal *Science* in 1887. Still not yet securely established as either an academic or a fieldwork discipline, anthropology was in a phase of its development that has sometimes been defined by its physical locus as its "museum period." The protagonist in the debate was the young German immigrant scholar Franz Boas, then seeking an anthropological career in the United States; his antagonist was Otis Mason, curator of anthropology at the U.S. National Museum and a member of the Washington anthropological establishment led by Major John Wesley Powell of the U.S. Bureau of Ethnology.[10]

Trained in geography and physics, Boas had spent a year among the Eskimo of Baffinland in 1883–84 and had visited the U.S. National Museum before returning to Germany, where he worked for a year under Adolf Bastian in Berlin at the newly established Royal Ethnographic Museum. In the fall of 1886, Boas had come to North America for ethnographic research among the Indians of British Columbia, after which he accepted a position in New York City as assistant editor of *Science*.[11] In May of 1887, in the guise of a discussion of museum arrangement, he ventured what was in effect a critique of the evolutionary theoretical assumptions of the dominant school of American anthropologists. The specific theoretical issue was how to explain "the occurrence of similar inventions in areas widely apart." As represented by Boas, Mason had offered three possible explanations for such similarities: "the migration of a certain race or people who made the invention"; the "migration of ideas" that might be taught or loaned; and the tendency, "in human culture as in nature elsewhere, [for] like causes [to] produce like effects"—for human beings, sharing a common mental makeup, to make the same inventions in similar environmental situations. Adopting the last alternative, which was in fact the fundamental assumption of what has since been called "classical evolutionism," Mason classified human inventions and other ethnological phenomena as "biological specimens," dividing them into "families, genera, and species," with each category arranged in a presumed sequence of "specific evolution out of natural objects serving human wants . . . up to the most delicate machine performing the same function." On this basis Mason had arranged the ethnological collections of the National Museum according to their "different species"—with the projectiles in one evolutionary sequence from simple to complex, and the means of navigation in another—without regard for "the tribes to whom they belong."[12]

Implicitly contrasting Mason's system with the one Bastian had adopted in Berlin, Boas granted that a typological arrangement might have a limited usefulness when applied to objects from a single "geographical province," where a "close connection between them" could be assumed. But as a general approach, it was subject to fundamental criticism on several grounds. Without venturing here into the German intellectual and philosophical context of Boas' position,[13] or to treat all the obscurities introduced by his still imperfect command of English, his position may be represented in a brief and schematically rationalized form, drawing on passages offered at several points in the continuing debate. In the first place, Boas argued that because "the elements affecting the human mind" were so complicated and "their influence so utterly unknown," one could not accept the maxim that "like causes have like effects" as a general principle in the human realm. By implication posing "function" against "meaning" as modes of explanation (or understanding), Boas suggested that objects that looked alike might in fact have quite different meanings that could be understood only by considering each object in its general cultural milieu: "The art and characteristic style of a people can be understood only by studying its productions as a whole." Given the system Mason had used in the National Museum, "the marked character of the North-west American tribes is almost lost, because the objects are scattered in different parts of the building, and are exhibited among those from other tribes."[14]

It is worth noting, as an echo of the German tradition in which Boas had been enculturated, that he took his two main examples from the realm of music. Thus a rattle could not be understood merely as "the outcome of the idea of making noise, and of the technical methods applied to reach this end." It was also "the outcome of religious conceptions," the goals of which might be accomplished by other means, since "any noise" could serve "to invoke or drive away spirits." Or it could simply be the "outcome of the pleasure children have in noise of any kind." Its specific form would depend on the characteristic art of the particular people.[15] At another point, Boas suggested that "from a collection of string instruments, flutes or drums of 'savage' tribes and the modern orchestra, we cannot derive any conclusion but that similar means have been applied by all peoples to make music. The character of their music, which is the only object worth studying, [and] which determines the form of the instruments, cannot be understood from the single instrument, but requires a complete collection of the single tribe."[16]

Although one hesitates to read Mason's response in stereotyped cross-cultural terms, its content suggests that, as a practical American democrat, Mason may have regarded Boas as an arrogant ethnological novice wrapping himself in the authoritative mantle of Germanic scholarship. The debate, however, clearly had both a pragmatic and a dogmatic aspect. In defending himself, Mason noted that he had been curator for only two years and had not yet gotten around to organizing the Northwest coast exhibits; he appealed also to the diversity of audiences, which included "archeologists, ceramists, musicians, [and] technologists of several kinds," as well as "students of war, religion, and the aesthetic arts," all of whom might want to see relevant specimens "in juxtaposition" to each other. Given charge of a "great collection," every "scientific anthropologist" had to choose one among several systems of "classific concepts"; if choosing one meant the loss of the virtues of another, well, other museums might be arranged on different systems. But "if all the museums in the world" had to be arranged on one plan, Mason "sincerely hoped" it would be "that of the national museum at Washington." "I think it is a growing conviction that inventions of both customs and things spring from prior inventions, just as life springs from life, and the sooner we recognize the fact that in the study of arts, institutions, language, knowledge, religions and races of men, we must always apply the methods and instrumentalities of the biologists, the sooner will our beloved science stand upon an immovable foundation."[17]

After firing one salvo, Mason quickly stepped aside in favor of his own authority, Major John Wesley Powell, who as director of the Bureau of [American] Ethnology was the intellectual and institutional leader of government anthropology. Responding to Boas' rebuttal of Mason's letter of self-defense, Powell, too, distinguished between pragmatic and theoretical issues, insisting that he would limit himself to the problem of methods of arrangement rather than that of interpreting similarities of culture. According to Powell, the "functions" of a museum were twofold: On the one hand, it served as a "repository of materials" for the researcher, a topic Powell disposed of simply by congratulating the National Museum on its excellence. He devoted himself instead to the second (and secondary) function: the "objective exemplification of some system of knowledge" pertaining to its collections, either for purposes of formal instruction or simply to enlighten the "passing observer." Here too Powell began by arguing pragmatically: In a great museum with "vast" collections, only "a small percent" of them could be shown to the public "with reasonable expenditure"; the rest must be reserved for specialist study.

The issue, then, was the principle of selection and of order in display: The museum administrator must "determine what is the most useful lesson to the general public which his materials can be made to teach." Since every investigator would have his favorites, and because foci would change with "the progress of research," the most important thing was flexibility. The technology of exhibit must be composed of "interchangeable parts," "easily adjustable to new conditions," to accommodate "new facts arising from the advancement of science and from the enrichment of the collections."[18]

Having made the case for pragmatic flexibility, Powell then offered his own theoretical remarks, by way of a critique of Boas' argument for a tribal arrangement. At this moment of the exchange, there is a disconcerting inversion of our retrospective intellectual historical expectations, which (quite appropriately) associate Boas with cultural relativism and Powell with evolutionary dogmatism. But just as there is a relativist potential in the evolutionary tradition, so is there an essentialist potential in the tradition of cultural relativism, notably in the work of Boas' intellectual ancestor Herder.[19] In attacking tribal arrangement, Powell argued in nominalist, historicist terms that, by implication, placed Boas in the position of essentializing human differences. Positing the existence of 25,000 tribes at "the discovery of America," Powell suggested that all of these tiny migrating "bodies politic" had, by virtue of division and coalescence, been in continuous historical flux, so that it was "not probable that there existed any one tribe which could claim to be the pure and simple descendant, without loss, admixture, or change, of any tribe existing at the time of the discovery." Similarly, intertribal and colonial trade contaminated the archeological record. On these and other grounds, Powell maintained that a tribal arrangement was in fact an impossibility. Extending the argument, he rejected also the very possibility of any "ethnic" classification of mankind, whether based on physical characteristics or linguistic differences. Although he was director of a Bureau of Ethnology, he insisted that there was no such thing as a "science of ethnology": "[T]he attempt to classify mankind in groups has failed on every hand," thereby establishing "the unity of mankind" as "the greatest induction of anthropology." Rejecting geographical arrangement on the grounds that it would require too many specimens at too great a cost, Powell by a process of elimination arrived at a classification of arts as the only viable basis of museum arrangement.[20]

Confronted thus by the most powerful single individual in the discipline in which he hoped to establish himself, Boas at this point quickly backed off from the dispute, although not without a final comment defending the viability of a tribal arrangement, which was already implemented in "numerous museums" in Europe. What was required was to exhibit a "full set" of specimens for one representative tribe within an "ethical group" and to show "slight peculiarities" of other related tribes "in small special sets" around it. Such groupings, however, were "not all intended to be classifications," as Powell had inferred. In a final parting shot, Boas suggested that the "principal difference" between the Mason-Powell plan and his own was that their classification was "not founded on the phenomenon, but in the mind of the student."[21]

To unpack here all the epistemological and methodological differences implicit in that final salvo would be impossible in the space available. Suffice it to say that they are profound, and that they do relate to some of the matters at hand. In the present context, the focus instead is on some knowledge-constructing issues more specifically implicated in the problem of museum arrangement. Behind the various pragmatic concerns expressed, and despite Powell's historicizing of the concept "tribe," the exchange reflects fundamental and perduring oppositions in anthropology, oppositions at once theoretical and ideological. All parties to the debate recognized that the ethnographic context of the empirical material available for display might constrain museum arrangement. They differed fundamentally, however, in how they proposed to respond to these limitations, especially in relation to the problem of comparison. For Mason and Powell, a maximally generalized—that is, evolutionary—comparison was the privileged goal of anthropology; that was what "science" was about, and museum arrangement should reflect this goal and contribute to its achievement and propagation. If ethnographic context constrained the comparative project, the solution was to narrow the focus of comparison in order to broaden the spectrum of evolutionary significance. For Boas, the nature of human cultural life was such as to place certain contextual constraints on any comparative project: He therefore chose to widen the contextual angle and narrow the spectrum of comparative significance, postponing to the indefinite future any attempt to derive universal laws of human cultural development.[22]

Furthermore, the theoretical differences implicated in museum arrangement had their ideological correlates: For Mason and Powell, the goal of museum exhibition was to demonstrate progress—the progress of anthropology, the progress of science, and the progress of human culture. For Boas, the goal of museum exhibition was to demonstrate for late Victorian Euro-Americans that "civilization is not something absolute, but that it is relative, and that our ideas and conceptions are true only so far as our civilization goes." If his notion of the tribe implied, as Powell by implication suggested, a certain essentialization of the culture of each ethnic group, it was to the end of defending a diversity of human value orientations that could not easily be arranged in a progressive sequence.[23]

As Mason pointed out, he had not invented the linear typological mode of museum arrangement; and as Boas indicated, his approach was widely practiced in contemporary European museums. When nineteenth-century curators of the various museums in which ethnographic specimens were exhibited began to put order into the apparent chaos of naturalia and artificialia inherited from the era of the cabinet of curiosities, typological and geographical arrangement were the two major options that presented themselves.[24] The former, however, was perhaps more reflective of a certain moment in the culture-history of ethnographic collection and exhibition, a moment when material was limited in quantity and had been somewhat randomly collected, with inadequate documentation as to its provenance, and when the existing technology of display constrained exhibition. At a slightly later moment—the 1890s—Mason, stimulated perhaps by Boas, also moved toward culture area displays, focusing on life figures representing different tribal groups, and this mode was by and large to triumph in the twentieth century—most notably, for present purposes, in the Northwest Coast Hall that Boas set up in the American Museum of Natural History.[25]

The history of anthropological museum exhibition after Boas severed his connection with the American Museum in 1906 lies beyond the scope of this paper.[26] During the succeeding years, many exhibits, often employing dioramic displays analogous to the life figure tribal groupings Boas favored, were based on physical anthropological assumptions that ran quite counter to those of Boas' anthropology.[27] But one twentieth-century development harks back directly to the Boas-Mason debate. Four years before that debate took place, General Augustus Lane Fox Pitt Rivers gave his collection of 14,000 ethnological and archeological artifacts to Oxford University with the stipulation that it be housed in a separate museum and arranged in evolutionary typological terms. As an aside, it is worth noting that Pitt Rivers expressed quite explicitly the presumed ideological implications of his evolutionary scheme: "The law that nature makes no jumps can be taught by the history of mechanical contrivances, in such a way as at least to make men cautious how they listen to scatterbrain revolutionary suggestions."[28] Given that the collection was housed in an annex at the rear of the Oxford University Museum, it seems unlikely that the conservative message of its artifacts was carried very widely among the British masses. But the basic arrangement, and many of the original exhibits, remained the same over nine decades in which the museum's collections multiplied fiftyfold. To cope with this mass of material, a plan was developed in the late 1960s for a new museum building that would also house all the various subdisciplinary components of anthropology at Oxford. A four-acre plot of land was chosen and a distinguished Italian architect (Pier Luigi Nervi) joined with the British firm of Philip Powell and Hidalgo Moya to create a museum that would preserve the general's typological scheme but at the same time accommodate other anthropological concerns.[29]

The principle adopted for the "New Pitt Rivers Museum and Proposed Centre for the Study of Anthropology and Human Environment" was that of a rotunda in which the exhibits were to be arranged in concentric rings around a central dome. (See figures 7.1 and 7.2.) Adapting to the environmentalist spirit of the later 1960s (as well as to the environmental determinism that could be read into both the typological and the geographical schemes), the dome was to be filled with tropical and subtropical plants, with the temperate zone represented on a surrounding roof garden. Around the dome would be the two levels of the major exhibit area—one devoted to archeology, the other to ethnology—on each of which the comparative typological principle was expressed concentrically, with the regional principle expressed radially. (See figures 7.3 and 7.4.) By walking in a circle, the visitor could examine all the different pottery vessels, or means of warfare, or making music, as they were expressed in different geographic zones. By walking from

7.1–7.2

Oxford University, Pitt Rivers Museum Project. Pier Luigi Nervi, Philip Powell, and Hidalgo Moya, 1969. Model of exterior and sections. The design of this museum was intended to accommodate not only the conditions stipulated in Pitt Rivers' nineteenth-century will, in which he bequeathed to Oxford his 14,000 artifacts, but also the concerns and interests of the twentieth-century anthropologists who studied those objects. The artifacts were to be displayed on several levels that would rise above and encircle a central dome filled with tropical plants. Source: "The New Pitt Rivers Museum and Proposed Centre for the Study of Anthropology and Human Environment," University of Oxford, n.d., pp. 24, 25A (unpaginated).

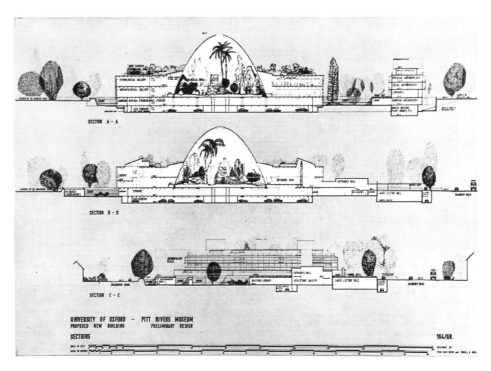

173 | THE SPACE OF CULTURAL REPRESENTATION, CIRCA 1887 AND 1969

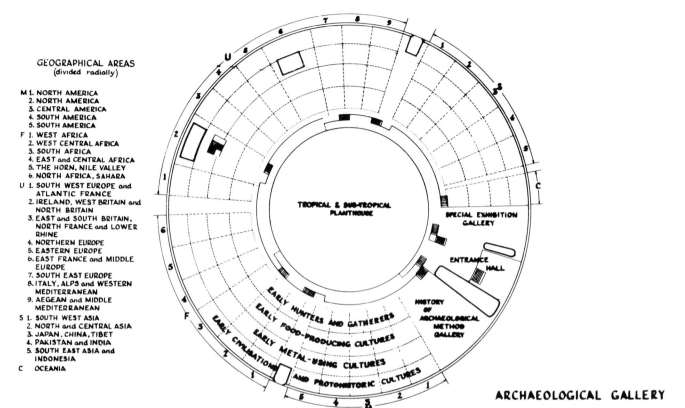

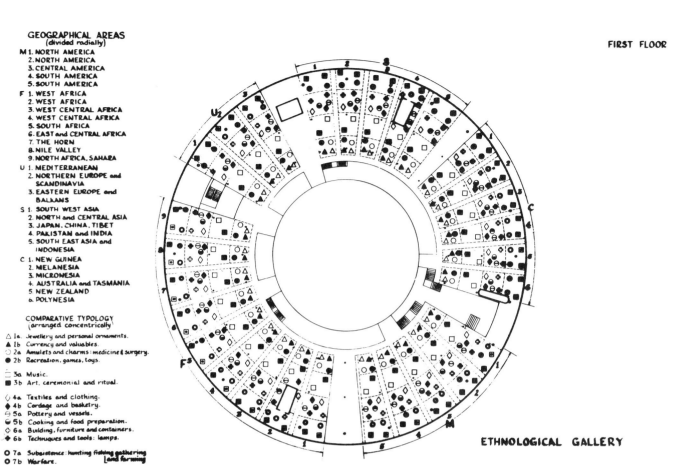

7.3–7.4

Oxford University, Pitt Rivers Museum Project. Pier Luigi Nervi, Philip Powell, and Hidalgo Moya, 1969. Plans of Archeological and Ethnological Galleries. Each concentric gallery was to incorporate both the typological principle of Mason and Powell and the regional principle of Boas. Typology would be expressed concentrically and regions represented radially. With the dual arrangement in place, a museum visitor would have been able to examine the artifacts in either typological or regional modes. By walking in a circle, one could view all the different pottery vessels, or musical instruments, or weapons in the collection. By walking outward from the dome, one could gain a sense of all the various cultural products—pottery, instruments, vessels, and so forth—of a particular geographic area. Source: "The New Pitt Rivers Museum and Proposed Centre for the Study of Anthropology and Human Environment," University of Oxford, n.d., pp. 16, 20 (unpaginated).

the ecological dome toward the outside rim, one could examine a particular geographical area's various cultural products, broken down, on the archeological floor, into a series of concentric cultural stages, from early hunter-gatherers on the inside to early civilizations at the rim. On the lower ground floor, a storage area for objects not on display was arranged in a similar fashion. Around the edges of the rotunda area and in an adjoining rectilinear structure were lecture halls and office and research areas for various subdepartments of a general anthropology department: social anthropology, physical anthropology, European archeology, general archeology, musicology, and the history of art.[30]

As the descriptive brochure describing the proposed museum suggested, it was in fact a plan that General Pitt Rivers had envisioned in 1888, in which concentric circles would be used for "the exhibition of the expanding varieties of an evolutionary arrangement," and "separate angles of the circle might be appropriated to geographical areas."[31] Otis Mason in fact proposed a similar combinatory principle in 1887: that in a "properly constructed" museum it would be possible "to arrange the cases in the form of a checkerboard," so that in one direction "the parallels of cases represent races or tribes or locations" and in another, "at right angles to the former," the "products of human activity in classes according to human wants."[32] Boas, however, was not taken with the idea, fearing that "long rows of cases" must remain empty, because "certain phenomena occur but in very few tribes."[33] And one suspects that he might also have reacted critically to the 1969 rotunda plan, despite the ideological correction implicit in segregating chronology on the archeological level. Given the rather broad definition of its geographical pie slices (North America, South Africa, Pakistan and India, and so forth) there was still no way the products of a single tribe or people might be seen in their cultural context.

In the event, however, the £3,000,000 estimated cost of the museum was not raised, and the Pitt Rivers Museum today looks very much the same as it did 25 or even 100 years ago. (See figure 7.5.) Late in that interim, new forms of anthropological or ethnographic museums have emerged, including regional cultural museums and those of specific native groups, that can be read as embodiments of Boas' concern for cultural contextualization.[34] Although comparative and evolutionary concerns may still be part of a general anthropological project, a comparative evolutionary typology, however ideologically corrected, does not at this point in the history of anthropology seem a likely guiding principle for museum arrangement.

If this somewhat arbitrarily microcosmic account of a single debate and its later architectural echo suggests a more general lesson about "the architecture of the science of anthropology," it is perhaps this: In the multifocal, rather ramshackle structure that is the discipline of anthropology, the relation of physical construction to knowledge construction has been, at the most obvious level (i.e., that of the buildings in which anthropological activity is carried on), a highly contingent one. The failed vision of the New Pitt Rivers Museum, smashed on the hard rock of financial reality, may perhaps be generalized. Architectural space has rarely been designed or realized with the knowledge-constructing activities of anthropology specifically in mind. Museums have, for the most

7.5

Pitt Rivers Extension to University Museum, Oxford. Photograph of galleries taken in 1960s. Funding for the proposed museum was never raised, thus it remains today much as it was 25 or even 100 years ago. Source: "The New Pitt Rivers Museum and Proposed Centre for the Study of Anthropology and Human Environment," University of Oxford, n.d., p. 29B (unpaginated).

part, been general purpose museums: academic structures, general purpose academic structures. Their design and construction have been motivated largely, one suspects, by a range of factors that are, in a significant sense, external to the demands of knowledge construction: financial constraints, construction codes, architectural fashions, cultural ideologies. Insofar as the demands of knowledge construction have been given weight, they have for the most part reflected generalized museological, academic, scientific, or educational concerns. The physical spaces left to be constituted specifically in terms of the construction or representation of anthropological knowledge have been interior spaces, already bounded and structured both at the margins and internally. There is no doubt a great deal more to be said about the ways in which these internal spaces have been employed in the construction and representation of anthropological knowledge. But these issues lie beyond the scope of the present argument. So also any reflections on the spaces of the field, the library, the archive, and the computer, each of which presents its own problems of construction and constraint.

NOTES

Lorraine Daston, Ira Jacknis, and William Sturtevant offered assistance and advice in the composition of this essay, which at an early point was conceived (but never realized) as a joint production with Jacknis. In addition to the comments of participants in the Architecture of Science symposium, it benefited from those of members of the Morris Fishbein Center for the History of Science and Medicine of the University of Chicago, to whom it was presented on May 10, 1994.

1. Stocking, *Ethnographer's Magic,* passim.
2. For example, Gupta and Ferguson, eds., *Anthropological Locations,* passim.
3. See, in general, Stocking, *Victorian Anthropology, Ethnographer's Magic,* and *After Tylor.*
4. Sanjek, ed., *Fieldnotes,* passim.
5. Erickson and Rice, *Strategies for Teaching Anthropology,* passim.
6. Stocking, "Delimiting Anthropology."
7. Stocking, "Essays on Museums."
8. Geertz, *Interpretation of Cultures,* p. 22.
9. Stocking, "Paradigmatic Traditions"; and Hinsley, *Smithsonian,* passim.
10. Anthropologists and historians of anthropology have frequently discussed this episode, most recently Bunzl, "Franz Boas and Humboldtian Tradition"; and Jacknis, "Ethnographic Object." The basic documents are contained in Stocking, ed., *Shaping of American Anthropology,* pp. 61–67, and "Dogmatism, Pragmatism, Essentialism." Further citations in this essay, however, are to the original publications.
11. Stocking, "From Physics to Ethnology"; Bunzl, "Franz Boas and Humboldtian Tradition"; and Liss, "German Culture and German Science."
12. Boas, "Occurrence," p. 485; cf. Stocking, *Victorian Anthropology,* pp. 169–170.
13. Bunzl, "Franz Boas and Humboldtian Tradition."
14. Boas, "Occurrence," p. 485.
15. Boas, "Museums"; cf. Bunzl, "Franz Boas and Humboldtian Tradition"; and Liss, "German Culture and German Science."

16. Boas, "Occurrence," p. 486.
17. Mason, "Occurrence," p. 534; cf. Stocking, "Dogmatism, Pragmatism, Essentialism."
18. Powell, "Museums," p. 612.
19. Bunzl, "Franz Boas and Humboldtian Tradition."
20. Powell, "Museums," pp. 613–614; cf. Stocking, "Dogmatism, Pragmatism, Essentialism."
21. Boas, "Museums," p. 614; cf. Bunzl, "Frans Boas and Humboldtian Tradition."
22. See Stocking, "Basic Assumptions."
23. Boas, "Museums," p. 589; cf. Bunzl, "Franz Boas and Humboldtian Tradition"; Hinsley, *Smithsonian;* and Stocking, "Basic Assumptions."
24. Daston, "Factual Sensibility"; Chapman, "Arranging Ethnology"; and Dias, *Trocadéro,* pp. 115–162.
25. Hinsley, *Smithsonian,* pp. 110–113; and Jacknis, "Franz Boas and Exhibits," pp. 80–81.
26. See, in general, the journal *Museum Anthropology.*
27. Teslow, "Representing Race."
28. In Chapman, "Arranging Ethnology," p. 39.
29. Fagg, *New Pitt Rivers.*
30. Chapman, "Arranging Ethnology," p. 39.
31. Fagg, *New Pitt Rivers.*
32. Mason, "Occurrence," p. 534.
33. Boas, "Museums," p. 589.
34. Clifford, "Northwest Coast Museums"; Karp and Lavine, eds., *Exhibiting Cultures;* Kaplan, ed., *Museums and Making of 'Ourselves.'*

BIBLIOGRAPHY

Boas, Franz. "The Occurrence of Similar Inventions in Areas Widely Apart." *Science* 9 (1887): 485–486.

———. "Museums of Ethnology and Their Classification." *Science* 9 (1887): 587–589, 614.

Bunzl, Matti. "Franz Boas and the Humboldtian Tradition: From *Volksgeist* and *Nationalcharakter* to an Anthropological Concept of Culture." In Stocking Jr., ed., *Volksgeist as Method and Ethic,* 17–78.

Chapman, William R. "Arranging Ethnology: A. H. L. F. Pitt Rivers and the Typological Tradition." In Stocking, Jr., ed., *Objects and Others,* 15–48.

Clifford, James. "Four Northwest Coast Museums: Travel Reflections." In Karp and Lavine, eds., *Exhibiting Cultures,* 212–254.

Daston, Lorraine. "The Factual Sensibility." [Review Essay]. *Isis* 79 (1988): 452–470.

Dias, Nélia. *Le Musée d'ethnographie du Trocadéro (1878–1908): Anthropologie et Muséologie en France.* Paris: Centre National de la Recherche Scientifique, 1991.

Erickson, Paul A., and Patricia Rice. *Strategies for Teaching Anthropology in the 1990s.* [Theme issue]. *Anthropology and Education Quarterly* 21 (June 1990).

Fagg, Bernard, ed. *The New Pitt Rivers Museum and Proposed Centre for the Study of Anthropology and Human Environment.* Oxford University, Oxford, 1969.

Geertz, Clifford. *The Interpretation of Cultures: Selected Essays.* New York: Basic Books, 1973.

Gupta, Akhil, and James Ferguson, eds. *Anthropological Locations: Boundaries and Grounds of a Field Science.* Berkeley and Los Angeles: University of California Press, 1997.

Hinsley, Curtis M. *The Smithsonian and the American Indian: Making a Moral Anthropology in Victorian America.* Washington, DC: Smithsonian Institution Press, 1994.

Jacknis, Ira. "The Ethnographic Object and the Object of Ethnology." In Stocking Jr., ed., *Volksgeist as Method and Ethic,* 185–214.

———. "Franz Boas and Exhibits: On the Limitations of the Museum Method of Anthropology." In Stocking Jr., ed., *Objects and Others,* 75–111.

Kaplan, Flora, ed. *Museums and the Making of 'Ourselves': The Role of Objects in National Identity.* London: Leicester University Press, 1994.

Karp, Ivan, and S. D. Lavine, eds. *Exhibiting Cultures: The Poetics and Politics of Museum Display.* Washington, DC: Smithsonian Institution Press, 1991.

Liss, Julia. "German Culture and German Science in the *Bildung* of Franz Boas." In Stocking Jr., ed., *Volksgeist as Method and Ethic,* 155–184.

Mason, Otis T. "The Occurrence of Similar Inventions in Areas Widely Apart." *Science* 9 (1887): 534–535.

Powell, John W. "Museums of Anthropology and Their Classification." *Science* 9 (1887): 612–614.

Sanjek, Roger, ed. *Fieldnotes: The Makings of Anthropology.* Ithaca, NY: Cornell University Press, 1990.

Stocking, George W., Jr. *After Tylor: British Social Anthropology, 1888–1951.* Madison: University of Wisconsin Press, 1995.

———. "The Basic Assumptions of Boasian Anthropology." In Stocking Jr., ed., *The Shaping of American Anthropology,* 1–19.

———. "Delimiting Anthropology: Historical Reflections on the Boundaries of a Boundless Discipline." *Social Research* 62 (1995): 933–966.

———. "Dogmatism, Pragmatism, Essentialism, Relativism: The Boas/Mason Museum Debate Revisited." *History of Anthropology Newsletter* 21 (1994): 3–12.

———. "Essays on Museums and Material Culture." In Stocking Jr., ed., *Objects and Others,* 3–14.

———. *The Ethnographer's Magic and Other Essays in the History of Anthropology.* Madison: University of Wisconsin Press, 1992.

———. "The Ethnographic Sensibility of the 1920s and the Dualism of the Anthropological Tradition." In Stocking Jr., ed., *Romantic Motives: Essays on Anthropological Sensibility,* 208–276. Vol. 6 of *History of Anthropology.* Madison: University of Wisconsin Press, 1992.

———. "From Physics to Ethnology." In *Race, Culture and Evolution: Essays in the History of Anthropology,* 133–160. New York: Free Press, 1968.

———, ed. *Objects and Others: Essays on Museums and Material Culture.* Vol. 3 of *History of Anthropology.* Madison: University of Wisconsin Press, 1985.

———. "Paradigmatic Traditions in the History of Anthropology." In *Ethnographer's Magic,* 342–361.

———, ed. *The Shaping of American Anthropology, 1883–1911: A Franz Boas Reader.* New York: Basic Books, 1974.

———. *Victorian Anthropology.* New York: Free Press, 1987.

———, ed. *Volksgeist as Method and Ethic: Essays on Boasian Ethnography and the German Anthropological Tradition.* Vol. 8 of *History of Anthropology.* Madison: University of Wisconsin Press, 1996.

Teslow, Tracy. "Representing Race: Artistic and Scientific Realism." *Science as Culture* 5 (1995): 12–38.

8

BRICKS AND BONES: ARCHITECTURE AND SCIENCE IN VICTORIAN BRITAIN

Sophie Forgan

Today there remains a chasm between the two cultures of architecture and science, and to talk about both in the same breath may well appear to invite confusion. In the nineteenth century, however, both professions were emerging from that broad and somewhat undifferentiated field that was termed in Britain the "arts and sciences". In the earlier part of the century the majority of practitioners in both fields were gentlemen amateurs rather than professional men, although their respective intellectual and cultural outlooks might be quite distinct. Professional identities were formed in response to different needs and aspirations. Nonetheless a number of interactions between the two fields throw into relief both complementarities and sharp distinctions. I propose to explore some of these interactions in two ways. Firstly, scientific knowledge had some direct input into architecture, in other words the early development of what is now usually called building science, which thus forms an obvious point of connection. After all, in a situation in which new types of building were required for quite new purposes, and in which construction technologies and mechanical services were developing fast, there were clearly opportunities for collaboration. The success of such attempts, however, depended on a variety of contingent factors, in particular the need for similar disciplinary practices. Collaborative ventures also reveal strikingly some of the problems in achieving and maintaining professional credibility. Secondly, I intend to examine certain buildings built specifically for scientific uses, purpose-built edifices, often constructed indeed with a great deal of collaboration between architects and scientists. In this respect the key area of interest is to try and analyze the effects that buildings might have on the shape of the discipline and on the face that discipline presented to the public. The buildings on which this paper focuses are museums, in part because these occupied a crucial if changing place in nineteenth-century scientific life, but also in part because the themes relate to those discussed in other papers in this volume, in particular the cultural spaces of representation analyzed by Stocking.

The scope of discussion is necessarily restricted to Britain. The architectural profession emerged and was practiced in very different ways in the United States or in Europe, with different forms of training and disciplinary outlook. The buildings discussed in detail are all in London to provide a more coherent focus. Moreover most professional institutions located their headquarters in London, and many of the great national museum collections were to be found there as well.

Building Science

The broader context of the introduction of building science into architectural training and practice relates to problems in the progress of professionalization, the story of which is well known in the case of science and needs little repetition here. From a situation dominated at the start of the nineteenth century by patronage and gentlemen amateurs, there rapidly developed an array of institutions—learned societies, professional associations, government bodies and surveys, together with numerous museums, and, in the latter decades of the century, schools and colleges for organized scientific and technical education—so that it became possible to speak of science as a profession.[1] Disciplinary fields became more clearly delineated, and while new subdivisions appeared regularly, the words "scientist," "physicist," "electrical engineer," and so on entered the language and served to define an occupation's parameters and its associated professional status. The process of professionalization was of course fraught with problems and marked by fierce intellectual and territorial battles. However, the process itself was visible enough.

For architects the process of professionalization was less straightforward. Nor indeed was it likely to follow the pattern of medicine or science, because the practice of architecture had a less obvious connection to exclusive and specialized sorts of knowledge, unless this knowledge was confined to an antiquarian appreciation of past styles and forms, unlikely in an age of massive expansion of the building industry and an eclectic approach to stylistic choice. The founding of the Royal Institute of British Architects (RIBA) in 1834 appeared to provide the institutional framework for the profession, but only a minority of architects belonged to it, and many of the best-known figures kept aloof.[2] There were not, however, the learned societies that underpinned the expansion of knowledge about scientific subjects, nor the schools to codify and disseminate an agreed education.[3] The main route into architectural education for the great majority was through the system of pupilage, in which the aspirant became the articled pupil of an established practitioner.

Pupilage varied enormously and did not generally include formal study of any of the relevant scientific subjects, but concentrated on architectural drawing, mensuration, site work and general office practice. The route toward compulsory examinations was long and tortuous.[4] There was nothing in Britain equivalent to the training Durand established at the École Polytechnique in Paris, though both University College London and King's College offered architectural courses. These courses, however, were generally regarded as a preliminary to entering pupilage.

Although the RIBA was dedicated to promoting the various arts and sciences relating to architecture and to securing the "uniformity and respectability of practice,"[5] the actual ways of doing so were problematic. Competition in the latter half of the nineteenth century grew ever more intense, and architects faced the problem of trying to exclude both amateurs and a growing army of builders, engineers, surveyors, and most threatening of all, large-scale firms of building contractors. They adopted a number of strategies to accomplish this, including a renewed emphasis on the architect's artistic skill, the art of his calling; a trend toward specialization in certain building types, such as hospitals or schools, that required detailed knowledge both of technical requirements and the needs of potential users, and thus promised to provide a better service to the client; and finally, specialization in particular technical areas, such as heating or lighting. None of these strategies were without risk, nor were they necessarily mutually exclusive. As the architect Reginald Blomfield said in 1892, the RIBA's view appeared to consider the "architect as a person of three parts, constructor, man of business, artist." In other words the architect's professional expertise included not only designing buildings but also being fully conversant with all technical questions of construction as well as the legal framework of building regulations and having the ability to run an office successfully. The balance accorded to each of these activities concerned many architects at the time. Blomfield went on to refer dismissively to the man who "was something of a scientific constructor, or 'a scientist' as the odious word is," who might even be preferred to the "simple designer of first-rate capacity."[6]

Nonetheless, many architects had broad general practices and managed to cope with demanding scientific requirements in their buildings. The best-known example is Alfred Waterhouse (1830–1905), architect of many university buildings and numerous offices and country houses and recipient of prestigious commissions such as Manchester Town Hall and the Natural History Museum.[7] He rose to the presidency of the RIBA and ran an enormous and very successful practice. In particular he was much in demand in the ancient universities as well as for new university colleges. His academic clients could be sure of getting an efficient building, constructed to the price required and generally within the given time. Waterhouse was also unusually prepared to listen to his scientific clients, and custom build fittings and furniture to their often demanding requirements. This was so at the Yorkshire College (Leeds), Owens College (Manchester), and particularly at the City and Guilds Central Institution in London, later incorporated into Imperial College. At the last of these, Waterhouse had the unenviable job of managing a large committee (who were difficult about the money), a number of eminent scientific experts called in (who had their own hobbyhorses), and the prospective professors of the new college, who inundated him with demands for fittings and details that had to be specially built. He was certainly not treated with any of the deference or cautious respect one might accord an artist but was somewhat pushed around by the unwieldy committee. It says much for Waterhouse's skill and tact, and perhaps a forbearance inculcated by his Quaker upbringing, that he concentrated on getting the job done without taking offense, a quality that we would probably now describe simply by the term "professionalism." He was also fortunate in that he found the scientific professors with whom he dealt to be people whose advice was technically unimpeachable and whose specially designed fittings actually worked.[8]

A contrasting sort of practice and career may be seen in the architect Edward Cookworthy Robins (1830–1918), who specialized in the technical requirements of scientific buildings and of laboratories in particular, publishing in 1887 *Technical School & College Building*. This manual included a great deal of information on foreign laboratories and provided a compendium of up-to-date wisdom. It continued to be consulted for lab design until the 1920s.[9] Robins had early developed an interest in public health questions and sanitary science, published papers and pamphlets on the subject, was a founding member of the Sanitary Institute (1877), and became a keen believer in technical education. He became involved in the design of Britain's first technical college, the Finsbury College, and was a devoted supporter of T. H. Huxley, whose fame and campaign for technical education was at its height in the late 1870s and early 1880s. Robins dedicated his book to Huxley in the most fulsome terms and arguably used his connections with scientists, and with Huxley in particular, to establish an unrivaled reputation for himself in this field.[10] Nevertheless, he could not afford to be seen as hostile to art and aesthetics, and gave the occasional lecture on such subjects as "The Ethics of Art" (1885), or "The Artistic Side of Sanitary Science" (1886), which was an ingenious attempt to marry science and art with the more practical requirements of buildings.

Turning now to the attitude of professional organizations toward science, the RIBA had always been uneasy about the role of science in architecture and what scientific principles, if any, underlay the profession of their art.[11] If art and aesthetic sensibility were the keys to producing creative, functional, and satisfying architecture, then science served merely in a subservient engineering role, as intimated in Blomfield's comments above. On occasion the RIBA had invited a prestigious scientist, such as Michael Faraday, to be an honorary member, and had long recognized a need to conduct tests on building materials in a scientific manner, but this remains a relatively unstudied aspect of their professional history. It was, however, an important area, in part because in the 1880s, as one historian has said, the RIBA was not "a flourishing institution," and ferocious aesthetic disputes as well as arguments about the nature of the profession and ways of excluding unwanted practitioners generally divided architectural culture.[12] Disputes revolved around the wisdom of controlling access to architecture by "registration," a necessary feature for any profession, but one that many architects regarded as a denial of the artistic nature of their calling. It was not the fact of restricting access that divided architects, but which particular approach to adopt to achieve that end. Pressure on architects was acute, because there was clearly a lucrative market in the many types of building that required up-to-date technical knowledge: colleges and technical schools, secondary schools with laboratory facilities, power stations, waterworks, sewage plants, post and telegraph offices, underground railway systems, and so on. The scientific content of RIBA examinations remained small. The real danger existed that people not trained as architects, such as builders and surveyors or even engineers, would move into this field. After all, Henry Cole at South Kensington had preferred using army engineers rather than architects to design buildings, notably of course Captain Francis Fowke.[13] In this sense one can interpret a renewed interest in the scientific and technical aspects of building as an attempt to fend off better technically qualified but less "artistic" practitioners. To do this effectively meant collaborating with elite scientists, as seen in E. C. Robins's cultivation of Huxley, not the ruck of everyday engineers. It was also no doubt a matter of self-protection, because architects who failed to master the scientific principles of areas in which they professed to be competent could face public humiliation when scientific experts challenged them.[14] In this context it is not surprising to find that a Standing Committee for Science was created in the RIBA's constitutional reorganization of the mid-1880s. After the reorganization, RIBA had four standing committees, those on Art, Science, Practice, and Literature, which neatly reflected the different areas of specialization of the sort of architect so disliked by Blomfield.

The story of the Standing Committee for Science reflects themes pertinent to the nature and practice of architecture as a profession, as well as the possibilities and limitations of collaboration with professional scientists. The Committee got off to a slow start.[15] It met first in June 1886, and six months passed before it was felt necessary to reconvene. For the next four to five years it was preoccupied with incorporating the acquired wisdom of a preceding committee (on Light and Air) and dealing with matters such as concreting over graves in disused churchyards or dismissing importunate inventors seeking endorsement for their products: "a new invention for fresco painting," or "Messrs. Merryweather's Microhlizer," few of which were considered "of sufficient importance for special notice." By 1891 the RIBA's Secretary, William White, actually had to suggest to the committee that they might contribute a column on "Scientific Notes" to the RIBA's *Journal.* Indeed the *Journal's* editor wrote (1892), virtually begging for notices of relevant books on "scientific" matters. The Committee decided, with much discussion, to organize an evening meeting of the Institute centered on a paper on the issues relating to architecture raised by the Congress on Hygiene and Demography taking place at the time (October 1891). They also discussed an evening on electric light, and the idea of a hygienic evening was abandoned when William Henry Preece, FRS, a well-known electrician and Engineer-in-Chief to the Post Office, was brought in to talk about "The Art of Internal Illuminating," which went off very well. Electric lighting was new, topical, and relevant to architects. In 1893 there was a brief flurry of interest in "heating domestic buildings by steam from central stations." Here clearly the United States was much more advanced, and the Committee wrote to Dankmar Adler at the American Institute of Architects, who replied at some length in reference to the Auditorium Building Tower in Chicago.[16]

The problem was that the Committee members were quite unclear as to what actually fell within their remit, what indeed was relevantly scientific in terms of architecture? In the jargon of today, the Committee was reactive rather than proactive. However, a subject emerged that scientifically minded architects could embrace without reservation. This concerned a laboratory-building project in which one Committee member was much involved. Consequently, in March 1894, an evening meeting was held on the new laboratories for physics and mechanical and electrical engineering then being built at University College London. University College, begun in 1827, was the oldest and architecturally most attractive of the colleges of London University. When the College had decided to build new labs, they naturally turned to their own Professor of Architecture, T. Roger Smith, to design them.[17] The three scientific professors were also then in an excellent position to make sure they got what they wanted. As a result, well-fitted laboratories, lecture rooms, workshops, and professorial quarters were designed to be built along the Gower Street frontage. T. Roger Smith arranged the evening meeting at the RIBA (with selected scientific guests), with talks by himself and the three professors and plans and pictures of the buildings, followed by a vigorous discussion (see figure 8.1). The chief dissenting voice in an evening of otherwise happy collaboration between science and architecture came from Henry Statham, editor of the important journal *The Builder*. Statham pointedly remarked that "University College were in possession of a building, by an eminent architect, which was one of the ornaments of London," yet not once had the original architect William Wilkins's name been mentioned, nor his original plans for the College.[18] Statham objected to the plans on aesthetic grounds—the quality of the original design by Wilkins, as well as faithfulness to the original architectural intention: whether Wilkins had intended to close the fourth side of the site. Surely the labs could have been placed somewhere else?[19] As another architect put it, in former times, after all, college buildings

> would have been designed in a very classical manner, and constructed in a very substantial manner, and the Professors of the several departments would then have been allowed to settle down in them, and to do the best they could. But [and it was implied, regrettably] that state of things was no longer tolerable [now].[20]

8.1

University College London: Completion of the Gower Street front as planned by T. Roger Smith and published in the *Journal of the RIBA* 1 (1894), facing p. 288. The view depicted gives a somewhat exaggerated impression of spaciousness, as in reality the hackney cab trotting briskly out of the college gateway would have crashed into the front door of University College Hospital immediately across the street unless it took a sharp turn left or right. Reproduced by courtesy of the British Architectural Library, Royal Institute of British Architects.

There was also a danger that the view of the cupola would be blotted out from the street. Urban vistas were important. This "valuable and remarkable building" should have been handled with more care.[21] Smith defended his handling of the building with its "very great" architectural claims, but revealed that the building committee, most of whom were Oxford or Cambridge men, had wanted to close the quadrangle entirely, arguing that the beauty of those ancient buildings was not apparent anyway until the quadrangle was entered.[22] Smith, though partly convinced, asserted that he had fought hard to retain an opening to the street, and the result was a compromise. Indeed a number of compromises were made, between urban vista and collegiate seclusion, between aesthetic integrity and the territorial demands of science, and between the departments themselves and the demands of the professors, who complained how badly housed they were.[23] Thus even apparent collaboration masked considerable differences among the various parties over the place of scientific education in an architectural work of art.

Statham pursued the matter for a time, the architectural press published reports of the meeting and discussion, but University College went ahead anyway. In any case the issues focused primarily on internal differences within the architectural profession (art and the sanctity of the architectural masterpiece) rather than any opposing disciplinary practices. The latter, however, did arise in an incident not long after. At the time, William Cawthorne Unwin was a member of the Science Standing Committee. (He had been appointed in 1893.) Unwin was a highly successful engineer, a Fellow of the Royal Society, and the professor of civil and mechanical engineering at the City and Guilds Central Institution in South Kensington. His research interests included work on the strength of materials. Who better then, than Unwin to take an active part in carrying forward a series of tests initiated by the Committee on bricks and brickwork, their relative load-bearing qualities and resistance to crushing? Two series of tests on different types of brick were carried out, and the results were presented at an RIBA meeting and published in the *Journal* (see figure 8.2).[24] A third series was carried out and also reported in the *Journal*. The Committee then decided to publish all the tests and their results in a separate booklet. Delays ensued, mostly on account of the cost, and as one member put it, "Up to this time everything had been harmonious, but now discords became apparent."[25] These discords concerned the tests' scientific validity (in some of the earlier tests, the bricklayers had cheated), the sort of conclusions that could be drawn from them (Unwin said they were "unsatisfactory and unreliable"), and the professional purity of the RIBA Council (who did not want to be associated with anything about which controversy might arise), as well as the conventions of publication.

Unwin himself did not feel the tests were really very significant. "My opinion is that the Brickwork Experiments are not of themselves important enough relatively to work done elsewhere to specially deserve separate publication,"[26] the "elsewhere" referred to work in Berlin and the United States. The tests, according to Unwin, could not provide a clear formula as to what loads brickwork could carry in the same way that one could

188 | *Sophie Forgan*

8.2
Tests on the load-bearing qualities of different bricks. Photographs were taken of the numbered columns before and after the hydraulic press had exerted pressure at an increasing rate. RIBA, *Report on Brickwork Tests,* p. 27. Reproduced by courtesy of the British Architectural Library, Royal Institute of British Architects.

Pier No. 11, Gauge at 103 lbs. pressure. Gaults in Portland Cement (p. 25).

Pier No. 12, after Compression. Front View. Gaults in Portland Cement (p. 28).

construct a formula for the strength of iron beams. Further, the tests had not been carried out in sufficiently controlled conditions and therefore were "unreliable." But, having helped to spend the RIBA's money, and still calling the results unsatisfactory and unreliable, Unwin then added insult to injury by summarizing them in the latest edition of his book, *Testing of Materials of Construction* (1883/1899) without asking any permissions, brushing aside complaints as follows:

> the ordinary course of scientific Societies is to publish researches in their transactions and then to allow very free re-publication in textbooks elsewhere. A scientific Society exists for the purpose of extending and spreading scientific knowledge and is in a wholly different position from a private person who rightly has a certain property in his work.[27]

Unwin was implying that the RIBA did not behave as a scientific society but rather was acting like a private person. Practices in carrying out such tests clearly needed to be tightened up, but in terms of his attitude to the dissemination of knowledge, Unwin took the stance of an academic scientist with scant regard for any of the professional sensitivities of architects. Unwin took no further part in the Committee for a number of years, and the report on the brickwork experiments was not published until 1905, six years later.[28] The Science Committee continued to consider all sorts of interesting questions, but not until it got interested in concrete did it attempt this sort of collaborative research effort again.

Buildings for Science

If joint research on the science relating to architecture was problematic, then a different perspective may be gained from examining architecture designed for science. Laboratory building, as seen above, was relatively straightforward, because in such an area the client's expertise was generally superior to that of the architect, at least as far as college or university labs were concerned. The architect might be free to design the exterior of the building appropriately, cost permitting, but internally crucial matters of planning and equipment dominated, and in Waterhouse's case, these were worked out in careful collaboration. But in an older and more general building type such as the museum, intended both for a scientific use and for a predominately nonscientific public, then the architect ostensibly had a freer hand. Museums were also commissions that carried a far greater prestige than laboratory buildings ever did and moreover acquired so to speak a certain weight of architectural meaning in addition to providing particular spaces for displaying scientific knowledge.

Museums had long been central to the growth and development of knowledge, above all in medicine and the sciences. Even the sixteenth-century scholar's room full of collections had a subtlety and complexity in the way its spaces were divided that related to notions about the proper place where knowledge was made, to appropriate social behavior, and to other contingent political and cultural factors as well (cf. Findlen, this volume). From the late eighteenth century onward, it is possible to argue that most scientific types of knowledge were conceived first and foremost in terms of collections.[29] Objects collected in the field were brought back to the museum to be analyzed, dissected, classified, and their significance and place in the order of things fixed and elaborated. Scientific disciplines emerged from these collections of specimens, and the professional structure of the subject was often located in museums, so that it is possible, for example, for George Stocking to talk of the "museum period" of anthropology. Furthermore, one of the factors that helped shape the construction of knowledge in particular fields was the building in which they were housed. Throughout the major part of the nineteenth century, even in the later years of the Victorian period, the museum was architecturally prestigious and scientifically authoritative. How its architecture might mediate that authority is the subject of the second part of this essay.

From midcentury onward, numerous museums were built, especially in the wake of the 1850 Act allowing town councils to establish public libraries and museums. In 1853, John and Wyatt Papworth brought out a prompt and timely architectural guide for architects and their potential clients to consider before embarking on building a new museum, library, or art gallery.[30] This guide covered most of the pertinent questions, such as site, accommodation required, maximum size of objects that might be exhibited, admission of the public, and conservation questions relating to ventilation and lighting. It included specimen plans for a "Museum for a small town" but also plans of that most prestigious of all museums, the Museum of Natural History in Paris. The text made much reference to the British Museum but also to others, such as the Museum of Practical Geology in Jermyn Street as well as several continental buildings. In all it was a practical guide to designing the smaller local museum, with sensible advice about planning and arranging facilities, height and depth of cases, and the troublesome problem of providing adequate light, which tended to result in confusing reflections off the glazed cases. Although due attention was paid to those who might want to use the museum for scholarly work, it was clearly conceived of first and foremost as a place where a less-than-scholarly public came for education and edification. There was in any case an enormous amount of rhetoric on the subject of the "usefulness" of museums in education, though not all Victorians were overly sanguine about the effectiveness of museum visiting in acquiring such education.[31] It was nevertheless taken for granted that all museum display should have an educative objective in view, and because museums provided men of science with an established career route, it is not surprising to find as curators men eager and capable of molding the visible shape of knowledge about a subject at that time. This was especially so in the great national museums, and to a lesser extent in the larger provincial and university museums, though these more often depended on miscellaneous donations for the bulk of their collections.

The vocabulary that scientists used about museums also indicated that museum displays shaped the knowledge of a discipline. Language and literary metaphor is one way of seeing how museums were represented and how they were regarded by an important sector of their users. The favorite analogy was with a library. This is not surprising given the common belief that museums contained, as it were, the sum of current knowledge, and furthermore, the assumption that they were built to contain "complete" collections. Collecting was not regarded as an open-ended, never-ending process, even though newly built museums demonstrated a regrettable tendency to fill up more rapidly than their creators had anticipated, and the "completion" of knowledge about a subject remained on the distant horizon. But museums, it was thought, could be "read" like a book or consulted like an encyclopedia. As a British Association committee on museums put it in 1887, "the museum is a book of plates close at hand to illustrate the volumes in the library," and the ideal provincial museum "ought, in the first place, to be a *fully illustrated monograph of its own district.*"[32] The museum was a text, in both a literal and a metaphorical sense. The committee's reports used the language of books throughout to explain and press home points about the proper use of display. For example, upright, hinged glazed cases were described as being "movable like the leaves of a book" or, with reference to good labeling practice, "A museum without labels is like an index torn out of a book; it may be amusing, but it teaches very little."[33] Such language fitted nicely inside that favorite metaphor, so often used, of the Book of Nature, which itself was a reference to that ultimate "book," the divinely created universe.

Furthermore, the architectural layout of museums was frequently similar to that of libraries. Libraries evolved in the early nineteenth century from a single large room, sometimes with a gallery above, with the bookcases ranged flat against the wall. It then became common to project cases at right angles at regular intervals from the wall to provide more shelf space. When the open space to be roofed over became larger, such projections were a good way to make use of, and in effect also to disguise, the upright piers needed to support the roof. Museum spaces and library spaces were thus relatively interchangeable. The Papworths, in their architectural guide, showed rooms that could equally well be used for libraries as for museums, and indeed in many local museums, the library became the dominant element and gradually pushed the museum out altogether. They also included examples of circular libraries, designed for "successive enlargement," and in the same breath referred to museums designed on a circular plan, such as the Scarborough Museum.[34] Analogies of museums with books or libraries therefore made a certain architectural sense as well as fitting the mind-set of the period.

On occasion, however, a scientist might go beyond the commonplace bibliological analogy and employ a more complex metaphor that revealed his own scientific beliefs. Richard Owen, the famous comparative anatomist and first director of the Natural History Museum, is a case in point. When Owen had finally succeeded in the creation of the Museum for which he had labored so long, he described the story of the building's creation to his contemporaries in the Biological Section of the British Association meeting in 1881. Owen concluded his narrative by calling on the language of phylogeny, in other words the language of classification, of archetypes, of the traceable evidences of ancestral structures. In the building just erected, Owen said, one could find only one characteristic retained of "the primitive and now extinct former museum," and that was the central hall.[35] In the new building, one could trace a "developmental advance" in museum design, which could be seen in the single-story galleries, improved admission of light, and adaptation of walls as well as floor to the needs of exhibition. To give added weight to his use of phylogenetic language in architectural metaphor, he added a footnote to Cuvier's debate in 1829 with Geoffroy St. Hilaire on the unity of composition or plan in cephalopods and vertebrates, in which Cuvier had distinguished between "*la composition d'une maison,* c'est le nombre d'appartemens [sic] ou de chambres qui s'y trouve; et son *plan,* c'est la disposition réciproque de ces appartemens et de ces chambres."[36] Thus for Owen, the language of archetypes and architecture reflected his own belief in slow developmental advance.[37]

Indeed, it is arguable that the Cuvierian argument about design in nature had a built mirror image in Owen's Museum of Natural History, in which the plan and composition constituted a scientific exposition of Owen's views about the natural world. The museum was entered through the central hall, Owen's original Index Hall, which formed a central reference point between the two wings, each of which contained one of the two major branches of natural history: zoology and collections relating to the living world in the western wing, and paleontology and extinct species in the eastern wing (with mineralogy and botany on the floors above). This conveniently matched the fiefdoms of the keepers under Owen. The spine of the great front galleries running along the facade also had a vertebrate, skeletal quality that related well to Cuvier's reciprocal disposition of parts. Architectural symmetry and design in nature blended together. The decoration of the facade and front galleries followed the same subject division, and Owen collaborated closely with Waterhouse's detailed renderings of animals and plants in terra cotta decoration. As a result, the visitor entered a magnificent building, impressed by the grand entrance appropriate to the subject's evident importance and prepared to be both delighted and instructed by the building's harmony, its decoration and the collections it displayed.[38] Here the architectural ensemble was clearly designed and realized on the basis of a particular construction of natural historical knowledge.

Other museums in London also shaped the relevant scientific disciplines in equally characteristic fashion. Again the building fabric was an integral part of the display system, in contrast to the more ramshackle structures that Stocking argues characterized anthropology in the later nineteenth and twentieth centuries. The Museum of Practical Geology in Jermyn Street provides a key example, being the earliest of purpose-built museums in which the building itself was intended to be as much a lesson as its contents. The Jermyn Street Museum, as it was often called, dates from the mid–nineteenth century. In 1845 the government agreed to Henry de la Beche's scheme to provide a building to house the Geological Survey, the Mining Records Office, and the collection of the Museum of Practical Geology. The building was started in 1847 and opened in 1851. It included by then the Royal School of Mines and provided teaching and laboratory instruction for prospective mining engineers. James Pennethorne designed the building, a veritable Palazzo Geologico, on the encyclopedic principle described above.[39] The ground floor was entered through a Hall of Marbles, which the visitor crossed to gain entrance to the lecture theater, which extended through the ground floor down to the basement. A broad staircase took the visitor up to the museum's main floor, on the *piano nobile,* around which two galleries rose to a top-lighted ceiling (see figures 8.3 and 8.4). The building also accommodated two laboratories, a large room for the Mining Records Office, rooms for models of mining machinery and equipment, a library, workshops, and some small, rather cramped rooms for the staff of the several organizations located there. Not surprisingly it rapidly became overcrowded, not only on account of the collections but also because of the teaching needs of the Royal School of Mines.

The Museum's main stated function was to instruct the visitor in the practical and economically profitable applications of geology, as the name implied. As such, it underpinned the practical utility of the Geological Survey. Its collections fell into two distinct categories, as Robert Hunt's early guidebook stated (1857): firstly, natural materials, which could be studied in terms of their lithological character, their geological order in time, and their mineral constitution. Secondly, the "Artificial Productions" of those natural materials, as Hunt put it, the results of human labor aided by the discoveries of science.[40] The latter, which occupied the museum's main floor, were arranged in a manner that juxtaposed cases containing the raw material, the processes of industrial production, and the finished product. As Henry Becker wrote enthusiastically,

> The specimens are admirably arranged in separate lines of cases placed in such juxtaposition that the progress of any one metalliferous mineral may be traced from the geological stratum whence the ore is extracted through the various processes of manufacture till the metal ultimately assumes the forms required for use or ornament.[41]

As a result, the Museum acquired a large collection of objects normally associated with the ornamental arts—pottery, glass, swords, metalwork, mosaics and so forth—to illustrate those forms required for use or ornament. The display also had a geographical dimension, in that the visitor proceeded from the geology to all the specimen's related characteristics, including its value and the geographical distribution of the raw material. This was most clearly organized in the rock collections, where the problems of trying to display collections in terms not only of their lithological character, but also of their stratigraphical and topographical characters, were carefully outlined and discussed in the descriptive catalogue published first in 1858 by A. C. Ramsay.[42] As such this became a model of the approach taken in many Victorian museums and was replicated in smaller provincial museums. Collections arranged thus came close to providing that "monograph" of the local district, its natural resources, and manufacturing potential the British Association recommended for the ideal museum thirty years later. At Jermyn Street the approach was enthusiastically supported by writers such as Henry Becker and seems to have been generally popular with visitors, judging by the sale of guidebooks. It was also an explicit justification, with its Peelite appeal to economy, utility, and improvement, for the annual demands on the Treasury for cash to support the Survey, the Royal School of Mines, and the Mining Records Office.

Moreover, the building itself was a lesson in stone, and one that predates the much better known surviving example of the Oxford University Museum. As Hunt explained, the Jermyn Street Museum "itself must be regarded as one of the illustrations of the main objects in view."[43] The guidebooks made much of the facades of Yorkshire dolomite or magnesian limestone, the steps of red Peterhead granite, the slab of Penrhyn granite at the doorway, the steps of Portland stone, the vestibule lined with Irish granite, polished Derbyshire alabaster, and pilasters of grey Peterhead granite.[44] The main hall housed columns and screens made up of British marbles, vases and tazzas, sculptures (including a replica of that contemporary favorite, the Dying Gladiator) and busts of famous British geologists, so that it laid out visibly both the material of the subject and its founding fathers, its British ancestry. The guidebooks reinforced the interpretation and often added helpful notes as to how to look at the exhibits and what conclusions to draw, for example, on the bust of "Professor John Playfair . . . by Noble, after Sir F. Chantrey R. A. A face marked by thought, penetration and care."[45] Even the physiognomy of scientific labor could thus be read.

However, the Museum was not just a museum to celebrate the country's rich natural resources and encourage their economic use in the "workshop of the world," though this may have been the view that many visitors took away with them. The Survey and Museum employed some of the most notable scientists of the day: Henry de la Beche; his successor as Director, Sir Roderick Murchison; A. C. Ramsay; and the young T. H. Huxley, who was appointed Naturalist to the Survey in 1854.[46] Among Huxley's duties was arranging the museum's upper galleries, which rose steeply around the open expanse of the main floor. These galleries contained a collection of British fossils, many of which had been acquired during the Survey's work. The Museum had moreover opened in its brand new building at the time that stratigraphical analysis was generally accepted as the major tool for establishing the geological record. The species and numbers of fossils and their distribution among strata were regarded as the key to unraveling the earth's history. The Devonian Controversy and the heated debates between Murchison and Adam Sedgwick and their respective adherents, which had occupied geologists in the later 1830s and 1840s, concerned boundary disputes between the various epochs of the earth's history as well as the correct interpretation of the fossil record.[47] It was therefore fortunate that the Museum opened at a time that allowed first de la Beche and his protégé John Phillips, followed by Murchison, to fix their version of the stratigraphic evidence in, if not exactly tablets of stone, then certainly the most authoritative setting available to anyone.

The building's form was well designed to allow such an authoritative display to be laid out. The main floor, around which the two galleries rose, was open to the ceiling. Giant cast-iron hoops, supported on pillars that allowed recesses for display cases, spanned the roof (see figures 8.3 and 8.4). The extensive continuous runs of display cases

8.3

The Museum of Practical Geology, Jermyn Street, in its heyday. Print of the principal floor from a watercolor painting by J. P. Emslie, 1876, IGS/897. Reproduced by permission of the Director, British Geological Survey. NERC copyright reserved. See also color plate 6.

8.4

Museum of Practical Geology: photograph of the main floor with galleries above, ca. 1910, notable by this time for the lack of visitors. The presence of policemen was normal in this period, and they fulfilled some of the duties now undertaken by attendants. Reproduced by permission of the Director, British Geological Survey. NERC copyright reserved.

and their containment visually within a single architectural volume allowed for a powerfully logical arrangement of specimens. So whereas the ground floor, the horizontal plane, was devoted to the uses of geology, to those processes that took place on the earth's surface, the vertical plane, the galleries above, displayed the earth's structure in terms of its geological record. They formed a sort of stratigraphical column, one that did not rise vertically, which would have been impractical given the number of fossils that had to be exhibited to demonstrate it, but was instead wrapped round the two galleries (see figures 8.5 and 8.6). The lower gallery started from the earliest—the Lingula Flags or Cambrian—at the lower west edge and moved to the Permian at the eastern corner, whereas the upper moved around from the Jurassic to the Eocene and Oligocene epochs.[48] The lower gallery therefore covered the older Palaeozoic era, and the gallery above

8.5

Plan of the Lower Gallery, Museum of Practical Geology, no date (probably late nineteenth century). The general layout of the galleries, over the course of the nineteenth century, did not differ substantially from this plan, with the exception of A. C. Ramsay's rock specimens, which were originally in wall cases in this gallery. New specimens were incorporated into the cases as they were acquired. Reproduced by permission of the Director, British Geological Survey. NERC copyright reserved.

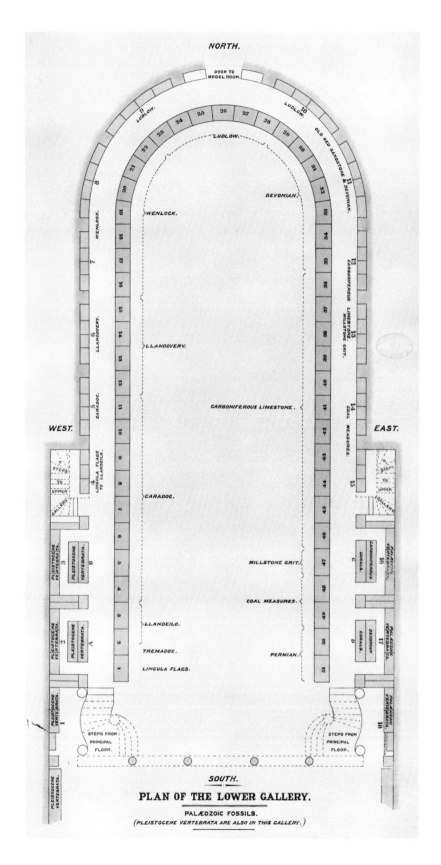

8.6

Plan of the Upper Gallery, no date. Larger specimens were displayed in the wall cases, and smaller ones in the glass-topped cases ringing the outer railing and in the drawers below. Reproduced by permission of the Director, British Geological Survey. NERC copyright reserved.

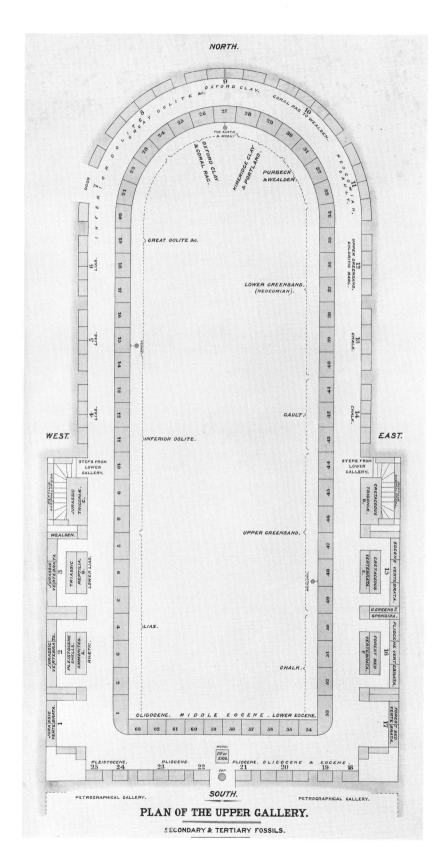

contained the more recent fossils of the secondary and tertiary eras. As Ramsay said, the fossil collections could be regarded as "a kind of stratigraphic collection of rocks."[49]

In addition to the stratigraphic arrangement, there was also a geographical one that accorded with the maps published by the survey. The earliest maps produced were of the western counties and of Wales, scene of so much geological controversy, and the fossils of Tremadoc, Llandeilo, Caradoc, Llandovery, and so on could be followed one after the other around the gallery. Each fossil was placed on a tablet with a label saying where it had been discovered, so each was therefore "sited" in its proper place, both in terms of its stratigraphical position and its topographical location. The fossils thus provided both the evidence for the geological maps produced by the Survey produced and a three-dimensional representation of those maps in what was in effect a gigantic geological section.[50] Huxley was quite clear about this, when he wrote in his *Catalogue* (1865), "In one sense, therefore, the collection of fossils is simply the product of, and key to, the maps of the Survey."[51] It was also the scientific evidence on which the map was based, which could be consulted at any time, as again Huxley wrote: ". . . it would be open to every one to examine for himself the evidence on which the map stood, and to satisfy himself of the accuracy of this part of the work of the surveyors."[52] The map's authority was embedded in the strata of galleries. There could be no clearer demonstration of the discipline of stratigraphic geology. No wonder that W. Stanley Jevons, visiting the Museum in the early 1880s, praised it for its logical arrangement, for the way that it gave an idea of natural order and succession and could be comprehended in a coup d'oeil.[53] After all, the eye takes in a map's general contours at a glance, before then reaching closer to examine the details.

However, by the end of the century, knowledge had moved on, and the Museum was no longer regarded as the exemplar of what a scientific collection should be. The building itself had become seriously overcrowded, and despite great hostility on the part of some of the staff, gradually all the teaching departments were removed to Henry Cole's growing empire at South Kensington, where T. H. Huxley was building up the Royal College of Science, a key part of what would later become Imperial College. The Museum of Practical Geology was one of the museums managed under the auspices of the Department of Science and Art, which after Cole's retirement was headed by Major-General Sir John Donnelly. But by the 1890s the Science and Art Department was itself under attack. Its museums were in a state of physical, financial, and indeed intellectual disorder and thus vulnerable to Parliamentary scrutiny and Treasury audit. The standard remedy of detailed scrutiny by Select Committee was applied, and such a committee was duly set up in 1897 to examine the Museums of the Science and Art Department.

One of the Select Committee's major recommendations was that the Jermyn Street Museum be removed to South Kensington and its building sold.[54] It is really rather extraordinary to find it proposed that a Government-funded, purpose-built museum building be sold, less than fifty years since it was opened. The proposal indeed raised a number of issues, some trickier than others. To begin with, many of the objects in the Jermyn Street collections, especially ceramics, would be better placed in the art collections of the South Kensington Museum, whose collections had grown quite enormous. Such ornamental objects were now seen primarily from an aesthetic viewpoint rather than as a product of a particular raw material and manufacturing process. There was little dispute from the scientists on that account. The collections also contained objects that the late-nineteenth-century museologist no longer considered it proper to collect and display; as the Select Committee derisively noted, "borax soap and Messrs. Truefitt's hair washes are gravely exhibited as 'technological specimens.'"[55] Weeding out was clearly required, and those exhibitions that had formerly demonstrated economic utility were clearly no longer regarded as useful, perhaps in part because the explosion of consumer goods available on the market and the multiplicity of different manufacturing processes made it more difficult to draw those simple lines of connection between raw material and finished product. But most difficult of all to resolve was the duplication that would result from placing another collection alongside that of the Natural History Museum, which included a palaeontological collection. Relocation would immediately highlight the overlap. However, as several witnesses vigorously protested, each collection had quite different underlying principles. Jermyn Street was not just a collection used for teaching geology: It was the only complete collection of British geology arranged so as to illustrate the stratigraphy of the British Isles.[56] It was, as suggested above, a map. And geologist J. W. Judd argued, in an identical way to Huxley a generation earlier, that a large part of the fossils and specimens in the collection were acquired in the constructing of that map: "They are the proofs of the correctness of the map, and they ought always to be placed so that they can be referred to."[57] Several witnesses at the Select Committee were deeply concerned that removal from its building would destroy the collection's value and that those authenticating specimens that underpinned the Geological Survey's maps, and even the general structure of knowledge about British geology, would be lost. The building and collections had together acquired a cohesive identity that could not survive removal to a different sort of building, they argued: the map was indivisible from the building's structure. Thus the Museum and the Survey's geologists defended the scientific shape of the discipline fixed in the 1850s and 60s, while elsewhere the center of geological attention had shifted to other areas such as extinct volcanoes or mountain building in the Scottish Highlands.

The Natural History Museum's collection, on the other hand, was essentially a palaeontological one that was not displayed stratigraphically.[58] There was duplication of specimens among the two collections, but the theoretical basis on which they had been collected was quite different. Having one collection did not do away with the necessity of the other, nor would it be a solution to let the Natural History Museum take most of the specimens and create "a continuous collection of palaeontology and geology."[59] Several witnesses found it easiest to describe the two museums in terms of one being a map, the other a biography.[60] The words used still call upon a bibliographical metaphor but imply a very different arrangement of the library shelves. And of course the architectural form of the Natural History Museum, with its axial symmetry, its regular spaces neatly marked off like the vertebra of a skeleton, was ideally suited to Owen's plan for displaying in distinct spaces all the members of a "family" and their relationship to each other.[61] The shape of the one discipline did not match the shape of the other. One can hardly be surprised that the integrity of the Jermyn Street building and collection was fiercely defended. In the end, of course, it was to no avail, though the Museum lingered on in a state of decline. No longer was it a place where the latest geological knowledge was mapped and laid out for public scrutiny, but simply a dusty collection of rocks and old bones. Finally, after many delays, it was removed to a new building at South Kensington in the 1930s. The Jermyn Street building was sold and demolished, and the site is now occupied by a superior department store.

Conclusion

In conclusion, two rather different perspectives have been offered in this paper on the architecture of science in nineteenth-century Britain. The first has examined some of the ways science and scientists were involved in the process of architectural professionalization, whether as practice or in building labs and testing bricks. Science indeed played a variety of roles in Victorian architectural practice and contributed to shaping the professional consciousness of architects, though on the whole building science was relatively unassimilated into architectural training and tended to remain the province of certain specialists. Collaboration among architects, engineers and scientists certainly lacked the long history that seems to have characterized the field in the United States (cf. Thompson). Certainly, however, the awareness of different disciplinary practices proved a salutary experience for architects and no doubt contributed to ensuring more successful instances of collaboration in the early twentieth century. A coda that exemplifies this more successful collaboration occurred in 1904 when the RIBA offered to assist the Geological Museum in the rearrangement of its collection of building stones, then became involved in tests on the weathering of stones carried out by the Museum.[62] Building science moreover provides a useful way of examining the variety and effectiveness of strategies adopted for professional advancement at a period of acute competition, as well as the limits of collaboration between two professions in very different fields.

The second perspective has attempted to analyze how buildings themselves might mould or constrain the construction of scientific disciplines. Buildings for science had an impact beyond their visible external appearance, which itself carried messages of importance and authority, in addition to great prestige for the architect in high-profile commissions. Certainly the physical construction of buildings is always contingent on a wide range of factors, and changes often take place during the course of building. However in many key instances in nineteenth-century Britain, museums were more than simply a shell within which scientists had to manage as best they might. Victorian museum buildings were intended to be "lessons in stone," and those architectural "lessons" could at the same time be the key textbooks for the disciplines they housed. How those scientific texts were written and displayed in exhibits and buildings can help us appreciate why the Victorian museum movement was so widespread and appealing, and how certain scientific disciplines came to be constructed within particular boundaries.

NOTES

I am particularly grateful to Adrian Forty, Graeme Gooday, Stephen Hayward, Jim Secord, Martin Rudwick, and John Thackray for their many helpful suggestions, and to the editors of this volume, Peter Galison and Emily Thompson. I also benefited much from the discussion at versions of this paper given at Harvard University and at the Cabinet of Natural Philosophy, Cambridge, UK. The archives of University College London, The Royal Institute of British Architects, and the British Geological Survey have been unfailingly helpful.

1. For the general development of nineteenth-century British science, see Russell, *Science and Social Change;* and Morrell and Thackray, *Gentlemen of Science.* Recent work also indicates that there was a great deal of invisible science by, for example, artisan botanists; Anne Secord, "Science in the Pub."
2. Saint, *Image of the Architect,* chap. 3; Crinson and Lubbock, *Architecture: Art or Profession?* chap. 2. I am also most grateful to Adrian Forty for allowing me to see his unpublished paper, "Problems in the History of a Profession: Architecture in Britain in the Nineteenth and Early Twentieth Centuries," and am indebted to his interpretation on some of the problems of professionalization for architects.
3. Societies concerned with ecclesiastical architecture, however, flourished in the midcentury, but Forty argues that architects generally regarded these as an unmitigated nuisance.
4. Crinson and Lubbock, *Architecture: Art or Profession?* pp. 184–192, includes a useful description of the RIBA examination syllabus and how it changed.
5. This was the principal objective of the RIBA's founders; cited in Saint, p. 61.
6. Shaw and Jackson, eds. *Architecture, A Profession or an Art?* p. 40. This volume brought together many of the best-known opponents of moves toward compulsory registration of architects and received much publicity.
7. Cunningham and Waterhouse, *Alfred Waterhouse.*
8. These included, among others, W. R. Ayrton, H. E. Armstrong, T. E. Thorpe, and Henry Roscoe; Brock, "Building England's First Technical College," p. 156.
9. Ibid., p. 156.
10. Robins, *Technical School,* dedication page: "To Thomas H. Huxley, Fellow of the Royal Society, these pages are inscribed in token of the author's reverence for his genius, and in grateful remembrance of his uniform and friendly counsel at all times." Robins also lived close to Huxley in London and no doubt exploited the opportunity for neighborly contact. For example, he hosted in his house in 1887 a meeting in support of technical education at which Huxley and other advocates of technical education were present.

11. Mace, *Royal Institute of British Architects,* pp. 189–190.
12. Power, "Arts and Crafts to Monumental Classic"; and Saint, *Image of the Architect,* chap. 3. Crinson and Lubbock, *Architecture: Art or Profession?* chap. 2, point out that many architects regarded knowledge of the building crafts as the key to aesthetics, rather than academic training in the principles of construction. Craft knowledge was thus regarded as superior to science.
13. Cole regarded army engineers as being more amenable to taking orders. Fowke's notable buildings included the 1862 International Exhibition Building and the Science Schools building of the South Kensington Museum, now known as the Henry Cole Wing of the Victoria and Albert Museum.
14. C. S. Peach, FRIBA, for example, suffered such a humiliation when giving evidence to a House of Commons Committee on how to measure vibration in buildings containing moving machinery; Forgan and Gooday, "A Fungoid Assemblage," pp. 179–180.
15. The following account is derived from the minute books of the Standing Science Committee, RIBA Archives, London.
16. The Auditorium Building was itself a technologically very advanced building, though Adler was unencouraging about the possibilities of heating buildings from a central station; *RIBA Minutes of the Standing Science Committee,* 1, pp. 279–281. Nonetheless the Committee spent some time trying to find an American engineer who would speak on this topic.
17. Smith was regarded as a sensible choice, a man "not of the highest genius, but with a great capacity for taking pains"; obituary in *Architect and Contract Reporter* (20 March 1903), p. 187. He had a genuine interest in the history and science of architecture and was author of numerous historical works. His scientific interests focused on architectural acoustics, on which he published *A Rudimentary Treatise on the Acoustics of Public Buildings* (1861, with subsequent editions ca. 1872, 1895).
18. Smith, "New Science Laboratories," p. 303.
19. Furthermore Statham pointed out that the center of the portico was not in line with the center of the gateway, nor with the central door of University College Hospital (an Alfred Waterhouse building), as may be seen today.
20. Smith, "New Science Laboratories," p. 301.
21. One may remember that cupolas had a long association with learning and ideal museums; Fabianski, "Iconography," pp. 95–134. C. Forster Hayward suggested that if carried out, the proposed plan would destroy a very important part of the architecture of London; Smith, "New Science Laboratories," p. 305.
22. Another architect, William Woodward, supported this view, arguing that University College London would thus achieve "a proper collegiate quadrangle." Ibid., p. 304.
23. The scientists got just about everything they wanted: insulation from the city's noise with double-glazed windows but direct connection with urban networks of electrical power. It should, however, be pointed out that Smith's plan was not immediately completed, and for lack of resources most of the Gower Street frontage was only built to first-floor level. The old lodges were not demolished and remained to be reconstructed in the final closing of the quadrangle by Hugh Casson in the early 1980s; Harte and North, *The World of University College London,* p. 97.
24. Street and Clarke, "Brickwork Tests: First Series" and "Second Series."

25. "Chronicle," p. 282.
26. *RIBA Minutes of the Science Standing Committee,* 2 (2 March 1899), p. 43. See also note of the committee's meeting on 8 June 1899, p. 50.
27. *RIBA Minutes of the Science Standing Committee,* 2 (2 March 1899), p. 43.
28. *RIBA, Report on Brickwork Tests,* 1905. This report included almost all the previous material, with no mention of who the authors were. Unwin is only mentioned once, as a member of the subcommittee, p. 1. The main author, Max Clarke, was forced to admit that the bricklayers had used some of the wrong bricks when they found the Leicester Reds and Staffordshire Blues very hard to cut, which meant that some of the test brickwork piers had to be rebuilt: "This is an unpleasant admission to make, but, if the experiments are to be of any value, everything in connection with them must be explained." P. 3.
29. Pickstone, "Museological Science?"
30. Papworth and Papworth, *Museums, Libraries and Picture Galleries.*
31. For a general overview of the role of scientific museums in the nineteenth century, see Forgan, "The Architecture of Display."
32. "Report of the Committee on the Provincial Museums," 1887, p. 119, and (1888), p. 126; emphasis in original.
33. Ibid. 1887, pp. 121, 127.
34. Papworth and Papworth, *Museums, Libraries and Picture Galleries,* pp. 58, 64–65, pls. 5, 6.
35. Owen, "Presidential Address," p. 660.
36. Ibid., p. 660.
37. If one follows the analysis of Nicolaas Rupke, one could also argue that Owen introduced into his architectural language the sort of duality between Cuvierian functionalism and transcendental morphology that characterized his oeuvre; Rupke, *Richard Owen,* chaps. 3, 4.
38. For an architectural analysis and description of the way the plan of the building developed, see Girouard, *Waterhouse and the Natural History Museum;* and Cunningham and Waterhouse, *Alfred Waterhouse,* chap. 9.
39. Sheppard, *Survey of London,* vol. 29, pp. 272–274, vol. 30, pp. 574–575, pls. 46–48; Tyack, *Sir James Pennethorne,* pp. 178–191.
40. Hunt, *A Descriptive Guide,* p. 1.
41. Becker, *Scientific London,* p. 251. There are parallels here with Stocking's discussion (this volume) on the argument between Mason and Boas over the arrangement of the ethnographic specimens in the U.S. National Museum. Mason justified his arrangement on grounds of needing to see relevant specimens "in juxtaposition" to each other. This was a familiar argument in mid–Victorian England, where "juxtaposition" was a heavily used term; see Forgan, "'But indifferently lodged . . . .'"
42. Ramsay, *A Descriptive Catalogue.* After its initial publication in 1858, this catalogue ran to a third edition by 1862.
43. Hunt, *A Descriptive Guide,* p. 10.
44. Becker, *Scientific London,* p. 251; but see also *Museum of Practical Geology: A Hand-Book Guide,* 1851: and Hunt, *A Descriptive Guide,* 1857.
45. *Museum of Practical Geology: A Hand-Book Guide,* p. 6.
46. On Huxley's appointment and move into paleontology, see Desmond, *Huxley: The Devil's Disciple,* pp. 218–219.

47. Rudwick, *Great Devonian Controversy;* and Secord, *Controversy in Victorian Geology.*
48. The term "Cambrian" tended to be used as little as possible while Murchison was Director of the Survey, no doubt because the term was Sedgwick's.
49. Ramsay, *Descriptive Catalogue,* p. ix.
50. The Survey also published geological sections, though not as many as the maps published. See Flett, *First Hundred Years,* p. 45.
51. Huxley and Etheridge, *A Catalogue,* p. ix.
52. Ibid., p. viii.
53. Jevons, "Use and Abuse," p. 62.
54. Select Committee on the Museums of the Science and Art Department, *Parliamentary Papers,* First Report, p. iii.
55. Ibid., First Report, p. xxvii.
56. Ibid., p. 39, evidence of Donnelly.
57. Ibid., p. 64, evidence of Judd.
58. There was, however, by this time a "limited but instructive" series of characteristic fossils in British strata, but it was not a very prominent display and was located behind the staircase on the ground floors; *General Guide,* p. 21.
59. Select Committee on the Museums of the Science and Art Department, *Parliamentary Papers,* First Report, p. 40, evidence of Donnelly.
60. For example, the evidence of Professor Hull, Director of the Irish Geological Survey, ibid. pp. 174–177.
61. For example, the arrangement of the minerals in the two museums differed. In the Jermyn Street Museum, the arrangement focused on economic utility and geological order, whereas in the Natural History Museum the arrangement was arguably "biographical": Minerals were arranged in their chemical families, such as oxides, carbonates, sulfides, and so on. My thanks to John Thackray for this information and for sharing with me his knowledge of the Museums.
62. *RIBA Minutes of the Science Standing Committee,* 2, pp. 151–152. The tests on weathering of stones started after much consultation in 1909.

Bibliography

Becker, Henry. *Scientific London.* London: Henry S. King, 1874.
Brock, W. H. "Building England's First Technical College: The Laboratories of Finsbury Technical College 1878–1926." In Frank A. J. L. James, ed., *The Development of the Laboratory,* 153–170. London: Macmillan, 1989.
"Chronicle." *Journal of the Royal Institute of British Architects* 6 (1899): 281–283.
Crinson, Mark, and Jules Lubbock. *Architecture: Art or Profession? Three Hundred Years of Architectural Education in Britain.* Manchester, UK: Manchester University Press, 1994.
Cunningham, Colin, and Prudence Waterhouse. *Alfred Waterhouse 1830–1905: Biography of a Practice.* Oxford, UK: Clarendon Press, 1992.
Desmond, Adrian. *Huxley: The Devil's Disciple.* London: Michael Joseph, 1994.
Fabianski, M. "Iconography of the Architecture of Ideal Museaea in the Fifteenth to Eighteenth Centuries." *Journal of the History of Collections* 2 (1990): 95–134.

Flett, Sir John Smith. *The First Hundred Years of the Geological Survey of Great Britain.* London: HMSO, 1937.

Forgan, Sophie. "The Architecture of Display: Museums, Universities and Objects in Nineteenth-Century Britain." *History of Science* 32 (June 1994): 139–162.

Forgan, Sophie. "'But indifferently lodged . . .': Perception and Place in Building for Science in Victorian London." In Crosbie W. Smith and Jon Agar, eds., *Making Space: Territorial Themes in the History of Science,* 195–214. London: Macmillan, 1998.

Forgan, Sophie, and Graeme Gooday. "'A Fungoid Assemblage of Buildings': Diversity and Adversity in the Development of College Architecture and Scientific Education in Nineteenth Century South Kensington." *History of Universities* 14 (1994–95): 153–192.

General Guide to the British Museum, Natural History, Cromwell Road, London SW. London: Trustees of the Museum, 1891.

Girouard, Mark. *Alfred Waterhouse and the Natural History Museum.* London: British Museum (Natural History), 1981.

Harte, Negley, and John North. *The World of University College London 1828–1978.* London: University College London, 1978.

Hunt, Robert. *A Descriptive Guide to the Museum of Practical Geology, etc.* London: HMSO, 1857.

Huxley, T. H., and R. Etheridge. *A Catalogue of the Collection of Fossils in the Museum of Practical Geology.* London: HMSO, 1865.

Jevons, W. Stanley. "The Use and Abuse of Museums." In *Methods of Social Reform,* 53–81. London: Macmillan, 1883.

Mace, Angela. *The Royal Institute of British Architects: A Guide to Its Archives and History.* London: Mansell Publishing, 1986.

Morrell, Jack, and Arnold Thackray. *Gentlemen of Science: Early Years of the British Association for the Advancement of Science.* Oxford, UK: Clarendon Press, 1981.

The Museum of Practical Geology: A Hand-Book Guide for Visitors. London: H. G. Clarke & Co., 1851.

Owen, Richard. "Presidential Address to Section D, Biology." *Report of the British Association for the Advancement of Science* (1881): 651–661.

Papworth, John W., and Wyatt Papworth. *Museums, Libraries and Picture Galleries, Public and Private: Their Establishment, Formation, Arrangement, and Architectural Construction etc.* London: Chapman and Hall, 1853.

Pickstone, John V. "Museological Science? The Place of the Analytical/Comparative in Nineteenth-Century Science, Technology and Medicine." *History of Science* 32 (1994): 111–138.

Power, A. "Arts and Crafts to Monumental Classic: The Institutionalisation of Architectural Education, 1900–1914." In N. Bingham, ed., *The Education of the Architect,* 34–38. Edinburgh: Proceedings of the 22nd Annual Symposium, Society of Architectural Historians of Great Britain, 1993.

Ramsay, A. C. *A Descriptive Catalogue of the Rock Specimens in the Museum of Practical Geology.* London: HMSO, 1859.

"Report of the Committee on the Provincial Museums of the United Kingdom." *Report of the British Association for the Advancement of Science* (1887): 97–130; and (1888): 124–132.

RIBA, *Minutes of the Standing Science Committee,* 1 (1886–1896), 2 (1896–1914). Archives of the Royal Institute of British Architects, London.

RIBA. *Report on Brickwork Tests Conducted by a Subcommittee of the Science Standing Committee of the Royal Institute of British Architects.* London: RIBA, 1905.

Robins, Edward Cookworthy. *Technical School and College Building.* London: Whittaker and Co., 1887.

Rudwick, Martin J. S. *The Great Devonian Controversy: The Shaping of Scientific Knowledge among Gentlemanly Specialists.* Chicago: University of Chicago Press, 1985.

Rupke, Nicolaas. *Richard Owen: Victorian Naturalist.* New Haven, CT: Yale University Press, 1994.

Russell, Colin. *Science and Social Change 1700–1900.* London: Macmillan, 1983.

Saint, Andrew. *The Image of the Architect.* New Haven, CT: Yale University Press, 1983.

Secord, Anne. "Science in the Pub: Artisan Botanists in Early Nineteenth Century Lancashire." *History of Science* 32 (1994): 269–315.

Secord, James A. *Controversy in Victorian Geology: The Cambrian-Silurian Dispute.* Princeton, NJ: Princeton University Press, 1986.

Select Committee on the Museums of the Science and Art Department. *Parliamentary Papers: Reports from Committees.* Vol. XI. London: HMSO, 1898.

Shaw, R. Norman and T. G. Jackson, eds. *Architecture, A Profession or an Art?* London: John Murray, 1892.

Sheppard, F. H. W., ed. *Survey of London: Parish of St. James, Westminster.* Vols. 29, 30. London: Athlone/GLC, 1960.

Smith, Professor T. Roger, et al. "The New Science Laboratories at University College London." *Journal of the Royal Institute of British Architects* 1 (1894): 281–308.

Street, W., and M. Clarke. "Brickwork Tests: Report on the First Series of Experiments." *Journal of the Royal Institute of British Architects* 3 (1896): 333–358; "Report on the Second Series of Experiments." *Journal of the Royal Institute of British Architects* 4 (1897): 73–103, 121–128.

Tyack, Geoffrey. *Sir James Pennethorne and the Making of Victorian London.* Cambridge: Cambridge University Press, 1992.

III

Modern Space

9

"Spatial Mechanics": Scientific Metaphors in Architecture

Adrian Forty

There are various ways to look at architecture's relationship to science, through objects, through practices, or through discourse. I want here to approach the relationship somewhat obliquely, through some aspects of the spoken and written language in which people have articulated their thoughts about architecture. Much of what is said about architecture, particularly about its aesthetic interest, is said (and can only be said) in metaphor; and when we ask where those metaphors come from, it is striking how many of the metaphors in the lexicon of day-to-day architectural speech have been drawn from science. Although architecture is hardly alone in its dependency upon scientific metaphors, and the reasons for this may indeed be obvious, insofar as science has become the dominant discourse of our times, we should not assume that a scientific term, just because it comes from science, is a successful metaphor for architecture. My purpose here, therefore, is to see whether, by asking what makes a good metaphor, we might learn anything about the relationship between architecture and science.

My first example is "circulation," in contemporary architectural parlance the conventional description for the means of movement, particularly human movement, within or around a building. This term—unquestionably a metaphor, drawn from physiology, as the early instances of its use confirm—has become virtually indispensable to the way we have come to think and talk about architecture: Indeed, one may recall Le Corbusier's "outrageous fundamental proposition: *architecture is circulation.*"[1] Yet although "circulation" might now appear to have an objective existence as a category within architecture, we must acknowledge that the term, used in its modern sense, was unknown before the second half of the nineteenth century. The first instance I know of "circulation" used to refer to the means for human movement within a building was by Viollet-le-Duc in the second volume of his *Lectures,* published in 1872, and although I am sure this was not the first occasion on which the word was used in this sense, I am also sure that it was not current before 1850. The appropriation of "circulation" as an architectural term coincides closely with the adoption during the latter half of the nineteenth century of most of the other scientific metaphors that have become commonplace in architectural vocabulary, terms such as "function," "structure," and others I discuss later.

The slightly surprising discovery that "circulation," now seemingly so indispensable to architectural vocabulary, was not part of architectural discourse prior to the 1850s at once raises two questions, the first of which is why it entered the lexicon just then, neither earlier, nor later? As a description for the movement of blood around the body, "circulation" was coined by Sir William Harvey in 1628 (see figure 9.1). Its potential as a metaphor to describe flow in other substances was seized upon almost at once by other disciplines, notably economics, where it was in general use by the second half of the seventeenth century; yet in architecture, despite its availability, the metaphor was not taken up for two and a half centuries. Why not? Part of the answer to this lies in the second and related question of whether, when architects and critics started using "circulation," they were simply renaming something already known by another name, or whether they were using it to articulate a genuinely new and previously nonexistent concept?

In considering our answers to these questions, we should bear in mind that in French, the language in which the metaphor was first applied to architecture, *circulation* also means vehicular traffic. That particular sense of the word appears to have been introduced, also as a metaphor, rather earlier, in the late eighteenth century. The first instance I know of its use is in Pierre Patte's *Mémoires sur les objects les plus importans de l'architecture* of 1769, in which he refers to *la libre circulation des voitures,*[2] but apparently this sense did not become widespread in French until the 1820s.[3] When French architects and critics started in the 1850s to use "circulation" in relation to buildings, they undoubtedly drew on this earlier meaning of the word, giving it a double signification.

To discover whether the concept of circulation existed previously, but by another name, we might look to the earlier architectural writers who could be expected to testify to some awareness of it. Of the various candidates, the most eligible witness must surely be J.-N.-L. Durand, whose *Précis des leçons d'architecture* was first published in 1802. Durand in fact only uses the word "circulation" once, in the most insignificant context: "an isolated support should generally be cylindrical, the form which best facilitates circulation."[4] Like earlier writers, Durand focused his—and his students'—attention upon the *distribution,* that is to say the arrangement of volumes so as to preserve the axes of

9.1

The Human Circulation System. From Pierre Larousse, *Grand Dictionnaire Universel du XIXme Siècle*, Paris: Larousse, 1869, vol. 2, p. 330, "Circulation." As a metaphor for human movement within buildings, this image did not enter currency until the 1850s.

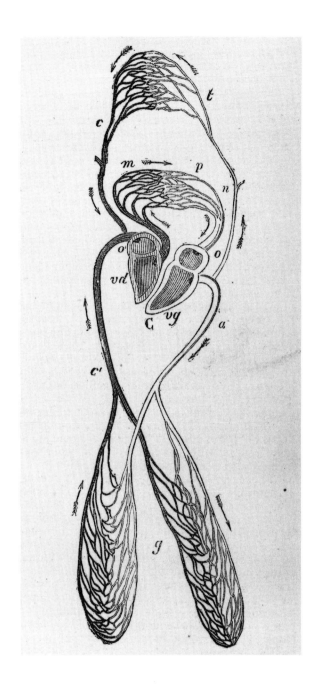

the plan, the *communications* between rooms and parts of a building, and the *dégagements,* or service communications. For Durand, *distribution* was the most important business of architecture, and there is no evidence that he ever considered the system of human movement around a building independently, or thought of it as having any particular significance for the design. If we go back a little farther, to the eighteenth century, we find evidence that a series of interconnecting rooms might permit people to "circulate"—Boullée commented on Versailles that "the public can circulate easily throughout the first part of the palace"[5]—but there is no evidence that architects ever thought of this arrangement as constituting a system of circulation. Although modern historians have described the interconnecting entertainment rooms of eighteenth-century aristocratic town and country houses as forming a "circuit," I am not aware that anyone at the time ever used this word or thought about it in this way: Contemporary accounts describe each room along the route in turn but do not make the leap to describe the entire route as a system of its own that could be considered apart from the individual rooms and staircases constituting the experience.[6] To call this arrangement a "circuit" is to superimpose a modern concept upon it. And if we go back even earlier, although we can certainly find physiological metaphors used to describe the relationship of different parts of buildings, they do not suggest that human movement was considered a separate component of the architecture. Thus in 1615, in a distinctly pre-Harveian metaphor, Venetian architect Vincenzo Scamozzi described staircases: "Of all the parts without doubt the stairs are the most necessary in buildings, like the veins and arteries in the human body; because just as these serve naturally to administer the blood to every part of the body, so do the principal stairs and the secret stairs reach to the most intimate parts of the building."[7]

If none of these examples suggest that architects before 1850 had an equivalent to the modern concept of "circulation," the point may be reinforced by looking at the distinctive use of the metaphor after 1850. One of the very first and most interesting post-1850 instances of the metaphor was its use by French critic César Daly in an 1857 article on Barry's Reform Club in London, a work that Daly regarded as prefiguring the architecture of the future (see figure 9.2): "This building," Daly wrote, "is no inert mass of stone, brick and iron; it is almost a living body with its own nervous system and cardiovascular circulation system."[8] Daly was referring not so much to the means for human movement, but to the largely invisible heating, ventilation, and mechanical communication systems buried within the walls. In Daly's demonstratively physiological metaphor, he suggested that each of these services was a system of its own, and could be considered quite apart from the materiality of the building it served: This sets his use of the metaphor apart from those earlier conceptions of architectural arrangement. Similarly, Viollet-le-Duc's use of the circulation metaphor in his remarks on domestic architecture is interesting in this light (as it is also for his use of the term "function" within the same passage):

9.2

Charles Barry, The Reform Club, London, 1839–41. Copyright Adrian Forty. Beneath the inert exterior of Barry's Reform Club, César Daly perceived a "living body with its own cardiovascular system."

There is in every building, I may say, one principal organ—one dominant part—and certain secondary orders or members, and the necessary appliances for supplying all these parts, by a system of circulation. Each of these organs has its own function; but it ought to be connected with the whole body in proportion to its requirements.[9]

For Viollet, "circulation" was a very fresh metaphor, and like Daly, he pointedly drew attention to its physiological origins. Also like Daly, he used the metaphor to stress the extent to which the circulation might be considered as a system of its own, independently of the house's other organs. Daly and Viollet were, it seems to me, expressing in "circulation" something not found in all those earlier terms for describing the arrangement of buildings. "*Distribution,*" "*communication,*" "*dégagements*" are all tied to the materiality of architecture: You have to have a building in view, or at least in mind, to understand what they mean. With "circulation," though, you do not, for what it describes refers not to the building's physicality, but to the possibility of flow within or about it. "Circulation" is special in that it describes not just an arrangement of parts, but as Daly stressed in his description of the Reform Club, a complete, self-contained system that can be considered independently of the building's inert, physical substance. This is indeed the way in which "circulation" was taken up and developed by architectural theorists and educators, so, for example, the fifth edition (1909) of Julien Guadet's *Eléments et Théories d'Architecture,* a classic text of the French Beaux Arts education, had a whole chapter on "Les circulations," dealing with them as an independent category within architectural composition. From here, it was but a short step to Le Corbusier's "outrageous proposition."

If, as I have suggested, the significance of "circulation" was to allow an aspect of architecture to be considered as a discrete system, its introduction must be seen as a symptom of the desire to bring scientific method into architecture. For architecture to approximate a scientific practice it had to be able to isolate and abstract specific features or properties from the complex phenomenal reality of the built work and to subject those abstractions to independent analysis. The concept covered by "circulation" fulfilled this criterion relatively well, and it is perhaps no surprise therefore that it has remained a favorite subject for architectural research.[10] But although we may say in general terms that introducing "circulation" had the effect of allowing architecture to proceed in a scientific manner, what exactly did those who coined the metaphor hope to express by it? It is in the nature of metaphors that they allow one to be precise about some things while remaining vague about others: "Circulation" was precise about presenting an aspect of architecture as a discrete system but vague about what flowed around that system: for Daly it was warm air and mechanical services, for Viollet it was people. As an instance of the uncertainty as to what sort of flow "circulation" could describe, we might look at what another writer, German art historian Paul Frankl, had to say about "flow" in his *Principles of Architectural History,* first published in 1914. Frankl insisted not only that human movement in buildings was bloodlike, but that knowledge of the system as a whole was necessary to understand a work of architecture:

> To understand a secular building we must get to know it as a whole by walking through it from end to end, from cellar to roof, through all its outstretching wings. The entrance, the vestibule or passage leading to courtyard or stair, the connections between several courtyards, the stairs themselves and the corridors leading away from them at each level, like the veins of our bodies—these are the pulsating arteries of a building. They are the passages that form the fixed circulation leading to individual rooms, to individual chambers, cells or loges. The organism of the house reaches as far as these arteries guide the circulation.[11]

However, physical movement, whether of people, or of things, or of energy—be it electrical currents or sound waves—is not the only sort of motion to be found in architecture. As well as the actual physical movement of matter or energy, there is also the perceived bodily movement through which we experience architecture. Indeed Merleau-Ponty argued that our ability to extend in imagination the body's power of movement lies at the origin of all knowledge: "motility in its pure state possesses the basic power of giving a meaning."[12] And more specifically, Edmund Husserl, in an article written a few years before Frankl's book, had argued that knowledge of space came about through the sense of movement present in the unmoving subject: "All spatiality is constituted through movement, the movement of the object itself and the movement of the 'I.'"[13] Frankl, in the *Principles of Architectural History,* was just as aware of movement as the agency of perception as he was of it as a functional property of architecture. There is in certain buildings, as he puts it, a "great flood of movement that urges us round and through the building."[14] In other words, movement, so to speak, is there in the building before anyone has arrived, and it is not the people who move, but space itself that goes round and round. Frankl was aware that he was dealing with two categories of circulation—one having to do with the experience of spatiality, the other with routes of human activity—and generally, but not always, he used the term "flow" to distinguish perceived spatial movement from the physical paths taken by people.[15] We should not forget that "circulation" may refer to more than one kind of movement in architecture.

I shall say at this point that I am not at all convinced that "circulation" is the most appropriate metaphor for human movement in buildings. When Sir William Harvey originally applied the term "circulation" to the flow of blood, he did so to draw attention to the fact that blood did not simply *move* in the body, but that a single fixed volume of fluid (not, as had previously been thought, two sorts of fluid) traveled around the body and always returned to the same point, the heart.[16] This does not correspond to what normally goes on in buildings: It is not always the same group of people who move about in a building, not all of them generally pass into all parts of it, nor (except perhaps for the guests at social gatherings in eighteenth-century townhouses) do they go round and round it always returning to the same point. On the other hand, it is quite a good description for the sort of spatial movement with which Frankl was concerned. The lack of correspondence between the circulation of the blood in the body and the movement of people in buildings does not seem to have stood in the way of the success of "circulation" as a metaphor, so it must have some other attraction. The most likely explanation for its appeal is to be found in the feature common to the use made of it by Daly, Viollet, and Frankl, all of whom invite the reader to think of the building as a sealed system, without orifices, and self-sufficient—in other words, a body, but a body perceived in the most clinical and unmetaphysical terms. For those many architects and building owners who persist, against all the evidence to the contrary, in seeing buildings as strictly bounded, self-contained entities, a metaphor reinforcing this delusion could not fail to be attractive.[17] It is quite possible, though, to think of other physiological metaphors that could express human movement in buildings with a rather closer correspondence to what goes on in them:—think, say, of "respiration," "breathing"—and interestingly the arrangement of Le Corbusier's Olivetti project of 1957 is much more like a respiratory system than a circulation system. (See figure 9.3.) But "respiration" has not caught on, I suspect because it would make buildings into open systems with indistinct boundaries, a prospect altogether too messy and too disturbing for most architects and building owners to want to be troubled with.

"Circulation" has without question been a very successful metaphor; indeed, I would say far too successful. What started as an innocent metaphor is now perceived as a fixed category: Architectural textbooks from late modernism took it for granted that "circulation" was a factor in the design of buildings,[18] and still today critics talk about it as if it were an absolute, objective property of architecture.[19] So deeply ingrained has this essentially modernist category become that for most of us it requires a positive act of mental effort to think about architecture *without* "circulation." When we ask to what we may attribute the success of this particular metaphor, it is clear that it has nothing to do with the exactness of any correspondence between the flow of substances around bodies and buildings. Instead, I suggest, its success is due to two more structural reasons: Firstly, it made architecture amenable to scientific method, and secondly, it satisfies a wish to see buildings as enclosed, self-contained systems against all the evidence to the contrary. It allows people to talk about what is untrue as if it were true and remain undisturbed by the contradiction.

The second set of scientific metaphors I want to discuss come from mechanics, both fluid and static. I am thinking of terms like compression, stress, tension, torsion, shear, equilibrium, centrifugal, centripetal, and so on. It is the use of these terms to describe not the actual stability of buildings, but their formal and spatial interest that concerns me. When I first started reading about architecture, I remember thinking it odd that terms from the most material of sciences, mechanics, should be chosen for the least material aspects of architecture, its spatiality, and I still find it strange.

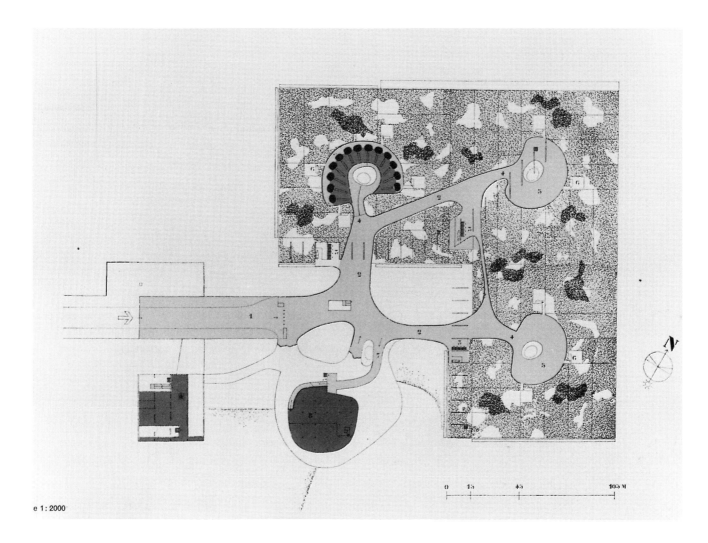

9.3

Le Corbusier, design for Olivetti electronic calculator plant at Rho-Milan, fourth- and fifth-floor plan. This project might better be described as based not on a *circulation* system, but upon a *respiration* system. From W. Boesiger, ed., *Le Corbusier: Oeuvre Complète 1957–1965,* London: Thames and Hudson, 1965, p. 123. © FLC/ADAGP, Paris and DACS, London, 1998.

9.4

Le Corbusier, Monastery Sainte-Marie de La Tourette, Evreux-sur-l'Arbresle, France, completed 1959. North facade, showing flank wall of chapel. Photo by Tim Benton. Reproduced by permission of Tim Benton. The oblique line of the cornice, which the eye normally "corrects" to the horizontal, inspired Colin Rowe's reading of the facade as rotating within a "tensile equilibrium."

These mechanical metaphors originated in German aesthetics, where there was tradition going back to Hegel and Schopenhauer of seeing architecture as the expression of its resistance to the force of gravity.[20] Later in the nineteenth century several writers developed this theme, and so, for example, the German aesthetic philosopher Robert Vischer argued that "art finds its highest goal in depicting a moving conflict of forces."[21] Undoubtedly the most imaginative exponent of the idea was Heinrich Wölfflin, who, in his doctoral thesis, "Prolegomena to a Psychology of Architecture" (1886), developed a theory of the empathetic experience of architecture, a theory he then applied in *Renaissance and Baroque* (1888) to the architecture of a particular place and time. Wölfflin's account of architecture in *Renaissance and Baroque* rests largely on "movement," that is to say the implied movement arising from "the opposition between matter and the force of form"[22] within things that are not themselves in motion. Although this could just as well be a description of the science of statics, we should not assume that is why these metaphors hold our interest. Wölfflin was not interested in the actual forces in the building, but in how the architecture gave the observer the sensation of feeling the compression on the columns, the thrust of the vaults, and so on. It was the way the architecture communicated these experiences to the viewer that mattered to him, rather than how the forces were actually transmitted through the structure of the building. In these descriptions, Wölfflin undoubtedly benefited from the use of the same set of metaphors—tension, stress, and so forth—that psychology had already appropriated to describe emotions, and indeed Wölfflin often exploited the coincidence, in remarks like "The baroque never offers us . . . the static calm of 'being,' only the unrest of change and the tension of transience."[23] I am suggesting here that the success of mechanics as a source of metaphors for the aesthetic experience of architecture has less to do with any direct correspondence it might have with architecture's tectonic aspects than with the availability of these same metaphors to describe states of feeling and emotion in the human subject.

Wölfflin's successor Frank took over the same set of metaphors, and his book extended them to the description of architectural space. By the 1950s, these metaphors from mechanics seem to have become sufficiently familiar for critic Colin Rowe to be able to assume that when he referred to "the spatial mechanics" of Le Corbusier's monastery of La Tourette, his readers would know what he meant.[24] (See figure 9.4.) Rowe's own writing made full play of these metaphors. He offers a wonderful description of the effect of the north side of the chapel at La Tourette, the first view the visitor has of the building, a description that shows us more than any image possibly could:

> Le Corbusier has . . . built into this frontal plane a depth which by no means exists in reality. The oblique cut of his parapet should now be noticed. It is a line so slightly out of the horizontal that the eye has an instinctive tendency to "correct" and translate it for what average experience suggests that it should be. For, being eager to see it as the normal termination of a vertical plane, the eye is consequently

willing to read it, not as the diagonal which physically it happens to be, but as the element in a perspective recession which psychologically it seems. Le Corbusier has established a "false right angle"; and this *fausse équerre,* which in itself infers depth, may also be seen as sporadically collaborating with the slope of the ground further to sponsor an intermittent illusion that the building is revolving.

Something of the vital animation of surface, the small but sudden tremor of mobility, in the area between bastion and belfry certainly derives from the torsion to which the wall is thus subjected; but, if this phenomenal warping of surface may be distinctly assisted by the real flexions of the bastion wall itself, then at this point it should also be observed how the three *canons à lumière* now introduce a counteractive stress.

For the spectacle of the building as seen on arrival is finally predicated on a basis, not of one spiral, but of two. One the one hand there are the pseudoorthogonals which, by the complement they provide to the genuine recession of the monastery's west facade, serve to stimulate an illusion of rotating and spinning. But, on the other, are those three, twisting, writhing, and even agonized light sources—they illuminate the Chapel of the Holy Sacrament—which cause a quite independent and equally powerful moment of convolution. A pictorial opportunism lies behind the one tendency. A sculptural opportunism lies behind the other. There is a spiral in two dimensions. There is a contradictory spiral in three. A corkscrew is in competition with a restlessly deflective plane. Their equivocal interplay makes the building. And, since the coiled, columnar vortex, implied by the space rising above the chapel, is a volume which, like all vortices, has the cyclonic power to suck the less energetic material in towards its axis of excitement, so the three *canons à lumière* conspire with the elements guaranteeing hallucination to act as a kind of tether securing a tensile equilibrium.[25]

In this remarkable description, in which the metaphor moves from static mechanics to fluid mechanics and back again to statics, the whole effect comes about through Rowe's allowing himself to see "voids . . . act as solids."[26] The La Tourette article is the most phenomenological of Rowe's criticism; in it, he is almost entirely concerned with the sensations experienced by the viewer as a result of the various optical illusions, but despite the great range of the metaphors, I think that their success comes from their ambiguity in referring both to psychology and to statics.

When, however, we turn to another exponent of spatial mechanics, Colin Rowe's one-time student Peter Eisenman, we find them used rather differently. Whereas for Rowe the metaphors describe what he could see, Eisenman, in his article on Terragni published in 1971, explicitly used them to describe what he could *not* see.[27] (See figure 9.5.) Eisenman distinguished between objects' sensual qualities—surface, texture, color, and shape—and a "deep aspect concerned with conceptual relationships which are not

9.5

Giuseppe Terragni, Casa Giuliani Frigerio, Como, Italy, 1939–40, North Elevation. From A. Sartoris, *Encyclopédie de l'architecture nouvelle,* Vol. 1. Milan: U. Hoepli, 1955. Reproduced by permission of the Royal Institute of British Architects, London. Peter Eisenman's analysis of this apartment block is one of the most elaborate investigations of a building's composition to be made through analogy with static mechanics.

sensually perceived; such as frontality, obliqueness, recession, elongation, compression and shear, which are understood in the mind."[28] An example of his application of these terms, which he does not acknowledge to be metaphors but sees as literal descriptions of the relationships of parts of buildings, is his analysis of the north facade of the Casa Giuliani Frigerio. (See figure 9.6.) Referring to the forward projection of the left-hand part of the facade, he writes:

> This volumetric extension seems purposely conceived as an element which does not carry across the entire facade, in order to create a condition of shear. This condition allows a dual reading: either the facade has been extended, in an additive manner, or the outer edge has been eroded to reveal an internal "solid" volume.[29]

The "shear" of which Eisenman writes is not a sensation experienced by the observer, nor is Eisenman interested in that possibility. On the contrary, it is something that belongs in the structure of the architecture, and it really takes Eisenman's analytical diagrams to understand it. The absence of any psychological dimension to his use of these terms, his entirely formalist interpretation of them, may have something to do with why Eisenman soon abandoned this kind of analysis.

To draw all this together, I have been pursuing so far the argument that the success of the most common scientific metaphors has not to do just with their being scientific, but with their reinforcing certain other perceptions of architecture, perceptions that may be rooted in social or psychological desires. The question I would like to reflect on now is what these, and the abundance of other scientific metaphors, can tell us about the relationship between architecture and science? On the face of it, architecture's adoption of words like "function," "structure," "circulation," or "transformation" would seem simply to be a symptom of a trend toward the scientization of the practice, a general characteristic of modernism. However, if we consider the matter in terms of what is known not about architecture, but about metaphors, the relationship turns out to be rather less straightforward. Although, as I have suggested, part of the attraction of these metaphors may have lain in their making architecture *seem* like a science, and so amenable to scientific procedures of analysis, what they really do, it seems to me, is—paradoxically—to confirm the opposite, that architecture is *not* a science, and indeed is not particularly like a science. Successful metaphors rely on the unlikeness of things, not upon their likenesses. An effective metaphor borrows an image from one schema of ideas and applies it to another, previously unrelated schema. As Nelson Goodman—in what is itself a wonderful metaphor—writes, a metaphor is "an expedition abroad," from one realm of thought to another; or, he says, it may be regarded as a "calculated category mistake."[30] But to be able to make that category mistake, there has to be a category distinction in place to begin with. To call a work of architecture "functional," which is unquestionably a metaphor, relies on an initial assumption that architectural objects are not natural organisms, nor mathematical equations, even though it may express a wish to see them as

9.6

Peter Eisenman, Analytical Isometric Diagrams of Casa Giuliani Frigerio, from *Perspecta* 13/14 (1971): 46, nos. 13, 14, 15, 16. Reproduced by permission of Peter Eisenman. One of Peter Eisenman's many drawings demonstrating the apparent static forces generated with the composition.

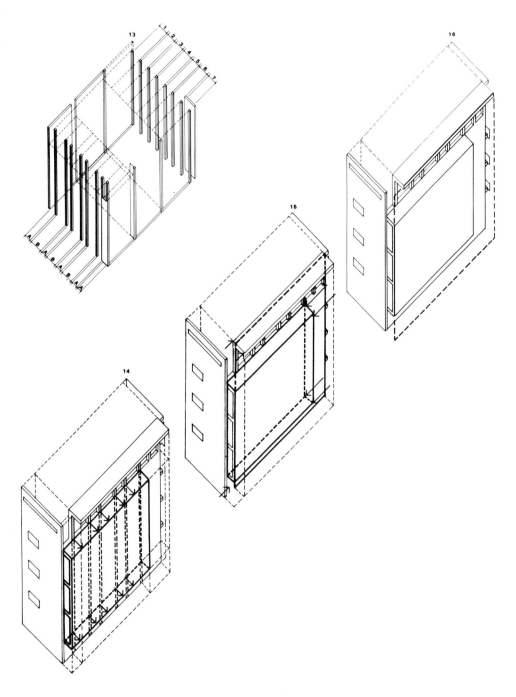

one or other of those things. The success of "functional" relies upon commonly accepted agreement that architecture is different from both biology and mathematics. And furthermore, since the scientific metaphors employed in architecture are drawn from such a diversity of scientific fields, from natural sciences as well as physical sciences and mathematics, the cumulative effect of their use is to suggest the unlikeness of architecture to science in general.

The most obvious thing we can say about these scientific metaphors in historical terms is that they belong exclusively to the modern era. Although the science of mechanics and the circulation of the blood have been known about since the seventeenth century and were therefore potentially available as metaphors all that time, nobody seems to have been interested in applying them to architecture until a century and a half ago. Why not? Those in other practices were: Economists took up "circulation" almost as soon as Harvey had discovered it in the body, and terms like "stress" and "tension" from mechanics featured in early-nineteenth-century psychology, so why not in architecture? There are two alternative sorts of answers to this question: Either, as I have already suggested was the case with "circulation," these metaphors expressed something that had not needed to be said before; or there was some obstacle to their use as architectural metaphors. Let us concentrate for a moment on the second possibility, that there may have been some structural reason why they could not be applied to architecture.

If indeed, as Alberto Pérez-Gómez suggests,[31] there was no conceptual distinction between science and architecture prior to the late eighteenth century, it is not surprising that no one was interested in scientific metaphors, for they would have been entirely ineffective. Only when science became a field of knowledge separate from architecture would there have been any appeal in seeing architecture as if it was a science. Metaphors are experiments with the possible likenesses of unlike things. Each one of the countless scientific metaphors in twentieth-century architecture is a little experiment, an attempt to find a relationship between architecture and one or another branch of science, but they all rely on our belief that really, at the bottom, architectural practice is not scientific. Were this epistemological separation ever to end and some kind of a reparation between architecture and science to occur, one sure sign that it had happened would be for every scientific metaphor to shrivel up and be discarded, and for no new ones to be invented. But until that happens, every scientific metaphor in architecture will continue to take its sustenance from our intransigent conviction, however misguided, that architecture cannot reach where science may.

NOTES

1. Le Corbusier, *Precisions,* p. 47. See also pp. 128–132, 136, and 230 for other remarks on the subject.
2. Patte, *Mémoire sur les objets,* p. 11.

3. "Circulation." In *Trésor de la langue française,* vol. 2, 1977.
4. Durand, *Précis des leçons d'architecture,* vol. 2, p. 9.
5. For example Boullée, *Architecture, essai sur l'art* (before 1799), p. 146r: "Il peut circuler facilement dans toute la première partie du Palais."
6. Girouard, *Life in the English Country House,* pp. 194–201. This is one of many examples of Girouard's modernist, functionalist interpretation of premodern architecture in this book.
7. Scamozzi, *L'idea della architettura universale,* Part 1, Book 3, chap. 20, p. 312.
8. Daly, "Reform Club," pp. 346–347: "C'est que cet édifice n'est pas une masse inerte de pierre, de brique et de fonte; c'est presque un corps vivant avec son système de circulation sanguine et nerveuse."
9. Viollet-le-Duc, *Lectures on Architecture,* vol. 2, p. 277.
10. See, for example, the studies of hospital ward layouts, Nuffield Provincial Hospitals Trust, *Studies in the Function and Design of Hospitals,* pp. 9–11; the early development of "space syntax" (see Hillier and Hanson, *The Social Logic of Space*) was also based upon the modeling of potential human movement within defined spaces.
11. Frankl, *Principles of Architectural History,* pp. 79–80.
12. Merleau-Ponty, *The Phenomenology of Perception,* p. 142. (Merleau-Ponty is in turn quoting from Grunbaum, *Aphasie und Motorik,* 1930, p. 394.)
13. E. Husserl, "Ding und Raum," 1907, quoted in Mallgrave and Ikonomou, *Empathy, Form and Space,* p. 84, footnote 222.
14. Frankl, *Principles of Architectural History,* p. 148.
15. Frankl, *Principles of Architectural History,* p. 17, refers to "circulation" as subjective perception of bodily movement. On p. 157, he distinguishes between "circulation" as purposive movement of people between places and the sense of movement implied by the spatial forms.
16. Harvey, "Movement of the Heart and Blood," p. 58. Harvey took the concept "circulation" from Aristotle's description of the cycle of water vapor and rain in the atmosphere: "We have as much right to call this movement of the blood circular as Aristotle had to say that the air and rain emulate the movement of the heavenly bodies."
17. Critics of the tight boundary conception of buildings include Groák, *The Idea of Building,* especially pp. 21–39; and Andrea Kahn, "Overlooking."
18. For example, Broadbent, *Design in Architecture,* pp. 393–399.
19. For example, Curtis, *Lasdun,* p. 196: "Circulation is always a driving force . . ."
20. Schopenhauer, *The World as Will and Idea,* vol. 1, p. 277.
21. Vischer, "On the Optical Sense of Form," p. 121.
22. Wölfflin, "Prolegomena," p. 189.
23. Wölfflin, *Renaissance and Baroque,* p. 62.
24. Rowe, *Mathematics of the Ideal Villa,* p. 186.
25. Rowe, *Mathematics of the Ideal Villa,* pp. 191–192.
26. Rowe, *Mathematics of the Ideal Villa,* p. 192.
27. Eisenman, "From Object to Relationship."
28. Eisenman, "From Object to Relationship," pp. 38–39.
29. Eisenman, "From Object to Relationship," p. 47.
30. Goodman, *Languages of Art,* p. 73.
31. Pérez-Gómez, *Architecture and the Crisis of Modern Science.*

Bibliography

Boullée, Etienne-Louis. *Architecture, essai sur l'art.* In Helen Rosenau, *Boullée and Visionary Architecture.* London: Academy Editions, 1976.

Broadbent, Geoffrey. *Design in Architecture.* London: John Wiley and Sons, 1973.

"Circulation." In *Trésor de la langue française.* Paris: Editions du Centre National de la Recherche Scientifique, 1977.

Curtis, William J. R. *Denys Lasdun: Architecture, City, Landscape.* London: Phaidon, 1994.

Daly, César. "Reform Club." *Revue generale de l'architecture et des travaux publics* 15 (1857): 342–348.

Durand, Jacques-Nicolas-Louis. *Précis des leçons d'architecture.* 2 vols. Paris: Chez l'auteur, 1819.

Eisenman, Peter. "From Object to Relationship II: Giuseppe Terragni Casa Giuliani Frigerio." *Perspecta* 13/14 (1971): 36–61.

Frankl, Paul. *Principles of Architectural History.* Translated by J. F. O'Gorman. Cambridge: MIT Press, 1968. (First published in Germany under the title *Die Entwicklungsphasen der neueren Baukunst,* 1914).

Girouard, Mark. *Life in the English Country House: A Social and Architectural History.* New Haven, CT, and London: Yale University Press, 1978.

Goodman, Nelson. *Languages of Art.* Brighton, UK: Harvester Press, 1981.

Groák, Steven. *The Idea of Building.* London: E & FN Spon, 1992.

Guadet, Julien. *Eléments et théories d'architecture.* 5th ed. Paris: Librairie de la Construction Moderne, 1909.

Harvey, Sir William. "The Movement of the Heart and Blood in Animals" (1635). In *The Circulation of the Blood and Other Writings,* translated by Kenneth J. Franklin. London: Dent, 1963.

Hillier, Bill, and Julienne Hanson. *The Social Logic of Space.* Cambridge: Cambridge University Press, 1984.

Kahn, Andrea. "Overlooking: A Look at How We Look at Site or . . . site as 'discrete object' of desire." In D. McCorquodale, K. Ruedi, and S. Wigglesworth, eds., *Desiring Practices. Architecture, Gender and the Interdisciplinary,* 174–185. London: Black Dog, 1996.

Le Corbusier. *Precisions.* Translated by E. S. Aujame. Cambridge: MIT Press, 1991. First published in Paris, 1930.

Mallgrave, Harry Francis, and Eleftherios Ikonomou, eds. *Empathy, Form and Space: Problems in German Aesthetics, 1873–1893.* Santa Monica, CA: Getty Center, 1994.

Merleau-Ponty, Maurice. *The Phenomenology of Perception.* Translated by Colin Smith. London: Routledge and Kegan Paul, 1962.

Nuffield Provincial Hospitals Trust. *Studies in the Function and Design of Hospitals.* London, New York, Toronto: Oxford University Press, 1955.

Patte, Pierre. *Mémoire sur les objets les plus importans de l'architecture.* Paris: Rozet, 1769.

Pérez-Gómez, Alberto. *Architecture and the Crisis of Modern Science.* Cambridge: MIT Press, 1983.

Rowe, Colin. "Le Corbusier's Dominican Monastery." *Architectural Review* 129 (June 1961): 401–410.

Rowe, Colin. *The Mathematics of the Ideal Villa and Other Essays.* Cambridge: MIT Press, 1976.

Scamozzi, Vincenzo. *L'idea della architettura universale.* Venice: expensis auctoris, 1615.

Schopenhauer, Arthur. *The World as Will and Idea.* Translated by R. B. Haldane and J. Kemp. 3 vols. London: Kegan Paul, Trench, Trübner & Co., no date.

Viollet-le-Duc, Eugène-Emmanuel. *Lectures on Architecture* (1881). Translated by B. Bucknall. 2 vols. New York: Dover Publications, 1987. First French edition, Paris: A. Morel, 1863–1872.

Vischer, Robert. "On the Optical Sense of Form" (1873). Translated in Mallgrave and Ikonomou, eds. *Empathy, Form and Space,* 89–123.

Wölfflin, Heinrich. "Prolegomena to a Psychology of Architecture" (1886). Translated in Mallgrave and Ikonomou, eds., *Empathy, Form and Space,* 149–190.

Wölfflin, Heinrich. *Renaissance and Baroque.* Translated by K. Simon. London: Collins, 1984. First published in German as *Renaissance und Barock,* Munich: Th. Ackermann, 1888.

10

Diagramming the New World, or Hannes Meyer's "Scientization" of Architecture

K. Michael Hays

> The blame for our blindness lies in the fact that the human being has quite a weak consciousness and poor instrument for understanding the structure of the picture; and a really poor sense of the functions. . . . Our sense of function, our feeling, goes to sleep in the shadow of the strongest sense of touch and sight. . . . We need an analysis and new synthesis of our whole way of life, under the spell of plant life, the basis of animal and human life.
>
> —Konrad von Meyenburg, "Kultur von Pflanzen, Tieren, Menschen," *Bauhaus* 1927

> Building is a biological event.
>
> —Hannes Meyer, *Bauhaus* 1928

Hannes Meyer has not been one of architecture history's most favored subjects. Known mostly as the "other" director of the Bauhaus, he has remained unpopular compared with both his predecessor there, Walter Gropius, and his successor, Mies van der Rohe. He has been denigrated for his architecture, which was dogmatically functionalist and technocratic, and for his politics, which were too far left even for liberal socialists. The label most often used to condemn Meyer is the same label that is most interesting in the present context, a label he used to describe himself: Meyer was a "scientific Marxist."[1]

In one sense, we can take this as a fairly routine modernist reference to both science and Marxism, especially among architects like Hans Schmidt, Mart Stam, and El Lissitzky. But less routine are the examples that articulate his position: his article "Die neue Welt," an analytic manifesto that took up most of the issue of *Das Werk* in which it was published in 1926; a project for an interior published in that article; and a project for a girls' school done the same year, which he referred to as his "first research into the 'scientization' (*Verwissenschaftlichung*) of architecture."[2] Together these items yield a particular understanding of the science of architecture in a series of mutual reductions and enfoldings: first, of architecture into technology and science; then, of science into a social and historical force field; and finally, of history into an unstoppable march toward a socialist future in which architecture as an art would have been negated but redeemed on a different plane as a "natural," scientific organization of life's needed and desired effects.

Meyer invalidated built form as architecture's highest achievement. Form, above all, must be utilized. Form is but the diagram for the production of effects: the arrangement and distribution of experiential and expressive contents whose domain extends from carefully fabricated building details intended to coax out the latent, contradictory, and marginal aesthetic effects of constructed materials to elementary geometrical systems that construct differentiated spaces and structures for programmatic activities. In Meyer's work, form is reconceptualized as a condition conducive to certain outcomes, certain possibilities of habitation; form is an instigator of responses and performances, a frame whose vocation is to suggest more than fix, to outline possibilities that will be realized only partly at any one time and only in occupation. Architecture in its fullest sense emerges as an *aprés-coup:* the enactment of a programmatic promise that, by form, had been deferred.

For Meyer science served as a rational, collective ideal antithetical to that of the private aesthetic life of the bourgeoisie. By 1926 science and technology were already showing that socialism was not an illusory prospect but an immanent reality obeying the very laws of nature, which science itself attempts to theorize. And primary among the signs of collective modernization were graphic traces that lay just underneath the surface of the images of industrialized landscapes and modern advertising: elementary signs, he said in "Die neue Welt," that "permeate" the modern environment and call for a refunctionalization of the human perceptual apparatus. Meyer begins "Die neue Welt" with an intense description of the psychological preconditions of modern subjectivity:

> The flight of the "Norge" to the North Pole, the Zeiss planetarium at Jena and Flettner's rotor ship represent the latest stages to be reported in the mechanization of our planet. Being the outcome of extreme precision in thought, they all provide striking evidence of the way in which science continues to permeate our environment. Thus in the diagram of the present age we find everywhere amidst the sinuous lines of its social and economic fields of force straight lines which are mechanical and sci-

10.1–10.2

Hannes Meyer, illustrations from "Die Neue Welt." Advertising as traces of an emergent perceptual apparatus.

entific in origin. They are cogent evidence of the victory of man the thinker over amorphous nature. This new knowledge undermines and transforms existing values. It gives our new world its shape.

Motor cars dash along our streets. On a traffic island in the Champs Elysées from 6 to 8 P.M. there rages round one metropolitan dynamicism at its most strident. "Ford" and "Rolls Royce" have burst open the core of the town, obliterating distance and effacing the boundaries between town and country. . . . Illuminated signs twinkle, loud-speakers screech, posters advertise, display windows shine forth.[3] [See figures 10.1 and 10.2.]

But Meyer reverses the traditional negative valence of the subjective consequences of such overstimulation, seeing its effects as rather expanding and sharpening our consciousness. The paragraph continues:

> The simultaneity of events enormously extends our concept of "space and time," it enriches our life. We live faster and therefore longer. We have a keener sense of speed than ever before, and speed records are a direct gain for all. Gliding, parachute descents and music hall acrobatics refine our desire for balance. The precise division into hours of the time we spend working in office and factory and the split-minute timing of railway timetables make us live more consciously.

Meyer's reportage of the objects of technological advance yields first to an account of their experiential effect, then to a laying bare of their psychovisual deep structure, a "diagram of the present age": technological objects as a kind of ideogram or gestalt of epistemologico-social potential.

> The picture the landscape presents to the eye today is more diversified than ever before; hangars and power houses are the cathedral of the spirit of the age. This picture has the power to influence through the specific shapes, colors, and lights of its modern elements: the wireless aerials, the dams, the lattice girders: through the parabola of the airship, the triangle of the traffic signs, the circle of the railway signal, the rectangle of the billboard; through the linear element of transmission lines. . . . [See figures 10.3 and 10.4.]
>
> [And later] The artist's studio has become a scientific and technical laboratory. . . . Burroughs's calculating machine sets free our brain, the dictaphone our hand, Ford's motor our placebound senses and Handley Page our earthbound spirits. Radio, marconigram and phototelegraphy liberate us from our national seclusion and make us part of a world community. The gramophone, microphone, orchestrion and pianola accustom our ears to the sound of impersonal-mechanized rhythms. . . . Psychoanalysis has burst open the all too narrow dwelling of the soul and graphology has laid bare the character of the individual. . . . National costume is giving way to fashion and the external masculinization of woman shows that inwardly the two sexes have equal rights. Biology, psychoanalysis, relativity and entomology are common intellectual property: Francé, Einstein, Freud and Fabre are the saints of this latterday. Our homes are more mobile than ever. Large blocks of flats, sleeping cars, house yachts and transatlantic liners undermine the local concept of the "homeland." The fatherland goes into a decline. We learn Esperanto. We become cosmopolitan.

Meyer's effort in these passages seems to be first, to turn subjective experience into objectivized form and images that then flow back into the space of experience thus left open; or in other words, to map the objective social field in terms of the conduct that it

10.3

Hannes Meyer, Co-op Foto, ca. 1926. Photograph of a building crane: elementary graph of the new world.

10.4

Hannes Meyer, Office of Meyer and Hans Wittwer, Basel. The artist's studio has become a scientific and technical laboratory.

imposes on its inhabitants—all of which demands that our cognitive and perceptual conventions be completely refunctioned to participate in the new world.

Meyer's interest in the graphic tracing of modernity, his research into "applied psychology,"[4] and the resultant trajectory toward radicalization of perception had begun as early as 1916, in preparation for his first built work, the Freidorf Cooperative Estate. He recounts that he had

> used my free time to draw all Palladio's plans on thirty standard sheets of paper (size 420/594) in common scale. This work on Palladio prompted me to design my first housing scheme, the Freidorf estate, on the modular system of an architectural order. By means of this system all the external spaces . . . and all public internal spaces . . . were laid out in an artistic pattern *which would be perceived by those living there as the spatial harmony of proportion.*[5]

Proportional harmony and the repetitive cellular structure are for Meyer the architecturalization of the harmony of socialism. Likewise the Siedlung's red color, what Adolf Behne called a "symphony in red,"[6] stands as a symbol of Freidorf's leftward-leaning social commitments. Already in this early project the surface-level image, color, together with an underlying geometry are perceived as an instrument of collective perceptual change. The standardization and serialization of the Palladian system evacuates the traditional, *Heimat* denotations of the buildings and attempts to reinstall a different sort of collectivity. Meyer acknowledges that

> although the co-operators no doubt appreciated the economic aspects of this standardization, it mostly ran counter to their sense of beauty. In regard to architectural simplification, the Freidorf standards go to the utmost limits of what the individualistic Swiss will tolerate in matters of taste and any further paring away of "architecture" will be branded as "prison and barrack" building and meet with an almost unbroken front of public resistance.[7]

But he insists that "Man looks small once he enters the temple of the community." The individual must be sacrificed to the collective, and geometric serialization was the ideogram of the new cooperative world. So if we must take pause at the contradiction that Meyer's plan for collective happiness must needs be installed by a manifest imposition of psychic serialism on all those individuals for whom liberation was intended, we must also register his understanding of form as a way of organizing ideological space. (See figure 10.5.)

The function of Meyer's repetitive module is to inscribe across the architecture the reiterative, serial building system of a collective society, to unfold architecture into the exteriority of mass technology and standardization. Meyer's conception of the relation of building and science here is based on a general leftist understanding of serialism as demanded by available machine technology, which is cogently summarized in the mani-

10.5

Hannes Meyer, Co-op Linocut (Abstract Architecture II), 1925. Diagram of the spatial effects of reiterative geometry.

festo "ABC fordert die Diktatur der Maschine," published by Hans Schmidt and Mart Stam in *ABC,* a radical journal with which Meyer was involved. Schmidt and Stam linked machine technology to an increased cooperative or collective labor already emerging in the interstices of capitalism. Moreover, they made the same demand for a corresponding transformation of culture that we saw in Meyer's efforts at Freidorf, a demand now put bluntly in the form of a "dictatorship" imposing on the masses an ironclad collective happiness.[8] Meyer, at this point in the trajectory of his own work, similarly sought to conjoin aesthetic and technical logics in order to force new conventions of mass perception adequate to the available conditions of mass production. To that end he restricted and selected his tools and materials to obliterate as nearly as possible the distinctions among the aesthetic sign, its functional operations, and advanced production procedures. But his concern, distinct from that of his comrades, is with what we might call the reception of production at its most advanced level more than with production itself, and he thus deploys mechanization as a latent network of possible new relations of perception more than a determining fact.

But when he reflected on the Freidorf project in 1925, he found its hierarchical classicism to be a "compromise, socially between the individual human being and the community, formally between town and country," preferring a new architectural form in which "there would arise in an asymmetric broken way an estate built only with justifiable respect for hygiene and return on capital, as garden and engineered dwelling machine, a sun-trap and at the same time conveying natural beauty and purity."[9] By the time Meyer had finished the project, he had questioned and inverted the tradition on which his *Siedlung* was based as a viable mode of revolutionary and transformatory signification, renouncing Freidorf as "inappropriate," "laughably ridiculous," the "product of an incomprehensible time."[10] He recognized that mechanization is not wedded only to the social relations that produced it; it could be brought into service for other modes of production. And indeed, architecture could precipitate them. In his later work the fundamental harmonizing principles of the Siedlung Freidorf would be exploded by the raw things of modernity itself, the work's nonidentity with its traditional physical and social surroundings asserted, with a certain dissonance as the result.

So one effect of this visual presence of science is to dissolve individual private life into collectivity. And it is here as well, among these traces of science and technological production, that Meyer's Co-op Zimmer, contra a comfortable bourgeois domesticity, makes its conceptual presence felt, but now with a more adequate iconography than Freidorf. (See figure 10.6.) "Co-op Zimmer" is something of a misnomer, for the project is, in fact, a photograph: not a photograph of an existing interior but of an interior that has been *always only* a photograph, published in "Die neue Welt" as example of the essay's aesthetic of standardization, repetition, mechanized media, advertisement, nomadism, impersonality, and collectivity. The "room" is represented by a mock up of white fabric, a folding wood-and-canvas chair, a cot raised on conical feet to allow air to

10.6

Page entitled "Der Standard," from "Die Neue Welt," showing the industrialized landscape and the new dwelling: Co-op Zimmer, a "diagram of the present age."

10.7

Hannes Meyer, Co-op Linocut (Self-Portrait), 1924. Diagram of the subjective effects of standardization, mechanized media, and nomadism.

flow underneath, and a phonograph on a collapsible stool; the uncropped version also shows shelves of food products. The reiterative circles of the gramophone and stool top, the double square of the chair hanging on the wall, the double triangles of the stool legs, and the rectangle and conical feet of the cot are combined in what could be taken as an elementarist geometrical construction virtually collapsed onto its canvas background. (See figure 10.7.) It is as if, across a graph of advanced technology and media, a membrane has been stretched, a membrane of everyday, domestic use objects that (immediate and familiar as they are) make actual the otherwise purely virtual graph.

By producing a kind of transverse communication between verbal and visual images in the essay, Meyer weaves a network of externalities into a cartography of the new world. In the chains of diverse references organized serially in declarative sentences, the reader cannot help feeling a kind of dispersion, as of tracers sent out in scattered directions registering functions of instruments, disciplines, modes of thought, habitats, and habits, all of which are constituent parts of the transformed life-world configured by the Co-op Zimmer. The Co-op Zimmer is an assemblage, a "conspicuous arrangement," Meyer called it, that still functions within some larger cultural machinery that includes a conceptualization of the nomadic mobility enabled by the portable furniture, the alimentary products, and the invasion of the bedroom by the jazz band whose sound is now severed from its instruments and flattened onto a reproducible disk.[11] Not an actual architectural space but a concept of space, not a fact so much as the possibility of a smoothly traversable world, this "diagram of the present age" is objectively determined by the imposition of new technical products and external "fields of force" that operate to dissolve established boundaries within various forms of experience and cognition.

The Co-op Zimmer attests to the possibility that forms of simultaneous collective reception, linked directly to the inexorable movement of science and technology, can afford a kind of protopolitical and practical apprenticeship for the collective society to come. The concrete experience of the visual products of science and technology, when understood as affording a symbolic and psychological mapping of the now vivid and tractable consequences of modernity, may be conceived as a functional diagram for the kind of mental retooling the human subject must undergo to divest itself of its historically conditioned defects and failures of development and begin its journey toward the classless future. "A diagram of the present age": The phrase describes not a utopian condition that one can only wish for, never to find in some actual place, but rather a machinery for producing utopian effects: possible new relations, pleasures, and freedoms. The machinery includes the liberative infrastructure of modern transportation and communication for certain but, as important, it includes a perceptual apparatus capable of scanning the surface of the modern industrial landscape for what I have called a graph of technology—gestalts that Meyer seems to understand as so many *Wunschbilder* ("the parabola of the airship, the triangle of the traffic signs, the circle of the railway signal, the rectangle of the billboard, . . . the linear element of transmission lines"). It does not simplify Meyer's

10.8

Hannes Meyer, Peterschule project, 1927, from *Bauhaus* 2, with H. H. Higbie's lighting diagram and "mathematical proof" of the lighting system's effects. Architecture as an abstract machine.

enterprise to insist on his concern for images, for what we understand as the significance of Meyer's pictorial reportage of the mass industrial and mass cultural landscape has less to do with the latter as a source of sheer aesthetic experimentation than it does with this picture's claim to cognitive and practical as well as visual and aesthetic status. The appropriation and presentation of the multiplicity of diverse images testify to Meyer's preoccupation not only with the industrialization process as a kind of "second nature," but also with the forms of experience that are the subjective consequences of such a process. This play of images—whose emblematic value is reconfirmed by the presence in Meyer's article of exemplary photographs including scenes of industry, its use objects, and its repetitive morphology—seeks to satisfy not only the appetite for form, but also the appetite for matter. The pictures stand as facts of a certain kind of seeing, as the actual forms of our collective knowledge of things, a kind of visual Esperanto. And their importance may therefore be recognized in terms of their ability to assimilate material and productive values to visual and psychological (even biological) effects, to convert the qualities of one into the forms of the other.[12]

The Co-op interior organizes one set of images that, by embodying in a concentrated package the rationalized formal and procedural necessities of mass technology, releases an entire range of potentials for inhabitation, use, and social practice not previously imaginable: Architecture as a *dispositif* or distribution apparatus for differential spaces, functions, contents, and expressions from the registers of industrial production and rationalized geometry now pressed together into a single tissue, all of which compels the viewer of this work to flex certain previously underused conceptual muscles to hold together possibilities for use and occupation that normally remain unthought. We find another diagram of possibilities in what Meyer calls his "Co-op building," the Petersschule project for Basel of 1926–1927, his more complete attempt at the "scientization" of architecture.[13] (See figure 10.8.)

The scientization of architecture in the case of the Petersschule is a matter of three basic tactics: First is the construction of potentials for a new function of learning-play: a composite activity based on Heinrich Pestalozzi's pedagogy, which rejected catechesis and memorization in favor of cultivating the child's natural capacities for observation, discovery, and experimentation. Meyer writes in the statement accompanying the 1926 competition project, "The goal: No commanded study but rather experienced knowledge!" He asserted that "the classroom, the decked and undecked play areas, and the toilets are the inseparable constructive units (or cells) of the school building," and designated that the school should make the "greatest possible separation of the teaching work from the ground level into the sunny, well ventilated and lit level."[14] And indeed, the block of classrooms, a vertical block of toilets attached to the classrooms, a lower block of ancillary functions, and the suspended decks constitute the building's basic configuration.

Meyer goes to some effort to document his lighting calculations relating to the depth of the room, the window area, and table height, which are based on Henry Harold

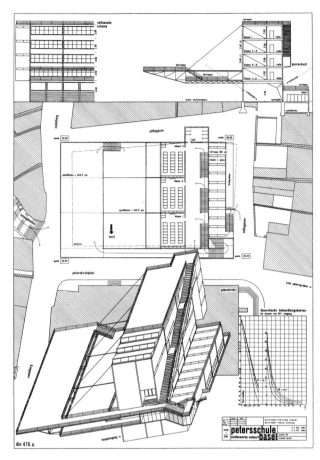

architekt hannes meyer basel/bauhaus-dessau
architekt hans wittwer basel

die petersschule basel
(wettbewerbsentwurf **1926**)

die aufgabe:

neubau einer 11 klassigen mädchen-volksschule mit turnhalle, zeichensaal, schulbad und suppenküche etc., 528 schülerinnen. sinnwidriger traditioneller schulhaus-bauplatz im altstadtgebiet von basel, im schatten hoher randbebauung, schlecht belüftet und im hinblick auf das umfangreiche bauprogramm mit 1240.0 qm gesamtfläche erheblich zu klein. übliche überbauung ergibt max. 500 qm schulhof, mithin 1.0 qm tummelfläche pro schulkind.

das ziel:

keine schulkrüppel! anzustreben wäre ausschließliche oberlichtbeleuchtung aller schulräume (vergleiche die resultate von fall 1 und 2 der beleuchtungsberechnung) und die bestimmung eines neuen baugeländes nach maßgabe planvoller stadtentwicklung. gegenwärtig erscheint die verwirklichung solcher forderungen aussichtslos, und es ergibt die auseinandersetzung mit dem alten schulhaus den umstehenden kompromiß.

der vorschlag:

größtmögliche entfernung des schulbetriebes von der erdoberfläche in die besonnte, durchlüftete und belichtete höhenlage.

im erdgeschoß nur schulbad und turnbetrieb im geschlossenen raum. die verbleibende hoffläche wird dem öffentlichen verkehr und dem „parking" freigegeben.

an stelle eines hofes sind 2 hängende freiflächen und alle oberflächen des gebäudekörpers der jugend als tummelplatz zugewiesen, im ganzen 1250 qm sonnige spielfläche, der altstadt entrückt.

freitreppe und verglaste treppe verbinden, parallel geführt, spielflächen und innenräume.

das eigengewicht des hauskörpers ist nutzbar verwendet und trägt an 4 drahtseilen die stützenlose eisenkonstruktion der 2 schwebenden freiflächen.

die gebäudekonstruktion als eisenfachwerkbau auf nur 8 stützen und mit diesem außenwand-querschnitt: aluminiumriffelblechverkleidung — bimsbetonplatten — luftlamelle — kieselgurplatten — luftlamelle — glanzeternitplatten.

bautechnische ausstattung: eiserne kippfenster, aluminiumblechtüren, stahlmöbel, flure und treppen mit gummibodenbelag.

rechnerischer nachweis der beleuchtungsstärke aller schulräume

fall **1**) östliches seitenlicht aller klassenzimmer.
fall **2**) shed-oberlicht des zeichensaales.
fall **3**) zweiseitiges seitenlicht der turnhalle.

berechnung der beleuchtungsstärke auf tischhöhe

fall **1**) klassenzimmer mit senkrechter fensterwand. (östliches seitenlicht.)
berechnet wird nur die beleuchtungsstärke für den ungünstigsten arbeitsplatz (P), dieser befindet sich in der vom fenster entferntesten reihe an der rückwand.
berechnungsverfahren nach higbie:
daten für die formel:
abstand des punktes P vom fenster $a = 5,1$ m
länge des fensters $m = 10,2$,,
abstand des oberen fensterrandes von der tischfläche $f = 2,4$,,
  ,,   ,, unteren   ,,   ,, $f' = -.-$,,
beleuchtungsstärke des fensters $b = 100,0$ ftcdl.

$$E_p = 50\left[\text{tg}^{-1}\left(\frac{10,2}{5,1}\right)\cdot\frac{5,1}{\sqrt{5,1^2+2,4^2}} - \text{tg}^{-1}\left(\frac{10,2}{\sqrt{5,1^2+2,4^2}}\right)\right] = 486,0 \text{ lx},$$

$$E_p' = 50\left[\text{tg}^{-1}\left(\frac{10,2}{5,1}\right) - \text{tg}^{-1}\left(\frac{10,2}{5,1^2}\right)\right] = 435,0 \text{ lx},$$

beleuchtungsstärke im punkte $P = E_p - E_p' = 41,0$ lx
(12 hefner-lux $/lx' = I$ footcandle.)

lichtverlust durch gegenüberliegende gebäude etc. wird auf grund empirischer werte festgestellt, hier beträgt er für alle stockwerke etwa 5 v. h.

die beleuchtungsstärke im punkte P an ort und stelle erreicht einen um etwa 40 v. h. höheren wert (zufolge der rückwürfe des lichtes an decke und wänden).

die leitsätze der D.B.G. verlangen für les- und schreibräume eine mittlere beleuchtung von 50–60 lx. die vorgesehene fensteröffnung gewährt also auch dem dunkelsten arbeitsplatz eine ausreichende beleuchtung. nahe der fensterwand ist die beleuchtung 10 mal stärker und in zimmermitte 4 mal stärker als im punkt P. die durchschnittliche beleuchtung beträgt etwa 180 lx, bei einer fensterfläche von etwas mehr als $1/_3$ der bodenfläche.

fall **2**) shed-oberlicht des zeichensaales.
berechnet wird die beleuchtung in jeder shed-axe.
berechnungsverfahren nach higbie und levin.
daten für die formeln:
abstand des punktes P_1 von der fensterfläche $a_1 = 2,5$ m
   ,,   ,,   ,, P_2 ,,   ,, $a_2 = 5,6$,,
   ,,   ,,   ,, P_3 ,,   ,, $a_3 = 8,6$,,
(diese abstände horizontal gemessen) ,,
länge des fensters $m = 11,0$ m
abstand des oberen fensterrandes von der tischfläche $f = 3,3$ m
   ,,   ,, unteren   ,,   ,, $f' = 2,6$ m
(diese abstände in der fensterebene gemessen).
beleuchtungsstärke des fensters $b = 100,0$ ftcdl.

$A_1 = \frac{a_1}{f} = 0,75,$ $A_1' = \frac{a_1}{f'} = 0,96,$ $A_2 = \frac{a_2}{f} = 1,70,$

$A_2' = \frac{a_2}{f'} = 2,15,$ $A_3 = \frac{a_3}{f} = 2,60,$ $A_3' = \frac{a_3}{f'} = 3,30,$

$B = \frac{m}{f} = 3,30,$ $B' = \frac{m}{f'} = 4,20.$

die beleuchtungsstärke in jeder shed-axe, erzeugt durch das zugehörige fenster, ist gleich dem unterschied zwischen den beleuchtungswerten von fenstern der höhe f und f'.

aus dem diagramm ergibt sich
beleuchtungsstärke in $P_1 = 56 - 39 = 17 \times 12 = 204$ lx. $= E_1$
       ,,   ,, $P_2 = 13 - 9 = 4 \times 12 = 58$,, $= E_2$
       ,,   ,, $P_3 = 5 - 3 = 2 \times 12 = 24$,, $= E_3$
die gesamtbeleuchtungsstärke in $P_1 = E_1 = 204$ lx.
    ,,   ,,   ,, $P_2 = E_1 + E_2 = 262$,,
    ,,   ,,   ,, $P_3 = E_1 + E_2 + E_3 = 286$,,

diese werte sind um weniger als $1/_3$ voneinander verschieden, gegenüber dem vielfachen beim seitenlicht. die durchschnittliche beleuchtung beträgt etwa 250 lx bei einer fensterfläche von etwa $1/_4$ der bodenfläche.

fall **3**) zweiseitiges seitenlicht der turnhalle.
berechnet wird die beleuchtung an den beiden längswänden und in der saalmitte.
beide längswände mit 2 m hohem fensterfries auf die ganze länge und unmittelbar unter der decke.
berechnungsverfahren nach higbie: (wie bei klassenzimmer mit seitenlicht).
daten für die formel: (P nahe längswand ost).
abstand des punktes P vom fenster (ost) $a_1 = 2,0$ m
   ,,   ,,   P   ,,   (west) $a_2 = 9,0$,,
länge des fensters $m = 23,0$,,
abstand des oberen fensterrandes von der tischfläche $f = 4,5$,,
   ,,   ,, unteren   ,,   ,, $f' = 2,5$,,
beleuchtungsstärke des fensters $b = 100,0$ ftcdl.
beleuchtungsstärke durch fenster (ost) $= 249$ lx
        ,,   ,,   ,,   (west) $= 29$ lx

lichtverlust durch gegenüberliegende gebäude, ostseite $= 5$ v. H.
       ,,   ,,   ,,   westseite $= 12$ v. H.
gesamtbeleuchtung in P $= 253$ lx.
daten für die formel: (P nahe längswand west).
abstand des punktes P vom fenster (ost) $a_1 = 9,0$ m
   ,,   ,,   P   ,,   (west) $a_2 = 2,0$,,
(die anderen werte wie oben).

beleuchtung durch fenster (ost) $= 29$ lx
     ,,   ,,   (west) $= 249$ lx
lichtverlust: ostseite $= 5$ v. h., westseite 27 v. h.
gesamtbeleuchtung in P $= 212$ lx.
daten für die formel: (P in saalmitte).
abstand des punktes P vom fenster (ost und west gleichviel) $a = 5,5$ m
(die anderen werte wie oben).

beleuchtung durch fenster (ost und west gleichviel) $= 110$ lx
lichtverlust: ostseite $= 5$ v. h., westseite 18 v. h.

gesamtbeleuchtung in P $= 195$ lx.

Higbie's experiments at the University of Michigan, first published, as best I can tell, the same year as the Petersschule project. Meyer even reproduces exactly the same diagram that Higbie illustrates as optimal for lighting a working surface.[15] The desire to maximize the separation from the ground also led him to propose the roof deck and the huge *Freiflächen* or open decks suspended from cables, directly accessible from the classrooms via outside gangways, all to be used as play space by the students. The suspended decks are held away from the classroom block by a dimension determined by the angle of light penetrating into the gymnasium and onto the playground. The entire arrangement of the basic units in this project can be explained in terms of the maximization of the area for outdoor recreation and the amount of light penetrating into the building, these coupled with the methods of its technical construction. The architecture is but an inscription of these functional conditions and relations.

"The building is built on a steel framework resting on only eight columns and with outside walls of this section: facing of chequered aluminum sheet—pumice concrete slabs—air space—kieselguhr slabs—air space—polished Eternit sheets. Fitting out [*Bautechnische Ausstattung*]: steel framed hopper-type windows, aluminum sheet doors, steel furniture, halls and stairs covered with rubber flooring."[16] Meyer's second tactic in the Petersschule is to intensify the raw materiality of the constructed thing—the glaring brightness, the hardness, one might even imagine the smell, the taste—and thrust the experience of that thing, previously indifferent and unimaginably external, toward the viewer with unpadded harshness. His materialism emphasizes the heterogeneous properties of things and their effects in real space and real time and induces a play of sensuous energies in the viewer, a compulsive pleasure taken in the quiddity of the building parts, but also in the contradictions, the disruptions, the gaps and silences, all of which explodes the received social meanings of those things.

In a later essay, Meyer would write,

> In line with the Marxist maxim that "being determines consciousness" the socialist building is a factor in mass psychology. Hence cities and their building components must be organized psychologically in keeping with the findings of a science in which psychology is kept constantly in the foreground. The individual pretensions to perceptions [*Empfindungsansprüche*] of the artist-architect must not be allowed to determine the psychological effect of the building. The elements in a building that have a telling psychological effect (poster area, loudspeaker, light dispenser, staircase, color, etc.) must be organically integrated so as to accord with our most profound insights into the laws of perception . . .[17]

Through such instrumentalization of built form to produce the desired psychophysical results, the Petersschule disenfranchises composition and vision as the dominant categories of architecture. Volumetric components are conceived in functional terms: simple adjacencies grouped according to use. And "elements . . . that have a telling psychological effect," such as, in this case, the stairs, walkways, and suspended platforms, are standardized or confiscated like so many found elements and affixed or grafted onto the basic unit of the building. All of this operates to negate the relational compositional strategies identified with traditional art of human facture and to substitute things untouched by personality.

Meyer's third tactic for the scientization of architecture in the Petersschule is a related urbanistic desire. The site of the Petersschule lies on the eastern periphery of the inner city wall of Basel, a former Roman fortification, adjacent to the Peterskirche. Meyer's project isolates itself on the site, holding the street line to the west and leaving more than half of the eastern part of the site free on the ground plane. The entry, which is an extended spatial and temporal sequence through the system of open and glazed stairs, begins at the western street, visible from the square in front of the church, and wraps around the building's north side. The passageway formed by the suspended platforms on the north of the building operates like an upper-level loggia; in concert with the deep entry door to the first level and the large window of the ground level, it describes a zone of circulation at the site's northern edge extending the space of the narrow passage that enters the site from the east and continues parallel to the south facade of Peterskirche. All of this furthers the preeminence of the diagonality so apparent in the perspective and axonometric drawings and blocks any frontal reading of a humanist building "face."

The suspended decks also determine the building's sectional organization: The ground level is left open for public circulation and parking; only the gymnasium, swimming pool, and kitchen are located at ground level or below. The building engages the specificities of its site but implies an entire sectional reorganization of the traditional city. The competition project was submitted under the "motto: 'compromise,'" the signal of Meyer's belief that the constrained building site in old Basel assigned by the competition was "absolutely unfit for a contemporary school" planned for more than 500 children.[18] In a sarcastic, boldfaced conclusion to his explanation of the project—"HOCH DIE DENKMALPFLEGE!" ("Cheers to the preservationists!")—Meyer implied that it would not be an unhappy consequence of the adventitious insertion of the new Petersschule if the surrounding old environment were allowed to wither away.

Like a prosthetic apparatus for a crippled and crippling city unable to function adequately on its own, the Petersschule organizes its elements in such a way as to reveal the present order as unsatisfactory, physically and socially, and to propose an antisocial response as a possible way out: The Petersschule would like to leave the old city behind. Short of that, it diagrams the concrete effects of what the city lacks. Like a prosthetic device that is both the mark of and compromised solution to a debilitation, the Petersschule produces a significant absence, that is to say an absence that it compensates for and at the same time represents.

But we need to move to another level of interpretation, other than representation, to fully explain the project. A provocative way of describing the Petersschule—one that, it will now be recognized, has remained in the background of my analysis of all Meyer's Co-op work—is suggested by Gilles Deleuze's concept of the "abstract machine" and the particular abstract machine he calls the diagram. Deleuze finds diagrams, for example, in Michel Foucault's analysis of disciplinary and punitive system like Panopticism that "*impose a particular conduct on a particular human multiplicity.* We need only insist that the multiplicity is reduced and confined to a tight space and that the imposition of a form of conduct is done by distributing in space, laying out and serializing in time, composing in space-time, and so on. . . . [A diagram] is a machine that is almost blind and mute, even though it makes others see and speak."[19] For Deleuze, the Panopticon, viewed as an abstract machine, has no form of its own but rather presides over relations between forces outside itself. Deleuze also finds diagrams in the painting of Francis Bacon. Here the diagram is "the operative set of lines and areas, of asignifying and nonrepresentative brushstrokes and daubs of color. And the operation of the diagram, its function, as Bacon says, is to 'suggest.' Or, more rigorously, it is the introduction of 'possibilities of fact.'"[20] Neither abstract in the sense of Mondrian nor expressionist in the sense of action painting, Bacon's painting breaks from traditional figuration in order, exactly, to produce a new figure: "It is like the emergence of another world."[21] In general, the abstract machine

> has no form of its own (much less substance) and makes no distinction within itself between content and expression, even though outside itself it presides over that distinction and distributes it in strata, domains, and territories. An abstract machine in itself is not physical or corporeal, any more than it is semiotic; it is *diagrammatic.* . . . It operates by *matter,* not by substance; by *function,* not by form. Substances and forms are of expression "or" of content. But functions are not yet "semiotically" formed, and matters are not yet "physically" formed. The abstract machine is pure Matter-Function—a diagram independent of the forms and substances, expressions and contents it will distribute.[22]

In one sense, perhaps any architecture is an abstract machine, insofar as any architecture enables certain functions and constrains others, produces certain effects and forecloses others. But the intensity of Meyer's understanding and pursuit of this dimension of architecture must be underscored. The format of Meyer's presentation of the Petersschule project as published in *Bauhaus* should be taken for what it is: Almost three-quarters of the single-page layout is devoted to diagrams and calculations; the "building itself" is only one component of the total architectural apparatus that includes these predictions of effect. In a letter to Walter Gropius of 1927, Meyer wrote about the illustration, "I have condensed into one drawing the design sent to L. Moholy. I should be pleased if the relevant lighting calculations were published with it. . . . I believe that we must if possible base our new designs which arise from functional building, on building science, in order to counter the otherwise justified complaint about the lack of objectivity."[23] Taking the format seriously means not only seeing it as a description of equipment, or as propaganda. Here is Deleuze again: "The diagrammatic or abstract machine . . . constructs a real that is yet to come, a new type of reality. Thus when it constitutes points of creation or potentiality it does not stand outside history but is instead always prior to history."[24] By organizing new spaces and new events, the Petersschule participates in the construction of the new world by reprogramming its inhabitants, training them in new perceptual habits, producing new categories of experience, delineating, by way of the architecture itself, new subject positions and hierarchies.

The scientization of architecture: the reconceptualization of architecture not as an aesthetic object—the organization of an organic whole that we normally mean by "good design"—but as a scientific commutation system—a program, a set of functions and procedures, a "biological process." Perhaps this claim can be further substantiated by way of one last analogy. Consider Meyer's assertion, "building is a biological event," together with his Co-op Construction of 1926, itself a kind of biological building. (See figure 10.9.) With its striated space, rectangular planes, diagonal placement within its frame, geometrical purity, and emphasis on visual layering and transparency as diagrammatic potentials of photographic technique, it resembles in its formal organization nothing so much as the Petersschule and its various attachments of suspended platforms, walkways,

10.9

Hannes Meyer, Co-op Construction I, 1926. Biological reproduction and visual effects.

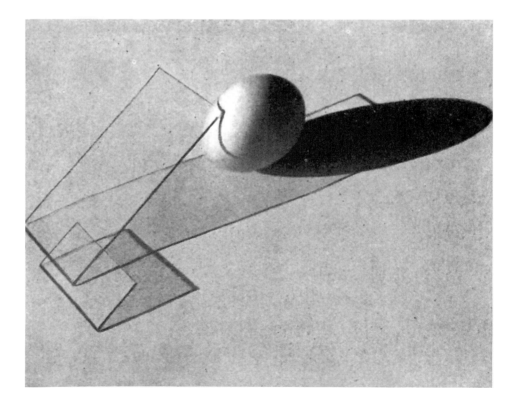

and stairs as techno-psychological expedients—appropriated industrial components organized in terms of intensified visual effect. Like the school's suspended decks, the glass fragments are unworked; they are palpably glass, propped up against the egg; they do not consent to a merely visual apprehension but, in their unmediated juxtaposition of material fragment to fragment, emphasize the full psychological effect of factural, technical construction. More important, the white ovoid is, after all, an *egg:* a biologically reproduced volume analogous to the mechanically reproduced body of the school's classroom block. Beyond its geometrical purity, the egg tends toward an identification with the alimentary products of cooperative societies like Freidorf and the utopian modes of production, distribution, and consumption they anticipate. As Jacques Gubler has written, "the co-op egg of 1926 is consumable, not by way of the oneiric, not in the sense of surrealism, but rather by way of the oral."[25] But then, the egg's alimentary aspect is just as surely canceled by the fact that the construction is *made out of* an egg and then photographed; it is assembled from selected pieces, not painted or carved, which links the activity of sign production to the activity of work, and it is distributed in a journal, which links it to available techniques of simultaneous collective reception. What we as viewers do with this piece must then inevitably oscillate between our aesthetic habits and the blockage of those by the sheer perceptual facts. The construction and distribution of the object begins to enter the process of collective-cooperative organization directed toward the socialization of all objects of consumption. And the Petersschule extends this functional proliferation to the multiple relations among bodies, equipment, movements, procedures, and technologies programmed by its spatiotemporal regime.

NOTES

1. Maldonado, "Preface," p. 7.
2. Meyer, "Biografische Angaben," p. 357.
3. Meyer, "Die neue Welt, p. 205." All subsequent excerpts are from this essay unless noted.
4. Letter from Meyer to Graf Dürckheim, 24 August 1930; in *Hannes Meyer: Bauen und Gesellschaft,* p. 75.
5. Meyer, "Wie ich arbeite," p. 21; emphasis added.
6. Behne, paraphrasing Meyer, in a review of Meyer's ADGB school, *Pädagogische Beilage,* p. 41.
7. Meyer, "Freidorf Housing Estate," p. 7.
8. "*The machine* is nothing more than the inexorable dictator of the possibilities and tasks common to all our lives.

    But we are still in a state of becoming, of transition. The machine has become the servant of bourgeois individualist culture born of the Renaissance. Just as the servant is paid and despised by the same master, so the machine is simultaneously used by the citizen and damned by his intellectual court, his artists, scholars and philosophers. The machine is not a servant, however, but a dictator—it dictates how we are to think and what we have to understand. As leader of the masses, who are inescapably bound up with it, it demands more insistently every year the transformation of our economy, our culture. . . .

> *We have taken the first step:* the transition from an individualistically producing society held together *ideally* by the concepts of the national State and a racially delimited religious outlook, to a capitalistically producing society *materially* organized in response to the need for industrialization and the international exchange of goods. . . .
>
> *We have to take the second step:* the transition from a society that is *compelled* to produce collectively but is still individualistically oriented to a society that *consciously* thinks and works collectively . . ."

  Schmidt and Stam, "ABC fordert die Diktatur der Maschine," pp. 115–116.
9. Meyer, "Die Siedlung Freidorf," p. 51.
10. Ibid.
11. In "Die neue Welt," p. 223, Meyer lists phonographic recordings "appropriate for the times."
12. Look again at the epigraph from the article by Konrad von Meyenburg in an issue of *Bauhaus* edited by Meyer: "The blame for our blindness lies in the fact that the human being has quite a . . . poor instrument for understanding the *structure of the picture (Struktur der Gebilde)* [emphasis added]."
13. There are actually two versions of the project, the original competition entry of 1926 (entry number 72, "motto: 'compromise'") and the revised presentation of the project in *Bauhaus* 1927, no. 2 (Dessau): 5. I blur the differences here.
14. Meyer and Wittwer, "Erläuterung zum Schulhaus," p. 81.
15. Higbie and Randall, "A Method for Predicting Daylight," p. 41. Emily Thompson helped me locate Higbie's work.
16. Meyer, "Projekt für die Petersschule," p. 17.
17. Meyer, "Über marxistische Architektur," p. 31.
18. Meyer and Wittwer, "Erläuterung zum Schulhaus," p. 81.
19. Deleuze, *Foucault,* p. 34; emphasis in original.
20. Deleuze, "The Diagram," p. 194.
21. Deleuze, "The Diagram," p. 194.
22. Deleuze and Guatarri, *A Thousand Plateaus,* p. 141.
23. Letter of 28 March 1927 from Hannes Meyer, Basel, to Walter Gropius, Dessau, in Getty Archives, Los Angeles.
24. Deleuze and Guatarri, *A Thousand Plateaus,* pp. 37, 142.
25. Gubler, ed., *ABC: Architettura e avanguardia,* p. 128.

Bibliography

Behne, Adolph. *Pädagogische Beilage zur Sächsischen Schulzeitung* 20 (June 1928): 41–42.
Deleuze, Gilles. "The Diagram." In Constantin V. Boundas, ed., The Deleuze Reader, 193–200. New York: Columbia University Press, 1993.
Deleuze, Gilles. *Foucault.* Translated by Seán Hand. Minneapolis: University of Minnesota Press, 1988.
Deleuze, Gilles, and Félix Guatarri. *A Thousand Plateaus.* Translated by Brian Massumi. Minneapolis: University of Minnesota Press, 1987.
Gubler, Jacques, ed. *ABC: Architettura e avanguardia, 1924–1928.* Milan: Electa, 1983.

Higbie, H. H., and W. C. Randall. "A Method for Predicting Daylight from Windows." *Engineering Research Bulletin* (Department of Engineering Research, University of Michigan, Ann Arbor) 6 (January 1927): 36–41.

Maldonado, Tomás. "Preface." In Schnaidt, *Hannes Meyer,* 7.

Meyer, Hannes. "Biografische Angaben." In *Hannes Meyer 1889–1954, Architekt Urbanist Lehrer,* 355–362. Berlin: Ernst & Sohn, 1989.

Meyer, Hannes. "Freidorf Housing Estate, near Basle, 1919–21." Translated in Schnaidt, *Hannes Meyer,* 2–15.

Meyer, Hannes. *Bauen und Gesellschaft: Schriften, Briefe, Projekte.* Dresden: VEB Verlag der Kunst, 1980.

Meyer, Hannes. "Die neue Welt." *Das Werk* 7 (1926): 205–224. Translated in Schnaidt, *Hannes Meyer,* 91–95.

Meyer, Hannes. "Projekt für die Petersschule, Basle, 1926." In Schnaidt, *Hannes Meyer,* 17.

Meyer, Hannes. "Die Siedlung Freidorf." *Das Werk* 12 (1925): 40–51.

Meyer, Hannes. "Über marxistische Architektur." In Meyer, *Bauen und Gesellschaft,* 92–97. Partial English translation in Schnaidt, *Hannes Meyer,* 31.

Meyer, Hannes. "Wie ich arbeite." *Architektura CCCP* 6 (1933): 103. Partial translation in Schnaidt, *Hannes Meyer,* 19–21.

Meyer, Hannes, and Hans Wittwer. "Erläuterung zum Schulhaus von heute." Published as facsimile in *Hannes Meyer 1889–1954 Architekt Urbanist Lehrer,* 81. Berlin: Ernst & Sohn, 1989.

Schmidt, Hans, and Mart Stam. "*ABC fordert die Diktatur der Maschine.*" *ABC* 4, vol. 2 (1927–1928). Translated in Ulrich Conrads, *Programs and Manifestoes on 20th-Century Architecture,* 115–116. Cambridge: MIT Press, 1970.

Schnaidt, Claude. *Hannes Meyer: Bauten, Projekte und Schriften. Buildings, Projects and Writings.* (English version by D. Q. Stephenson.) New York: Architectural Book Publishing Co., ca. 1965.

11

Listening to/for Modernity: Architectural Acoustics and the Development of Modern Spaces in America

Emily Thompson

Introduction

In 1914, the architect Bertram Grosvenor Goodhue was having trouble with a client. Goodhue was working on the design for St. Bartholomew's Church in New York, and the church rector, Dr. Leighton Parks, was proving hard to please. Goodhue wrote of his difficulties to his friend, colleague, and occasional collaborator, Wallace Sabine:

> Dr. Parks has asked me to design a new exterior based on the present plan—a new exterior that shall be "*modern*"—again whatever this term may mean; so as you probably have already gathered I am almost at my wits' end. I was, however, clever enough to say that I thought in such a matter you should be the designer as much if not more than I. Perhaps if you were here you could shed light on the whole subject, which I must admit is now the most vexed one that I have ever had to do with, so please arrange to see me as soon as you can.[1]

Sabine was not a fellow architect but a scientist, a professor of physics at Harvard University who had applied his research in architectural acoustics to numerous designs by Goodhue and other important architects of the era, including Goodhue's former partner, Ralph Adams Cram.

Goodhue's frustration over the meaning of "modern" and his plea to Sabine to resolve this problem for him suggests that historians faced with the same problem of understanding "modernity" would do well to consider the connections between architects and scientists as these groups worked together to create the new architecture of the early twentieth century. Goodhue understood that the scientific point of view was central to the idea of a modern architecture, and he called upon Sabine to provide this perspective in an attempt to modernize his design. Sabine did not "design" St. Bartholomew's Church,[2] nor any other architectural structure, but his science nonetheless played an important role in the development of American architecture in the century's first few decades. Sabine's science, the science of architectural acoustics, thus provides a case study for understanding more generally the role of science in the development of modern architecture.

In assuming that Sabine possessed the power to define what constituted the modern, Goodhue articulated the widely held belief that science was the driving force behind the historical process of modernization. Goodhue and Sabine lived in a world captivated by a perception of the ever increasing intellectual power of science and the ever widespread application of scientific knowledge to transform the natural world through technological processes. Although the roots of modern science extend back to the sixteenth century, modernization as a potent historical force is typically associated with the employment of new industrial technologies such as the steam engine and the factory in the eighteenth century. The process continued to build steam, literally as well as figuratively, throughout the nineteenth century, and was an integral element of what is identified in the Anglo-American sphere as Victorian culture.[3]

Whereas modernization constitutes the economic and physical substrate of Victorianism, modernism constitutes its cultural antithesis. The twentieth-century intellectual and aesthetic culture of modernism is generally characterized as a rejection of Victorian values.[4] Whether considering fashion, sexual mores, art, literature, music, or architecture, the historian confronts a deep chasm in the early twentieth century, a chasm that separates the Victorians from the Moderns: Dickens from Dos Passos, Schumann from Stravinsky, Cram from Le Corbusier. This gap is, however, bridged by the continuous technological forces of modernization. The machines of the "Machine Age" were built by engineers whose work constituted an uninterrupted progression from the accomplishments of their nineteenth-century predecessors. As countless histories have told us, the avant-garde artists and intellectuals who formulated the new culture of modernism were those willing to trust the bridge provided by science and technology, to cross that bridge into a new cultural realm. Indeed, the "moderns" themselves were the first to present this account of the origins of their ideas and works. The Italian Futurist school of painters proclaimed in 1910:

> Comrades, we tell you now that the triumphant progress of science makes changes in humanity inevitable, changes that are hacking an abyss between those docile slaves of tradition and us free moderns who are confident in the radiant splendor of our future.[5]

This type of connection between modernizing scientists and modernist artists is, although compelling, unsatisfyingly simple. It assumes a unidirectional process in which the technological accomplishments of modernizers stimulated an aesthetic response from modernists. For example, Terry Smith's study of industry, visual art, and design in America, *Making the Modern,* opens with an account of the Ford Motor Company's Highland Park automobile factory, with all subsequent analysis—of the architecture of Albert Kahn, the paintings of Charles Sheeler, and the graphic design of *Fortune* magazine—implicitly presented as products of that industrial environment, generated by the machines of production as were the Model T's themselves. In a more general study of the modern culture of time and space, Stephen Kern resorts to a similar model of "the impact of technology on human experience."[6]

In this kind of account of modernism, the "impact of technology" is at times literal: The Italian Futurists gleefully plunge their fast new cars into roadside ditches, crawl out of the wreckage, and stumble home to transform the world of art.[7] The train annihilates space, the telegraph annihilates time, the cinema annihilates single-point perspective, and avant-garde artists—from painters to poets to architects—step in to fill the void with something new, something "modern." Their sources of inspiration are the very machines that have seemingly obliterated the world of the past.

"Let us listen to the counsels of American engineers," Le Corbusier proclaimed in 1923.[8] European architects did listen, but many architectural historians would have us believe that their American counterparts covered their ears. Even as American engineers were busy creating the new machines, their architectural compatriots have predominantly been portrayed as being stuck in a historicist rut, rejecting the new and endlessly reiterating the old, filling the American landscape with phony-colonial mansions and pseudogothic campuses. Henry-Russell Hitchcock spoke of the "sluggish life, sunk in inertia and conservatism" of twentieth-century "traditional" architecture,[9] and even those who find much beauty in the works of Gothic and Renaissance revivalists are still inclined to agree with Sigfried Giedion's view that this architecture used "the past as a means of escape from its own time by masking itself with the shells of bygone periods."[10] The essay that follows does not, in fact, contest Giedion's claim, but instead offers an alternative way of perceiving the history of American architecture in this period. By stripping away the mask, by peering inside the shell, by holding the shell up to the ear and *listening* to the building within, a new story is heard.

In this story, technology does not simply stimulate a new architecture; instead, architects, scientists, and engineers work together to transform both technology and architecture. Exploration of this interaction subsequently emphasizes certain affinities between

11.1

Saint Thomas's Church, exterior. Cram, Goodhue & Ferguson, 1908–14, New York. Although visually evoking a distant, Gothic past, St. Thomas's Church constituted a very modern kind of acoustical environment. The architects employed new, special-purpose acoustical building materials that resulted in a very untraditional sounding space within. *Architectural Record* 35 (February 1914): facing 101. The Athenaeum of Philadelphia. Copyright The McGraw-Hill Companies. All rights reserved. Reproduced with the permission of the publisher.

11.2

Philadelphia Savings Fund Society Tower, exterior. Howe & Lascaze, 1929–32, Philadelphia, Pennsylvania. Herald of the new aesthetic of Modern architecture in America, the PSFS tower celebrated visually and embodied acoustically the technological enthusiasm of the early twentieth century. *Architectural Forum* 57 (December 1932): 487. The Athenaeum of Philadelphia. Copyright BPI Communications Inc.

revivalists and the moderns who appeared on the American scene circa 1930. Orthodox histories of American architecture must always cross the Atlantic at some point in the late 1920s to pick up the new modern style that first developed in Europe and then transplant this aesthetic into American soil. The story that follows instead emphasizes the continuity between such seemingly antimodern American structures as St. Thomas's Church (Cram, Goodhue & Ferguson, New York, 1908–1913; see figure 11.1) and the herald of the new aesthetic, the Philadelphia Savings Fund Society Tower (Howe & Lascaze, Philadelphia, 1929–32; see figure 11.2).[11]

When Le Corbusier promoted listening to American engineers, he simultaneously warned "beware of American architects."[12] His justification for this warning was visual, however: a photograph of an American building. If Le Corbusier had listened to American architects or to their buildings, he might have rescinded his warning and perhaps found inspiration in their engagement with new technologies.[13] Similarly, if historians of architecture begin to listen to buildings as well as scrutinize them visually, they might formulate new narratives of the type that this essay attempts to present. And if historians of modern culture resist evoking "The Machine" and instead analyze particular machines, in the way that this essay analyzes the machine of acoustical science and technology, a richer understanding of the culture of American modernism, including architectural modernism, will surely result.

A Brief History of Architectural Acoustics[14]

A body of published literature addressing the problem of controlling the behavior of sound in rooms first appeared in Europe during the late eighteenth century.[15] Like the neoclassical architecture of that era, these treatments were rooted in geometrical analysis. European writers and builders such as Pierre Patte, Francesco Algarotti, and others sought to determine the one best form, a simple geometric form, that would optimize the propagation of rays of sound through space. The appearance of new, commercial theaters stimulated the concern over acoustics articulated in these texts. These spaces were much larger than the royal opera houses that had preceded them, and it was believed that the voices of the singers who performed there would be insufficient to fill these rooms with sound. The employment of one best geometrical form was seen as a means to support, even multiply or amplify, the rays of sound to ensure that everyone could hear. To determine this form, the authors traced, with pen and straightedge, the propagation of rays of sound as they reflected off the variously shaped surfaces represented on the plans of rooms.[16]

In America, the architect Benjamin Latrobe and his student Robert Mills represent this approach to architectural acoustics. In 1803, Latrobe carried out what he called "a very minute and laborious mathematical investigation of the number of echoes produced by different formed rooms." He concluded that the sphere was the best form, as it produced beneficial "rings" of echoes that reinforced the voice.[17] Mills followed a similar theory in 1830 when he was asked by Congress to improve the faulty acoustics of the Hall of the House of Representatives in the Capitol Building in Washington. He proposed modifying the chamber's irregularly shaped rear wall to approximate a uniform circular form, to transform what he called "disconsonant echo" into "consonant echo."[18]

The technique of tracing the propagation of rays of sound remained popular throughout the nineteenth century, but most practitioners admitted that such analysis offered no guarantee of success. Architectural writers on the subject of acoustics unanimously complained of the lack of scientific attention to this problem. In 1803, Latrobe had complained that "[t]here is still wanted . . . a system by which an architect could be guided in his design."[19] At century's end, architects and their clients were still seeking such a system. In 1895, the Science Standing Committee of the Royal Institute of British Architects looked into the "question of acoustics,"[20] hoping to shed light on what was still considered to be "a very obscure subject."[21] Committee member T. Roger Smith had published in 1861 an oft-cited treatise on the topic, but when the third edition of this text appeared in 1895, Smith could still only confess that architects depended more on instinct than on "a code of laws" when attempting to plan for acoustics, and Smith's Science Standing Committee effort similarly failed to answer the question of acoustics.[22] Yet in 1895, a young American physicist was commencing work that would, for the first time, provide architects with the kind of system that they had long sought.

In 1898, the philanthropist Henry Higginson was preparing to build a new home for his Boston Symphony Orchestra. Higginson required the advice of one who could ensure with the perceived authority of scientific laws that his hall would do justice to the great symphonic works of the Romantic era, particularly those of his favorite composer, Ludwig van Beethoven. He wanted specifically to re-create the acoustical quality of the Neues Gewandhaus at Leipzig.[23] Wallace Sabine, an assistant professor of physics at Harvard University, was recommended to Higginson by the University's president Charles Eliot to provide the necessary expertise. Sabine had spent the previous several years studying the behavior of sound in rooms. Initially asked by President Eliot to recommend alterations to a poor-sounding lecture room in the university's new Fogg Art Museum, Sabine turned this simple, almost casual request into the impetus for a lengthy and meticulous study of architectural acoustics.[24]

Operating within the intellectual framework of late-nineteenth-century physics, Sabine characterized sound not as geometric rays, but as a body of energy, capable of not just reflection but also absorption.[25] He chose not to manipulate sound through the geometrical *form* of reflective surfaces, but instead through the sound-absorbing qualities of the *materials* that constituted those surfaces. His technique was simple: A tank of compressed air was employed to sound an organ pipe in a room. After shutting off the air supply, Sabine would listen to the residual sound, or reverberation, until it was no longer audible. He measured this interval, the reverberation time, with a chronograph accurate to hundredths of a second. Sabine measured the reverberation times of numerous rooms on the Harvard campus and in Cambridge and Boston, and he additionally manipulated the reverberation times of these rooms by introducing to them sound-absorbing seat cushions borrowed from Harvard's Sanders Theater.

By 1898, Sabine had spent three years collecting data and had no practical result to show for his efforts; the Fogg lecture hall remained unusable and unused. President Eliot eventually ran out of patience with the zealously precise young scientist, and he responded to Sabine's request for several more years of research time with the ultimatum: "You have made sufficient progress to be able to prescribe for the Fogg Lecture Room, and you are going to make that prescription."[26] Thus forced, Sabine had panels of sound-absorbing felt attached to certain wall surfaces in the lecture room, and the auditorium was finally marginally usable.

Higginson's subsequent request for acoustical expertise regarding his new music hall led Sabine to review his research notebooks, and at this time he was able to perceive a mathematical relationship amidst the mass of data that he had so carefully accumulated. He developed a formula that related a room's reverberation time to its volume and to the sound-absorbing power of the materials that constituted its surfaces:

$$T = \frac{.164 \; V}{S \; \alpha_a},$$

where

T = reverberation time (in seconds),

$.164$ = a hyperbolic constant,

V = the volume of the room (in cubic meters),

S = the surface area of the room (in square meters), and

α_a = the average coefficient of absorption in the room.

With this mathematical relationship in hand, Sabine was now able to quantify the sound-absorbing properties of different architectural materials, such as glass, plaster, and wood. Sabine's formula, in tandem with these new "coefficients of absorption," could then not only be used to determine how to alter the acoustical properties of extant rooms, it could also be applied to the designs of rooms not yet built, to predict and then manipulate their acoustical qualities.[27]

Sabine applied his formula to the plan for Higginson's music hall that had been developed by Charles McKim of McKim, Mead & White. His calculations indicated that the new hall's reverberation time would significantly exceed that of the Neues Gewandhaus at Leipzig. Sabine thus suggested changes to achieve the acoustic result that Higginson sought. Together, McKim and Sabine developed a revised plan with a reduced overall volume and a new seating arrangement (a second balcony was added to maintain the au-

11.3

Symphony Hall, section. McKim, Mead & White, 1900, Boston, Massachusetts. Harvard University physicist Wallace Sabine worked with the architects during the design and construction of the hall to ensure its acoustical quality, employing a mathematical formula that he had derived from experimental data collected over the preceding five years. *Architectural Review* 7 (1900): plate LXV. Princeton University Library. Copyright The McGraw-Hill Companies. All rights reserved. Reproduced with the permission of the publisher.

dience capacity of the now smaller hall). Sabine was consulted numerous times as the hall was constructed over the next two years and offered scientifically informed advice on such items as the placement of the organ pipes, the best type of wood to use for paneling, and the type of seating to employ.[28] The Boston Symphony Orchestra presented its first concert in the new hall in October 1900, and the result was judged an acoustical success.[29] (See figure 11.3.)

Perhaps more significant than the actual sound of Symphony Hall, however, was the power and simplicity of the formula that Sabine had applied to it. Architects embraced Sabine's scientific answer to the "question of acoustics," and they welcomed the opportunity to employ his formula, or to hire an acoustical consultant who wielded that formula, to inform their designs for churches, concert halls, and legislative assembly halls.

After Symphony Hall, Sabine continued to work with McKim, Mead & White, and by 1916, more than eighty additional firms, including Cram, Goodhue & Ferguson, had sought his services.[30] Other scientists and engineers, too, were easily able to pick up Sabine's method and develop his program of measuring the absorbing powers of various building materials. They could then, like Sabine, utilize this data to modify extant spaces or to predict the reverberation of projected rooms through analysis of architects' plans and application of the reverberation formula. Sabine initially employed traditional building materials such as wood, plaster, and glass to control sound. Not long after his work on Symphony Hall, however, he began to develop new, special-purpose acoustical materials. The scientists and engineers who followed Sabine increasingly employed these new

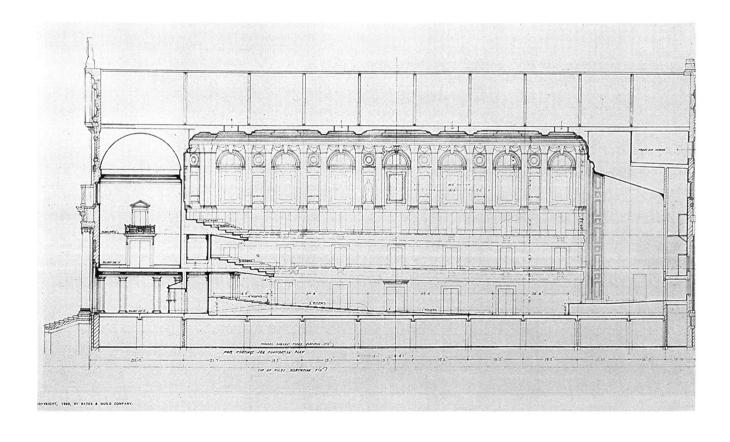

kinds of building materials for solving their problems of acoustical design, and the employment of these new materials introduced a new sound, a modern sound, to American architecture.

The Sound of Modernity

The proliferation of new, sound-absorbing building materials only reflects the larger material bounty of the early decades of the twentieth century. This was a time in which Louis Mumford chose to organize an entire history of technology and civilization around the materials that constituted human construction, and Mumford identified his own era as one characterized by the promise of new kinds of man-made materials.[31] The origins of the new acoustical materials, however, are located in the very structures that visually seemed most resistant to the promise of the future, buildings that appeared to cling longingly to the distant past, buildings of the Gothic revivalists Cram, Goodhue & Ferguson.

The medievalism of Ralph Adams Cram was motivated by his belief in the symbolic impoverishment of contemporary American culture. Religious ceremony, as practiced in pre-Reformation England, was, for Cram, the key to rejuvenation, and his neo-Gothic churches provided environments in which worshipers could escape from the secularized world of the surrounding city. They offered what historian Jackson Lears has called "therapeutic antidotes to feverish modern haste."[32]

Cram considered the Gothic style appropriate only for those institutions that could trace their history back to pre-Reformation days, specifically, churches and universities. Yet he recognized that these institutions did not survive unchanged. The intoned Latin chants of the medieval mass had been replaced, for Cram's clients, by the protestant sermon, which emphasized intellectual as well as spiritual engagement on the part of the congregation. Although Cram insisted that a church's primary function remained that its inhabitants "be filled with the righteous sense of awe and mystery and devotion,"[33] he also recognized that an ideal church must be a place "where a congregation may conveniently listen to the instruction of its spiritual leaders."[34] Cram's medieval aesthetic conflicted with the acoustical necessities of the modern service, and he turned to Wallace Sabine to resolve this conflict.[35]

In 1911, Cram introduced Sabine to Raphael Guastavino, a ceramic craftsman whose family had brought to America from Spain the technique of thin-shelled, or timbrel, vaulting. The Guastavino system of vault construction utilized multiple layers of terra cotta tiles to build large domes and arches in buildings such as the Boston Public Library, Penn Station in New York City, and the churches and chapels of Cram, Goodhue & Ferguson.[36] Cram introduced Guastavino to Sabine with hopes that the two would collaborate to develop a vaulting tile that would absorb sound to an extent that would render his churches suitable for sermons as well as for high masses. The result of their collaboration, a porous ceramic material named Rumford tile, was first utilized in St. Thomas's Church, New York. (See figures 11.1 and 11.4.) Rumford presented a surface that absorbed 26 percent of incident sound energy, much more than the 3 percent typically absorbed by traditional stone surfaces. Rumford's successor, Akoustolith, absorbed even more: as much as 60 percent of incident sound energy.[37]

Rumford and Akoustolith were utilized in hundreds of churches and chapels, temples and secular buildings all over the country. (See figure 11.5.) By drastically reducing reverberation times, these materials profoundly affected the sound of the spaces that

11.4

Saint Thomas's Church, interior. Cram, Goodhue & Ferguson, 1908–14, New York. Interior surfaces were covered with Rumford Tile, a porous ceramic material developed by Wallace Sabine and Raphael Guastavino. Rumford absorbed far more sound energy than did traditional stone or ceramic tiles, so although St. Thomas's looked traditional, its sound was not. *Architectural Record* 35 (February 1914): 105. The Athenaeum of Philadelphia. Copyright The McGraw-Hill Companies. All rights reserved. Reproduced with the permission of the publisher.

11.5

Guastavino Company advertisement, Temple B'Nai Jeshurun, interior with Rumford tile. Albert Gottleib, 1915, Newark, New Jersey. One of hundreds of installations utilizing the company's new sound-absorbing acoustical tiles. *Brickbuilder* 24 (December 1915): 16. Princeton University Library. Copyright BPI Communications Inc.

they enclosed: They transformed the traditional relationship between sound and space. Reverberation is a way to experience space through time, thus these new materials effected a transformation of traditional space-time relationships. Like other contemporary reformulations of space and time, such as Cubist art and non-Euclidean geometry, these materials constitute signposts of the new culture of modernism.[38] In this way, structures like St. Thomas's should not be seen as pseudovestiges of the past but should instead be placed at the forefront of cultural change.

Rumford and Akoustolith initiated a rapid proliferation of new materials for controlling the relationship between sound and space. By the 1920s, architects could choose from an endless variety of products to compose interior surfaces. Acousti-Celotex, Sanacoustic Tile, Acoustone, Sabinite, and Sprayo-Flake Acoustical Plaster represent only a sampling of what was available. These materials were made of seemingly anything and everything: mineral wool; fibers from flax, wood, and sugarcane; disinfected cattle hair; and asbestos. There were insulating papers, rigid wall boards, ceramic tiles, sprayed-on plaster, and all sorts of mechanical devices for structurally isolating floors, walls, and ceilings.[39]

Whereas Guastavino tile construction drew upon a craft tradition of hand production and skilled installation, those materials that followed were mass produced by modern industrial corporations. One of the largest, the Celotex Corporation, started in 1920 by producing 60,000 square feet of building board per day. Less than a decade later, the daily output was 1,600,000 square feet.[40] Materials like Acousti-Celotex not only were mass produced by new techniques, but additionally were employed in installations very different from those where Rumford and Akoustolith had been used. Guastavino acoustical tiles were typically used in places such as churches, chapels, and synagogues—places that most people visited but did not inhabit continually. Celotex, in contrast, promoted the use of its acoustical products in such quotidian spaces as offices, factories, and shops. By 1925, not only had architectural acoustics become big business, but the locus of activity itself increasingly moved into the commercial world.

Cass Gilbert's New York Life Insurance Building (see figure 11.6), for example, constituted in its day the world's largest single installation of sound-absorbing materials.[41] "To safeguard and promote the health of its 3,500 home office employees," the *American Architect* reported, "the New York Life Insurance Company carries on an extensive welfare program that required consideration in connection with the planning of the home office building."[42] The New York Life Insurance Company provided its employees with "healthful working conditions," including controlled heating and ventilation, sterilized drinking water, and quiet. The building was acoustically treated throughout; offices, cafeterias, lounges, even the pneumatic mail tube system were all insulated to prevent the accumulation and transmission of unwanted sounds.

Most of the acoustical treatment consisted of a thick felt of asbestos and sanitized cattle hair cemented to walls and ceilings. The felt was covered with decorative fabric chosen to suit the location. In addition, heavy window glass kept out external noise, forced ventilation ensured that the number of open windows would be kept to a minimum, and hardware and plumbing were selected for "quiet operation."[43]

By 1930, this kind of "soundproof construction" was identified as a primary feature to seek in apartments as well as in office buildings.[44] Such construction might consist of structural barriers like the Herringbone Rigid Metal Lath. (See figure 11.7.) Advertisements claimed that Herringbone was "proven an effective barrier to the most penetrating sound."[45] Also available were wall coverings like Sprayo-Flake Acoustical Plaster. Sprayo-Flake advertisements described the product as an "insulating blanket," "sprayed on with guns" to form a "thick blanket of insulation covering the surface and sealing all cracks and crevices."[46]

The acoustical ideal represented in these advertisements contrasts sharply with Henry Higginson's earlier criterion for "good acoustics." Whereas Higginson's goal had been to create a musical temple that would do justice to the great symphonic masterpieces of his idol Beethoven, the new goal of acoustical construction was something quite different:

> The fact that noise has become a problem necessitating control indicates a fundamental and important change in the life of society. This change relates to the mushroom growth in complexities of existence. It is noticed in pressure and confusion. The stream of business and industrial life swirls daily into new nervous whirlpools; and the demand is growing that more compensatory measures for private life be developed. . . . The isolation of sound is one such measure.[47]

11.6

New York Life Insurance Building, exterior. Cass Gilbert, 1929, New York. Not only did the NYLI Tower house one of the most famous examples of the modern American corporation, the building itself was just as innovative. When built, it constituted the world's largest single installation of sound-absorbing materials. These materials were employed to reduce noise levels within the building to reduce workers' stress and increase their productivity. *American Architect* 135 (20 March 1929): 357. The Athenaeum of Philadelphia. Copyright The McGraw-Hill Companies. All rights reserved. Reproduced with the permission of the publisher.

11.7

Herringbone Rigid Metal Lath advertisement. By 1930, many apartment buildings offered tenants "soundproof construction," which was designed not only to prevent the transmission of sound between rooms, but also to seal the interior off from the increasing cacophony of the city outside. *Architectural Forum* 39 (July 1923): 25 (advertisements section). Princeton University Library. Copyright BPI Communications Inc.

265 | LISTENING TO/FOR MODERNITY

Concern over the din of urban life had, of course, existed in the nineteenth century and no doubt long before, but the difference between earlier concerns and those voiced in the 1920s is one of substance as well as of degree. Acoustically controlled architecture had traditionally consisted of places that offered temporary refuge from the turmoil without, places like churches and concert halls. As the scale and pervasiveness of industry grew, however, and as the nature and landscape of city living changed, escape from the noise no longer seemed possible. Acoustical materials were thus increasingly used to attempt to eliminate the noise of the world.

In 1930, for example, New York City appointed a Noise Abatement Commission to study the sources, amount, and effect of noise in New York. The city health commissioner explained, "I am confident that we can live these lives of ours more efficiently if we do away with the unnecessary noises that surround us."[48] This movement to eliminate noise was driven by hopes for a particular kind of noise-free world. Noise abaters did not conjure up visions of pastoral quiet, of peaceful, preindustrial landscapes. Quiet was instead a means to better mental health, greater concentration, and more focused activity. The goal of this quest for quiet was a more efficient, productive, modern industrial society.

The harmful effects of noise on production were demonstrated by "scientific" findings, for example, those of Donald Laird, an industrial psychologist at Colgate University who studied the effect of the aural environment on office workers' productivity. In 1929, Laird measured the difference in energy expended by typists working under both quiet and noisy conditions. His experiments indicated that an average typist expended almost 20 percent more energy when working under noisy conditions than when in a quiet environment.[49] Laird's results were cited in advertisements, articles, and textbooks describing acoustical treatments and products, and they helped stimulate the application of sound-absorbing materials in countless offices across the United States.[50]

Though the Noise Abatement Commission ultimately failed to achieve its goal of silencing the streets of New York, the campaign did succeed in promoting the employment of acoustical building materials to isolate commercial offices from the noise without. And, the noise of business itself was increasingly attacked and eliminated. The acoustical technology employed within the New York Life Insurance Building isolated the interior of the structure from the surrounding cacophony of the city, and it eliminated the unnecessary sounds of typewriters, adding machines, and other office technologies generated within. Gilbert's skyscraper was at the forefront of acoustical design circa 1929, and in this sense its design was as modern as the large business organization that it housed. Yet stylistically, the skyscraper was far from innovative. Its neo-Gothic exterior was complemented on the interior with medievalesque murals depicting tales of knights, princesses, and fairies. At least one observer perceived the incongruity of a "Gothic" insurance office:

The problem of furnishing the special rooms of the New York Life Insurance Company Building presented many interesting aspects. To begin with, the period of the building designed by Cass Gilbert was, to use his own term, "American Perpendicular," but on examination one could see Mr. Gilbert's interpretation of Tudor or Gothic motives, connected with English tradition, adapted to our skyscraper form. In this twentieth century . . . it is necessary to select a period of interiors and furnishings wholly different from those used by our Mediaeval ancestors. To be more in accord with our present day business life, then, it was decided that the English period of the eighteenth century would be generally featured in the principal rooms.[51]

The interior designer, while leaping aesthetically forward several centuries by foregoing Gothic for Georgian, still landed far short of his own day and age. One space within the building did, however, fully and unselfconsciously express visually the modernity of the organization housed in the tower.

The kitchen of the NYLI Building, located deep in the basement of the structure, was little noticed by the critics who intensely scrutinized the architecturally legitimated structure that rose above it. This room constituted, not an aesthetically designed space, but instead, the solution to a particularly complicated "engineering problem," that of providing 7,000 hot, nourishing meals to employees between the hours of 11 A.M. and 2 P.M.[52] The solution to this problem was not cloaked in tapestry or tracery; instead, it was a sleek, steel space, filled with the efficient technology of polished metal machines, ducts, and hoodwork. (See Figure 11.8.) American architects and architectural critics generally did not recognize the beauty of this type of engineered environment in the 1920s. In Europe, however, this look—the look of American engineering—was already contributing to a transformation of architectural design.

11.8

New York Life Insurance Building, Dishwashing Room. Upstairs, the NYLI Tower concealed its modernity beneath a screen of neo-Gothic imagery and iconography. But here in the basement, the engineered environment of the modern commercial kitchen was left "as is": an efficient space filled with steel machines and polished metal ductwork. This was the "look" that European architects were finding increasingly compelling. *American Architect* 135 (20 March 1929): 399. The Athenaeum of Philadelphia. Copyright The McGraw-Hill Companies. All rights reserved. Reproduced with the permission of the publisher.

267 | LISTENING TO/FOR MODERNITY

A new aesthetic was developing in Europe, but it developed without the acoustical technologies available in America. Sound-controlling materials weren't as prevalent in Europe, nor were the techniques of their employment as widely known there.[53] Mies van der Rohe had called for such materials in 1924 when he proclaimed:

> Our technology must and will succeed in inventing a building material that can be manufactured technologically and utilized industrially, that is solid, weather-resistant, soundproof, and possessed of good insulating properties.[54]

Mies's call went largely unheeded abroad, but when the new aesthetic arrived in America, modern architects found both a market full of the very kinds of materials that Mies had sought and a tradition of employment of those materials that helped establish the new style in this country.

That new style was formally presented to the American public in 1932. Within the walls of the Museum of Modern Art and on the streets of Philadelphia, the new principles of design were set forth. The MOMA exhibit on "The International Style" catalogued the principles that the Philadelphia Savings Fund Society Tower of Howe & Lascaze exemplified: a new emphasis on architectonic volume rather than mass, regularity rather than symmetry, and a vehement proscription of "arbitrary" applied decoration.[55]

The functionally differentiated spaces of the PSFS Tower—the main banking room, the office block above, and the elevator bank that serviced those offices—were all distinguished by the different volumes that constituted the structure. The asymmetrical arrangement was efficient and orderly, and ornament was "conspicuous by its absence." (See figure 11.2.) "The surfaces of machine production are inherently beautiful," one reviewer wrote, and they produce "a natural aesthetic movement supplied in other buildings by sculptured swags and terra cotta gargoyles."[56] Customers were conveyed by sleek steel escalators (see figure 11.9) to a banking room filled with gleaming curved surfaces of chrome and glass (see figure 11.10). The Society, usually conservative in matters of style as well as of finance, had "gone Gershwin,"[57] and it was Ira's brother, not some long-dead king of England, who inspired the "Georgian" interior of this office building. With the PSFS Tower, the engineered environment of the kitchen of the New York Life Insurance Building moved upstairs to become an architecturally legitimated space.

Although the PSFS Tower heralded a new, machined look for architecture in America, the acoustical technology employed within was not revolutionary but instead followed a continuous line of development that connected it historically to the New York Life Insurance Building, St. Thomas's Church, and Symphony Hall. The Savings Fund Society's banking room and offices were all treated with sound-absorbing materials to maximize the efficiency of the bank's clerical workers and to minimize noise-induced stress experienced by customers and bank executives.

11.9

Philadelphia Savings Fund Society Tower, escalators. Howe & Lascaze, 1929–32, Philadelphia, Pennsylvania. The "look" of machines and bare steel moves upstairs to become aesthetically legitimate in America. *Architectural Forum* 57 (December 1932): 489. The Athenaeum of Philadelphia. Copyright BPI Communications Inc.

11.10

Philadelphia Savings Fund Society Tower, main banking room. Howe & Lascaze, 1929–32, Philadelphia, Pennsylvania. Although not fully visible in this reproduction, the original photograph indicates clearly the gridded surface of the banking room ceiling, a grid created by the system of sound-absorbing tiles specially developed for this project. In the PSFS tower, acoustical and visual modernity converge. *Architectural Review* (London) 73 (March 1933): 104. Princeton University Library. Reproduced courtesy of the Architectural Review.

The acoustical treatment employed throughout the PSFS Tower consisted of arrays of sound-absorbing ceiling tiles fastened onto suspended metal frameworks. The acoustical treatment of office ceilings was, by 1932, standard practice,[58] but the innovation of a suspended tile grid fully integrated the acoustical technology with the new aesthetic.[59] The mass-produced tiles fit the industrial ideology of the new modern style, and the regular, rectangular patterns that resulted from their use modularized the ceiling in a way that purists found pleasing. In their definitive catalogue of new style, MOMA curators Henry-Russell Hitchcock and Philip Johnson referred to the "geometrical web of imaginary lines," the "scheme of proportions" that "integrates and informs a thoroughly designed modern building."[60] Photographs of the interiors of the PSFS tower indicate that this imaginary web, in fact, became real on the ceilings of the various rooms.[61] (See figure 11.10.)

Conclusion

Hitchcock and Johnson missed the material reality of the PSFS ceiling web. Perhaps if they had looked up, or if they had *listened* to the architectural spaces within the building, they might have perceived this web as a network of strands that not only organized the structure, but additionally connected it to its American past. They might then have bridged the historical "breach"[62] that seemed to separate the new, modern buildings from those that had immediately preceded them; they might have developed a "plot" for the story of early-twentieth-century architecture, a plot that continued to elude Hitchcock for the next fifty years.[63]

Such a plot might sound something like this: At the turn of the new century, the architect Charles McKim collaborated with the scientist Wallace Sabine to create what was heralded as the world's first "scientifically designed" concert hall.[64] Soon thereafter, Sabine applied his scientific understanding of the behavior of sound in rooms to the work of Cram, Goodhue & Ferguson, to enable those architects to manipulate the sound of the spaces that they created. Sabine worked with Raphael Guastavino to create traditional-looking building materials with very untraditional acoustical properties. These new materials were employed in structures like St. Thomas's Church to create spaces in which traditional relationships of sound and space, relationships of time and space, no longer applied. Cram believed that such innovative technological practice was true to the spirit of the Middle Ages, which he hoped to recreate in twentieth-century America. By striving in this way to reawaken the spirit of the past, he perhaps paradoxically stimulated the modern spirit of his own age, a spirit he would vociferously condemn.[65]

Cram's transformation was limited to the sacred space of the church and chapel; Cass Gilbert, through his New York Life Insurance Building, transmitted that transformation into the secular world of office skyscrapers. The sound-absorbing materials in his structure were put to work on working people to reinforce the productive values of a modern, industrial society. Howe & Lascaze, with their Philadelphia Savings Fund Society Tower, continued this process, and additionally celebrated it visually by means of the new European style.

Perhaps American architecture in the years between 1900 and 1930 can be thought of as a sort of chrysalis. The (to some, homely) caterpillar of late-nineteenth-century architecture sequestered itself for several decades. While externally appearing static, while presenting no growth or change to the observer, significant transformations were occurring within that concealing outer shell. Circa 1932, the shell cracked open and fell away to reveal a change as spectacular as the pattern of colors adorning the butterfly that replaces the caterpillar.

In order best to appreciate that final result, however, one needs to understand the internal, invisible developments and transformations that brought it about. By listening to the transformations that occurred within the shell, the invisible has become knowable. This knowledge indicates that the appearance of Modern architecture in America was not simply a reaction or a response to new scientific and technological conditions; it was instead the result of architects, scientists, and engineers working together and stimulating one another to change the intellectual circumstances of their existence and the material substance of their environment.

When, in 1914, Bertram Grosvenor Goodhue struggled to understand what his clients wanted when they called for something "modern," he knew the answer would be found by turning to his friend Wallace Sabine. He urged the scientist to come see him and discuss the subject in person. Though the specific conversation that ensued left no immediate historical record, we can still listen in on such dialogues between architects and scientists by listening to the buildings that resulted from their collaboration. By listening to these buildings, we enhance our own understanding of what it means for a building to be "modern."

Notes

Earlier versions of this essay were presented at the 1994 conference on the Architecture of Science at Harvard University, the 1994 meeting of the Society for the History of Technology, and the Workshop Series of the Department of History and Sociology of Science at the University of Pennsylvania. I thank the participants at all of these events for helpful questions and comments. Thanks also to the referees and editors at The MIT Press, and to Peter Galison, for careful, critical, and constructive readings. Finally, special thanks to Charles Gillispie, who supervised—with equal measures of rigor and enthusiasm—the doctoral dissertation from which this essay has evolved.

1. Papers of Wallace Clement Sabine, 1899–1919, Correspondence, Bertram Grosvenor Goodhue to Sabine (October 22, 1914) [HUG 1761.xx], Courtesy of the Harvard University Archives (hereafter "Sabine Correspondence"). Emphasis in original. The Pusey Library at Harvard holds photocopies of the Sabine correspondence; the originals are located at the

Riverbank Acoustical Laboratories, Illinois Institute of Technology Research Institute, Geneva, IL.

2. Goodhue consulted Sabine numerous times as he designed and built St. Bartholomew's. Most queries concerned the effect of different aspects of the architect's design upon the acoustics of the church. No response to Goodhue's letter concerning a "modern" exterior has been located. See Goodhue to Sabine (February 19, 1916) and Sabine to Goodhue (February 21, 1916), Sabine Correspondence; Smith, *St. Bartholomew's Church;* and Oliver, *Bertram Grosvenor Goodhue.*

3. Marshall Berman also employs this general chronological organization in a treatise on modernization and modernity that emphasizes European culture: *All That Is Solid Melts Into Air: The Experience of Modernity,* pp. 16–17. See also Harvey, *Condition of Postmodernity,* chap. 2, "Modernity and Modernism," pp. 10–38. For more on the American context of Victorian culture and modernization, see Howe, "American Victorianism"; Brown, "Modernization: A Victorian Climax"; and Ward and Zunz, eds., *Landscape of Modernity.* See also the special issue, "Focus on American Modernism," of *Modernism/Modernity* 3 (September 1996), which emphasizes literary modernism.

4. Coben, "Assault on Victorianism."

5. Umberto Boccioni et al., "Manifesto of the Futurist Painters," quoted in Berman, *All That Is Solid,* pp. 24–25. For more on the Italian Futurist painters and their acoustical collaboration with modern musicians such as Luigi Russolo, see Morgan, "'A New Musical Reality.'"

6. Smith, *Making the Modern;* Kern, *Culture of Time and Space,* p. 119.

7. Kern, *Culture of Time and Space,* p. 98.

8. Le Corbusier, *Towards a New Architecture,* p. 42. While I have cited an English translation of the 13th French edition, the passage also appeared in the first (French) edition of 1923, on p. 29.

9. Hitchcock, *Architecture,* p. 531.

10. Giedion, *Space, Time and Architecture,* p. xliii.

11. While the modern architects themselves loudly proclaimed the ahistorical character of their work, as early as 1929 architectural historians had begun to construct a historical context for the new style of building. From Henry-Russell Hitchcock's 1929 *Modern Architecture: Romanticism and Reintegration* to current histories on bookstore shelves today, the work of the moderns has been traced back to the late-nineteenth-century ideas and designs of the Art Nouveau movement, Henry Hobson Richardson, Louis Sullivan, the Chicago School, and Frank Lloyd Wright. Yet, the turn of the century gulf, especially in the American context, remains. Most often, the precipice appears when Daniel Burnham takes over the construction of the buildings for the Chicago World's Columbian Exposition of 1893. The White City at Jackson Park is portrayed as a mausoleum for the innovative architectures of Sullivan and Wright, and in their place rises the historical revivalism of McKim, Mead & White et al. The title of William Jordy's account of American architecture, *The Impact of European Modernism in the Mid-Twentieth Century,* indicates the emphasis in contemporary histories on the European origins of modern American architecture. This storyline similarly dictates the organization of material and chapters in Hitchcock's *Architecture.* See also Pevsner, *Sources of Modern Architecture;* Frampton, *Modern Architecture;* Handlin, *American Architecture;* Roth, *American Architecture;* and Curtis, *Modern Architecture since 1900.* Curtis does begin to create a place

for the "older tradition" of revivalism in his account, but I believe there is a much more significant connection between the revivalist tradition and the new modern aesthetic. See Curtis, chap. 17, "The Continuity of Older Traditions."

12. Le Corbusier, *Towards a New Architecture,* p. 42.
13. Le Corbusier would become increasingly aware of the acoustical aspects of architectural space. He worked with the acoustical consultant Gustave Lyon on a 1927 entry in the design competition for the Large Assembly Hall for the League of Nations, and he experimented in his Pavillon Suisse of 1931 to remedy problems of sound transmission through the partition walls. His collaboration with the avant-garde musician Edgard Varèse on the Philips Pavilion at the 1958 Brussels World's Fair represents the architect's full appreciation of the acoustical aspect of space and its aesthetic possibilities. See Banham, *Architecture of the Well-Tempered Environment,* pp. 153–155; Osswald, "Acoustics of the Large Assembly Hall"; and Treib, *Space Calculated in Seconds.*
14. For an extended treatment of the history that is briefly sketched out in this section, see Thompson; "'Mysteries of the Acoustic.'" See also Forsyth, *Buildings for Music.* For architects' consideration of the acoustical aspects of building, see Rasmussen, *Experiencing Architecture,* pp. 224–237; Conrads and Leitner, "Audible Space" in *Daidalos* 17 (September 1985), a special issue on sound and architecture; Fitch, *American Building 2,* pp. 131–157; and Elliott, *Technics and Architecture,* pp. 407–431.
15. Although earlier texts treating architecture and acoustics exist, most notably the writings of Vitruvius and the sixteenth-century Jesuit Athanasius Kircher, not until the second half of the eighteenth century did a sustained textual treatment of the subject develop. For more on premodern building technology, see Fitchen, *Building Construction Before Mechanization;* and Mark, ed., *Architectural Technology.*
16. Though these authors did not agree upon which particular form was best, all followed the same general approach to the problem. See Algarotti, *An Essay on the Opera;* Patte, *Essai sur l'architecture théâtrale;* Saunders, *A Treatise on Theatres;* and Wyatt, *Observations on the Design for the Theatre Royal.* For a more general treatment of the works of these and other late-eighteenth-century architects, see Pérez-Gómez, *Architecture and the Crisis of Modern Science;* and Rykwert, *The First Moderns.* Forsyth's *Buildings for Music* describes the architectural transformation of music rooms, and Johnson's *Listening in Paris* charts the social and cultural changes that led to new demographics of concert going in eighteenth- and early-nineteenth-century Paris.
17. Latrobe, "Remarks on the Best Form of a Room for Hearing and Speaking," pp. 403, 404.
18. Mills, "Memorial: House of Representatives," p. 3.
19. Latrobe, "Acoustics," p. 120.
20. Burrows, "Sound in its Relation to Buildings," p. 65.
21. "Discussion on Sound in Its Relation to Buildings," p. 78.
22. Smith, *Acoustics,* p. 3; and *Royal Institute of British Architects Journal* (3rd series) 4 (6 May 1897): 323. See Forgan (this volume) for more on Smith and the Science Standing Committee.
23. The Neues Gewandhaus at Leipzig, completed in 1886, was designed by Martin Gropius and Heinrich Schmieden. In his account of this concert hall, Michael Forsyth notes, "The reputation of the Neues Gewandhaus became established at once, and the main hall was re-

garded from the outset as a model of acoustical excellence." Yet, he presents no evidence of a conscious effort at acoustical design by the architects. See Forsyth, *Buildings for Music,* pp. 208–214 (quotation from p. 214).
24. The following summary of Sabine's work is elaborated upon in Thompson, "'Mysteries of the Acoustic'" and "Dead Rooms and Live Wires." See also Beranek, "Notebooks of Wallace C. Sabine" and "Wallace Clement Sabine and Acoustics"; and Orcutt, *Wallace Clement Sabine.*
25. This approach stems directly from Sabine's background in studying other kinds of physical energy, including electromagnetic energy. See, for example, Trowbridge and Sabine, "Selective Absorption of Metals" and "Electrical Oscillations in Air."
26. Charles Eliot to Wallace Sabine, 3 November 1897, quoted in Orcutt, *Wallace Clement Sabine,* p. 125.
27. Sabine's own account of his research and a well-detailed development of the reverberation equation is found in Sabine, "Reverberation." See also Thompson, "'Mysteries of the Acoustic.'"
28. The correspondence among Higginson, McKim, and Sabine preserves the process whereby the design was worked out. See, for example, Henry Lee Higginson to Charles Follen McKim: 26 January 1899 and 5 May 1899, Papers of McKim, Mead & White, New-York Historical Society, Folder M-10: Boston Music Hall. See also Wallace Sabine to Higginson: 26 February 1899, 13 November 1899, and 8 March 1900; and McKim to Higginson: 27 February 1899, 17 March 1899, and 8 November 1899, Henry Lee Higginson Collection, Historical Collections, Baker Library, Harvard Business School.
29. Early reviews of the hall offered a mixture of praise and criticism, but a consensus regarding the hall's acoustical excellence soon emerged, and it remains in place today. For more on the critics' reception of Symphony Hall, see Thompson, "'Mysteries of the Acoustic.'"
30. For more on Sabine's consulting career, see Thompson, "'Mysteries of the Acoustic'" and "A Tale of Two Physicists." See also Beranek and Kopec, "Wallace C. Sabine, Acoustical Consultant."
31. Mumford, *Technics and Civilization.*
32. Lears, *No Place of Grace,* p. 194.
33. Cram, *Gothic Quest,* p. 101.
34. Cram, *Church Building,* p. 10.
35. Cram's willingness to apply the technological tools of the modern world to improve upon the medieval is identified (in an analysis of Cram's writings, rather than of his buildings) in Clark, "Ralph Adams Cram." Still, Clark shows how Cram saw such technological innovation as a primary characteristic of the Middle Ages rather than of his own era. See also Muccigrosso, *American Gothic.*
36. A detailed account of the history of Guastavino construction is found in Collins, "The Transfer of Thin Masonry Vaulting from Spain to America." See also Milkovich, "Guastavino Tile Construction." For a contemporary account, see Wight, "The Works of Raphael Guastavino."
37. Rumford was made of a mixture of clay, feldspar, and "vegetable bearing earth," or peat. During firing, the peat burned out and left sound-absorbing pores on the tile's surface and throughout its body. Akoustolith was a similarly porous aggregate of pumice particles loosely

bonded with Portland cement. United States Patent Office, W. C. Sabine and R. Guastavino, "Wall and Ceiling of Auditoriums and the Like," #1,119,543 (1 December 1914); and W. C. Sabine and R. Guastavino, "Sound Absorbing Material for Walls and Ceiling," U.S. Patent #1,197,956 (12 September 1916).

38. For more on these transformations, see Kern, *Culture of Time and Space,* chap. 6, "The Nature of Space"; Harvey, *Condition of Postmodernity,* chap. 16, "Time-Space Compression and the Rise of Modernism as a Cultural Force"; and Friedland and Boden, eds., *NowHere.*

39. Knudsen, *Architectural Acoustics,* is a valuable source for data, physical descriptions, and photographs of the many acoustical products available by 1932. See also annual editions of *Sweet's Architectural Trade Catalogue.*

40. Production figures come from biographical entries of Celotex founders: "Bror Dahlberg," *National Cyclopedia of American Biography* (New York: James T. White and Co., 1930) C: 327; and "Carl Muench," *National Cyclopedia of American Biography* (Clifton, NJ: James T. White and Co., 1979) 58: 532.

41. Green, "Soundproofing," p. 411.

42. "Planning for Employees' Welfare," p. 397.

43. Green, "Soundproofing," p. 412.

44. Dahl, "Check List," p. 371.

45. Herringbone advertisement, *Architectural Forum* 39 (July 1923): 25 (advertisements section).

46. Sprayo-Flake Company advertisement, *Sweet's Architectural Trade Catalogue* (1931): B2513–B2514.

47. Sherman, "Sound Insulation," p. 373. For more on the changing criteria of architectural acoustics in the early twentieth century, see Thompson, "Dead Rooms and Live Wires."

48. Brown et al., eds., *City Noise,* p. 218.

49. Laird, "Measurement of Effects of Noise," p. 432.

50. See, for example, the 1931 entry in *Sweet's Architectural Trade Catalogue* for the United States Gypsum Company, p. B2675; Swan, "Noise Problems in Banks"; Knudsen, *Architectural Acoustics,* p. 447; and Paul Sabine, *Acoustics and Architecture,* p. 224.

51. von Ezdorf, "Design of the Interior of the New Home Office Building," p. 369.

52. "Planning for Employees' Welfare," p. 398.

53. One acoustical consultant active in France in the 1920s was Gustav Lyon. Lyon, head of the Pleyel piano company, contributed to the 1927 design of the Salle Pleyel, a concert hall in Paris. Also in 1927, he worked with Le Corbusier on a design for the League of Nations Assembly Hall competition. In both these projects, Lyon neglected the sound-absorbing properties of materials and instead focused on the reflective patterns established by curved wall surfaces. Forsyth characterizes the acoustics of the Salle Pleyel as "a notorious disappointment." See Forsyth, *Buildings for Music,* pp. 262–267 (quotation on p. 263); Calfas, *La Nouvelle Salle;* and Lyon, *L'acoustique architecturale.*

54. Ludwig Mies van der Rohe, "Industrialized Building," p. 82, in Conrads, ed., *Programs and Manifestoes.*

55. Hitchcock and Johnson, *International Style,* p. 20.

56. "New Shelter for Savings," p. 488.

57. "New Shelter for Savings," p. 483.

58. In 1932, Vern Knudsen wrote, "The absorptive treatment of offices is nearly always accomplished by treating the ceiling with a highly absorptive material . . ." Knudsen, *Architectural Acoustics,* p. 453.
59. William Jordy has described the acoustical technology as follows: "These tiles were cast, perforated, plaster of Paris panels with rock-wool pellets poured in place above, making the tiles four inches thick. PSFS ordered its tiles from a small Philadelphia inventor and manufacturer who went bankrupt during the construction period. His bankruptcy forced the Society to take over his plant and act as its own supplier in order to finish the job." Jordy, "PSFS," p. 53. This inventor was probably M. C. Rosenblatt. In *Sweet's* 1931 catalogue, the Acoustical Corporation of America, of Philadelphia, "successors to M. C. Rosenblatt, Inc." advertised its "Silent-Ceal," a "complete suspended ceiling construction," and claimed to have completed the "largest single contract ever awarded in this field." This claim may refer to the PSFS installation. Silent-Ceal was described as "a revolutionary development in acoustical engineering. It is a *complete* suspended ceiling, not a superficial acoustical treatment. It gives the highest sound absorption possible—above 70 per cent for four principle octaves." *Sweet's Architectural Trade Catalogue,* 1931: B2650. Emphasis in original.
60. Hitchcock and Johnson, *International Style,* p. 61.
61. The grid formed by the acoustical tiles on the ceiling of the Main Banking Room may not be fully evident in figure 11.10 as reproduced in this volume, but the lines are clearly evident in the original photograph as it appears in the *Architectural Review.* See also additional photographs of PSFS interiors in *Architectural Forum* 57 (December 1932); *Fortune* 6 (December 1932); and *Architectural Review* 73 (March 1933).
62. Hitchcock and Johnson, *International Style,* p. 26.
63. Hitchcock, *Architecture,* p. 532. Referring generally to the "traditional" architecture of the early twentieth century and specifically to the Woolworth building of Cass Gilbert, Hitchcock wrote that "the story is not an easy one to tell because it seems—at least to most scholars today—to lack plot." Hitchcock could only characterize the PSFS tower as a "Sullivanian slab," treated "along the lines that the leading European exponents of the new architecture had adumbrated in the previous ten years" (p. 514).
64. A plaque dedicated to Sabine was installed in the lobby of Symphony Hall circa 1950. It reads: "Symphony Hall, the first auditorium in the world to be built in known conformity with acoustical laws, was designed in accordance with his specifications and mathematical formulae, the fruit of long and arduous research."
65. In 1931, reflecting on New York's new Radio City complex (which symbolized as well as embodied the continued development of modern acoustical technologies), Cram complained: "From these belatedly truncated towers will go out even to the ends of the earth the nourishing vitamins of the chosen culture of this climacteric age. Amos 'n' Andy, Mr. Wrigley's Musical Hour, the intimate and revealing details of the latest *crime passionnel* and, in the hours that cannot be profitably disposed of to the exponents of super-salesmanship, such varied propaganda as may covet the high privilege of being 'on the air.'" Cram, "Radio City—And After," p. 295.

Bibliography

Algarotti, Count Francesco. *An Essay on the Opera* (1762, Italian). London: Davis and Reymers, 1767.

Banham, Reyner. *The Architecture of the Well-Tempered Environment.* 2nd ed. Chicago: University of Chicago Press, 1984.

Beranek, Leo L. "The Notebooks of Wallace C. Sabine." *Journal of the Acoustical Society of America* 61 (March 1977): 629–639.

Beranek, Leo L. "Wallace Clement Sabine and Acoustics." *Physics Today* (February 1985): 44–51.

Beranek, Leo L., and John W. Kopec. "Wallace C. Sabine, Acoustical Consultant." *Journal of the Acoustical Society of America* 69 (January 1981): 1–16.

Berman, Marshall. *All That Is Solid Melts Into Air: The Experience of Modernity.* New York: Penguin Books, 1988.

Brown, Edward F., et al., eds. *City Noise: The Report of the Commission Appointed by Dr. Shirley W. Wynne, Commissioner of Health, to Study Noise in New York City and to Develop Means of Abating It.* New York: Department of Health, 1930.

Brown, Richard D. "Modernization: A Victorian Climax." *American Quarterly* 27 (December 1975): 533–548.

Burrows, H. W. "Sound in its Relation to Buildings." *American Architect and Building News* 48 (18 May 1895): 65–70. Originally appeared in the *Journal of the Royal Institute of British Architects* (3rd Series) 2 (28 March 1895): 353–375.

Calfas, Paul. *La Nouvelle Salle de Concert Pleyel à Paris.* Paris: Publications du journal le Génie Civil, 1927.

Clark, Michael D. "Ralph Adams Cram and the Americanization of the Middle Ages." *Journal of American Studies* 23 (August 1989): 195–213.

Coben, Stanley. "The Assault on Victorianism in the Twentieth Century." *American Quarterly* 27 (December 1975): 604–625.

Collins, George R. "The Transfer of Thin Masonry Vaulting from Spain to America." *Journal of the Society of Architectural Historians* 27 (October 1968): 176–201.

Conrads, Ulrich, ed. *Programs and Manifestoes on 20th-Century Architecture.* Translated by Michael Bullock. Cambridge: MIT Press, 1970.

Conrads, Ulrich, and Bernhard Leitner. "Audible Space: Experiences and Conjectures." *Daidalos* 17 (15 September 1985): 28–45.

Cram, Ralph Adams. *Church Building.* Boston: Small, Maynard and Co., 1901.

Cram, Ralph Adams. *The Gothic Quest.* New York: Baker and Taylor Co., 1907.

Cram, Ralph Adams. "Radio City—And After." *American Mercury* 23 (July 1931): 291–296.

Curtis, William J. R. *Modern Architecture Since 1900.* 3rd ed. Upper Saddle River, NJ: Prentice Hall, 1996.

Dahl, J. O. "A Check List of Features That Make Apartments Popular." *Architectural Forum* 53 (September 1930): 371–372.

"Discussion on Sound in Its Relation to Buildings." *American Architect and Building News* 48 (25 May 1895): 78–81.

Elliott, Cecil D. *Technics and Architecture: The Development of Materials and Systems for Buildings.* Cambridge: MIT Press, 1992.

Fitch, James Marston. *American Building 1: The Historical Forces That Shaped It.* 2nd ed. New York: Schocken Books, 1973.

Fitch, James Marston. *American Building 2: The Environmental Forces That Shape It.* 2nd ed. New York: Schocken Books, 1972.

Fitchen, John. *Building Construction Before Mechanization.* Cambridge: MIT Press, 1986.

Forsyth, Michael. *Buildings for Music: The Architect, the Musician and the Listener from the Seventeenth Century to the Present Day.* Cambridge: MIT Press, 1985.

Frampton, Kenneth. *Modern Architecture: A Critical History.* 3rd ed. London: Thames and Hudson, 1992.

Frampton, Kenneth. *Studies in Tectonic Culture: The Poetics of Construction in Nineteenth and Twentieth Century Architecture.* Edited by John Cava. Cambridge: MIT Press, 1995.

Friedland, Roger, and Dierdre Boden, eds. *NowHere: Space, Time and Modernity.* Berkeley and Los Angeles: University of California Press, 1994.

Giedion, Sigfried. *Space, Time and Architecture: The Growth of a New Tradition.* 5th ed. Cambridge: Harvard University Press, 1967.

Green, Jr., L. "Soundproofing the New York Life Insurance Company Building." *American Architect* 135 (20 March 1929): 411–412.

Handlin, David P. *American Architecture.* London: Thames and Hudson, 1985.

Harvey, David. *The Condition of Postmodernity: An Enquiry into the Origins of Cultural Change.* Cambridge, MA and Oxford, UK: Blackwell, 1990.

Hitchcock, Henry-Russell. *Architecture: Nineteenth and Twentieth Centuries.* 4th ed. London: Penguin Books, 1977.

Hitchcock, Henry-Russell. *Modern Architecture: Romanticism and Reintegration* (1929). New York: Da Capo Press, 1993.

Hitchcock, Henry-Russell, and Philip Johnson. *The International Style* (1932). New York: W. W. Norton and Co., 1966.

Howe, Daniel Walker. "American Victorianism as a Culture." *American Quarterly* 27 (December 1975): 507–532.

Johnson, James H. *Listening in Paris: A Cultural History.* Berkeley and Los Angeles: University of California Press, 1995.

Jordy, William H. *The Impact of European Modernism in the Mid-Twentieth Century.* Vol. 5 of *American Buildings and Their Architects.* New York: Oxford University Press, 1972.

Jordy, William H. "PSFS: Its Development and Its Significance in Modern Architecture." *Journal of the Society of Architectural Historians* 21 (May 1962): 47–83.

Kern, Stephen. *The Culture of Time and Space, 1880–1918.* Cambridge: Harvard University Press, 1983.

Knudsen, Vern O. *Architectural Acoustics.* New York: John Wiley and Sons, Inc., 1932.

Laird, Donald A. "The Measurement of the Effects of Noise on Working Efficiency." *Journal of Industrial Hygiene* 9 (October 1927): 431–434.

Latrobe, Benjamin Henry. "Acoustics" (appendix to Archibald Campbell's article on Acoustics). In David Brewster, ed., *Edinburgh Encyclopedia,* American ed., Vol. 1. 120–124. Philadelphia: Joseph and Edward Parker, 1832.

Latrobe, Benjamin Henry. "Remarks on the Best Form of a Room for Hearing and Speaking" (letter to Thomas Parker, ca. 1803). In John C. Van Horne and Lee W. Formwalt, eds., *The Correspondence and Miscellaneous Papers of Benjamin Henry Latrobe,* Vol. 1: 1784–1804, 400–408. New Haven, CT: Yale University Press, 1984.

Lears, T. J. Jackson. *No Place of Grace: Antimodernism and the Transformation of American Culture 1880–1920.* New York: Pantheon Books, 1981.

Le Corbusier (Charles-Edouard Jeanneret). *Towards a New Architecture.* Translated by Frederick Etchells from the 13th French edition (1931). New York: Dover Publications Inc., 1986.

Lyon, Gustave. *L'acoustique architecturale.* Paris: Bibliothèque Technique du Cinéma, 1932.

Mark, Robert, ed. *Architectural Technology up to the Scientific Revolution: The Art and Structure of Large-Scale Buildings.* Cambridge: MIT Press, 1993.

Milkovich, Ann Katherine. "Guastavino Tile Construction: An Analysis of a Modern Cohesive Construction Technique." Master's thesis, University of Pennsylvania, 1992.

Mills, Robert. "Memorial: Hall of the House of Representatives U.S." 21st Cong., 1st sess., 1830. H. Rept. 83.

Morgan, Robert P. "'A New Musical Reality': Futurism, Modernism and 'The Art of Noises.'" *Modernism/Modernity* 1 (September 1994): 129–151.

Muccigrosso, Robert. *American Gothic: The Mind and Art of Ralph Adams Cram.* Washington, DC: University Press of America, 1980.

Mumford, Lewis. *Technics and Civilization* (1934). New York: Harcourt Brace Jovanovich, 1963.

"A New Shelter for Savings." *Architectural Forum* 57 (December 1932): 483–498.

Oliver, Richard. *Bertram Grosvenor Goodhue.* New York: The Architectural History Foundation; Cambridge: MIT Press, 1983.

Orcutt, William Dana. *Wallace Clement Sabine: A Study in Achievement.* Norwood, MA: Plimpton Press, 1933.

Osswald, F. M. "Acoustics of the Large Assembly Hall of the League of Nations, at Geneva, Switzerland." *American Architect* 134 (20 December 1928): 833–842.

Patte, Pierre. *Essai sur l'architecture théâtrale. Ou de l'ordonnance la plus avantageuse à une salle de spectacles, relativement aux principes d'optique et de l'acoustique.* Paris: Chez Moutard, 1782.

Pérez-Gómez, Alberto. *Architecture and the Crisis of Modern Science.* Cambridge: MIT Press, 1983.

Peters, Tom F. *Building the Nineteenth Century.* Cambridge: MIT Press, 1996.

Pevsner, Nikolaus. *The Sources of Modern Architecture and Design.* London: Thames and Hudson, 1968.

"Planning for Employees' Welfare in the Design of the New York Life Insurance Company Building." *American Architect* 135 (20 March 1929): 397–401.

Rasmussen, Steen Eiler. *Experiencing Architecture.* Cambridge: MIT Press, 1962.

Roth, Leland M. *A Concise History of American Architecture.* New York: Harper and Row, 1979.

Rykwert, Joseph. *The First Moderns: The Architects of the Eighteenth Century.* Cambridge: MIT Press, 1980.

Sabine, Paul E. *Acoustics and Architecture.* New York: McGraw Hill, 1932.

Sabine, Wallace. "Reverberation." *American Architect and Building News* 68 (April–June 1900): 3, 19, 35, 43, 59, 75, 83. This article, which appeared serially, also appeared in *Engineering Record* 1 (1900), and it is reproduced in Sabine's *Collected Papers on Acoustics,* Cambridge: Harvard University Press, 1922 (reprinted by Dover, 1964, and Peninsula Press, 1993).

Saunders, George. *A Treatise on Theatres*. London: I. and J. Taylor, 1790.

Sherman, Roger W. "Sound Insulation in Apartments." *Architectural Forum* 53 (September 1930): 373–378.

Smith, Christine. *St. Bartholomew's Church in the City of New York*. New York: Oxford University Press, 1988.

Smith, T. Roger. *Acoustics in Relation to Architecture and Building*. 3rd ed. London: Crosby Lockwood, 1895.

Smith, Terry. *Making the Modern: Industry, Art, and Design in America*. Chicago: University of Chicago Press, 1993.

Swan, Clifford Melville. "Noise Problems in Banks." *Architectural Forum* 48 (June 1928): 913–916.

Thompson, Emily. "Dead Rooms and Live Wires: Harvard, Hollywood, and the Deconstruction of Architectural Acoustics, 1900–1930." *Isis* 88 (December 1997): 597–626.

Thompson, Emily. "'Mysteries of the Acoustic': Architectural Acoustics in America, 1800–1932." Ph.D. diss., Princeton University, 1992.

Thompson, Emily. "A Tale of Two Physicists: The Origins of Acoustical Consulting." *Sound and Video Contractor* (20 March 1997): 14–22.

Treib, Marc. *Space Calculated in Seconds: The Philips Pavilion, Le Corbusier, Edgard Varèse*. Princeton, NJ: Princeton University Press, 1996.

Trowbridge, John, and W. C. Sabine. "Electrical Oscillations in Air." *Proceedings of the American Academy of Arts and Sciences* 25 (1890): 109–123.

Trowbridge, John, and W. C. Sabine. "Selective Absorption of Metals for Ultra Violet Light." *Proceedings of the American Academy of Arts and Sciences* 23 (1888): 299–300.

von Ezdorf, Robert. "The Design of the Interior of the New Home Office Building of the New York Life Insurance Company." *American Architect* 135 (20 March 1929): 369–372.

Ward, David, and Olivier Zunz, eds. *The Landscape of Modernity: New York City, 1900–1940*. Baltimore: The Johns Hopkins University Press, 1992.

Wight, Peter B. "The Works of Raphael Guastavino." *Brickbuilder* 10 (1901): 79–81 (April); 100–102 (May); 184–188 (September); 211–214 (October).

Willis, Carol. *Form Follows Finance: Skyscrapers and Skylines in New York and Chicago*. New York: Princeton Architectural Press, 1995.

Wyatt, Benjamin Dean. *Observations on the Design for the Theatre Royal, Drury Lane*. London: J. Taylor, 1813.

12

Of Beds and Benches: Building the Modern American Hospital

Allan M. Brandt and David C. Sloane

The modern American hospital sits squarely between the world of science and public culture. This institution, in the course of the twentieth century, has gained preeminent authority in a complex social process of taking problems of disease, debility, and suffering and redefining them in the language and practices of modern medicine and science. This, of course, was not always the principal function of the hospital, which only a century ago had a distinctively different ethic and ethos.[1] This chapter suggests that the dominant cultural values and expectations of American medicine are written on the hospital's facades and spatial arrangements. Tracing the dramatic changes in hospital architecture offers one critical avenue toward understanding not only this particular institution, but also the broader history of medicine, science, and popular culture in the twentieth century. Ultimately, embedded in hospital design—in the nature of the physical plant and its aesthetic—are notions of normative doctor-patient relations, physician-staff relations, and medical and scientific authority, as well as specific explanatory models for understanding and treating disease.[2]

As recently as 1900, most Americans received their health care, when they received health care at all, in their homes. The hospital, deeply stigmatized as an institution of last resort, remained an unusual site for the delivery of care with few attractions. Not only was it viewed as a home for the down and out, the destitute without resource to family or friends. It was also perceived as a dangerous place in which problems of disease could be exacerbated by visits to the institution itself. For the better part of the nineteenth century, hospitalization meant not only that disease or disability had intruded into the nature of everyday life, but also that the normal course of seeking help had in some way broken down. Why else would people seek, as historian Charles Rosenberg has so aptly put it, the care of strangers?[3]

The designers of America's early hospitals in the late eighteenth and early nineteenth centuries understood that they had to contend with such negative images. Architecture was crucial to the meaning and practice of the early-nineteenth-century hospital. Hospital lay trustees, who were responsible for the institutions' construction and operation, consistently felt a need to justify them. The grand facades that adorned many of the early hospitals reflected this desire to reassure the public that the hospital had an eminent purpose and role in the society.

Emulating other public buildings of the period, designers of hospitals attempted to evoke their benefactors' generosity and public spirit.[4] As historian David Rosner has written, the hospital was, above all else, a charitable enterprise, dominated by its patrons' moral and sectarian ideals.[5] Its structure reflected a particular view of the social order more than any specific notion of the nature of disease, its transmission, or medical practice. Opened in 1821 explicitly to help the poor, Massachusetts General Hospital was a typical example of an early American "voluntary" hospital.[6] Just as the new city halls, courthouses, and jails were intended to manifest the virtue of republican order, the hospitals were built to express the contemporary belief that patients were "moral minors" in need of assistance.[7]

Medicine and science, important to the hospital's mission, nonetheless could not dominate its culture or for that matter its organization and design. Lay trustees, for example, controlled all admission decisions in these institutions, emphasizing the potential patients' rectitude and often overruling physicians' desire for more interesting "clinical material," as patients were often called.[8] The acutely ill and contagious were shuffled off to the public almshouse or pesthouse. The others, especially Blacks and immigrants, were often simply turned away. By "choosing" their patients, hospital lay trustees felt they could insure that hospitals instilled virtue as well as restored health.

The hospital brought together a set of social and medical concerns in an age when anxieties about poverty, morality, and disease were difficult to differentiate. The vitiation of moral norms was considered as compelling a view of disease causality as more specific theories of miasmas or contagions. In this respect, the hospital's scientific rationale, or even its medical rationales, did not necessarily dominate in the organization and design of these buildings. In an institution so fundamentally associated with the disorder of disease—the uncontrolled aspects of disability, suffering, and death—order and control became paramount values in design.

Typically, the nineteenth-century hospital was characterized by a central structure that radiated long and straight wings to house inmates (as administrators referred to pa-

12.1

Massachusetts General Hospital, 1821, designed by Charles Bulfinch, is representative of the early civic hospital. Courtesy Countway Library, Harvard Medical School.

tients). The main edifice was usually the most ornate, often with columned porticos and a cupola or dome. Symmetrical wings were unrelieved in their uniformity and regularity.[9] The central facade reflected the donors' generosity; the wings, moral order and regularity. Massachusetts General Hospital, designed by noted Boston architect Charles Bulfinch, demonstrated the powerful public and moral expectations invested in these institutions. The hospital was built in the shadow of the State House, an earlier architectural success of Bulfinch's.[10] (See Figure 12.1.)

The rigid physical structure of these hospitals mirrored the strict daily life of the wards. In repetition came the order and discipline required for healing, but more importantly, moral regeneration. Moreover, similar to the nation's schoolhouses, almshouses, workhouses, and prisons, such forms of organization offered opportunities for central oversight, authority, and discipline. Relatively small staffs of nurses typically supervised large wards of up to thirty or forty beds with dozens of patients. Strict rules dictated appropriate behavior on these communal wards.[11]

MASSACHUSETTS GENERAL HOSPITAL, BOSTON, IN 1821.

The spatial organization of the nineteenth-century hospital reflected fundamental beliefs about the place of the patient within the hierarchy of medical discipline and practice.[12] The hospital's facade and structure reflected its larger goals of bringing order to the disordered, health to the sick, morality to those who had lost their way. The structural organization permitted oversight and clarified an increasingly articulated medical hierarchy.[13] According to Assistant Surgeon General John S. Billings, "To be an 'inmate' was to barter independence for security, to subject oneself to the physical and moral authority of trustees, administrators, and attending physicians."[14]

Nineteenth-century hospitals also reflected architecturally the miasmatic theory of disease. Along with primary concerns about order and charity, concerns about the space between patients, the flow of air through the wards, and ventilation between the wards dominated discussions of hospital design well into this century.[15] Most famous for her part in the emergence of modern nursing, Florence Nightingale was also involved in hospital design after her experience in the Crimean War convinced her of the need for ventilation. Nightingale was a principal proponent of what was known as the zymotic theory of disease, which characterized the spread of disease as a process of fermentation.[16] According to this popular theory, a small amount of infectious disease in the air, just like one bad apple in a bushel, would ferment the whole atmosphere.[17] Along with most of her generation, she believed that air, space, and isolation were crucial because they reduced exposure to the miasmas transmitting disease. She calculated precise recommendations for the distribution and allocation of space inside the hospital, suggesting 1,500 cubic feet per patient and 100 square feet for each bed. These standards persisted well into the twentieth century, long after bacteriology replaced zymotic theories of disease.[18] The later use of verandas and sunrooms continued to reflect a related desire to expose patients to the elements.

Even as designers attempted to counteract miasmas, however, the concentration of illness under one roof in hospitals continued to be a major cause for concern. Deep and persistent worries about the dangers of hospitals were voiced as the institution grew in stature and function. "Hospitalism"—diseases transmitted in hospitals—drew particular attention to these anxieties. The hospital was the preeminent "sick building"; constructed for the sick, it emanated sickness to the well. One physician proposed razing hospitals every fifty years and rebuilding elsewhere to avoid miasmatic contaminants, and many others echoed his concern that the buildings became "hospitalized" and therefore unsafe over time.[19] Despite efforts to make hospitals appear to be safe havens for the worthy poor, they continued to be perceived as dangerous places, institutions of last resort. As late as 1876 one acute observer concluded, "the truth is, the majority of our hospitals . . . are liable to do more harm than good."[20] The middle class and the wealthy continued to receive their care at home.

By the late nineteenth century, hospital design began to reflect a new set of historical contingencies. To be sure, architects continued to focus attention on theories of contagion and the requisite needs for ventilation. But new hospitals reflected other needs considered to be medically important as well. These hospitals were constructed with the explicit goal of bringing a new, expansive science to medical care. Now the hospital would come to be fundamentally redefined from a civic institution serving the working poor to a scientific institution consolidating a newly formed professional authority. In this respect the hospital drew together the growing aspirations of the profession, science and an increasingly expectant public. At the same time new designs expressed long-standing concerns for miasmatic theory, they looked forward to a new medical science dominated by professional authority, diverse patients, and public confidence in new and dramatic therapeutic rationales.

Although these new buildings were seen as housing the latest advances of the new medical science, much remained familiar to the stolid structures of antebellum American medicine. In particular, long common wards remained the principal approach to housing patients. But now hospitals offered radiating wings called pavilions attached in skeleton-like form to central administrative buildings. Typically justified as aiding in ventilation, these pavilions also offered a series of important advantages to institutions eager to expand their appeal to an increasingly heterogeneous group of patients.

The pavilions' isolation was seen as protection against contagion, contamination, and contact. But the logic of pavilion design was how explicitly it met sociocultural goals: One institution could bring together the diverse constituents of medical practice and research. The "clinical material" for education and investigation rested in close (but rigorously separate) proximity to prestigious pay wards.

The Johns Hopkins Hospital, built between 1877 and 1885, marked the preeminent example of these transformations in design and function. Hopkins Hospital offered both pay and public pavilions and space for diagnostic labs and research, as well as medical student instruction. Its high Victorian facade announced that it was well prepared to "administrate" among these sometimes complimentary, sometimes competitive constituencies. (See figures 12.2 and 12.3.) John Shaw Billings supervised the building project, bringing with him experience gained as surgeon and administrator during the Civil War and also as Assistant Surgeon General; in the latter position, he led an extensive study of military hospitals following the war. The entire building was designed symmetrically. The administration building stood at the center of the design, with the apothecary directly behind it. Separate wards not only isolated patients from possible contagion; they also separated them by social class and ethnicity. Flanking the administration building symmetrically were female and male pay wards and the kitchen and nurses' home. Offset and extending behind the administration building were the common wards. Each building was reached through a corridor, and each was isolated from the others by these corridors.[21]

12.2

Johns Hopkins Hospital exemplifies the hospital as it became an authoritative and scientific institution in the late nineteenth century. A. McGehee Harvey et al., *A Model of Its Kind*. Vol. 2, *A Pictorial History of Medicine at Johns Hopkins*. Baltimore: The Johns Hopkins University Press, 1989, p. 12.

12.3

John Shaw Billings' 1875 sketch for the Johns Hopkins Hospital used a pavilion style, relying on the miasmatic theory of disease. "Sketch Plan of Arrangement for Johns Hopkins Hospital, with One Story Pavilions, Temporary or Permanent. By Dr. J. S. Billings U.S.A." Trustees of the Johns Hopkins Hospital, *Hospital Plans,* New York: William Wood, 1875. Reproduced from Charles E. Rosenberg, *The Care of Strangers: The Rise of America's Hospital System.* New York: Basic Books, 1987.

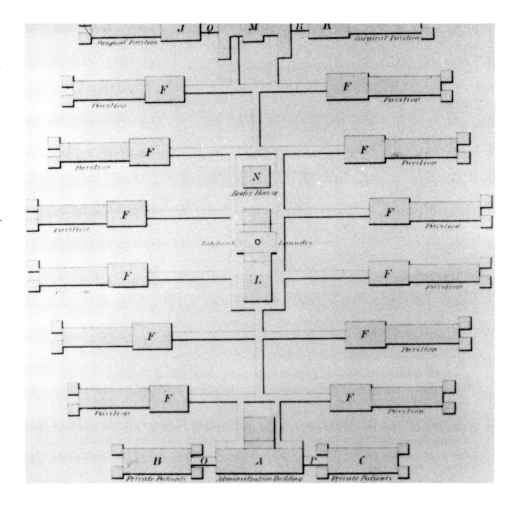

As hospitals like Johns Hopkins began to appear, changes in medical theory, technology, and education were setting the stage for a new generation of hospitals more generally accepted by the public and the profession. The introduction of anesthesia and antisepsis and the triumph of the germ theory created a new set of exigencies for hospital design and construction, as well as a new set of cultural and social expectations of medical intervention. Rates of surgical procedures and, for that matter, surgical mortality shot up in the last years of the nineteenth century, as physicians in hospitals were emboldened to attempt new surgical cures.[22] The new proclivity for surgical intervention forced old hospitals to add surgical suites, often on the back of their existing buildings, and no new hospital would dare open its doors without touting its modern operating rooms. In the first decades of the twentieth century, the hospital moved from the cultural periphery to become a core institution, crucial to developments in both medicine and science.

More individuals would seek the care of strangers with a new set of rationales. From a charitable institution dominated by notions of social welfare and moral order, the new hospital of the early twentieth century promised secular redemption from disease itself. The compelling forces of science and medical expertise now promised benefits unavailable elsewhere. This view of the hospital as the house of science drew a new clientele to its doors. The desire to attract paying patients to the hospital, often for the first time, required buildings that evoked precision, technology, and the promise of a new science, even when that science sometimes had little to offer patients in need. To bring in paying patients, creature comforts were also emphasized: feather beds, fine china, antique furniture. The hospital created a model of a two-class system of care. Pavilions now became a structural means of isolating patients—not from microbes, but rather to separate those of means from those of lower social status. Harkness Pavilion at Presbyterian Hospital and Klingenstein Pavilion at Mount Sinai are two examples of prestigious pay wards in which patients with means would receive treatment.[23]

In this respect, though scientific developments no doubt had an important impact on the hospital's reconstruction, it would be a misimpression to suggest that it was purely science that guided the building of the new hospital. A wide range of social and cultural forces, such as changes in the constitution of families and domestic service, as well as the interests of the medical profession typically dictated aspects of hospital design and construction.

But even though cultural changes affected the hospital's structure, its perceived purpose was scientific. The hospital quickly became the modern temple of science. What began with Johns Hopkins Hospital soon solidified; here science was brought to the public. Here the esoterica of the laboratory, the wonders and magic of a brave new world would be brought to the bedside, turning despair and disability to hope and sometimes (although less frequently perhaps than was purported) cure. The modern hospital, in its architectural form and spatial arrangements, reflected the values of this new biomedical paradigm, which emphasized the specificity of diagnosis and treatment. Disease was no longer considered principally in the environments that produced it. It was brought to the hospital to be isolated, identified, and when possible, treated.

With this new social acceptance and compelling cultural rationale came a dramatic increase in the number of American hospitals. As late as 1873, it was estimated that the United States had only 178 hospitals. Even accepting that a variety of other institutions provided many of the same services in the nineteenth century, the rapid expansion of the hospital after 1890 is truly remarkable. By 1904, the U.S. Department of Commerce reported almost 1,500 hospitals. Forty years later, at the end of World War II, the number had risen to 4,445. With the passage of the Hill-Burton Act in 1946, which provided federal funding for the development of hospitals nationwide, the numbers continued to climb right through 1960, when 5,736 hospitals were counted nationwide.[24] Eventually, the number of hospitals would peak at more than 7,000, a manifestation of the enormous rise in the significance and importance of tertiary health care in American society.

The new cultural understanding of childbirth that emerged in the 1920s illustrates how hospital care had become sanctioned. Historian Judith Leavitt describes how popular magazines of the 1920s and 1930s were full of "heart-rending personal examples" of women who died in childbirth because they did not deliver in hospitals.[25] Boston's Lying-In Hospital opened a new, much larger building in 1922, indicating this change in the public perception of pregnancy and labor.[26] The rise of such maternity hospitals played a critical role in the medicalization of childbirth. (See figure 12.4.)

12.4

As hospital births came to be sanctioned in the early twentieth century, new buildings, such as Boston Lying-In Hospital (completed in 1922), met both consumer and medical demands. Courtesy of Countway Library, Harvard Medical School.

The many new hospitals of the early twentieth century contained a wide range of new scientific machinery that appealed particularly to patients seeking the advantages of the "new" medicine. Medical technology and the hospital became synonymous.[27] The hospital became the one social institution in which members of the general public came in contact with cutting-edge science and technology. A hospital visit for most was not unlike visiting a World's Fair in which future technological achievements would be displayed, the key difference, of course, being that the lay person now utilized (or was subjected to) these technologies, namely, X-rays, laboratories, IV treatments, and technical monitoring. The exigencies of the machine increasingly dictated the hospital's organization and structure.[28]

The development of these new medical technologies was but one way hospitals were becoming more complex. During the early years of the twentieth century, medical practice, education, and research came to be consolidated in the hospital. As medicine rose in status and prominence, these buildings became the center of scientific stature and power. The hospital soon emerged as the central locus for the training of physicians as well as the center for biomedical research. Not just clinical research that required the presence of "clinical material," but basic research as well came to be situated in the contemporary academic hospital. These various uses of the hospital forced administrators to reconsider their allocation of space and resources. By 1939, *Architectural Record* asserted that the new hospital was "a far cry from the comparatively simple structures that, a few decades ago, cared for 'charity cases' in large wards." Instead of these "simple" hospitals, "nearly two-thirds of the hospital's floor area is devoted to services and one-third to patients."[29] For example, the rising significance of specialties like pathology created new demands for laboratories and space.

With the expansion of hospitals' purpose came major changes in the physical plant. Some commentators began arguing for a new style of hospital, the skyscraper. In 1905, Dr. Albert Ochsner proposed a multileveled hospital, arguing that in the city it was a necessity because of the lack of space.[30] At Massachusetts General Hospital, the new Robert White Memorial Building opened in 1939. (See figure 12.5.) In their discussion of why the building was constructed, the institution's trustees cited both the need for more space and a shift away from older scientific beliefs that had led to the construction of pavilions. They explained that the White Building, with its thirteen floors, would "make it possible to assemble the scattered wards and bring them under one roof, near the facilities which they should use."[31]

Sleek new skyscrapers appealed to architects for their almost weightless appearance, but they also conveyed a sense of efficiency and functionalism, which made them ideally suited for new hospital designs. The science of medicine was not only practiced in the laboratory, it was also modeled in the wards and private rooms. These structures mirrored new "scientific management" techniques of architects eager to rid the hospital of its previous "old, haphazard, unbusiness-like methods" of operation (as one hospital consul-

12.5

Completed in 1939, the Robert White Memorial Building, a thirteen-story, towering addition to Massachusetts General Hospital, reflected a sleek, new multilevel design as well as an increased need for hospital beds. Courtesy of Countway Library, Harvard Medical School.

tant characterized them).[32] Discipline in the hospital shifted from moral to scientific rationales. Buildings shifted from the long horizontal structures of pavilions to towering monuments to scientific ingenuity.

The buildings themselves reflected the emphasis on efficiency and effectiveness. In 1949, the American Institute of Architects held a seminar on hospitals during their annual convention. One architect noted at that conference that "functional design best correlates these varied activities such as nursing education, intern and resident physician education, research and various adjunct or service facilities of the hospital into a unit offering ultimately, proper care for the patient."[33] A design that emphasized purpose and medical effectiveness to the exclusion of ornament had become the new norm. The sterility of the mid-twentieth-century hospital was structural, medical, and aesthetic.

In the immediate post–World War II period, the rise of the National Institute of Health, private employer-based insurance, and federal funds to support the building of new hospitals led to a remarkable surge in hospital construction. The hospital was soon to become a megalith.[34] Massachusetts General Hospital, for example, had been transformed from the statuesque facade of 1821 to a maze of buildings in 1992, dwarfing the original. (See figure 12.6.) Constant growth and renovation compromised the hospital's architectural integrity. Medical sprawl often surrounded—and overwhelmed—the stately institutions of nineteenth-century medicine. Not surprisingly, the aesthetic of the twentieth-century hospital reflected this emphasis on the institution's dominant scientific qualities. If the nineteenth-century hospital's architecture reflected aspects of beneficence and charity, the twentieth-century hospital's architecture paid homage to the religion of science.[35]

The hospital's complexity of purpose and structure that developed in the first half of the twentieth century is still with us today. Distinguished from its predecessors by its wide range of constituencies and functions, the modern hospital has become one of the most diverse and complex of all social institutions, encompassing a daunting range of activities and arguably serving the broadest range of constituencies of any institution in modern life. Almost all Americans are born in hospitals, and, of course, there too, most shall die. Virtually all of us will make interim visits to these institutions, both as patients seeking care and as visitors. The hospital remains the locus of care for the seriously ill and dying, but nonetheless, care for the chronically ill continues to be centered in the hospital as well. Moreover, especially in our cities, the hospital has in the last two decades additionally become the center for the delivery of primary care and ambulatory care as basic social support has come to be compromised. The very evolution of the term "ambulatory" was to distinguish those hospital users who come and go from those actually hospitalized. In this respect, the hospital continues to draw upon its nineteenth-century lineage as a social welfare institution.[36]

12.6

By 1992, Massachusetts General Hospital had become a complex maze of buildings spread out over a large urban area. Find the original Bulfinch building of 1821! Courtesy of Massachusetts General Hospital, Public Affairs Office.

It is a remarkable phenomenon, and one that can be explained only by the nature of certain historical contingencies, that the same institution in which the urban poor seek care at the door of the emergency room has, at the same moment, researchers conducting studies of the new molecular biology. At the core of this apparent paradox has been the powerful notion of the twentieth-century hospital, that science and medicine are fundamentally interpenetrated. Basic research would be translated quickly and efficiently into clinical research and eventually integrated into state of the art patient care. The hospital would be the center of this "translational activity" of science, demonstrating on a daily basis the utility of investigation, even at the molecular level.[37]

In some ways, situating these laboratories in hospitals reflects an anachronism, a time in which it was assumed that the scientist would also be a physician, easily moving from bench to bedside; within academic medicine such careers have become increasingly unlikely. Nonetheless, just as vestiges of the nineteenth-century hospital are found today, so this ideal of the physician-investigator is no small part of the academic hospital's structure and design. Academic medical centers promised to combine effectively teaching, research, and patient care, each goal benefitting the other.

Also, the very qualities of the twentieth-century hospital that made it so appealing, so dominant an aspect of twentieth-century life have in the last two decades become the focus of an intense criticism of the institution. Beginning in the 1960s, criticism of the hospital began to coalesce; hospitals were identified as impersonal, bureaucratic institutions in which anonymous physicians and health care providers treated anonymous patients. The triumph of specialization was now seen as fracturing care among a plethora of experts who never made contact with the "whole" patient. Hospital "transport" became a crucial aspect of building design as patients were shuttled from one diagnostic technology to another, rarely ever seeing "the doctor." Patients were "turfed" from service to service.[38] In the maze of corridors and departments, patients and their visitors often found themselves lost.

Impersonal and bureaucratic, the hospital, in its effort to cure, had, it was argued, lost the capacity to care. Rather than seeing hospitals as institutions in which miracle cures took place, some critics even went so far as to suggest that hospitals, like their nineteenth-century predecessors, were typically dangerous places where the sick would go only to be further injured. Critics, like Ivan Illich, suggested that iatrogenesis was a ubiquitous feature of modern health care institutions.[39] Hospitals came to be subject to a broader sociopolitical critique that viewed powerful social institutions as essentially exploitative and the hospital in particular as serving the interests of the medical profession and elite science rather than patients' needs.

By the 1970s, many observers began to argue not only that the hospital was impersonal but that it also had failed to achieve the lofty expectations of the early to mid–twentieth century. This critique was part of a larger attack on biomedicine and the biomedical model: Medicine had failed to control disease; it was relatively ineffective in addressing the predominance of chronic disease; and the hospital was poorly suited to address the medical problems implicit in the epidemiological transition from infectious disease to the systemic chronic diseases of the second half of the twentieth century.[40] Hospital costs had skyrocketed while basic health indicators changed little, or as was the case with infant mortality in some locales, actually worsened. Often towering over inner cities, the hospital, many suggested, had become isolated from the communities in which it stood, a symbol of elite science and medicine that rarely contacted the social world of disease and illness. Much to their dismay, surrounding communities saw neighborhoods razed to make room for new hospital buildings as the physical plant expanded. One commentator noted that the hospital had come to be viewed as a "mixed blessing, a technological and bureaucratic brontosaurus with an enormous appetite, an inadequate heart, and a minute social brain."[41] These critics articulated a deep public ambivalence about the hospital and more particularly, the reductionist, technologically centered medicine that it had come to embody. An expectant public both eagerly sought scientific medicine and, at the same time, perceived it to be deeply alienating. Once symbols of scientific prowess, these same glass and concrete structures now were viewed as alienating, cold, insular, and oppressive.

Hospitals and their designers have responded to these critiques in recent years, attempting to build "patient-centered" institutions. The hotel and the shopping mall present the two most compelling and noted architectural paradigms for the contemporary hospital.[42] Typically, these structures are praised as appropriate analogues for the hospital given their inviting environments, their broad accessibility, and their integral relationship with the public community.[43] Their customer-oriented approach has replaced the moral meaning of the civic building and the functional sterility of the scientific building as paradigms for constructing a symbolic relationship between the public and the hospital.

Recent attempts to humanize the hospital, to take the hard edge off the machine aesthetic of the postwar period, have often led to considerable architectural successes. Patients and physicians alike have praised the return to scale in hospital buildings. In this perspective, the patient is viewed as a consumer of health care; just as retail architecture centered attention on the patron's care and comfort, drawing consumers into the marketplace, so now the new hospital architecture is organized around the values of a consumer culture.[44] As one architectural journal recently noted in describing a new, patient-centered structure:

> Among the happy consequences of Baptist Hospital's attempt to attract patients and gain the allegiance of people visiting its new emergency care department is that they are treated like royalty, or at least like hotel guests. Staff members are put through a "guest relations" program emphasizing courtesy, friendliness, and the like. . . . All in all, this is the least institutional of emergency departments and looks far more enticing than most examples of the building it emulates, the American hotel.[45]

Just as architects of the nineteenth century had suggested the therapeutic values and assumptions in the organization of hospitals, so now again in the late twentieth century, the actual design of hospitals is suggested to have powerful therapeutic implications. As one architect explained: "These 'high-push' environments will help ease patient stress, reduce medication levels, and promote shorter hospital stays."[46]

More-inviting public areas, including the introduction of fast-food restaurants and mall-like shops as well as atriums and sitting areas that could easily pass as hotel lobbies, make the hospital less "sterile" and imposing. Dartmouth-Hitchcock Medical Center in Lebanon, New Hampshire, a principal example of the new hospital architecture, uses a mall design to orient patients. In 1992, the new hospital replaced a building originally constructed a century earlier and incrementally expanded on many occasions afterwards. Colored tape on the floors had been used to guide patients and visitors along the old hospital's mazelike tunnels and corridors. The new structure's central mall now provides patients clear guides to the various departments; for the medical consumer, this functions much like the traditional elevator listing of department store "departments."[47] The hospital closely resembles commercially successful shopping malls like the Chestnut Hill Mall, built in the early 1970s.[48] (See figures 12.7 and 12.8.)

While contemporary architects have worked diligently to make the hospital more hospitable, their clients have also made them aware of the intensely competitive aspects of the medical marketplace. The new hospital must draw patients/customers if they are to succeed in this market environment. These new hospitals, manifesting heightened aesthetic concerns and patient comfort, are especially targeted at those aspects of the medical market most likely to be profitable. Sophisticated entrepreneurs are seeking to attract patients who are both well insured and willing to pay additional out-of-pocket expenses for special medical services.[49] These range from increasingly popular cosmetic surgeries to cutting-edge cancer therapies yet to be approved for general use and therefore not covered by insurance. Although the general hospital with its complex constituencies remains, new specialized hospitals are being built, eager to solicit patients/consumers for particularly profitable services. Embedded in such hospitals is a historically shifting definition of medical "need" and the provision of care.

Architects have been enlisted in this new consumer orientation. "The patient/customer now shops for cost, quality, and convenience in health care services," writes one hospital architect. "Responding to this new marketplace, hospitals of the future will transform themselves into a resemblance of successful commercial and retail centers."[50] As he goes on to predict:

> The new age health center will be the connecting point for a variety of different services including laser clinics, regeneration centers, biotech research, birthing centers, hospice and self-care. Patients, staff, and equipment will shuttle between buildings and over parking lots on Disneyesque skywalks. Convenient self-service medical

12.7–12.8

The architecture of Dartmouth-Hitchcock Medical Center (right, completed in 1992) owed much to the commercial architecture of shopping malls designed to appeal to consumers, such as the Chestnut Hill Mall of Newton, Massachusetts (left). Chestnut Hill Mall, Newton, Massachusetts: Photo by Allan Brandt, 1994. Dartmouth-Hitchcock Medical Center: Courtesy Shepley Bulfinch Richardson and Abbot, Architects. Photo by Jean M. Smith.

malls will encourage out-of-pocket spending for a wide array of ambulatory care services at health shops, discount pharmacies, and wellness programs. These new health centers will adopt standard features from the airline and hotel industries, introducing computerized, curbside admitting services and building inpatient towers around large greenhouse atrium spaces.

Such a merging of medicine and the market is occurring in Disney's Celebration development in central Florida. Medical corporate interests in Florida have developed plans for a "hospital of the future" on the Walt Disney World property near Orlando. In January 1998, Disney opened a medical complex under the banner of "Celebration Health."[51] Preliminary plans for Celebration Health included an ambulatory surgery facility, a diagnostic imaging center, a wellness center, and a sports medicine facility. Residents exercising in their homes can be connected to the wellness center, where data is collected to monitor health measures such as blood pressure.[52] The hospital is touted as part of the larger planned community, which boasts neotraditional urban plans intended to construct a community environment for the roughly 20,000 residents. Developing the town is integrally related to the construction of the medical center; both are pieces of the celebration, inseparable in the plan for appealing to the consumer.

New designs like Celebration Health are not, however, merely deployed to be attractive to a medical consumer; they also promise greater efficiencies and reduced costs to health care financial interests, now increasingly eager to expand profit margins in the delivery of care. When Pru-Care, a health maintenance organization and subsidiary of Prudential Insurance Company of America, built a new primary care center in Nashville in the early 1990s, they followed the trend of creating a medical mall with attractive, open waiting areas for a range of patient services. According to architect Earl Swensson, the client was eager to avoid "unnecessary patient/doctor cross traffic." This problem was solved by creating a second corridor reserved for doctors and nurses with access to examining rooms, "allowing both to enter and leave their work areas without patient contact."[53] Mall designs, inviting and attractive, may also reduce costs by managing the time that patients and their caregivers spend together.

A consumer orientation to health care and hospital design inevitably reflects the broader commodification of health care that has characterized the economy of medical services since the mid-1970s and has dramatically intensified in the 1990s with the rise of managed care. In this postmodern medical world, doctors have been redefined as "providers," and patients are "customers," purchasing "product lines" that have been rigorously evaluated through statistical techniques not only for their medical effectiveness but for customer satisfaction as well. These "product lines" must reap profits to gain a place in the medical market. The hospital, broadly construed, is the commercial site for such transactions.

Since this new consumerist aesthetic has emerged, it makes sense to inquire about its historical implications and significance. In part, no doubt, it reflects a responsiveness to an ongoing critique of medical authority and paternalism in the patient-doctor relationship and new desires to "empower" patients. The development of new, user-friendly hospitals that seem neither imposing nor threatening is clearly unobjectionable. Nonetheless, it is important to recognize that these buildings do not reflect a deeper critique of the biomedical paradigm that has dominated Western medicine since the turn of the century. Nor do they offer alternative visions of disease causality, risk, or responsibility.

The new hospital architecture is an explicit attempt to "civilize the machine."[54] And although current designs do reflect a respect for patient comfort, more significantly they reflect the further commodification of health care within the contemporary market of health care services. In the explicit desire to obfuscate the architectural boundaries between the shopping mall and the hospital we see the powerful and implicit message that medicine is a business, medical care a commodity. Certainly this is not a new development, but only recently has it enlisted the powerful symbolic logic of the architect.

Reading the architecture of the hospital offers one possibility for coming to a more sophisticated understanding of both medical theory and practice, as well as the nature of the medical economy and culture. The history of space and spatial divisions, the history of the hospital aesthetic are revealing of not only core medical values and ideals, but also broader cultural notions of disease, health, and health care. Even the new consumer-oriented hospital cannot ultimately mask the pervasive uncertainties inherent in the experience of and response to human disease.

NOTES

1. In recent years, the history of the American hospital has attracted considerable interest. See especially Rosenberg, *The Care of Strangers;* Rosner, *A Once Charitable Enterprise;* Vogel, *The Invention of the Modern Hospital;* and Stevens, *In Sickness and In Wealth.*
2. The most important work on the history of hospital architecture remains Thompson and Goldin, *The Hospital.* Although this volume treats twentieth-century hospital architecture only briefly, we have, nonetheless, relied heavily on it. For contemporary discussions, see also Stevens, *The American Hospital;* Woodworth, "Hospitals and Their Construction"; and Hornsby and Schmidt, *The Modern Hospital.*
3. Rosenberg, *The Care of Strangers.*
4. Rothman, *The Discovery of the Asylum,* pp. 42–45.
5. Rosner, *A Once Charitable Enterprise,* p. 20.
6. Thompson and Goldin, *The Hospital,* p. 102. See also Vogel, *The Invention of the Modern Hospital.*
7. Rosenberg, *The Care of Strangers,* p. 35.
8. Rosenberg, *The Care of Strangers,* pp. 303, 315. See also Stevens, *In Sickness and In Wealth,* p. 137.
9. Thompson and Goldin, *The Hospital,* pp. 118, 125–169. See also Rothman, *The Discovery of the Asylum,* pp. 152–154.
10. Kirker, *The Architecture of Charles Bulfinch.*

11. Rosenberg, *The Care of Strangers,* pp. 36–37.
12. Rosenberg, "And Heal the Sick." Although Rothman's focus is on the mental asylum, a similar generalization about the desire for order and control may be extended to the general hospital of the nineteenth century; Rothman, *The Discovery of the Asylum.* See also Rosner, *A Once Charitable Enterprise,* pp. 19–22.

    Many of the nation's nearly 200 hospitals in the early 1870s had been established as private charitable institutions for the urban poor; paternalistic and moral assumptions of stewardship and charity dominated the organization and practices of these institutions. Whereas the cities' acutely ill poor ended up in publicly managed almshouses and pesthouses, only those worthy poor—not acutely ill, not suffering from a contagious or morally unacceptable disease like syphilis—ended up in these voluntary hospitals.
13. Foucault discusses the architectural figure of the Panopticon, based on a supervisor's being able constantly to view and therefore control inmates. See Foucault, *Discipline and Punish,* pp. 200–203.
14. Billings, "Notes on Hospital Construction," p. 387.
15. Rosenberg, *The Care of Strangers,* pp. 127–128. See also Thompson and Goldin, *The Hospital,* pp. 208–209.
16. See Rosenberg, *Florence Nightingale on Hospital Reform.* See Woodham-Smith, *Florence Nightingale,* for a biographical account of Nightingale.
17. The "bad apple" is an example shared by Martin Pernick in an e-mail sent to the history of science and medicine list, January 16, 1998.
18. Prior, "The Architecture of the Hospital," p. 94.
19. See Vogel, *The Invention of the Modern Hospital.* See also Bordley and Harvey, *Two Centuries of American Medicine, 1776–1976,* pp. 62–63.
20. Wylie, W. G. "Hospitals: Their History, Organization, and Construction," p. 60.
21. See Thompson and Goldin, *The Hospital,* pp. 183–187; and Harvey, *A Model of Its Kind.* Vol. 2, *A Pictorial History of Medicine at Johns Hopkins.*
22. Pernick, *A Calculus of Suffering,* pp. 208–223.
23. Lamb, *The Presbyterian Hospital and the Columbia-Presbyterian Medical Center, 1868–1943,* pp. 187, 249 and 378; and Hirsch and Doherty, *The First Hundred Years of the Mount Sinai Hospital of New York, 1852–1952,* pp. 181 and 294.
24. Stevens, *In Sickness and In Wealth,* pp. 24 and 229; and Fox, *Health Policies, Health Politics.*
25. Leavitt, *Brought to Bed,* p. 175.
26. Boston Lying-In Hospital Trustees, *Condensed Annual Report, 1922,* pp. 1–2. The new building was designed by Coolidge and Shattuck, Architects (see *The Report of the Boston Lying-In Hospital, 1917–1920,* p. 118).
27. Howell, "Early Use of X-Ray Machines and Electrocardiographs at The Pennsylvania Hospital."
28. On the rise of medical technologies in the hospital, see Reiser, *Medicine and the Reign of Technology;* and Rothman, *Beginnings Count.* See also Howell, *Technology in the Hospital.*
29. Neergaard, "Planning the Small General Hospital," p. 77.
30. Kostof, *History of Architecture,* p. 661.
31. Massachusetts General Hospital Trustees, *125th Annual Report,* p. 7. The building was designed by Coolidge, Shepley, Bulfinch and Abbot, Architects. Harry K. Shepley was the main architect associated with the project. See Faxon, *The Massachusetts General Hospital, 1935–1955,* p. 163.

32. McCalmont and Associates, "National Hospital Bureau."
33. Norman, "Administrative Aspects of Hospital Design," p. 45.
34. Stevens, *In Sickness and In Wealth,* p. 339.
35. For a powerful account of this ethos, see Sinclair Lewis's novel *Arrowsmith,* published in 1925.
36. Stevens, *In Sickness and In Wealth,* pp. 351–352.
37. The National Cancer Institute has been particularly active in funding translational research efforts. Recent discussions of translation research include Lerman, "Translational Behavioral Research in Cancer Genetics"; and Karp and McCaffrey, "New Avenues of Translational Research in Leukemia and Lymphoma."
38. Shem, *The House of God: A Novel.* See also Rosenberg, *The Care of Strangers;* and Knowles, ed., *Doing Better, Feeling Worse.*
39. Illich, *Medical Nemesis,* esp. pp. 32–34.
40. See Omran, "The Epidemiological Transition: A Theory of the Epidemiology of Population Change."
41. Rosenberg, "Community and Communities," p. 4.
42. Sloane, "Scientific Paragon to Hospital Mall." See also Sloane, *Mall Medicine.*
43. Leach, *Land of Desire.*
44. Sloane, "Scientific Paragon to Hospital Mall."
45. A. O. D., "Emergency Unit Puts a Welcoming Face on a Hospital," p. 65. See also Kelly and Sanchez, "The Space of the Ethical Practice of Emergency Medicine"; and Malkin, *Hospital Interior Architecture.*
46. McKahan, "The Healing Environment of the Future," p. 37. See also Horsburgh, "Healing by Design."
47. Sloane, "In Search for the Hospitable Hospital." Shepley, Bulfinch, Richardson, and Abbott, Architects, designed the new hospital. See also Gregerson, "Medical Center Adds Human Touch to Health Care."
48. Chestnut Hill Mall was designed by Sumner Schein, Architects.
49. See Herzlinger, *Market-Driven Health Care.*
50. McKahan, "The Healing Environment of the Future," p. 36. On the current issues confronting health care, see Ginzberg, *Tomorrow's Hospital.*
51. See Celebration Health Website (28 January 1998 Newsletter, copyright 1998, Florida Hospital) at http://www.celebrationhealth.com/.
52. Greene, "Orlando Hospital Picked to Create State-of-the-Art Facility for Disney," p. 18.
53. Swensson, "Mall Concept: Ambulatory Care," p. 128. See also Miller and Swensson, *New Directions.*
54. Kasson, *Civilizing the Machine.*

Bibliography

A.O.D. "Emergency Unit Puts a Welcoming Face on a Hospital: Baptist Hospital, Miami, The Ritchie Organization." *Architecture* (April 1986): 65–67.

Abel-Smith, Brian. *The Hospitals in England and Wales 1800–1948.* Cambridge: Harvard University Press, 1964.

Billings, John Shaw. "Notes on Hospital Construction." *Public Health Reports and Papers of the American Public Health Association Meeting, 1874–75,* 385–388. New York: Hurd and Houghton, 1876.

Bordley, James, and A. McGehee Harvey. *Two Centuries of American Medicine, 1776–1976.* Philadelphia: W. B. Saunders Company, 1976.

Boston Lying-In Hospital. *Condensed Annual Report, 1922.* Boston: Boston Lying-In Hospital, 1923.

Boston Lying-In Hospital. *Report of the Boston Lying-In Hospital, 1917–1920.* Boston: Boston Lying-In Hospital, 1922.

Faxon, Nathaniel W. *The Massachusetts General Hospital, 1935–1955.* Cambridge: Harvard University Press, 1959.

Fiset, Martin. "Architecture and the Art of Healing." *The Canadian Architect* 35 (March 1990).

Forty, Adrian. "The Modern Hospital in England and France: The Social and Medical Uses of Architecture." In Anthony D. King, ed., *Buildings and Society: Essays on the Social Development of the Built Environment.* London: Routledge & Kegan Paul, 1980.

Foucault, Michel. *Discipline and Punish: The Birth of the Prison.* New York: Vintage Books, 1977.

Fox, Daniel M. *Health Policies, Health Politics: The British and American Experience, 1911–1965.* Princeton, NJ: Princeton University Press, 1986.

Fox, Daniel M. *Power and Illness: The Failure and Future of American Health Policy.* Berkeley and Los Angeles: University of California Press, 1993.

Ginzberg, Eli. *Tomorrow's Hospital: A Look into the Twenty-First Century.* New Haven, CT: Yale University Press, 1996.

Golden, J., and C. E. Rosenberg. *Pictures of Health: A Photographic History of Health Care in Philadelphia, 1860–1945.* Philadelphia: University of Pennsylvania Press, 1991.

Greene, Jay. "Orlando Hospital Picked to Create State-of-the-Art Facility for Disney." *Modern Healthcare* (November 1992): 18.

Gregerson, J. "Medical Center Adds Human Touch to Health Care." *Building Design and Construction* (August 1992): 24–29.

Harvey, A. McGehee, et al. *A Model of Its Kind.* Vol. 2. *A Pictorial History of Medicine at Johns Hopkins.* Baltimore: Johns Hopkins University Press, 1989.

Herzlinger, Regina E. *Market-Driven Health Care: Who Wins, Who Loses in the Transformation of America's Largest Service Industry.* Reading, MA: Addison-Wesley Publishing Co., 1997.

Hirsch, Joseph, and Beka Doherty. *The First Hundred Years of the Mount Sinai Hospital of New York, 1852–1952.* New York: Random House, 1952.

Holmes, Christian. "The Planning of a Modern Hospital: An Address Delivered Before the Department of Nursing and Health, Teacher's College, Columbia University, February 21, 1911." Detroit, MI: National Hospital Record Publishing Company, 1911.

Hornsby, John Allan, and Richard E. Schmidt. *The Modern Hospital: Its Inspiration, Its Architecture, Its Equipment, Its Operation.* Philadelphia: W. B. Saunders Company, 1913.

Horsburgh, Jr., C. Robert. "Occasional Notes: Healing by Design." *The New England Journal of Medicine* 333 (September 14, 1995): 735–740.

Howell, Joel D. "Early Use of X-ray Machines and Electrocardiographs at The Pennsylvania Hospital." *Journal of the American Medical Association* 255 (May 2, 1986): 2320–2323.

Howell, Joel D. *Technology in the Hospital: Transforming Patient Care in the Early Twentieth Century.* Baltimore: Johns Hopkins University Press, 1995.

Plates 1-4

(1) Panorama of *Glienicke,* August C. Hahn, after Wilhelm Schirmer, ca. 1845, Staatliche Schlösser und Gärten Berlins, Charlottenburg, with details of (2) the *Teufelsbrücke;* (3) the *Glienicke* engine house; and (4) the steamboat *Alexandra* with the *Babelsberg* engine house behind the bridge.

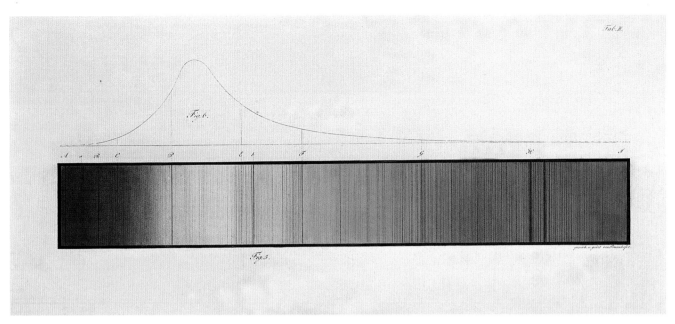

Plate 5

Joseph von Fraunhofer's 1814 depiction of the solar spectrum and the spectral lines. Reprinted with the kind permission of the Deutsches Museum, Munich, photograph 43952.

Plate 6

The Museum of Practical Geology, Jermyn Street, from a watercolor painting by J.P. Emslie, 1876, IGS/897. Reproduced by permission of the Director, British Geological Survey. NERC copyright reserved.

Plate 7

Lewis Thomas Laboratory for Molecular Biology, Princeton University. Photo by Matt Wargo.

Plate 8

Color-coded early program study for the organization of space in the Lewis Thomas Laboratory for Molecular Biology at Princeton University.

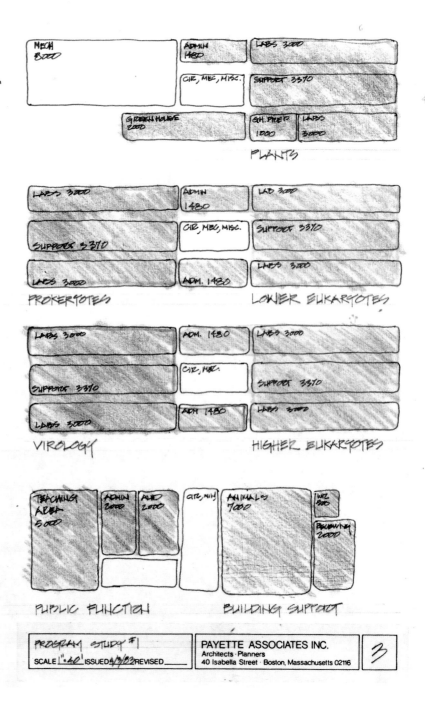

Plate 9

Central staircase of the Lewis Thomas Laboratory for Molecular Biology at Princeton University.

Plate 10

Central foyer and stairwell of the Lewis Thomas Laboratory for Molecular Biology at Princeton University.

Plate 11

Fermi National Accelerator Laboratory (Fermilab), Batavia, Illinois. Photo copyright Fermilab Visual Media Services.

Plate 12

Design for the Superconducting Super Collider Laboratories, Waxahachie, Texas. Photo copyright Michal Ronnen Safdie.

Illich, Ivan. *Medical Nemesis: The Expropriation of Health.* New York: Random House, 1976.

Karp, J. E., and R. P. McCaffrey. "New Avenues of Translational Research in Leukemia and Lymphoma: Outgrowth of a Leukemia Society of America–National Cancer Institute Workshop." *Journal of the National Cancer Institute* 86: 16 (August 17, 1994): 1196–1201.

Kasson, John. *Civilizing the Machine: Technology and Republican Values in America, 1776–1900.* New York: Penguin Books, 1976.

Kelly, Michael, and Ricardo Sanchez. "The Space of the Ethical Practice of Emergency Medicine." *Science in Context* 4 (1991): 79–100.

Kirker, Harold. *The Architecture of Charles Bulfinch.* Cambridge: Harvard University Press, 1969.

Knowles, John H., ed. *Doing Better and Feeling Worse: Health in the United States.* New York: Norton, 1977.

Knowles, John H., ed. *Hospitals, Doctors, and the Public Interest.* Cambridge: Harvard University Press, 1965.

Kostof, Spiro. *A History of Architecture: Settings and Rituals.* New York and Oxford: Oxford University Press, 1995.

Lamb, Albert R. *The Presbyterian Hospital and the Columbia-Presbyterian Medical Center, 1868–1943.* New York: Columbia University Press, 1955.

Leach, William. *Land of Desire: Merchants, Power, and the Rise of a New American Culture.* New York: Pantheon, 1993.

Leavitt, Judith W. *Brought to Bed: Childbearing in America, 1750 to 1950.* New York: Oxford University Press, 1986.

Leavitt, Judith W., and Ronald L. Numbers, eds. *Sickness and Health in America: Readings in the History of Medicine and Public Health.* Madison: University of Wisconsin Press, 1985.

Lerman, Caryn. "Translational Behavioral Research in Cancer Genetics." *Preventive Medicine* 26 (September-October 1997): S65–S69.

Long, Diane E., and Janet Golden, eds. *The American General Hospital: Communities and Social Contexts.* Ithaca, NY: Cornell University Press, 1989.

Ludlow, William. "Why Not Homelike Hospitals?" *The Literary Digest* 60 (January 18, 1919): 19. (Condensed version of article with the same title in *Hospital Management.*)

Malkin, Jain. *Hospital Interior Architecture: Creating Healing Environments for Special Patient Populations.* New York: Van Nostrand Reinhold, 1992.

Massachusetts General Hospital Trustees. *125th Annual Report.* Boston: Massachusetts General Hospital, 1938.

McCalmont, M. E., and Associates. "National Hospital Bureau." Pamphlet from the New York Academy of Medicine Collections, New York, 1916.

McKahan, Donald C. "The Healing Environment of the Future." *Healthcare Forum* (May/June 1990): 36–39.

Miller, Richard L., and Earl S. Swensson. *New Directions in Hospital and Healthcare Facility Design.* New York: McGraw-Hill, 1995.

Neergaard, Charles F. "Planning the Small General Hospital." *Architectural Record* 86 (December 1939): 75–106.

Norman, Jacque B. "Administrative Aspects of Hospital Design." In American Institute of Architects, *Hospitals: Convention Seminar Addresses.* Washington, DC: American Institute of Architects, 1947.

Nuland, Sherwin B. *How We Die: Reflections on Life's Final Chapter.* New York: Alfred A. Knopf, 1994.

Omran, Abdel R. "The Epidemiological Transition: A Theory of the Epidemiology of Population Change." *Milbank Memorial Fund Quarterly* 49 (October 1971): 509–538.

Pernick, Martin S. *A Calculus of Suffering: Pain, Professionalism, and Anesthesia in Nineteenth-Century America.* New York: Columbia University Press, 1985.

Prior, Lindsay. "The Architecture of the Hospital: A Study of Spatial Organization and Medical Knowledge." *The British Journal of Sociology* 1 (March 1988).

Reiser, Stanley Joel. *Medicine and the Reign of Technology.* Cambridge: Cambridge University Press, 1978.

Rosen, George. "The Hospital: Historical Sociology of a Community Institution." In E. Freidson, ed., *The Hospital in Modern Society,* 1–36. New York: The Free Press of Glencoe, 1963.

Rosenberg, Charles E. "And Heal the Sick: The Hospital and Patient in Nineteenth Century America." *Journal of Social History* 10 (1977): 428–447.

Rosenberg, Charles E. *The Care of Strangers: The Rise of America's Hospital System.* New York: Basic Books, 1987.

Rosenberg, Charles E. "Community and Communities: The Evolution of the American Hospital." In Long and Golden, eds., *The American General Hospital: Communities and Social Context.*

Rosenberg, Charles E. "Florence Nightingale on Contagion: The Hospital as Moral Universe." In Charles E. Rosenberg, ed., *Healing and History: Essays for George Rosen.* Kent, UK: Dawson and Sons, 1979.

Rosenberg, Charles E., ed. *Florence Nightingale on Hospital Reform.* New York: Garland Publishers, 1989.

Rosner, David. *A Once Charitable Enterprise: Hospitals and Health Care in Brooklyn and New York 1885–1915.* Cambridge: Cambridge University Press, 1982.

Rothman, David J. *Beginnings Count: The Technological Imperative in American Health Care.* New York: Oxford University Press, 1997.

Rothman, David J. *The Discovery of the Asylum: Social Order and Disorder in the New Republic.* Boston: Little, Brown and Company, 1971.

Shem, Samuel. *The House of God: A Novel.* New York: R. Marek, 1978.

Sloane, David C. "In Search for the Hospitable Hospital." *Dartmouth Medicine* 18 (Fall 1993): 23–31.

Sloane, David C. *Mall Medicine: The Evolving Landscape of American Health Care.* Baltimore: Johns Hopkins University Press, forthcoming.

Sloane, David C. "Scientific Paragon to Hospital Mall: The Evolving Design of the Hospital 1900–1990." *Journal of Architectural Education* 48 (November 1994): 82–98.

Sloane, David C. "Twilight Homes: Life and Death in America's Hospitals, Nursing Homes, and Hospices, 1900–1990." Paper presented at the meeting of the *American Association of the History of Medicine,* Buffalo, New York (April 1994).

Sommer, Robert, and Robert Dewar. "The Physical Environment of the Ward." In E. Freidson, ed., *The Hospital in Modern Society,* 319–342. New York: The Free Press of Glencoe, 1963.

Stevens, Edward F. *The American Hospital of the Twentieth Century.* New York: F. W. Dodge Corporation, 1928.

Stevens, Rosemary. *In Sickness and In Wealth: American Hospitals in the Twentieth Century.* New York: Basic Books, 1989.

Swensson, Earl S. "Mall Concept: Ambulatory Care." In Albert Bush-Brown and Dianne Davis, eds., *Hospitable Design for Healthcare and Senior Communities,* New York: Van Nostrand Reinhold, 1992.

Thompson, John D., and Grace Goldin. *The Hospital: A Social and Architectural History.* New Haven, CT: Yale University Press, 1975.

Vogel, Morris J. *The Invention of the Modern Hospital: Boston 1870–1930.* Chicago: University of Chicago Press, 1980.

Woodham-Smith, Cecil. *Florence Nightingale.* New York: McGraw-Hill Book Company, 1951.

Woodworth, John M. "Hospitals and Their Construction: The Principles Which Should Govern in the Location, Design, Material, General Management, and Duration of Use of Hospitals." *Public Health Reports and Papers of the American Public Health Association Meeting, 1874–75,* 389–395. New York: Hurd and Houghton, 1876.

Wylie, W. G. "Hospitals: Their History, Organization, and Construction." (Boylston Prize Essay of Harvard University for 1876.) New York: D. Appleton and Company, 1877.

IV

Is Architecture Science?

13

Architecture, Science, and Technology

Antoine Picon

History of Architecture—History of Science and Technology: A Productive Crossing

"Since the beginning of the century, architects and engineers are staring at each other with more bewilderment than true sympathy. It is as if god A+B and goddess Imagination were looking hard at each other,"[1] declared the architectural critic César Daly at a dinner organized by the French Civil Engineers Society in 1877. Throughout the nineteenth and twentieth centuries, a whole range of discourses and writings have been devoted to the divorce between architecture and engineering, as well as to the more general split between architecture and science and technology. Most of those discourses and writings have stressed the importance of bridging the gap between the architectural field and the scientific and technological one. But until recently the fundamental assumption that a divorce had taken place between architecture and science and technology has not been seriously discussed.

More recently however, theoreticians and practitioners of architecture as well as historians have challenged this vision of a radical divorce between architecture and science and technology. In the historical field, one is especially struck by the renewal of the question of the relations between the disciplines. To begin with the history of architecture, books such as Alberto Pérez-Gómez's *Architecture and the Crisis of Modern Science* and

Kenneth Frampton's *Studies in Tectonic Culture* are representative of this renewal.[2] Simultaneously, architecture has begun to interest historians of science and technology. A historian of science such as Peter Galison has written, for instance, on the status of the architectural metaphor in early-twentieth-century epistemology,[3] whereas the celebrated historian of technological systems, Thomas Hughes, is more and more curious about architecture.

Adopting the same perspective, I would like to present here some evidence of the productive character of the crossing between the history of architecture and the history of science and technology. But the adoption of this point of view doesn't imply that one has to abandon the idea of an estrangement between architecture and science and technology. It is rather based on the hypothesis that this estrangement is not as radical as it may appear at first sight, and that furthermore, the boundary between the two domains is still permeable.

Before entering into some details about the materials, it is perhaps interesting to consider a few supplementary reflections on the renewal of the question of the relations between architecture and science and technology. What underlying preoccupations can account for this renewal?

Some immediate reasons can be given to explain such a phenomenon. If architecture is no longer situated near the cutting edge of scientific and technological research, it is still a practice involving some scientific and technological knowledge. Its practical dimension is also stimulating for historians who have become more and more interested in the material basis of intellectual production as well as in the complexity of design processes. Many historians of science have repudiated the idealistic conception of scientific knowledge as a pure system of ideas, whereas historians of technology now endeavor to decipher the ways that technological artifacts are designed with a certain degree of arbitrariness (just as are buildings or furniture), a degree of arbitrariness often previously neglected.

A more general reason for the interest in the relations between architecture and science and technology can certainly be found in the widespread nostalgia for the former unity of culture. Until the eighteenth century, architecture was close to science and technology. Because it has especially suffered from the estrangement between the sciences and the arts created by the first industrial revolution, some may find it interesting to have a closer look at the relations that have remained between those two worlds through the mediation of an intermediary practice such as architecture. But nostalgia is only one part of the general answer.

The main factor is perhaps the change in status that science and technology have recently experienced. Science especially is no longer revered as the only true form of knowledge. In relation to a context of growing disillusion regarding its social impact, science is more and more often considered a cultural production, just as are the arts. For technology, the change is of course less radical because technology and society seem to share so many determinations. But the consequence is in both cases the same: Because science and technology are cultural productions, they must certainly bear some relation to other cultural fields.

Taking now the point of view of the architectural historian, what can be the interest in crossing the history of architecture with the history of science and technology? Once again, I don't want to deliver a radical message. My aim is not to reinterpret entirely the history of architecture but rather to contribute to a better understanding of architecture's place within the culture of its time. In other words, the aim is to contribute to the study of architectural meaning.

What I would like to do now is to return to the period when architecture and science had strong and evident links to one another, from the Renaissance to the mid–eighteenth century. Then I underline some of the difficulties that one encounters when dealing with those links in the nineteenth and twentieth centuries. As I illustrate, taking the example of Eugène Emmanuel Viollet-le-Duc and the life sciences of his time, the main danger is perhaps the false analogy. If one wants to exhibit real relations between contemporary architecture and science and technology, one must, in my opinion, reposition all of them within rather large cultural patterns. The influence of these patterns can be traced both in the architectural field and in the scientific and technological one. I give two examples of the type of broader frame of interpretation I have in mind, a deeper way to locate architectural and scientific and technological developments within a certain historically specific set of concerns. The first one deals with the relation between French architecture and the scientific trends of the late eighteenth century. The other deals with more contemporary architectural production and the relation it has to some of the technological mutations we are experiencing today.

Architecture, Science, and Technology in the Vitruvian Tradition

From the Renaissance to the second half of the eighteenth century, European architectural theory and practice remained relatively faithful to the architectonic principles Vitruvius espoused in his *Ten Books on Architecture.* The Vitruvian tradition was of course no monolithic body of knowledge but rather a range of basic assumptions that one could discuss and even challenge.[4]

In the Vitruvian frame of thought, architecture was considered an art much closer to the sciences than it has ever been considered since. This connection was well illustrated by the existence of people such as Simon Stevin in the Netherlands, François Blondel or Claude Perrault in France, and Christopher Wren in England, who were all both architects and scientists.[5] Beside those rather extreme examples, it was not uncommon to see architects reading scientific treatises or scientists interested in problems of architecture.

The interesting scientific issues raised by architectural theory and practice accounted for this situation. Some of these issues were self evident. Structural problems had always been a scientific puzzle. Since Galileo and his *Discourse Concerning Two New Sciences,* strength of materials had become an interesting field for scientific investigation.[6] At the end of the seventeenth century, understanding of the mechanism of solidification of quicklime began also to attract the attention of scientists.[7] Such examples could be multiplied easily. Architecture and construction were a source of stimulating questions for science.

Architecture was also the source of other, less obvious scientific questions. Since the Renaissance, the practice of optical correction recommended by Vitruvius involved what we would today call the physiology and psychology of vision. An illustration of the question of optical correction is found in the first French translation of Vitruvius: Is it necessary to incline statues situated near the edge of a building to avoid an impression of disequilibrium? This type of question was part of a broader range of inquiries regarding the relation between architecture and the life sciences.

An intricate network of metaphors and analogies linked the study of architectural structures and the anatomical research of the Classical Age. To understand the use of iron reinforcement in the Louvre Colonnade, a technique still exceptional at the time of its construction in the 1670s, one must for instance take into account the very peculiar theory of muscular contraction or rather muscular relaxation of its main designer, Claude Perrault.[8] For Descartes, the action of muscles was carried out through their contraction, which was the result of a flow of "esprit animaux," or animal spirits, very small particles produced by the brain and sent throughout the body to make the body move. Thus, muscles inflated and contracted just like pneumatic devices. This principle is well illustrated in Figure 13.1, taken from Descartes' treatise on the human body. Here the eye turns because the muscle to the left is inflated and shortened.

In Perrault's theory, on the contrary, muscles were naturally contracted, and the action of animal spirits was to expand them. (See figure 13.2.) Rather than being provoked by a local tension, movement was the result of a local relaxation. To illustrate this general idea, Perrault gave the example of a mast held by two ropes in tension. Cutting one rope made the mast move in the opposite direction.

The relation between this theory and the Colonnade had to do with the conception of naturally tensioned structures. In the Colonnade, the iron reinforcement was analogous to the muscles giving tone to the soft flesh.

Architecture was additionally studied typologically, as in the works of Serlio and Le Muet, and could thus also provide precious methodological orientation for the problems of classification encountered in natural history or the study of living beings. This approach was implicit in Perrault's use of a common type of representation for Roman temples and animals in his translation of Vitruvius and in his anatomical memoirs. Both series of engravings were based on the distinction between static structures or anatomy and dynamic functioning.

13.1

René Descartes, *Traité de l'homme* (1664). Explanation of the contraction of muscles. The eye turns because one of the muscles is inflated and shortened as a result of the flow of animal spirits.

13.2

Claude Perrault, *Essais de physique* (1680–1688). The flexion and extension of the arm explained through a mechanical analogy. According to Perrault, each muscle responsible for a certain movement has an antagonist responsible for the opposite movement. Thus, the motion is provoked by the relaxation of one of the two muscles. The situation is similar to what happens when a mast held by two ropes in tension moves because one of the ropes is suddenly cut.

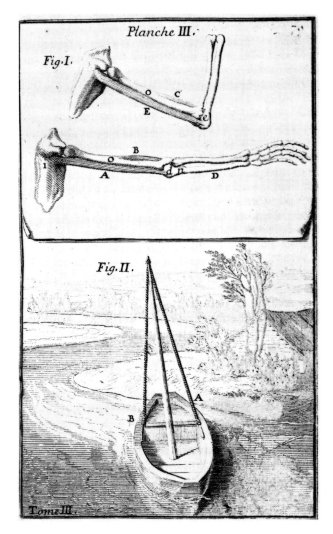

Throughout the Classical Age, the main relation between architecture and science was based on the widespread belief in an architectonic world ruled by proportion. Although the new modern science had begun to challenge this vision, architecture was still considered a discipline embodying very essential natural principles. This belief explains the fortune of speculations regarding the Temple of Jerusalem, from the Spanish Jesuit Juan Bautista Villalpando to Isaac Newton, since the proportions used by God in this sacred building were certainly expressing some esoteric truth about the universe.[9]

Returning to Perrault's work, the numerous connections between science and architecture help us to understand a project such as the Observatory of Paris. In this building, the architecture was not simply meant to convey the symbolic importance of astronomy. Actually, Perrault designed the monument so that it could be used as a giant scientific instrument. Its main axis corresponded to the meridian, whereas the facades of its polygonal towers gave the direction of the sun at the solstices and equinoxes. Furthermore, a central shaft penetrating the entire building was supposed to help astronomers to make zenithal observations.[10]

During the second half of the eighteenth century, all these very immediate relations between architecture and science gradually weakened. Structural engineering split from architecture, and strength of materials was transformed into a new science based on calculus rather than on geometrical figures. The study of living beings adopted models other than the mechanical ones that bore so many analogies to architecture. Above all, a new vision of nature seriously challenged the faith in an architectonic world. To appreciate the extent of the split between architecture and science, one can compare Perrault's Observatory with the famous Cenotaph for Newton by Etienne Louis Boullée.[11] Whereas the first project aimed to use architecture as an efficient scientific tool, Boullée's ambition was simply to provide a striking symbol.

THE DANGERS OF ANALOGY

Since the beginning of the nineteenth century, architecture and science have seemed to belong to two different worlds. This does not mean that all possible relations have been abolished. There are, for example, links between the vision of structure developed by the nineteenth-century rationalist architects and the conceptions of organism prevailing around the same time in the life sciences. Some themes of twentieth-century modern architecture can also be related to their scientific context. But one must be very cautious when making these connections. The main danger, in my opinion, is that one is often tempted to construct superficial analogies.

An example of such a superficial analogy is found in the case of Eugène Emmanuel Viollet-le-Duc and the alleged direct and determining relation between his vision of structure and the conceptions of living organism that were discussed at the time.[12] Dwelling on famous illustrations such as those taken from his *Histoire d'un dessinateur* and on some passages of his theoretical work, one is very often tempted to draw a direct comparison between his structural intuitions and those, let us say, of Georges Cuvier. (See figure 13.3.) After all, Viollet-le-Duc mentioned Cuvier in his *Dictionnaire raisonné de l'architecture française,* and his conception of structural coherence seems to echo one of Cuvier's most famous achievements. One day Cuvier was brought a few bones of an unknown fossil, a kind of prehistoric opossum, that had just been found in the gypsum quarries surrounding Paris. During the night, Cuvier reconstructed the entire skeleton, starting from the few bones he had at his disposal. Then, in the following days, a well-preserved skeleton was found in the quarries. Cuvier's reconstruction and the skeleton matched perfectly. "The architecture of the Middle Ages proceeds with the type of logical order we discover in the works of nature. Therefore, just as from the leaf of a plant one can deduce the entire plant, from the bone of an animal the entire animal, seeing a single profile is sufficient to deduce the architectural member to which it belongs, and from the member to reconstruct the monument,"[13] wrote Viollet-le-Duc in the entry "Style" of his *Dictionnaire.*

Combined with Geoffroy Saint-Hilaire's reflections on types and variations, the reference to Cuvier is supposed to shed light on some of Viollet-le-Duc's most daring reconstructions, such as his ideal Gothic cathedral. I don't think that the analogy, though extremely seducing, plays such a fundamental role. Two reasons, at least, make me doubtful. The first is the scarcity of the direct references made by Viollet-le-Duc to scientists

13.3

E. E. Viollet-le-Duc, *Mémoires d'un dessinateur* (1879). Application of the articulation of bones to engineering.

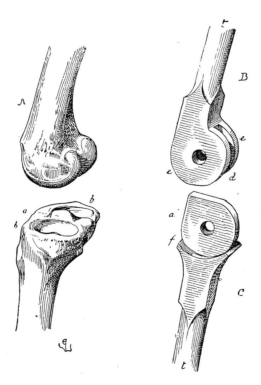

such as Cuvier or Saint-Hilaire in his major theoretical books such as the *Dictionnaire* or the *Entretiens sur l'architecture.* The second is the fact that, for Viollet-le-Duc, structures were social products, that is to say they were not pure natural developments but principles defined at the crossing of nature and society. In his opinion, for example, the Gothic vault emerged at the crossing of a social need for large gatherings of people, an economic context of expansive labor, and a reflection on weight and thrust.

There is certainly a reference to natural achievements in Viollet-le-Duc's theoretical work. But I think that its best expression is to be found rather in his study of the Alps, in his belief in geometrical patterns of growth, which explains his astounding reconstruction of Mont-Blanc.[14] In this perspective, Viollet-le-Duc is perhaps closer to D'Arcy Thompson's celebrated book, *On Growth and Form,* than to the life sciences of his own time. With his fascination for the triangle and the tetrahedron, I would even like to suggest that he offered a kind of a precursor of Buckminster Fuller's Synergetics. Of course, the belief in very general patterns of growth has something to do with the study of life, but it is more significantly a more general concern with the dynamism of nature.

To avoid the dangers of superficial analogies, I would like to suggest now the necessity of building broader frames of interpretation, cultural frames that are somewhat anterior or rather deeper than architectural doctrines or scientific theories. The crossing between the history of architecture and the history of science and technology is interesting if one does not reduce architecture to a consequence of the scientific and technological context, or the reverse, though this reverse approach is certainly harder to sustain.

I now give two examples of the type of broader frame of interpretation one can suggest to understand certain phases of architectural development. I deal first with late-eighteenth-century French architecture, then with a more contemporary episode. The first of my two case studies focuses on science; the second emphasizes technology.

"Revolutionary" Architecture and Scientific Analysis

In the last decades of the eighteenth century, French architecture evolved in a striking manner. Architects such as Etienne-Louis Boullée and Claude-Nicolas Ledoux abandoned the traditional compositional techniques for a clear-cut geometry of simple shapes, for spatial dispositions based on distinct and often isolated volumes. The historian of art Emil Kauffmann was among the first to study this evolution, which led to neoclassicism, and he coined various phrases to describe it. In his famous *Von Ledoux bis Le Corbusier* and *Three Revolutionary Architects: Boullée, Ledoux, Lequeu,* published in the 1930s and 1950s, he spoke of a "pavilion system" of composition as well as of a "revolutionary architecture."[15] The first phrase referred to the abandonment of the traditional hierarchies of the French classical tradition, the second to the possible relation of this evolution to the political issues of the time.

Kauffmann's views have been challenged often since the publication of his major books.[16] Contemporary historians such as Anthony Vidler have argued quite convincingly that Ledoux was no radical revolutionary at all.[17] Still, there is something to be understood in the evolution embodied in Boullée's and Ledoux's work. In his books on Ledoux, Vidler has explained quite subtly the true political meaning of this evolution. What I try to do next is to relate it to some scientific as well as technological issues of the time. Those scientific and technological issues are intertwined, but as I have said, I emphasize science.

In his introduction to the French translation of *Three Revolutionary Architects,* Georges Teyssot was among the first to relate this episode of architectural history to its scientific and technological context.[18] I follow the path he has opened. A motive for this pursuit can certainly be found in the success of Boullée's and Ledoux's conceptions among the engineers of the time. At the Ecole des Ponts et Chaussées, for instance, Boullean principles very often directly inspired the architectural designs produced in the 1780s.[19] Figure 13.4 is a Ponts et Chaussées design for a church that is clearly inspired by Boullée's famous project for a cathedral. There is certainly a relation between Boullée's and Ledoux's principles and the scientific and technological issues of the time.

13.4

F. Forestier de Villeneuve. Design of a cathedral for the 1782 architectural competition of the Ecole des Ponts et Chaussées, cross section. Ecole Nationale des Ponts et Chaussées, Ms. 105. A typically Boullean composition.

My starting point is the urge for new theoretical foundations expressed in the writings of Boullée and Ledoux as well as in the works of less famous architects such as Nicolas Le Camus de Mézières. The context of those theoretical reflections was, of course, the crisis of the Vitruvian tradition, a tradition challenged by factors ranking from the influence of travel and the discovery of the relative value of Vitruvian principles to the growing autonomy of construction and engineering.

For Boullée, architecture's new theoretical foundations were to be discovered in the systematic exploration of the relations between sensations and elementary shapes and volumes, in a reflection on the various types of architectural programs.[20] The same orientation characterized Ledoux's approach. In his major treatise, *L'architecture considérée sous le rapport de l'art, des mœurs et de la législation,* Ledoux wrote, for example, that "the circle and the square are the alphabet of the best conceived designs."[21] Given the pervasive influence of the sensationalist philosophy of Locke and Condillac at the time, one is not surprised to find the same basic assumption, namely that architecture is to be grounded on the sensations created by elementary shapes and volumes, in Nicolas Le Camus de Mézière's *Le génie de l'architecture.*[22] As did Boullée, Le Camus de Mézières also reflected on the various architectural programs and their ultimate components in his *Le génie de l'architecture,* which was mainly devoted to a thorough study of the different types of rooms in housing projects.

Beyond the sensationalist influence, the common feature of all these theoretical attempts was their ambition to identify elementary components of architectural practice. Besides the elementary shapes, volumes, and functionalities, there were also constructive components such as those described by the architect Pierre Patte in his *Mémoire sur les objets les plus importants de l'architecture.*[23]

The most striking feature is, however, the analogy between those reflections in terms of elements and the emphasis put, around the same time, by philosophers and scientists on the elements of the sciences. Standing in contrast to the brevity of its article on the natural elements, the *Encyclopédie* had devoted pages to the "éléments des sciences." "Generally speaking, elements are the primitive parts of a whole," wrote d'Alembert in the latter article. He then observed that the sciences were cumulative, that they could be taught and learned because they were at bottom combinations of elements.[24] Through this seemingly trivial remark, one can see immediately the appeal for an art such as architecture to be a combination of elements as well.

In the same article, d'Alembert made another crucial remark, namely that the elements he had in mind were not the product of an absolute reduction but rather the expression of the relations between man and nature. Announcing nineteenth-century positivism, this meant that elements were facts rather than substances. Among the important scientific elements of the time is, for example, Newton's inverse-square law, which Coulomb was to transpose so successfully to electrostatics,[25] as well as the chemical elements of Lavoisier. But again, those latter elements were not to be considered ultimate substances, but rather the result of human operations. As Chaptal was to state it in his *Eléments de chimie,* published in 1803, chemical elements were only provisional results of the chemical processes provoked by man.[26]

If elements were not ultimate laws or substances, what then was their true characterization? The term "operation," often employed by authors such as d'Alembert, Lavoisier, and Chaptal, provides a hint. Elements could be considered as such because they combined in a dynamic way, because they enabled man to approach the true dynamism of nature. Behind that conception, a profound change in the vision of nature was taking place. Nature was no longer seen as a stable architectonic structure but rather as a productive power, as a complex set of intertwined processes. To describe it, one had to pass from an approach in terms of structure to the consideration of operations and functions.[27]

In eighteenth-century philosophical culture, the identification of elements and the understanding of the way they combined bore a name. Analysis was the method that consisted in the identification of elements, followed by the study of their various combinations. As Condillac stated it in his *Cours d'études* of 1775, "analysis is the entire decomposition of an object and the arranging of its components so that generation becomes both easy and understandable."[28] Applied to the human mind, the analytical method was supposed to explain how abstract knowledge and complex judgments derived from primitive sensations. There again, what was at stake was the understanding of a dynamic process. Applied to chemistry, it meant a systematic decomposition of complex substances such as the atmosphere, as well as their reproduction starting from their elementary components. In mathematics, calculus was the true analytical tool, because it enabled man to relate elementary variations to global phenomena. There again, the main utility of calculus, especially in fields such as astronomy or hydrodynamics, was to provide a better understanding of dynamic processes.

Technology was imbued with the same kind of concern for rational decomposition into elements followed by a rational recomposition.[29] Engineers applied analysis to their designs as well as to the construction process. An engineer such as Jean Rodolphe Perronet, who was both director of the Ecole des Ponts et Chaussées and the most famous bridge builder of the time, explained that bridge design had to be founded on the analysis of the bridge's forms and functions.[30] Ideally, bridges were to be composed of vertical piers and entirely horizontal decks. This analysis thus gave birth to an ideal model which we can see in figure 13.5. Inspired by manufacturing processes such as pin making,[31] engineers also analyzed all the stages involved in bridge construction. Their analytical approach reached its climax in their designs for ideal harbors and arsenals based on a thorough analysis of functions and operations, both during the construction process and in normal use after the design's completion.[32]

Dealing with territorial planning, which was part of their responsibilities, Ponts et Chaussées engineers also made use of the analytical approach. The drawings made by the

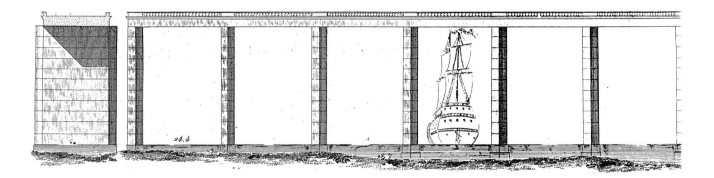

13.5

The Loyang Bridge in China, according to E.-M. Gauthey, *Oeuvres* (1809–1816). For Perronet as well as for Gauthey, his former student at the Ecole des Ponts et Chaussées, this bridge is almost ideal, with its vertical piers and its entirely horizontal deck. The bridge's size, as indicated by the travelers of the time is, of course, greatly exaggerated.

Ponts et Chaussées students for the annual map competition organized by the Ecole expressed well these engineers' conception of territory.[33] On those drawings depicting an imaginary territory, planning was based on the systematic identification of types of natural land and obstacles, plains, rivers, swamps, and mountains. The various kinds of infrastructures were envisioned in the same way, from roads and bridges to canals and harbors. These components were then combined to produce an ideal territory. The analogy between this ideal territory and the picturesque gardens of the time was, by the way, no accident, for picturesque gardens themselves were based, according to theoreticians and practitioners such as René Louis de Girardin or Jean Marie Morel, on an analysis of natural elements and ambiances.[34] Furthermore, according to the same authors, gardening was definitely related to agricultural and territorial planning.

In this same analytical perspective, engineers who were often in charge of architectural designs made use of Boullée's and Ledoux's compositional principles. Figures 13.6 and 13.7 illustrate a court of justice designed by a Ponts et Chaussées student that is based on the identification of elementary functions and volumes, an identification followed by their combination through the consideration of the various circulations linking them. Such an approach is clearly analogous to that of Ledoux at his famous salt factory of Arc-et-Senans.

Boullée's and Ledoux's attempts at a radical renewal of architectural foundations were thus related to a broad cultural frame, a frame that also shaped science and technology. Part of their ambition was to adapt architecture to the new vision of nature embodied in this frame. The importance of dynamism, natural and human, was clearly one of their preoccupations, from the symbolic alternance of day and night in Boullée's Cenotaph for Newton to the emphasis Ledoux placed on movement and travel. His *Architecture* opens with an account of a traveler visiting his ideal city of Chaux. Buildings such as his famous house traversed by a waterfall are part of the same inspiration, just as is the systematic use of oblique views to convey an impression of dynamism. Just as with engineers, part of the ambition of the so-called revolutionary architects was to transform

13.6

M. Sevestre, design for a court of justice for the 1782 architectural competition of the Ecole des Ponts et Chaussées, plan. Ecole Nationale des Ponts et Chaussées, Ms. 105. Boullean principles of composition combine with a rational analysis of functions and circulations.

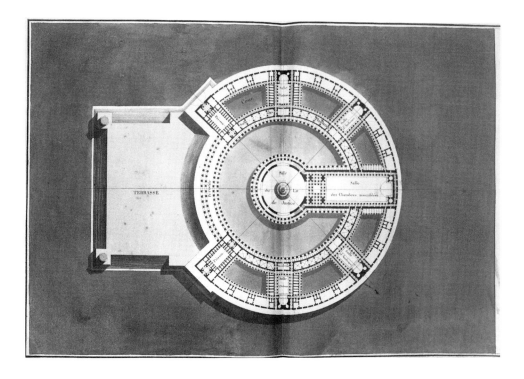

13.7

M. Sevestre, design for a court of justice for the 1782 architectural competition of the Ecole des Ponts et Chaussées, cross-section and elevation. Ecole Nationale des Ponts et Chaussées, Ms. 105.

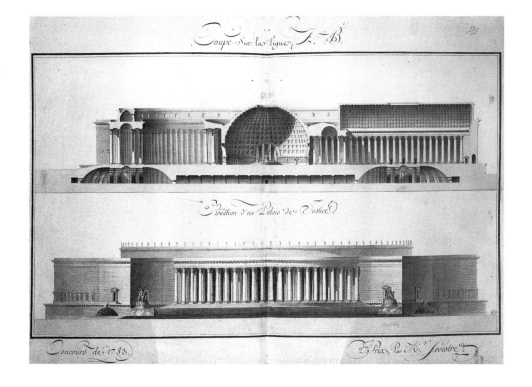

their constructions into regulating devices. Whereas engineers' bridges were to regulate the road traffic as well as the flow of water, architects' buildings had to regulate human activity.

An obsession for control is clearly at stake in this concern for regulation. Ledoux's ideal salt factory and city are no panopticon, but they announce Bentham's design. His preliminary study of the various operations involved in the production of salt is just as authoritarian in its principle as the specifications made by engineers concerning the workers employed on their construction sites. Both cases involve a typically eighteenth-century utopian belief, namely that this control benefits the people who submit to it, that the specifications governing their actions can be considered a kind of education. Ledoux's architecture was to possess an educational value, just as were the rationalized sites of the engineers.

Architects such as Boullée, Ledoux, and Le Camus de Mézières were at the time conscious that the new analytical foundation they tried to build was not the solution to all the problems encountered by architecture. Because architecture was primarily an art, it could not be reduced to the rationale of elements and their rational composition, an attitude that Durand was later to adopt.[35] To preserve what Boullée called the "poetry of art,"[36] one had to find other dimensions, expressive ones, outside the elementary sensations provoked by volumes.

Various answers were proposed: the dramatic play with light and shadow or the sublimity of bare walls contrasted with more ornate surfaces. All converged toward an urge for symbolization. Architecture was not only to be rational and thus useful, a key work of eighteenth-century philosophy, it was also to symbolize the process transforming an answer to primitive needs into a more elaborate message. In other words, architecture had to make tangible the process leading from nature to the human language. It had to speak, or rather to describe the path leading from what Condillac called the "language of action," that is to say the initial, uncoordinated attempts at communication, into the elaborate sphere of meaning that characterized civilization. Such a meaning was achieved by Boullée and Ledoux through the research of highly symbolic plan composition as well as by a decoration referring to the process of symbolization. Figure 13.8 offers a good example of this process of symbolization with the opposition between the rocks and the bare columns at the entrance of the salt factory, an opposition referring to the contrast between nature and civilization.[37] The rearchitecturing of Perronet's type of bridges attempted by both Boullée and Ledoux was based on a similar reference to the alleged origin of bridges, boats put one beside another to cross a river.

What was emerging thus at the end of the eighteenth century was the modern idea of expression. Architecture was no longer assimilable to the very structure of the natural world, but it could express the process of transformation of this world into a sphere of social meaning and communication. The growing importance of techniques of visual representation was part of this redefinition of the status and role of the architectural

13.8
C.-N. Ledoux, entrance to the salt factory of Arc-et-Senans. Ledoux symbolizes here the contrast between nature and civilization with the opposition between the rocks and the bare columns.

discipline, a discipline that could no longer be considered independent from society. Boullée's or Ledoux's spectacular drawings and engravings were inseparable from the conception of architecture as an essential dimension of the "public sphere."[38] In this perspective, architectural design had to be communicated and discussed just as political and social proposals.[39]

Architecture and the Crisis of Modern Technology

The second case study, which is much shorter than the previous one, could be entitled, paraphrasing Pérez-Gómez, "Architecture and the Crisis of Modern Technology." What I would convey through this study is, however, quite different from Pérez-Gómez's message. I am certainly more optimistic than he is.

A crisis of modern technology is not meant here as a threat that would endanger technological development. On the contrary, the idea is rather to suggest that are living in a more and more technological world, but in a world in which technology has become so pervasive that its very success alters part of its traditional meaning, at least in the developed countries. To deal with this change, I adopt a starting point that may seem strange at first: I begin with the growing ambiguities attached to the notion of technological object or artifact.

Traditionally, artifacts were rather easy to characterize. They lacked the plenitude of natural beings, even if they provided useful tools for understanding those beings. This view was well expressed by the Aristotelian distinction between a natural being, whose finality was inherent to its very existence, and an artifact (such as a table), whose finality remained exterior to it (since it lay in the mind of the craftsman who created it).[40] Later on, even if the boundary between the natural and the artificial became somewhat permeable, beings and objects could still be ranked following a single progression, from the entirely natural to the absolutely artificial.

Such is no longer the case, for two reasons at least. The first is the emergence of a lot of phenomena which are rather hard to classify. Is a registered molecule or a transgenic mouse natural or artificial? A second reason is the multiplication of points of view that have replaced the former single progression from the natural to the artificial. To take an example, a computer, because of its materials and its complexity, is in many ways more artificial than, let us say, a wooden table. But in a world in which the notion of information is omnipresent, a computer structure has certainly more to do with nature than the wooden table. On the other hand, as an information-based device, the brain is in many respects a machine.

Another classical characterization of artifacts was through reference to their to technological lineages. Such was the approach adopted by the French philosopher Georges Simondon in *Du mode d'existence des objets techniques,* a book published in 1969 that was and is still rather influential in France.[41] In this perspective, artifacts were the product of a rather clear evolution, such as the one leading from the nineteenth century's internal combustion engines to the modern automobile engine. But there again, two difficulties have arisen. The first is linked to increased historical attention to social factors, factors that very often confuse the clear lineage one is tempted to construct.[42] To take an example, when the Frenchman Cugnot put a steam engine on a four-wheeled frame at the end of the eighteenth century, was this the precursor of the locomotive or of the automobile? If you are interested in steam engines, you choose the locomotive, but if you are more interested in the type of use aimed at, the automobile is a better answer. Second, the hybridization of artifacts is today so important that lineages are no longer so illuminating as they used to be. A contemporary car has, for instance, so much electronics in it that one can wonder whether cars are not transforming themselves gradually into computers on wheels.

A third possible characterization of artifacts is their belonging to large technological systems.[43] They are in coherence with other artifacts, as many historians of technology have emphasized. Thus, the Ford automobiles were part of a technological system based on mass production. But there is a growing problem with the characterization of technological systems. Whereas a historian such as Bertrand Gille could quite easily describe the technological system of the first industrial revolution as based on the triad of iron, coal, and steam engine, such a simple description is simply impossible when dealing with our world, a world made of so thoroughly interconnected technologies that it seems to have no center and no periphery. In other words, our technological world is rather a densely interwoven web than a discrete system based on a few elements.

The point at which I wanted to arrive is the following: The status of artifacts is highly paradoxical today. We have never been surrounded by so many of them. But on the other hand, are these artifacts still objects, machines, in the traditional sense? Nothing is more uncertain. The most striking feature of contemporary technological development is the multiplication of quasi-objects: huge and complex networks with their terminals, which seem to lack either the precise spatial definition of the traditional machine or its relative autonomy. A network is not a machine with its dramatic appearance; neither is a terminal.

Such an evolution has, of course, something to do with the gradual replacement, in many fields, of mechanical devices by electronic ones that do not function as traditional machines. The electronic "revolution" is not a mere technological change; it implies also a transformation in our perception of technology. This transformation is probably as important as those that took place at the beginning of the Renaissance or on the eve of the first industrial revolution.

What I want to suggest is that we are entering a new phase in which technological development, at least in industrially developed countries, is no longer to be synonymous with a kind of dramatic confrontation between human and nonhuman actors, men and machines, as in Fritz Lang's *Metropolis* or Chaplin's *Modern Times*. Technology is rather assimilated into a kind of pervasive environment, a kind of landscape. We are perhaps entering also an age of fetishism with objects as well as an age of growing disillusionment about them, since the former ideal of "dialogue" between man and machine will probably lose its importance in a context of innumerable and rather disorderly connections.

The crisis of technology to which I have referred is a change in its meaning, from a drama on a stage to an environment, to a landscape with a more fluid dynamism. To design objects in such a context cannot have exactly the same meaning as in former times. To come back to architecture, I have the impression that some contemporary positions can be interpreted as a response to this new context, a context in which objects are at the same time fetishized and put under suspicion.

Architectural attempts to address this issue have also something to do with the evolution of our cities, as they transform themselves into strange and disordered landscapes.[44] Cities can be no longer assimilated as a sequence of various urban stages. With their more and more suburban character, they convey strangely sublime and picturesque impressions. They are also characterized by strange short circuits between centers and peripheries, by discontinuities, and by radical contrasts in scale and appearance. In many cases, the gradual replacement of distance by the time of accessibility as the key factor is responsible for those short circuits, discontinuities, and contrasts.[45] How is one to produce convincing architecture in the context of a fractured urban space?

Deconstruction, as a play with traditional building appearances as well as with the boundaries between order and disorder, rationalism, and lyricism, can perhaps be seen as a possible answer to this challenge.[46] It provides at least a rather interesting way to produce objects while underlining their limits, as a kind of motionless explosion in which something is to be seen that is neither the anatomy of the object, nor its final annihilation, though the effect produced has something to do with both.

In the same line as his former apologies for contemporary urban chaos, Rem Koolhaas's recent obsession with big objects seems also to deal with the problem.[47] The answer given in this case is, however, fundamentally different, since the boundaries are reached through a scale intermediate between traditional architecture and urban design. It can be seen as a play on the possibilities of architecture as an urban infrastructure, as architecture extending beyond its traditional limits to become a landscape in itself because of its dimension.

Santiago Calatrava's answer is quite different. Here the limits of the architectural object are reached through a disciplinary blurring between architecture and engineering.[48] Designs such as his recent train station for Satolas near Lyons are quite representative of his play with the traditional boundaries between architecture and engineering. (See figure 13.9.) On one hand, such a design conveys an impression of structural achievement typical of engineering. On the other hand, the structure's general thickness clearly adopts some distance from the engineering concern with performance, returning to a more architectural preoccupation with plasticity.

13.9

Santiago Calatrava, the Satolas railway station, near Lyon, France, during construction. A realization that seems to blur the traditional distinction between architecture and engineering.

13.10

Jean Nouvel, Fondation Cartier, Paris, France. Nature and technology seem to merge in a new kind of landscape.

In the French context, Jean Nouvel is probably the architect who addresses with the greatest clarity this issue of the architectural object's limits, hence his influence on so many French students and young graduates. Fascinated with contemporary technology, Nouvel is at the same time playing with the meaning it conveys, rather than with its immediate capacities of expression. Thus, he is far from the high-tech approach. His idea is rather to use architecture as a means to crystallize various key images of contemporary culture than to try to be scientifically and technologically sophisticated. His buildings are objects, objects that try to escape their condition through the mobilization of some of the salient features of the contemporary technological landscape. His projects, from the Institut du Monde Arabe to his more recent reshaping of the opera of Lyon or the Foundation Cartier, convey technological images without really dealing with immediate technological questions.

If we consider Nouvel a bit more closely, his emphasis on bidimensionality, on effects of light and texture are certainly the most pronounced salient features of the contemporary technological landscape. (See figure 13.10.) In a recent article, Nouvel prophesied the gradual disappearance of the Albertian concern with perspective and space.[49] He may be wrong, but the interesting thing is that this intuition is rooted in his assessment of our technological environment. I don't want to suggest that this environment is the key to the architectural developments, rather that both this environment and the contemporary architectural evolution reflect a more general cultural change.

This emphasis on surfaces, lights, and textures can be observed not only in Nouvel's work but also in Koolhaas's and others' productions. Contemporary landscaping often follows the same path, with a renewal of the interest in questions of texture. The French journal *Pages paysages* is representative of this interest.

Such an approach is, in my opinion, related to the growing impact of computer images. Contrary to the common opinion, I think that computer images exert a much more powerful impact through the textures and lights that they propose than through the possibilities of three-dimensional simulation that they provide.

Bidimensionality, moving lights and textures as on computer screens, has perhaps a deeper meaning. It is tempting to mention here Svetlana Alpers's *The Art of Describing: Dutch Art in the Seventeenth Century,* in which she contrasts the three-dimensional Italian art of the time with a Northern painting more concerned with surfaces, lights, and contrasts in materials and textures.[50] The distinction I would make between Alpers's type of opposition and the one we are experiencing now is that in Alpers's perspective, a strong narrative character accompanies three-dimensional art, whereas now the new bidimensionality of technology as well as of Nouvel's architecture is related to the concern for simulation and/or narration.[51]

In the same line of thought, it would be tempting to refer to Tschumi's concern with "events."[52] Technology is, by the way, more and more concerned with events. Computers are systems of programmed events, from the single bit of information that is nothing but an elementary event to more complex occurrences. The new industrial management techniques that are gradually replacing the scientific management methods are based on the identification of key events in the production process, from machine breakdowns to human communication.[53] Events are perhaps one supplementary link between contemporary architecture and technology.

Conclusion

As I have already said, I am rather optimistic. If one is not, one can very well regret the nonimmediate character of the relations between architecture and the cutting edge of sciences and technology. Another possibility is to believe that architecture is entirely subject to scientific and technological rationality, so much for the worse.[54] Far away from those two attitudes, my position is rather to be interested in what is still circulating from the sciences to the arts and vice versa. I think that the position occupied by architecture today, if one makes the assumption that it is not at the innovative frontier of science and technology, is truly interesting. First (and that is not to disdain) because such a position can be truly critical; second, because one of the characteristics of post-modernity is the possibility of avoiding the clear-cut distinction between the modern, the brand new, the innovative, on one hand, and the old, the archaic, on the other. Architecture is precisely neither the former nor the latter. Part of its lasting ascendance comes from its capacity to be contemporary without being always heroically modern.

Notes

1. Daly, *Ingénieurs et architectes,* p. 4.
2. Pérez-Gómez, *Architecture and the Crisis of Modern Science;* and Frampton, *Studies in Tectonic Culture.*
3. Galison, "Aufbau/Bauhaus."
4. On the Vitruvian tradition, see for instance Kruft, *Geschichte der Architekturtheorie;* Szambien, *Symétrie goût caractère;* and Germann, *Vitruve et le vitruvianisme.*
5. Cf. Picon, *Claude Perrault;* and Bennett, *The Mathematical Science of Christopher Wren.*
6. See for instance Timoshenko, *History of Strength of Materials;* and Benvenuto, *La scienza delle costruzioni e il suo sviluppo storico.*
7. Cf. Simonnet, *Matériau et architecture.*
8. On Perrault's theory, see Azouvi, "Entre Descartes et Leibniz"; and Picon, *Claude Perrault.*
9. Cf. Rykwert, *La maison d'Adam au paradis.*
10. Petzet, "Claude Perrault als Architekt des Pariser Observatoriums"; and Picon, *Claude Perrault.*
11. On Boullée's work, the basic reference remains Pérouse de Montclos, *Etienne-Louis Boullée.*
12. Among the recent studies devoted to this theme, see Baridon, *L'imaginaire scientifique de Viollet-le-Duc;* and above all Bressani, *Science, histoire et archéologie.* Bressani's dissertation is by far the best-informed work.

13. Viollet-le-Duc, *Dictionnaire raisonné,* tome 8, p. 486.
14. Cf. Frey, ed., *E. Viollet-le-Duc et le massif du Mont-Blanc 1868–1879,* especially Véry, "A propos d'un dessin de Viollet-le-Duc."
15. Kaufmann, *De Ledoux à Le Corbusier,* and *Trois architectes révolutionnaires.*
16. For a critical reassessment of Kaufman's interpretation, see Mosser, "Situation d'Emil K."
17. Vidler, *Ledoux* and *Claude-Nicolas Ledoux.* See also Vidler, *The Writing of the World.*
18. Teyssot, "Klassizismus et 'Architecture révolutionnaire'."
19. Cf. Picon, *French architects and engineers* and *L'invention de l'ingénieur moderne.*
20. Boullée's exploration of the relations between sensations and elementary shapes and volumes as well as his reflection on architectural programs can be found in his *Architecture: Essai sur l'art.* On Boullée's typological approach, see Szambien, "Notes sur le recueil d'architecture."
21. Ledoux, *L'architecture considérée sous le rapport de l'art, des mœurs et de la législation,* p. 135.
22. Le Camus de Mézières, *Le génie de l'architecture,* pp. 1–3 in particular.
23. Patte, *Mémoire sur les objets.* On the theoretical importance of Pierre Patte, see Choay, *La règle et le modèle;* and Picon, *French Architects and Engineers.*
24. d'Alembert, "Eléments des sciences," p. 491 in particular.
25. On the elementary character of Newton's law that encouraged Coulomb to look for its transposition to electrostatics, see Blondel and Dörries, eds., *Restaging Coulomb.*
26. Chaptal, *Eléments de chimie,* p. 55.
27. About Buffon's natural history, Jacques Roger wrote in the same perspective that "if there is a natural order of the world, this is not a structural order. . . . It is an order of the natural operations, an order of the processes that condition life and its renewal." Roger, *Buffon,* p. 130.
28. de Condillac, "Cours d'études pour le prince de Parme. V: De l'art de penser," p. 769.
29. Cf. Picon, *L'invention de l'ingénieur moderne.*
30. *Ibid.*
31. At the beginning of his career, Perronet studied pin making at a factory in Normandy. The two memoirs he wrote on this occasion were used later for the *Encyclopédie* and the *Encyclopédie méthodique.* Perronet, *Explication de la façon dont on réduit le fil de laiton à différentes grosseurs dans la ville de Laigle,* 1739, manuscript in the library of the Ecole Nationale des Ponts et Chaussées, Ms. 2383; *Description de la façon dont on fait les épingles à Laigle en Normandie,* 1740, manuscript in the library of the Ecole Nationale des Ponts et Chaussées, Ms. 2385.
32. On these designs, see Demangeon and Fortier, *Les vaisseaux et les villes;* and Morachiello and Teyssot, *Nascita delle città di stato.*
33. On these map competitions, see Picon, *French Architects and Engineers.*
34. de Girardin, *De la composition des paysages;* and Morel, *Théorie des jardins.* On the relations between eighteenth-century engineering and gardening, cf. Picon, "Le Naturel et l'efficace."
35. Cf. Szambien, *Jean-Nicolas-Louis Durand;* and Villari, *J.-N.-L. Durand.*
36. The expression is used by Boullée at the beginning of his *Essai sur l'art* to differentiate architecture from the building activity.
37. On this theme, see Mosser, "Le rocher et la colonne."
38. On the emergence of the "public sphere," cf. Habermas, *L'espace public.*
39. Public architectural competitions were organized in the same perspective during the French revolution. Cf. Szambien, *Les projets de l'an.*

40. Cf. Lloyd, *Les débuts de la science grecque de Thalès à Aristote,* pp. 126–127; and Séris, *La technique,* pp. 24–28.
41. Simondon, *Du mode d'existence des objets techniques.* On Simondon's reflections on technology, see as well Simondon, *Gilbert Simondon: Une pensée de l'individuation de la technique.*
42. The development of the social construction of technology approach has played a role in this blurring of the notion of technological lineage. Such a blurring goes with the idea that there is nothing "natural" in technological evolution, so that the passage from one artifact to another is always socially constructed.
43. Bertrand Gille in France and Thomas Hughes in the United States have developed this notion; Gille, "Prolégomènes à une histoire des techniques"; and Hughes, "The Evolution of Large Technological Systems." For a comparison between the two definitions of technological systems given by these authors, see Picon, "Towards a History of Technological Thought," pp. 37–39 in particular.
44. See, for instance, Ascher, *Métapolis ou l'avenir des villes.*
45. Cf. Desportes, "Liaisons, nœuds et déliaisons"; and Veltz, *Mondialisation, villes et territoires.*
46. On the deconstructivist attitude, see for instance Noever, ed., *The End of Architecture?* This attitude's ambiguous relation to philosophy has been scrutinized in Wigley, *The Architecture of Deconstruction.*
47. Koolhaas, "Bigness," and *S,M,L,XL.*
48. Cf. Picon, "Santiago Calatrava."
49. Nouvel, "A venir," p. 50.
50. Alpers, *The Art of Describing.*
51. The growing importance of narration in our contemporary world may perhaps account for the growing interest taken in the reflections of a philosopher such as Paul Ricœur. On the strategic importance of narration in the industrial field, see, for instance, De Coninck, *Travail intégré, société éclatée.*
52. See Tschumi, *Architecture and Disjunction.*
53. Cf. Coriat, *L'atelier et le robot;* and Picon and Veltz, "L'informatique et les nouveaux modèles d'organisation dans l'industrie."
54. This is for instance Alberto Pérez-Gómez's position, a position I am far from sharing.

Bibliography

Alpers, Svetlana. *The Art of Describing: Dutch Art in the Seventeenth Century.* Chicago: The University of Chicago Press, 1983.

Ascher, François. *Métapolis ou l'avenir des villes.* Paris: Odile Jacob, 1995.

Azouvi, François. "Entre Descartes et Leibniz: L'animisme dans les *Essais de physique* de Claude Perrault." *Recherches sur le XVIIe siècle* 5 (1982): 9–19.

Baridon, L. *L'imaginaire scientifique de Viollet-le-Duc.* Paris: L'Harmattan, 1996.

Bennett, J. A. *The Mathematical Science of Christopher Wren.* Cambridge: Cambridge University Press, 1982.

Benvenuto, Eduardo. *La scienza delle costruzioni e il suo sviluppo storico.* Florence, Italy: Sansoni, 1981.

Blondel, Christine, and Matthias Dörries, eds. *Restaging Coulomb. Usages, controverses et réplications autour de la balance de torsion.* Florence, Italy: L.-S. Olschki, 1994.

Boullée, Etienne-Louis. *Architecture: Essai sur l'art.* Edited by J. M. Pérouse de Montclos. Paris: Hermann, 1968.

Bressani, Martin. *Science, histoire et archéologie. Sources et généalogie de la pensée organiciste de Viollet-le-Duc.* Ph.D. diss., Université de Paris IV-Sorbonne, 1997.

Chaptal, Jean Antoine. *Eléments de chimie.* Paris: Deterville, 1803.

Choay, Françoise. *La règle et le modèle.* Paris: Seuil, 1980.

Coriat, Benjamin. *L'atelier et le robot. Essai sur le fordisme et la production de masse à l'âge de l'électronique.* Paris: Christian Bourgeois, 1990.

d'Alembert, Jean le Rond. "Eléments des sciences." In *Encyclopédie, ou dictionnaire raisonné des sciences, des arts et des métiers,* tome 5: 491–497. Paris: Briasson, 1751–1772.

Daly, César. *Ingénieurs et architectes (un toast et son commentaire),* extrait de la *Revue générale de l'architecture et des travaux publics.* Paris: Ducher et cie, 1877.

de Condillac, E. Bonnot. "Cours d'études pour le prince de Parme. V: De l'art de penser." In *Œuvres philosophiques de Condillac,* tome 1: 769. Paris: Presses Universitaires de France, 1947–1951.

De Coninck, Frédéric. *Travail intégré, société éclatée.* Paris: Presses Universitaires de France, 1995.

de Girardin, René Louis. *De la composition des paysages.* Geneva and Paris: P.-M. Delaguette, 1777.

Demangeon, Alain, and Bruno Fortier. *Les vaisseaux et les villes: L'arsenal de Cherbourg.* Brussels: Mardaga, 1978.

Desportes, Marc. "Liaisons, nœuds et déliaisons. La Ville modelée par les transports." *Le Débat* 80 (1994): 123–139.

Frampton, Kenneth. *Studies in Tectonic Culture: The Poetics of Construction in Nineteenth and Twentieth Century Architecture.* Cambridge: MIT Press, 1995.

Frey, Pierre, ed. *E. Viollet-le-Duc et le massif du Mont-Blanc 1868–1879.* Lausanne, Switzerland: Payot, 1988.

Galison, Peter. "Aufbau/Bauhaus: Logical Positivism and Architectural Modernism." *Critical Inquiry* 16 (1990): 709–752.

Germann, Georg. *Vitruve et le vitruvianisme. Introduction à l'histoire de la théorie architecturale.* Darmstadt: 1987. French translation, Lausanne, Switzerland: Presses polytechniques et universitaires romandes, 1991.

Gille, Bertrand. "Prolégomènes à une histoire des techniques." In Bertrand Gille, ed., *Histoire des techniques,* 1–118. Paris: Gallimard, 1978.

Habermas, Jürgen. *L'espace public. Archéologie de la publicité comme dimension constitutive de la société bourgeoise* (1962). French translation, Paris: Payot, 1978.

Hughes, Thomas P. "The Evolution of Large Technological Systems." In Wiebe Bijker, Thomas Hughes, and Trevor Pinch, eds., *The Social Construction of Technological Systems,* 51–82. Cambridge: MIT Press, 1987.

Kaufmann, Emil. *De Ledoux à Le Corbusier. Origine et développement de l'architecture autonome* (1933). French translation, Paris: L'Equerre, 1981.

Kaufmann, Emil. *Trois architectes révolutionnaires. Boullée, Ledoux, Lequeu* (1952). French translation, Paris: Société des Architectes Diplômés pa le Gouvernement, 1978.

Koolhaas, Rem. "Bigness." *L'architecture d'aujourd'hui* 298 (1994): 84.

Koolhaas, Rem. *S,M,L,XL.* Rotterdam, the Netherlands: 010 Publishers, 1995.

Kruft, Hanno Walter. *Geschichte der Architekturtheorie. Von der Antike bis zum Gegenwart.* Munich: C.-H. Beck, 1985.

Le Camus de Mézières, Nicolas. *Le génie de l'architecture, ou l'analogie de cet art avec nos sensations.* Paris: L'auteur et B. Morin, 1780.

Ledoux, Claude-Nicholas. *L'architecture considérée sous le rapport de l'art, des mœurs et de la législation.* Paris: L'auteur, 1804.

Lloyd, G. E. R. *Les débuts de la science grecque de Thalès à Aristote.* (1970). French translation, Paris: La Découverte, 1990.

Morachiello, Paolo, and Georges Teyssot. *Nascita delle città di stato: Ingegneri e architetti sotto il Consolato e l'Impero.* Rome: Officina, 1983.

Morel, Jean Marie. *Théorie des jardins.* Paris: Pissot, 1776.

Mosser, Monique. "Le rocher et la colonne, un thème d'iconographie architecturale au XVIIIe siècle." *Revue de l'art* 58–59 (1982–1983): 55–74.

Mosser, Monique. "Situation d'Emil K." In *De Ledoux à Le Corbusier. Origines de l'architecture moderne,* 84–89. Arc-et-Senans: Fondation C. N. Ledoux, 1987.

Noever, Peter, ed. *The End of Architecture?* Munich and Vienna: Prestel, 1993.

Nouvel, Jean. "A venir." *L'architecture d'aujourd'hui* 296 (1994): 50.

Patte, Pierre. *Mémoire sur les objets les plus importants de l'architecture.* Paris: Rozet, 1769.

Pérez-Gómez, Alberto. *Architecture and the Crisis of Modern Science.* Cambridge: MIT Press, 1983.

Pérouse de Montclos, Jean-Marie. *Etienne-Louis Boullée (1728–1799). De l'architecture classique à l'architecture révolutionnaire.* Paris: Arts et Métiers Graphiques, 1969.

Petzet, Michael. "Claude Perrault als Architekt des Pariser Observatoriums." *Zeitschrift für Kunstgeschichte* 30 (1967): 1–54.

Picon, Antoine. *Claude Perrault 1613–1688, ou la curiosité d'un classique.* Paris: Picard, 1988.

Picon, Antoine. *French Architects and Engineers in the Age of the Enlightenment* (1988). English translation, Cambridge: Cambridge University Press, 1992.

Picon, Antoine. *L'invention de l'ingénieur moderne. L'Ecole des Ponts et Chaussées 1747–1851.* Paris: Presses de l'Ecole Nationale des Ponts et Chaussées, 1992.

Picon, Antoine. "Le naturel et l'efficace. Art des jardins et culture technologique." In M. Mosser and Ph. Nys, eds., *Le Jardin, art et lieu de mémoire,* 367–396. Paris: Les éditions de l'imprimeur, 1995.

Picon, Antoine. "Santiago Calatrava: Tettonica o architettura?" *Casabella* 615 (September 1994): 24–29.

Picon, Antoine. "Towards a History of Technological Thought." In Robert Fox, ed., *Technological Change: Methods and Themes in the History of Technology,* 37–49. London: Harwood Academic Publishers, 1996.

Picon, Antoine, and Pierre Veltz. "L'informatique et les nouveaux modèles d'organisation dans l'industrie: Information, événements, communication." *Annales des Ponts et Chaussées* 69–70 (1994): 14–20.

Roger, Jacques. *Buffon.* Paris: Fayard, 1989.

Rykwert, Joseph. *La maison d'Adam au paradis* (1972). French translation, Paris: Le Seuil, 1976.

Séris, Jean-Pierre. *La technique.* Paris: Presses Universitaires de France, 1994.

Simondon, Georges. *Du mode d'existence des objets techniques.* Paris: Aubier, 1969.

Simondon, Georges. *Gilbert Simondon: Une pensée de l'individuation de la technique.* Paris: Albin Michel, 1994.

Simonnet, Cyrille. *Matériau et architecture. Le Béton armé: origine, invention, esthétique.* Ph.D. diss., Ecole des Hautes Etudes en Sciences Sociales, Paris, 1994.

Szambien, Werner. *Jean-Nicolas-Louis Durand 1760–1834. De l'imitation à la norme*. Paris: Picard, 1984.

Szambien, Werner. "Notes sur le recueil d'architecture privée de Boullée (1792–1796)." *Gazette des Beaux-Arts* (March 1981): 111–124.

Szambien, Werner. *Les projets de l'an. II: Concours d'architecture de la période révolutionnaire*. Paris: Ecole Nationale Supérieure des Beaux Arts, 1986.

Szambien, Werner. *Symétrie goût caractère. Théorie et terminologie de l'architecture à l'âge classique 1500–1800*. Paris: Picard, 1986.

Teyssot, Georges. "Klassizismus et 'Architecture révolutionnaire.'" Introduction to Kaufmann, *Trois Architectes révolutionnaires,* 12–31.

Timoshenko, Stephen P. *History of strength of materials* (1953). New York: Dover, 1983.

Tschumi, Bernard. *Architecture and Disjunction*. Cambridge: MIT Press, 1994.

Veltz, Pierre. *Mondialisation, villes et territoires*. Paris: Presses Universitaires de France, 1996.

Véry, Françoise. "A propos d'un dessin de Viollet-le-Duc." In Frey, ed., *E. Viollet-le-Duc,* 109–118.

Vidler, Anthony. *Claude-Nicolas Ledoux*. Cambridge: MIT Press, 1990.

Vidler, Anthony. *Ledoux*. Paris: Hazan, 1987.

Vidler, Anthony. *The Writing of the World*. Princeton, NJ: Princeton University Press, 1987.

Villari, Sergio. *J.-N.-L. Durand (1760–1834). Arte e scienza dell'architettura*. Rome: Officina, 1987.

Viollet-le-Duc, Eugène Emmanuel. *Dictionnaire raisonné de l'architecture française du XIe au XVIe siècle* (1854–1868). New edition, Paris: Morel, 1875.

Wigley, Mark. *The Architecture of Deconstruction: Derrida's Haunt*. Cambridge: MIT Press, 1993.

14

Architecture *as* Science: Analogy or Disjunction?

Alberto Pérez-Gómez

Running the risk of being unfashionable, I do not, in this article, simply describe some structural or formal analogy between architectural and scientific thought or between the modes of production of science and architecture. My interest in this topic is motivated by a hope that discourse may still orient action, and that therefore a discussion of the old question of architecture's status as science, as a legitimate form of knowledge, may have a bearing on practice. Indeed, given the state of affairs in contemporary theory and practice, it is important to discuss whether and how scientific discourse may be normative for architecture. I approach this complex problem through a series of historical narratives aimed at clarifying the differences among traditional, classical, and contemporary science and examining its changing relationships to architectural theory. I conclude with some remarks about architecture's relationship with its own technological modes of production and its potential place in technoculture.

Scientific metaphors "applied" through instrumental thinking have been common in architecture during the last two centuries, from "functionalism" itself, a mathematical metaphor with its origins in differential calculus and the laws of maxima and minima, to more specific biological or mechanical metaphors used to describe buildings' internal efficiency or aesthetic character. In his contribution to this volume, Adrian Forty has reminded us of Viollet-le-Duc's use of "circulation" to express a building as a complete system (first employed around the 1870s), and of the many terms drawn from statics and dynamics applied to the articulation of formal strategies of buildings *cum* machines in the period of high modernism.[1] The degree to which these metaphors may have functioned poetically is, of course, highly debatable. It is not a simple matter to determine the extent to which they may have revealed something of the essence of architecture, something hitherto unknown in our experience that may have contributed to architecture's meaning. Regardless of our answer to this question, prior to the late eighteenth century scientific metaphors were seldom used in architectural discourse. Recalling metaphor's linguistic function, which is to reveal a tacit analogy of elements explicitly disconnected, we must accept Forty's conclusion that this difference reveals, if nothing else, at least a problematic relationship between science and architecture in our times, and not a merely natural, working situation.

Indeed, rather than assuming that science and architecture have become linked only as a result of recent "revolutions," such as the end of metaphysics, logocentricity, classical authorship, or whatever, it is important to understand that architecture and science were linked at the very inception of our Western tradition. Their aims always ran in parallel. Philosophy and science, the crowning jewels of the *bios theoreticos,* aimed at revealing truth, a truth understood since Plato's *Timaeus* as a mathematical correspondence. Plato's *Timaeus* became not only the model for science until its culmination in Newtonian physics but also the model for architectural theory. The demiurge as an architect, creating the world out of geometry from the primordial gap/primordial matter, *chasho/chaos/chora,* was a commonplace in classical theory. The architect's cosmos is Plato's cosmos, and the philosopher's "cosmobiology" underlines all "revelations" of architectural meaning in traditional architectural writing. Architecture disclosed truth by revealing the order of the cosmos in the sublunar world. It was a form of precise knowledge humanity implemented to frame the rhythms of human action, of political and religious rituals, guaranteeing the efficacy and reality of the human experience. The metaphors for architecture were, obviously, that which architecture *was not,* but that it revealed by analogy, the created orders of the cosmos, of nature, and the human, live body. One could argue that architectural theory, therefore, was science; it had the same status as *scientia,* while being in a noninstrumental relationship with practice. *Scientia* named that which should be contemplated, the proportional order that architecture embodied, not only as a building, but as a human situation, in the space-time of experience. Not surprisingly, Plato's Socrates evoked Daedalus as his most important ancestor.[2]

As I have tried to show elsewhere, this *status quo* starts to change during the seventeenth century, although the transformations sometimes evident in theoretical treatises don't affect architectural practice until the nineteenth century.[3] In the mid-1600s Girard Desargues developed an instrumental theory of perspective and stonecutting that practitioners never accepted, and toward the end of the same century, Claude Perrault extrapolated his understanding of biology and physics into a controversial architectural theory. In his *Ordonnance for the Five Kinds of Columns,* Perrault questioned the traditional role of proportions to guarantee the relationship between the microcosmos and the macrocosmos and the importance of optical corrections that had been always regarded as the reason for any observed discrepancies between the proportional prescriptions in theoretical treatises and building practice.[4] These two notions had invariably been present in the theoretical literature on architecture from Alberti (in the fifteenth century) to the late seventeenth century. Perrault couldn't understand the traditional priority of practice and the power of architecture to demonstrate perfect measurement for embodied, synesthetic experience; he couldn't believe that the architect's task hinged on his ability to adjust such proportions according to the site and the "program" at hand. For Perrault, the status of theory was no longer that of absolute truth but rather, as in inductive physical systems, the "most probable" and mathematically precise. Its purpose was merely to be as easily "applicable" as possible, a set of recipes to control an architectural practice in his view always prone to error and subject to the clumsiness of craftsmanship. Perrault conceived architecture and its applied theory as a discipline participating in a progressive history, in all likelihood bound to be perfected in the future.

In a certain sense, Perrault was merely continuing the tradition of architecture as science. Yet he radically transformed the nature of architectural theory and practice. This transformation heralds the "beginning of the end" of traditional architecture, to paraphrase Peter Eisenman, the end of the "classical" way of conceiving and making buildings, related to a cosmological "picture" that served as an ultimate, intersubjective framework for meaningful human action. The beginning of our architectural crisis does not date back a few years to "the end of the avant-garde," or even to the inception of panopticism and the industrial revolution, or to the demise of the Beaux-Arts in the early twentieth century. Rather, it must be seen in parallel with the beginning of modern science itself and its impact on architectural discourse. After Perrault, but particularly after Jacques-Nicolas-Louis Durand, the popular teacher of architecture whose early-nineteenth-century work contains *in nuce* all the theoretical presuppositions and stylistic debates that still plague us, the legitimacy of architectural theory and practice, predicated on its "scientificity," was reduced to pure instrumentality. The value of architectural theories henceforth depended on their applicability. Other well-known forms of deterministic theory followed suit, from Eugène-Emmanuel Viollet-le-Duc's structural paradigms to Buckminster Fuller's technological dreams and more recent behavioral and sociological models. Even today, after Jean-François Lyotard's well-publicized critique of the grand narratives of science,[5] architects and theorists still view this issue quite uncritically.

These misunderstandings are compounded by a disregard for the history of architecture as a complex, multifaceted cultural order with epistemological connections, embodied in a diversity of artifacts, impossible to reduce to a typology or sociology of buildings, to a single, progressive and continuous line, or to discontinuous, hermetic moments. History need not be a burden for practice. In his seminal essay *On the Uses and the Disadvantages of History for Life,* Friedrich Nietzsche articulated both the dangers and the possibilities history has opened up for a new man, particularly for the creative and responsible individual in the postcosmological era.[6] There are, of course, useless and problematic forms of history, particularly pseudo-objective progressive narratives, but this should not result in an unwillingness on our part to pay attention to what we *are,* which is, indeed, what we have been. As I elaborate later, there is a particular way to understand and "use" history as a framework for ethical creation. Lacking a living tradition for architectural practice since the nineteenth century, we are in fact called to reconstruct it, visiting and interpreting the traces and documents of our past, invariably with fresh eyes, to discover hitherto hidden potentialities for the future, as one recovers coral from the bottom of the ocean or extracts pearls out of ordinary-looking mollusks.

Much recent writing on architecture of diverse ideological filiation, ranging from scientistic and methodological approaches to more carefully considered attempts to continue the project of critical rationality from the Enlightenment (often under the influence of Jürgen Habermas and the Frankfurt School), reiterates the view of history as merely an accumulation of uninteresting shells, quite dead and unyielding. This common disregard for history, easily embraced because it coincides with popular assumptions about linear temporality, progress, and the past as an alien, closed book, goes hand in hand with the embracing of "alternative" scientific models. Chaos and catastrophe theory, for example, often irresponsibly extrapolated into architectural theory, are made to suggest formal strategies for architecture, metaphorical connections in themselves merely a mannerism of modernity. Identifying truth with science and science with applied science, that is, the theory of technology, the result is an incapacity to consider truly radical alternative modes of thinking in architectural theory. Witness the irrelevant results of the recent interest in chaos and catastrophe theory in architecture schools, and of the return of semiotics in theoretical discussions in Europe and North America. Indeed, these strategies seem to offer no new possibilities beyond the relationship between theory as applied science and practice as technology inaugurated by Durand almost two centuries ago.

In order to examine this problem and contemplate potential alternatives, we must first consider science itself. Are there other ways pure science may constitute itself as an epistemological model for architecture? Ancient, medieval, and early modern science were always about the "whole"; they functioned on the assumption that it was possible to reveal a meaningful order for man through a kind of discourse in which mathematical reason would not be at odds with narrative reason. Even Newtonian natural philosophy, despite its often explicit distinction between "true causes" and "laws," is grounded in an implicit metaphysics that includes assumptions such as the identification of absolute space, the space of gravitation, with God.[7] Furthermore, Newton spent long hours working on an elaborate reconstruction of the Temple of Solomon in Jerusalem.[8] Since the late sixteenth century, this "project" had been undertaken many times by philosophers, theologians, scientists, and architects in the hope of understanding God's "own" architecture.

(God dictated the Temple's proportions to Ezekiel in a prophetic dream related in the Old Testament.) The reconstructed versions of the Temple were then understood as exemplary for the architecture of the present, reconciling the (human) history of architecture (always assumed in the reconstructions, regardless of their specific cultural provenance) with the (divine) order of the universe. Newton's interest in this problem seems to reveal his desire to reconcile human creations (including his own natural philosophy) with the true order of the universe, a concern that must be understood as significantly qualifying his famous statement: *Hypotheses non fingo.*

Starting from the unquestioned assumption of a rational cosmological order, presumed to be the intelligible foundation of an intersubjective value system contained in the laws of nature, traditional science could reconcile the Platonic truth-as-correspondence with the human experience of truth as a permanently shifting reality in everyday life. *Mythos* and *logos* could coexist in the same universe of discourse. Newtonian natural science's impact on eighteenth-century architectural discourse seems to have contributed even to the delay in the process of instrumentalizing of theory that I mentioned previously. It also precipitated the consideration of alternative strategies to classical architecture. Once the coincidence between history and cosmology was no longer self-evident, theoretical discourses started to associate architecture with language and custom, rather than nature, and the possibility of meaning with a hermeneutic approach to history through the notion of character.

After the time of Laplace and Lagrange, in the early nineteenth century, science was finally in a position to "purge" itself from speculation. This is well documented in the transformations that Newtonian cosmology underwent through Laplace's work.[9] Philosophy, we may remember, tried to follow suit and establish its own self-evidence. Here we may simply recall Kant's indictment of traditional narrative structures in his *Prolegomena to Any Future Metaphysics.* Put simply, the issue was to avoid discourse outside of mathematical logic's strict parameters, particularly, of course, metaphysical and theological speculation, recognizing that if science alone can offer man effective truths, these are in fact only partial truths, probabilities, or instrumental devices. Operating in tension with the scientism often driving the modern popular imagination, science does not concern itself with either the whole "picture" or with essences, however much it would like to do so. Although generalizations are always dangerous, it could be argued that this assumption has not changed, whether we examine the premises of fundamental physics like quantum electrodynamics or chaos theory.

For this reason, a basic aspect of Edmund Husserl's famous diagnosis of the crisis of European sciences is still crucial.[10] Husserl believed that although increasing specialization and mathematization of all disciplines would result in a greater instrumentality and more effective control of practical tasks, the discourse and its products would be alienated from the expectations of lived experience. Human beings need language and artifacts to make sense of their world in the present, here and now. This is indeed the nature of truth, even if it must be "weak" in our historical epoch, devoid of a socially shared mythical or cosmological picture. Truth must designate a discourse about reality that can be of consequence for ethical thought and action. Husserl justly feared that specialized and positivistic scientific discourses would never be able to reconstitute the place for human dwelling, the poetic instant that may bring life and death to a single point of incandescent purpose. Architecture, concerned as it is with the life-world, has to pay special attention to this observation. Husserl was originally a mathematician. His obsession with making a rigorous science out of philosophy, continuing the philosophical vocation of the Western world, suggested to his disciples the possibility of developing a discourse for the exceptional, a *logos* to account for the thick vivid present of our everyday experience.

The phenomenology that stemmed from Husserl's thought recognizes that experience, always mutable and changing, is "given" with a framework of categories. In the disorder of appearance there is order that makes meaning possible in the first place. We don't merely constitute meanings in our minds, nor are they simply "there," they appear literally in between. This is the nature of our human reality, and its mystery should be celebrated. This observation, it must be emphasized, has taken place in a mental space totally free of mystical, theological, or "metaphysical" presuppositions. The world is neither chaos nor cosmos, it is indeed *chaosmos,* the world revealed by James Joyce and Marcel Duchamp.

Some time after philosophers and artists had recognized reality's unfathomable nature, speculative scientists seem to have arrived at analogous descriptions. We should not be surprised that there are points of contact between chaos theory's questioning of classical physics by examining concrete, ordinary phenomena and phenomenology, or between Heidegger and Heisenberg. It is impossible not to acknowledge a relationship between the space-time of Feynman's quantum electrodynamics, in which particles seem to have a will and to behave differently in view of the experiment being performed, in which time is reversible and light's essence must be proclaimed as utterly strange, and the thick temporality of experience described by Merleau-Ponty as the flesh of the world. Even the Platonic *chora,* that third element of reality described in the *Timaeus*—establishing an ontological continuum between being and becoming that the Greek philosopher acknowledged as "very hard to grasp," the substance of dreams, yet not immediately accessible to the senses[11]—connotes a mysteriously dense space-time (the depth of art and architecture) that seemed to become transparent and irrelevant only since the advent of applied science. For phenomenology, the elasticity of time and dimensionality are evident through the primary engagement of embodied being in the world: the verticality of an object, a facade for example, is an inherent component of its size and significance: It can be demonstrated that in the primacy of experience vertical dimension is always greater than the identical horizontal measurement. Yet as I try to demonstrate, despite these interesting connections between science and phenomenology, there are important differences, particularly when considering the problem of language in relation to creative or productive action. Is it possible for science to proclaim that mystery is the meaning? That the most wonderful thing about life is its uncertainty?

Before elaborating on the differences between scientific theory and phenomenological hermeneutics, I would like to touch on a related question that has fascinated architects and critics in recent years. The potentially fascinating consequences of chaos theory for architecture are indeed a popular topic these days. In its expectation to find complex behavior as a result of simple systems, and in its understanding that complex systems give rise to simple behavior, in holding that the laws of complexity hold universally, caring not at all for the details of a system's constituent parts, chaos is indeed different from classical physics.[12] Chaos theory demonstrates self-similarity, retrieving Leibniz's notion that "the world is in a drop of water" to understand, finally, that everything is connected. This is a formidable and exciting realization. We have at last "discovered" that the ancient analogical assumptions that drove traditional architecture and science were not merely stupid dreams. I cherish stories about a living world and the life of minerals, about the body without organs, about nature as a machine without parts. Mandelbrot's fractals have demonstrated how the structures of nature's different orders are analogous, despite scalar changes. Living in a similar world, but back in the seventeenth century, Leibniz could therefore postulate a formal discipline, a "universal calculus" as the point of departure for action. This is obviously a seductive argument for architects: We make forms and we also make content; it is all on the surface because depth is an illusion; we can play and expect to be responsible, because, supposedly, there is no alternative. Architecture can therefore be circumscribed to a quest for original form, intelligible for the initiates, thundering with mystical reverberations.

But we must examine this very carefully. It is symptomatic that Mandelbrot himself expressed some opinions about architecture.[13] He believed that the architecture of the Bauhaus, with its emphasis on spare, orderly, linear shapes, was particularly hostile. He claimed not to be surprised by architects' desire to stop building skyscrapers like the Seagram Building in New York, supposedly "inhuman" works because of their simple geometry, whereas the complex fractal geometries were supposed to "resonate with the way nature organizes itself." Curiously, Mandelbrot argued that the plurality of scales in a Beaux-Arts building with its volutes and dentils was therefore much more appealing. Hardly to his credit, Mandelbrot's view is not very different from the Prince of Wales's ultraconservative and rather naive opinion about modern architecture. Yet in contemporary architectural criticism, the fascination with fractals is often cited as a justification for a complex aesthetic of fragmented forms, a "deconstructive" style of architecture, as distant as one could imagine from a classical revival.

I share with many contemporary critics a fascination with the unexpected formal consequences of these relations, now manipulable in practice through computer design programs. Interesting buildings have undeniably started to populate some of our cities, buildings endowed with a prodigious ability for structural gymnastics. Many are fun, certainly less boring than your average glass box. Nevertheless, one should not lose sight of the fact that the relationship between geometry and architecture Mandelbrot and some of his architectural fans imagined is thoroughly classical, simply mimetic in the traditional sense. How can this "formula" for architectural meaning be appropriate for our epoch of incomplete nihilism?

I will be reminded that the disjunction of form and content in aesthetics is an invention of the seventeenth century, particularly evident since the advent of Baroque architecture. Indeed, elsewhere I have myself described anamorphosis in these terms, demonstrating this initial disjunction of presence and representation.[14] But the splitting of art into form and content is also the result of our civilization's being "thrown" into history. As long as we, as a civilization, may not be completely beyond historicity, we have to be careful with our assumptions. In other words, however I may share a dislike of this problematic split as expressed by postmodern critics and philosophers, to pretend it doesn't exist in our intersubjective, political reality is a dangerous delusion. Leibniz could start from the mathematical and operate on his *clavis universalis* because of his theological a priori. God had ordered the world and chosen the best possible. God was at the end (and the beginning) of it all. Leibniz imagined our free will as a ferryboat in a river; we all, individual monads, go our own chosen ways, though we are still loosely guided by Divine Providence. This sort of human action, however intelligently articulated, was still operating in a traditional world. Only the eighteenth century saw the beginning of history in the sense that is familiar to us when we hear in the news that a political figure, signing a peace agreement, just "made history." History as human-generated change is not "natural," it is part of the modern Western consciousness, with its obsession with scientific progress and material improvement. It could be argued that before the Enlightenment, particularly before the works of Vico and Rousseau, human actions were more or less irrelevant vis-à-vis the explicit order of creation. Renaissance architecture, for example, turned its eyes toward the past, but only to confirm its actions of reconciliation with a cosmological order that was perceived as absolutely transhistorical, just as history was unquestionably the sacred narrative of the church—with salvation, and therefore apocalypse, just around the corner. Modern history, on the other hand, starts from the assumption that human actions truly matter, that they can effectively change things, like the French Revolution, that there is potential progress—obvious in modern science and technology—and that the present differs qualitatively from the past. This "vector" has indeed characterized modernity, questioned for the first time by Nietzsche, and most recently by postmodern cultural critics.

I share Gianni Vattimo's perception that although history as the grand narrative of progress and the avant-garde may have ended, we must yet accept our historicity.[15] We can never simply overcome modernity and leave it behind: Rather we can convalesce, heal ourselves of resentment, and reconcile our present with our past. In other words, it is time to embrace, rather than try to resolve, the *aporias* associated with our human condition since the nineteenth century. We cannot act as if we lived in a cosmological epoch, in a perpetual present, where there would indeed be no distinction between architectural form and content, leading us to abdicate responsibility for our actions; nor can we merely pretend to continue the project of modernity with its future orientation, its absurd disjunctions of form and content, and its deferral of responsibility. All we can do is modify the terms of our relationship to historicity, accepting the multiplicity of discourses and traditions while assuming our personal responsibility for projecting a better future. This is what a hermeneutic discourse aims to accomplish. Clearly today we control, individually, very little; yet our actions, even those as apparently minute as a decision to recycle paper, have a phenomenal importance. This absurd situation is itself a consequence of our technological reality, our wholly constructed world. This is why, I would argue, formal attitudes in architecture, optimistically related to chaos theory and other recent developments in the theory of physics or biology, may be dangerously irresponsible.

Rather, I would emphasize that lacking as we do a theological a priori, we must start from our experience and its historical roots to construct a normative theory. Leibniz's insight should not be lost.[16] We must engage a perceptual faith aiming to discover the exceptional coincidences we call order; to discover, through our making, that connections do exist, and that their significance may be shared with other human beings—in the case of architecture, with the occupants and participants of projects and buildings. The world of our experience includes the artifacts that make up our artistic tradition, the revelatory moments we call architecture, moments of recognition in spatiotemporal forms that are completely new, yet strangely familiar. Understanding these forms of specific embodiment and articulating their lessons in view of our own tasks, we will have a greater chance to construe an appropriate architecture, an intersubjective reality that might fulfill its social and political task as an affirmation of culture. The issue for architecture is the disclosure of a social and political order from the *chaosmos* of experience, starting from the perceptions of meaning that our culture has shared, embodied in historical traces, and projecting imaginative alternatives that go beyond stifling and repressive inherited institutions. The architect's narratives and programs must begin by accounting for these experiences of value, thus articulating an ethical practice. I return to this question at the conclusion of my article. For now, it should be emphasized that this theoretical position has its origins in the eighteenth century, in the works of many architects and philosophers who were already discussing alternatives to the crisis of classical architecture. The issue is to ground architecture and its meanings through its relationship to language, to understand history (stories) as the true normative discipline of humanity and therefore as the appropriate discourse of architectural theory. This is a tradition that develops from Giambattista Vico's *New Science* to the hermeneutics of Hans-Georg Gadamer, Paul Ricoeur, and Gianni Vattimo.

In our obsessive search for a scientific theory of architecture, we stumble upon one remarkable model that seems indeed appropriate for the architect's search for form: Alfred Jarry's science of pataphysics. Jarry (1873–1907), one of the most remarkable French writers of the avant-garde, whose controversial work has long been considered precursor of the most significant landmarks in twentieth-century theatre and prose, invented this new "synthetic" discipline, a paradoxical "poetic science." The program of pataphysics remains relevant, and today is even fashionable, though I would argue that its content is perhaps still too radical for most architects. Pataphysics is the science of the exceptional, a science of imaginary solutions, a celebration of mystery, truly analogous to art. Its master key is irony: What science clarifies is what remains obscure. Pataphysics enacts discovery through making, it celebrates technical processes and architecture as a verb. The artist's life is the paradigmatic work of art: In this sense the aim to deconstruct the distance between form and content is truly radicalized, in the only way possible for the twentieth century: Jarry became *Ubu Roi* (the main character in his most celebrated set of plays, an absurd antihero), a process of self-transformation that takes precedence over formal products. Pataphysics demands a different relationship between thinking and making, where thinking and values are crucial and the act "calculated," yet there is never a method or instrumental theory.

In fact what Jarry described was art's potential to embody a different kind of truth, one that is never positively given once and for all and thus may be called objective, nor merely makes sense to the one that utters it and thus may be labeled subjective. Rather art's potential is to reveal a truth both unique and universal, personal and socially relevant, culturally specific and transhistoric. This is an "experience," our capacity to perceive meaning in privileged human works that reveal our mortal lives as purposeful, a "revealing/concealing" of being rather than its objectification, making shockingly evident to all those participating in the recreation of a work of art how human mortality and our perception of order are not in contradiction, but rather depend on their copresence for their reality. This "reality" of art cannot be theorized or logically demonstrated. The evidence is in the traces left by Mozart, Michelangelo, or Le Corbusier. We either participate in this proposed meaning (sense, direction), or we don't. Whereas contemporary science still pursues the dream of immortality, a more or less explicit future orientation towards an objective (final!) disclosure of the order of things, art accepts the possibility of transcendence only in the thick, always evanescent present moment. Art and architecture communicate, but what they "say" cannot be transcribed. It is what they are. The truth art conveys can be experienced only in its own medium, in the specificity of the work at hand.

As the ironic embodiment of the scientific project, we can understand pataphysics as the culmination of Western science and as a potential model for architecture. If the positive sciences demythified traditional natural science and philosophy, pataphysics demythifies the demythification of the world supposedly accomplished by positivism and technology. This may be particularly relevant as we discover that architectural theory is not science, but that architecture as a mode of production is necessarily technology. Rather than a practice furthering the transparency and correspondence of signifiers and signifieds, a pataphysical architecture would possess a theoretical discourse in the form of ontological hermeneutics (historical narratives) and a practice in the form of a poetics (fictional narratives), both exploding and celebrating the gap between the two terms of metaphorical constructions as the effective place of meaning.

Technology, we know well, is more than machines or neutral processes, it is our world, the historical reality that we have fabricated, qualitatively different from a world of traditional techniques. As Octavio Paz has suggested, ours is a world of artifacts that no longer are a bridge between our consciousness and the external realities that we have not created, as human technical products certainly were prior to the nineteenth century, but rather appear to constitute a wall, impossible to escape, surrounding us with our own dreams of control, self-referentiality, and cyberspace. Although retrospectively it may be argued that this difference is merely a question of degree, the fact is that in its desire to answer the same old questions about the measure of man and place amidst the immeasurable and alien, technology opted for control and domination, for arrogant and efficient action on a reduced and objectified picture of the world. Success within this instrumental mode of action has reinforced humanity's ambition to exploit a universe reduced to "natural resources," one that, we strongly suspect, may not have been created only for that purpose, but whose final opacity, we may believe, gives us license to act in this fashion.

The story I told earlier about architectural theory assimilating itself with instrumental methodologies is hardly surprising in this context. This all-encompassing technology has been bound up with the tools of architectural ideation, representation, and practice since the early nineteenth century. Architecture, the art of mediation par excellence, has tended to become a vehicle of political or economic domination, repression and control. Technology problematizes architecture, a space-time of human situations whose meaning coexists with embodied consciousness. This realization is of particular import for the architect as he or she realizes that tools of representation are indeed not value neutral, a complex topic that cannot be developed here.

It is important to emphasize, though, that technology cannot simply be dismissed from the point of view of the traditional fine arts, metaphysics, or humanism. The thrust behind its accomplishments (and failures) is indeed the human thirst for transcendence. Once God and man were removed from the center of the world, technology became a morally justifiable response to the inveterate human lack in the face of the cosmos. Obsessed by instrumentality's accomplishments and its own assertive will to power, technology has led postmodern humanity into a gnostic trap in which the preexisting "world out there" is forgotten. Today we are often controlled by forces that we thought we were capable of mastering.

Within the wall we have built it is difficult to understand that there is no human creation that is not also a mimesis. As Walter Benjamin has pointed out, the mimetic function allows for human culture to become our nature. In the process, through works of art (and architecture), the nightmare of constructionism is shattered.[17] We cannot name the transhistorical ground, but we cannot do without it. However mutable and historically determined this world of primary experience may be, phenomenology discloses it as one where the universal and the specific are given simultaneously in the mystery of perception, in the space between being and becoming. This ground gives us immediate access to historical artifacts (always, of course, reinterpreted), and it must be presupposed by any narrative, understanding narrative as the fundamental structure of human truth. Merleau-Ponty reminds us that "what consciousness does not see is what makes it see . . . as the retina is blind where the fibers that will permit vision spread out into it."[18]

Despite all this evidence, it is still easier to believe that all reality is constructed and that we have little or no option but to embrace our "destiny": the *jouissance* of giddiness, intertextuality, immaterial bodies, and the like. The city is dead: Long live the cybernetic megalopolis. Under this assumption it is easy to affirm that architecture and design as technology are all-encompassing, that in this mode architecture offers a vision of life "yet to come" for a post-Romantic, decentered, and fully social-democratic self. Can it really be that simple? What about our present contradictions and pathologies, evident as we contemplate the political realities of the postmodern world?

We must recognize that there are dangers present in the technological world that are more subtle, yet more serious than humanity's potential for self-destruction and the threat of ecological disaster. The messianic denial of death as a positive limit and essential qualifier of life may indeed end up jeopardizing the possibility that human existence can remain open to meaning and thus perpetuate civilization through its cultural institutions and symbols. A nihilism of despair may become a powerfully destructive force, overriding the balance that we must maintain when we engage, with Nietzsche, in an indispensable affirmative nihilism. Though questioning the absolute truth of Western metaphysics as something outside experience (the Platonic truth-as-correspondence), this form of affirmative nihilism enables humanity to remain open to the powerful silence of art and architecture and to the dark radiance of light, enhancing our sensitivity to the sound of an angel's flight, in case it happens to come by. The issue, therefore, rather than blindly embracing technology and the tools it offers, is to recognize its mysterious origin and its historical transformations. Diagnosing the enframing without resentment is a crucial step: Technology itself is founded on "weak truths." God and "man" may be irretrievable, but our unique planet earth and its sky are present (finally!) as a ground for our full, embodied experience, one through which we are now capable of questioning the hegemony of abstract constructs. In view of this realization, we are called to transform our own individual, often arrogant, relationship to the world and explore other options.[19]

For architecture, the end of historical progress, itself bound up with technology, places the hitherto unquestionable value of innovation in a different light. The issue, however, is not merely to assume the collapse of differences between technological and aesthetic cultures, the often expressed dream of the whole world's becoming a work of art. . . . This might express the desired end for an accomplished nihilistic epoch, were it ever to arrive. Our own time, however, is hardly at that stage. We live in a time of incomplete nihilism, in which repressive economical, political, or pseudoreligious values occupy the place of the strong values of the *ancien régime,* not only in Bosnia or Northern Ireland, but in Europe and North America as well. The identification of aesthetic and technological values could be dangerous. We should not readily give up the possibility for art and architecture to destructure technology and other hegemonic institutions. It is important to maintain the erotic distance that has constituted the effective human depth in our tradition, the space of presentation *and* representation, rather than accepting the collapse of desire into the four dimensions of cyberspace, or the dissolution of the function (or the content) in the architectural form. This sort of sophisticated formalism, despite its good intentions, could effectively exacerbate the barbaric aspects of our human nature: Rather than designing a world for the *übermann,* a democracy of responsible individuals beyond good and evil and other assorted dualisms, we may end up with a society of violently territorial fish.

If nihilism is epochal, as Heidegger himself suggested, we must be patient. The last thing we should promote, however, is the abdication of the individual imagination. The imagining self is not the punctual, self-centered, and fully coincident Cartesian ego, it is our vehicle for compassion and thus, for ethical action. By means of the ethical imagination, oriented historically yet not bound by history, the architect must find effective connections. As I suggested previously, Nietzsche articulated this task beautifully in his "untimely meditation" on history.[20] It is often stated that true scientists also have great imaginations: Logic, mathematical virtuosity, cleverness, and hard work are never sufficient. Couldn't it be argued that it was imagination that Oppenheimer lacked? In the end, the personal imagination is the issue; without it man cannot act ethically.

The architect is called to destructure the perception of technology as a strong value, to make embodied consciousness spin faster than the electrons in a computer and demonstrate that embodied perception is always more mysterious than simulations. The aim is to use the tools in order to subvert them, to evidence technology's weakness and celebrate its mysterious origin. As suggested, irony is an invaluable tool for architecture in dealing with this delicate situation, particularly when language must be engaged. And this is most important. The work of architecture must not end in silence, even if it must evoke it. Silence is politically dangerous, given the risks I have described. For this reason I argue that the program in architecture must not remain disengaged. The program as narrative, as a vision of life put forward by the architect, is a crucial part of the project. We have to envision, through language, a projection of an appropriate, poetic life on earth. Of course we know there are ambivalences associated with changing use, just as the building itself changes in time, deteriorates, and dies. In the end we are never the judges of our own acts: Immortality is for others to confer. Yet, the architect's ethical responsibility, a tall order of business as I hope to have demonstrated, is carried beyond formal strategies, basically through the recognition and subversion that the building allows through its program. To construe utopic situations, to open possibilities for our present society's having the courage to question institutional imperatives, reconstructing our tradition through narratives and thus hopefully truly acting outside history and its repressive values—such is the mission that I see emerging from architecture's complex relationship with science and technology.

NOTES

1. See Forty, this volume, "'Spatial Mechanics': Scientific Metaphors in Architecture."
2. Plato, *Euthypro,* II c–e.
3. This relationship between architectural theory and practice remained essentially unchanged until the end of the Renaissance. For more details about the transformation of this relation and of the very nature of theory and practice after the scientific revolution, see my own *Architecture and the Crisis of Modern Science.*
4. See Perrault, *Ordonnance for the Five Kinds of Columns,* "Introduction," pp. 1–38.
5. Lyotard, *The Postmodern Condition,* esp. chaps. 5 and 9.
6. Nietzsche, *Untimely Meditations,* pp. 57 ff.
7. See Burtt, *The Metaphysical Foundations of Modern Physical Science.*
8. See Ramirez, ed., *Dios Arquitecto;* also my review of it in *Design Book Review* 34 (1994): 49–53.
9. Laplace, *Exposition du Système du Monde,* p. 443.
10. Husserl, *Phenomenology and the Crisis of Philosophy,* "Introduction" and "Philosophy and the Crisis of European Man."
11. Plato, *Timaeus and Critias,* p. 70.
12. Gleick, *Chaos,* p. 304.
13. Gleick, *Chaos,* pp. 116–117.
14. Pérez-Gómez, *Architecture and the Crisis of Modern Science,* chap. 5.
15. Vattimo, *End of Modernity,* esp. Part 3, pp. 113 ff.
16. Long before Gilles Deleuze took it upon himself to emphasize the relevance of the seventeenth-century philosopher, Leibniz was read as a proto-phenomenologist by Anna Teresa Tymieniecka, a student of Husserl. See her *Leibniz' Cosmological Synthesis.*
17. See Taussig, *Mimesis and Alterity,* "Report to the Academy."
18. See Merleau-Ponty, *The Visible and the Invisible,* pp. 74 ff.
19. I have attempted to explore options for architecture in a technological world in my recent book *Polyphilo or the Dark Forest Revisited.* By necessity, the form of the text in this work is narrative rather than "scientific." The words are analogous to the structure of stones and mortar in conventional architecture, weaving a nonlinear discourse that discloses a "dramatic" space for potential human action.
20. Nietzsche, *Untimely Meditations,* pp. 57 ff.

Bibliography

Burtt, Edwin A. *The Metaphysical Foundations of Modern Physical Science.* London: K. Paul, Trench, Trubner and Co. Ltd., 1927.

Feynman, Richard P. *QED: The Strange Theory of Light and Matter.* Princeton, NJ: Princeton University Press, 1985.

Gadamer, Hans-Georg. *Reason in the Age of Science.* Cambridge: MIT Press, 1981.

Gleick, James. *Chaos: Making a New Science.* New York: Penguin Books, 1987.

Husserl, Edmund. *Phenomenology and the Crisis of Philosophy.* New York: Harper & Row, 1965.

Kagis McEwen, Indra. *Socrates' Ancestor: An Essay on Architectural Beginnings.* Cambridge: MIT Press, 1993.

Kant, Immanuel. *Prolegomena to Any Future Metaphysics.* Translated and edited by Gary Hatfield. Cambridge: Cambridge University Press, 1997.

Laplace, Pierre-Simon. *Exposition du système du monde.* 4th ed., Paris, 1813.

Lyotard, Jean-François. *The Postmodern Condition: A Report on Knowledge.* Manchester, UK: Manchester University Press, 1986.

Merleau-Ponty, M. *The Visible and the Invisible.* Evanston, IL: Northwestern University Press.

Nietzsche, Friedrich. *Untimely Meditations.* Cambridge: Cambridge University Press, 1983.

Pérez-Gómez, Alberto. *Architecture and the Crisis of Modern Science.* Cambridge: MIT Press, 1983.

Pérez-Gómez, Alberto. *Polyphilo or the Dark Forest Revisited.* Cambridge: MIT Press, 1994.

Perrault, Claude. *Ordonnance for the Five Kinds of Columns after the Method of the Ancients.* Introduction by Alberto Pérez-Gómez. Translated by Indra Kagis McEwen. Santa Monica, CA: The Getty Center for the History of Art and the Humanities, 1993.

Plato. *Euthypro.* New York: Viking Penguin, 1954.

Plato. *Timaeus and Critias.* Middlesex, UK: Penguin Books Ltd., 1971.

Ramirez, J. A., ed. *Dios Arquitecto: Juan Bautista Villalpando y el Templo de Salomon.* 3 vols. Madrid: Ediciones Siruela, 1991.

Taussig, Michael. *Mimesis and Alterity: A Particular History of the Senses.* New York: Routledge, 1993.

Tymieniecka, Anna Teresa. *Leibniz' Cosmological Synthesis.* Assen, Netherlands: Van Gorcum & Comp. N.V., 1964.

Vattimo, Gianni. *The End of Modernity.* Baltimore: Johns Hopkins University Press, 1988.

15

THE MUTUAL LIMITS OF ARCHITECTURE AND SCIENCE

Kenneth Frampton

> If a critical theory is a kind of knowledge (and if we reject naturalism), it seems obvious that it won't be a kind of *scientific* knowledge: How would one go about examining instances of normative beliefs? How would one apply the hypothetico-deductive method? Ideological beliefs and attitudes aren't refuted by pointing out observed negative instances, but by inducing reflection, i.e. by making the agents who hold these beliefs and attitudes aware of how they could have acquired them.
>
> —Raymond Guess, *The Idea of a Critical Theory*

The first thing to be observed about the relationship between architecture and science is the numerous attempts that have been made during the course of the century to reconstitute architectural practice so as to bring it into line with the methods of science. This impulse is by no means unprecedented, for there have been times, dating back to the last century and even beyond, when architects were trained as structural engineers, as though their responsibilities as professionals had to be extended into this area if they were to retain their status within an emerging technocratic society. This has been particularly true in Germany, Switzerland, Spain, and the Netherlands, where architects were invariably trained in technical universities. In the Technical University Delft, architects still graduate with the title of "building engineer," and in Spain, by law, architects are responsible for the calculation of the structure of a building up to six floors in height.

In Anglo-Saxon countries, in which the technical university has remained relatively undeveloped but the tradition of technological pragmatism has been strong, rather futile attempts were made in the 1960s to recast architecture as applied ergonomics. This was particularly true at the universities of California at Berkeley and of London, which, although remote geographically, were closely linked ideologically, so much so that they both opted for the term "environmental" as the key adjective with which to qualify their status as institutions of design. It is a measure of the capriciousness of fashion that the Bartlett School of Architecture at the University of London had its name changed to the School of Environmental Studies, only to have it revert to its original title at the end of the 1970s. Its director in the 1960s, Richard Llewelyn Davies, also insisted that entering graduates should have high mathematical scores, until it was discovered that the synthetic capacity demanded of architects did not necessarily favor minds with mathematical ability. We may be touching here, however, on inexplicable cultural differences and traditions, for in the case of the Madrid School of Architecture, mathematical ability has always been a prerequisite for entering the technical university, and this qualification does not seem to have compromised its graduates' architectural capacity.

A comparable effort to reconstitute architecture as an applied science was equally evident at Cambridge University, England, in the 1960s, where a group of architect-mathematicians, Marcial Echenique, Lionel March, and Philip Steadman, were able to establish the so-called Center for Built Form and Land Use Studies, a research institution still operative today, although its publications have been curtailed and its studies no longer seem to possess the relevance that they once did. One should note that these relatively short-lived epistemological forays were not mere isolated incidents within the evolution of twentieth-century architecture. The unfinished modern project, in a scientific sense, has made itself manifest in multiple ways over the past century, and I briefly cite here other attempts made during the century to rationalize the practice of architecture in its name.

One may begin with Alexander Klein's ergonomic, econometric housing studies that were pursued during the heyday of the Weimar Republic between 1923 and 1931.[1] We may pass from this to Hannes Meyer's polemical use of the term *building* rather than architecture in his inaugural Bauhaus address of 1928 when he concluded with the words: "Building is only organization: social, technical, economic, physiological organization."[2] Meyer's own teaching in the Bauhaus (both before and after his promotion to the directorship) was technocratic in tone, focusing as it did upon the rationalization of prefabricated production, on the normalization of furniture design, and on the optimization of sun exposure. His highly polemical antipathy to composition in architecture is notorious. In 1928 he wrote:

> All things in this world are the product of the formula: function times economics.
> So none of these things are works of art:
> All art is composition and hence unsuited to a particular end.
> All life is function and therefore not artistic.
> The idea of the composition of a dock is enough to make a cat laugh.

> But how is a town plan designed? or a plan of a dwelling?
> Composition of function? Art or life?[3]

Meyer was hardly alone in expounding such views, and architects as various as Fred Forbat, Arthur Korn, Otto Häsler, and Ernst May, the city architect of Frankfurt, subscribed to very similar materialist values.[4]

Despite the "degree-zero" functionalist consensus of the Weimar Republic, the so-called International Style varied much more widely in the first half of this century than we generally accept, ranging from the extreme rationalization of on-line production as we find this in Walter Gropius's prefabricated Törten Housing of 1926, in which the track of a rail-mounted, mobile, tower crane exclusively determined the housing layout, to the much more *biotechnical,* overall approach evident in the work of such revisionist modernists as Richard Neutra and Alvar Aalto. Thus we find Neutra writing, in 1954, in *Survival Through Design:*

> It has become imperative that in designing our physical environment we should consciously raise the fundamental question of survival, in the broadest sense of this term. Any design that impairs and imposes excessive strain on the natural human equipment should be eliminated, or modified in accordance with the requirements of our nervous and, more generally, our total physiological functioning.[5]

In a similar vein, Aalto qualified the very idea of rationality when he wrote in 1940:

> It is not the rationalization itself which was wrong in the first and now past period of modern architecture. The wrongness lies in the fact that the rationalization has not gone far enough. Instead of fighting the rational mentality, the newest phase of modern architecture tries to project rational methods from the technical field out to human and psychological fields.[6]

Where Klein had been oriented toward finding the most efficient and economical way of laying out a typical minimal dwelling, and where Meyer devised a scientific method for assuring the optimum penetration of sunlight into his Trades Union School built in Bernau in 1930, Neutra and Aalto attempted to provide environments more attuned to their clients' psychosocial needs, with Neutra making extensive use of landscape as a way of integrating the living volume into the site and Aalto employing warm materials instead of chromium-plated metal as a way of providing a modern biotechnical environment that was more sympathetic to the user's tactile sensibilities. In spite of these excursions into the psychological susceptibilities of different classes of users, bioclimatic design methodology remained the prevalent scientific means for introducing regional, not to say local, inflection into the otherwise daunting uniformity of international modernism, as we may judge from a relatively late compendium on this subject, Victor Olgyay's *Design With Climate* of 1963[7] (figure 15.1). In his *Owens Valley Study Section* of virtually the same date, Ralph Knowles will apply the Olgyay method to the plotting of

15.1

Victor and Aldar Olgay. A typical sun/prevailing diagram showing optimum conditions for a given locale. Source: Victor and Aldar Olgay, *Design with Climate*.

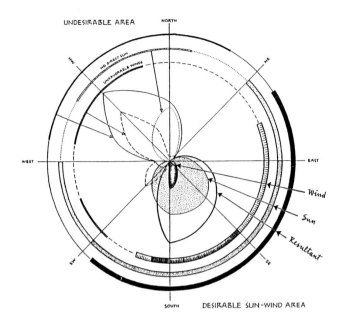

the entire region's passive energy potential, not only identifying the most ecologically favorable sites for future development but also indicating the building envelop profile that would be able to take the greatest advantage of these latent thermal and prevailing wind capacities.[8] However, it has been long since realized that general orientation and sun exposure are only two variables in the design of an appropriate environment.

We may turn from a totally different standpoint to the various attempts made over the years to integrate perceptual and behavioral psychology into the conception and appropriation of architectural form. Among the more encyclopedic attempts in this regard was Sven Hesselgren's *The Language of Modern Architecture* of 1959.[9] One notes that this metalinguistic approach seems to have anticipated the semiotic paradigm's then-imminent emergence in all the arts. This Saussurian science of culture would soon become manifest in the architectural field in such publications as Charles Jencks and George Baird's anthology *Meaning in Architecture* of 1965,[10] and later, at a more rigorous level, in Umberto Eco's *A Theory of Semiotics* of 1976.[11]

Hesselgren, for his part, was clearly conditioned by the welfare-state culture-politics that prevailed in Sweden at the time. "The architect," he wrote, "must take his basic facts from several realms of knowledge. As part of his task is to create healthy buildings for humans he must know something of biology. . . ."[12] Hesselgren went on to assert that architects must also be familiar with physics and sociology, only in these instances he readily concedes the need for specialists. He then makes the rather sweeping claim that "In applied psychology the situation is somewhat different, for here the architect is the only

356 | *Kenneth Frampton*

specialist and must therefore himself be an expert, especially in applied perception psychology."[13] Then as now the close connection between the science of perception and visual culture is something of a truism, as we may judge from Paul C. Vitz and Arnold B. Glimcher's *Modern Art and Modern Science* of 1984,[14] even if this study doesn't treat architecture as such.

In 1971, Lionel March and Philip Steadman of the Built Form and Land Use Studies unit at Cambridge published their seminal work, *The Geometry of the Environment*,[15] as an elaboration of a design methodology pioneered by the mathematician-architect Christopher Alexander in his *Notes on a Synthesis of Form* of 1966.[16] Their stress on geometry as a spatial discipline was hardly new, aside from the fact that, in this instance, it was logarithmically articulated and ostensibly not geared to any particular form. The mathematical procedures demonstrated therein were mainly topological, illustrating such procedures as mapping, stacking, and nesting and such spatially manipulative operations as transformation, rotation, reflection, and the like. They went on to demonstrate the use of other normative, mathematical devices such as matrices, vectors, networks, mosaic formulations, symmetrical patterns, Venn diagrams, graph theory, and different kinds of proportional systems. One of their more challenging expositions devolved upon the application of a permutational method for optimizing proxemic relationships, in which they conceded at the outset that "the job of measuring Cp [optimal circulation path] is certainly one for a computer. With ten rooms the number of possibilities is up to over 3½ million, and the addition of only one more room increases that number by a further 36 million. For a problem with any realistically large number of locations, the permutations become quite unmanageably numerous, even using computers."[17]

Here we see that British logarithmic research of the 1960s attempted to extend Klein's proxemical methodology to much larger and more complex architectural tasks. The ostensible goal was to find the single most economical way for arranging a given spacial cluster from the standpoint of overall convenience. Even so, as March and Steadman admit, the kind of perimeter resulting from such a method would need to be manipulated by hand afterward to yield an integrated form from the point of view of general architectural criteria. In other words, once other variables were allowed to impinge on other aspects of the problem, such as lighting, compositional form, production, sense of identity, and so forth, then the entire method seems only to have had limited application.[18]

One needs to mention, just for the record, two other approaches from the 1960s that attempted to fuse the practice of industrial design and architecture with the methodology of science. The first of these took place under direction of Herbert Ohl in the Building Department of the Hochschule für Gestaltung in Ulm, where the heuristic methods adopted were invariably addressed to arriving at an economic mode for the construction of building components and the prefabrication and stacking of optimized cellular units.[19] This approach returned once more to the promise of industrialized building, notwithstanding the fact that by this date it was already known that one of the impediments to capitalizing on the potential economies of mass-produced dwelling units was the marked resistance on the part of the real estate and banking industries to market and fund such units, let alone the fact that without an assured demand on a large scale one could not really approach the levels of prototypical refinement and cost reduction commonly achieved in the automobile industry, as Alexander Pike would point out in his seminal essay of 1962.[20]

A more pragmatic and, indeed, more romantic technoscientific approach dating from the same period was the geodesic methodology of Richard Buckminster Fuller, first celebrated as a critical alternative to the received functionalism of the prewar modern movement in Reyner Banham's *Theory and Design in the First Machine Age* of 1960.[21] Fuller propagated the notion of a "new nature" in the form of a universal, tetrahedral cosmos that he justified partly on the grounds that this particular matrix corresponded to matter's molecular structure, namely to the geometrical order that provided for spheres in closest packing.[22] Fuller's generic solution to any building task was always the same, and in that sense, it bordered on the simplistic, for it invariably comprised the throwing of a geodesic dome over the required accommodation, irrespective of whether it was a botanical garden, a railway shed, or a private house (figure 15.2). Fuller's solution was conceived, it would seem, to meet three particular criteria: efficient, lightweight structural assembly; optimal openness of both structure and space; and last but not least, an Archimedean geometrical form capable of representing the universe's underlying structure.

15.2

R. Buckminster Fuller. A typical application for a geodesic dome for the instant storage of military supplies. Source: *Architectural Design,* July 1961.

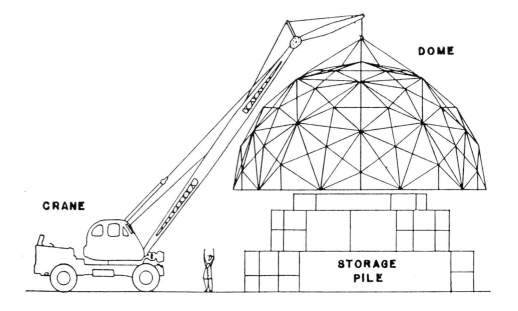

15.3

Konrad Wachsmann and Walter Gropius. General Panel Systems House, 1945–47. Wachsmann's idealized timber and metal coupling. Source: Konrad Wachsmann, *The Turning Point of Building*.

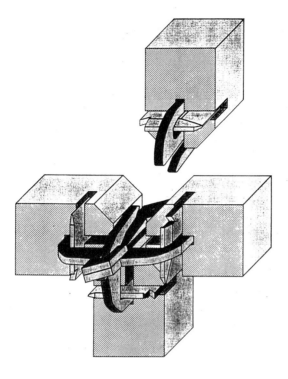

The German emigré architect Konrad Wachsmann, whose work in certain respects paralleled that of Fuller, concentrated his energies conceiving vast cantilevered hangar structures for the U.S. Air Force and on devising suspended structural systems capable of yielding unprecedented span-to-depth ratios. Immediately after the Second World War, Wachsmann collaborated with Walter Gropius on the design of a prefabricated, prepackaged suburban housing system known as the General Panel Systems house, for which he would devise an ideal dry joint in timber and in timber and metal (figure 15.3) such as would later be achieved exclusively in metal in Fuller's geodesic domes and in the tensegrity masts of American sculptor Kenneth Snelson. Thus although repudiating the International Style for its formalist aesthetics, both Fuller and Wachsmann ended up assuming an equally formal/technical posture except for the fact that they projected their works as artificial inventions comparable in their authority to the natural organisms that were first posited as "design models" in D'Arcy Thompson's *On Growth and Form* of 1917.[23] Certainly Wachsmann was more focused on the constraints of machine production and on the logic of practical, lightweight, machine-tool assembly than either Snelson or Fuller, as we may judge from his *The Turning Point of Building* of 1961. On the other hand, Fuller's comprehensive "science fiction" world view, as epitomized by his dymaxion Air-Ocean World Map, was better able to capture the imagination of an entire generation, particularly in England, where his definition of the dymaxion principle as the "maximum advantage gain for minimum energy input" would be applied by the Sixties generation to romantic ends as we find this, say, in the work of the Archigram Group or in the equally visionary projects of Cedric Price. Price's Potteries Thinkbelt of 1966 was an ingenious "ecological" proposition for the re-employment of existing, underused rolling stock as classrooms and other kinds of educational facilities[24] (figure 15.4).

Production-oriented functionalists such as Konrad Wachsmann and Hannes Meyer were to have a major impact on the work of the Swiss architect Max Bill, who in fact had been one of Meyer's pupils at the Bauhaus and later, after the Second World War, became the founder and director of the Hochschule für Gestaltung in Ulm. One is indebted to Bill for his coinage of the term *Produktform,* wherein the form that an object or a building assumes is primarily determined by an interplay between its program's functional requirements and the logic and the sequential elegance of its economic production. This is still the main heuristic method that lies behind many of the achievements of today's high-tech architects, this plus Cedric Price's hypothetical ideal of the 1960s of providing "well serviced anonymity," which has since been brought to perfection, both culturally and technically, by architects as accomplished as Richard Rogers, Norman Foster, and Renzo Piano.

In no single work has Piano proven his prowess as a technocratic builder than in the 1.7 kilometers long Kansai Airport (1988–94), built on an artificial island in the midst of the sea, close to Osaka in Japan. Capable of handling 100,000 passengers a day, this building is a technological tour de force. That a new genre of team-design and productive reasoning was used in its generation is evident from Piano's description of the way in which the cross-section of the roof was devised (figure 15.5):

15.4

Cedric Price, Potteries Thinkbelt, 1966. An ingenious approach to the reuse of existing rail infrastructure and rolling stock for educational purposes. Source: Charles Jencks and George Baird, *Meaning in Architecture.*

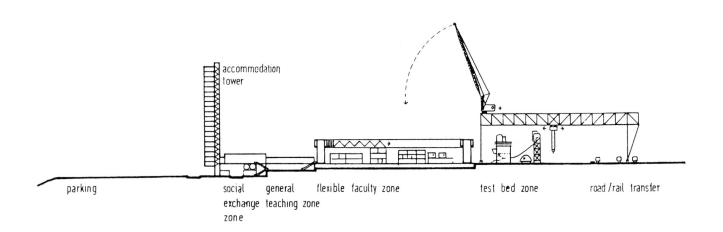

15.5

Renzo Piano, Kansai Airport, Japan, 1988–94. Interior perspective showing cowels blowing onto distribution baffles, which are suspended within the roof structure. Source: Renzo Piano RPBW Genoa.

It has been given this shape to channel air from the passenger side to the runway side without the need for closed ducts. Baffles resembling blades, not set in pipes, but left open to view, guide the flow of air along the ceiling and reflect the light coming from above. In this way all the elements that would have prevented people from seeing the structure have been eliminated. We regulated the movement of the air by creating a ceiling that is aerodynamic, but the "other way up," for the flows we are interested in are on the inside not on the outside. We did this by entrusting the work of calculation to excellent designers and the computer, which gave us the speed and precision we needed . . .[25]

It is surely no accident that what one may call the *Produktform* approach has yielded its most substantial and convincing results where the works have been relatively expensive and intrinsically technological, that is to say, where the building in question has been construed, in one way or another, as a large machine, as in say an airport, a factory, a laboratory, or an office building. This technocratic stance has been rather less successful in dealing with programs such as housing or civil architecture, that is, with tasks involving psychologically complex human needs or with buildings having a public-symbolic character.

The original modern movement (1918–1939) justified its structural syntax and its programmatic method in terms of technology. This was a plausible position to adopt at the time given the multiple technological advances perfected between 1870 and 1918.

This period's pioneer architects argued that their architecture was a direct consequence of such innovations as reinforced concrete and steel construction combined with plate glass and a complementary array of electromechanical installations of every conceivable kind, the entire mythic modernizing story, in fact, that Sigfried Giedion recorded in his two canonical texts, *Space, Time and Architecture* of 1940 and *Mechanization Takes Command* of 1948.[26] As we have seen, many other innovations and technical procedures have since greatly augmented this technical repertoire, ranging from space frames to tented assemblies and from pneumatic membranes to climatic servomechanisms of the most sophisticated kind, all of which returns us once again to the conflation of engineering with architecture.[27]

Despite this convergence, architecture, unlike technology, is categorically value laden, and as such it corresponds to the role assigned to artistic discourse in Charles Morris's seminal essay of 1939, "Science, Art and Technology." Morris's schema of three distinctly different primary discourses, each having an entirely different mode of address and method, retains its relevance inasmuch as it still seems plausible to regard science, art, and technology as being essentially limited to *predictive, evaluative,* and *instrumental* practices. As he was to put it, now nearly sixty years ago:

> Scientific discourse is, in summary, statemental or predictive in character, and the statements are either confirmable (or disconfirmable) in terms of empirical evidence. . . . Men have, however, other needs than that of accurate prediction. As beings with needs and values, they are concerned with the vivid portrayal of what they value, and in devices by which their needs can be satisfied. Esthetic discourse ministers to the first interest; technological discourse to the second. By esthetic discourse is meant a specialized type of language which is the actual work of art. . . . The view proposed is that the esthetic sign designates value properties of actual or possible situations and that it is an iconic sign ("an image") in that it embodies these values in some medium where they may be directly inspected. . . . it must be recognized that objects have value properties . . . (and that) in dealing with them esthetic discourse is concerned with the same world with which science and technology are concerned. . . . the artist is himself a technologist in that he must work his will upon some material or other.[28]

Although Morris did not address himself to architecture as such, the implications of his thesis are clear, namely that architecture, aside from representing value, also embodies it concretely by virtue of the physical attributes integral to its spatial organization and material detailing. One may cite, by way of an example, Herman Hertzberger's Montessori School in Delft of 1966, where the articulation and inflection of the microspace is redolent throughout with sociocultural values consummated through the way the building is appropriated and used. With this work in mind, one may assert that architecture is initially metamorphic and pragmatic in its operations rather than cognitive or inventive in a scientific or techno-scientific sense. One needs to recall in this regard that architecture became a liberal profession relatively late with the consequence that it is still, in essence, as much a craft-based *metier* as it is an applied technique. For this reason, value and myth may be readily incorporated into its procedures rather than the pure instrumentality of techno-science or the various activities that may be subsumed today under the rubric of art.

The traditional roles of the sciences and of the humanities were radically reformulated in 1979 by Jean François Lyotard's *La condition postmoderne,* wherein he argued that the multiplicities of contemporary science and the skepticism of poststructuralist philosophy were such that the Enlightenment project of a unified master narrative could no longer be sustained. Lyotard's recognition of the succeeding metanarratives and the proliferation of language games in all disciplines was substantiated at the time by the poststructuralist deconstruction of language, that is, by the reduction of all texts to undecidable differences and by the destabilization of the subject through post-Freudian psychology. Despite the opening up of new lines of thought and our seeming ability to transcend both the received syntax of the modern movement and the legacy of history, we are nonetheless distracted today by abstruse speculation as to the nature of architecture, combined with a compulsion to aestheticize its surface effects, as though these reactions were sufficient to compensate for our growing incapacity to articulate environmental space and to resolve the production of built form at an adequate level of detail. This degenerative tendency is not just a consequence of rethinking architecture in terms of the postmodern condition and poststructuralist philosophy. It also appears to be a direct consequence of the globalization of production and consumption and of the role now played by the media in the worldwide marketing of commodity.

This is now so much the case that it seems that we may be aroused from our laissez-faire somnambulism only by acknowledging that we do indeed know what practices, scientifically speaking, are conducive to the maintenance of the ecosystem and, by the same token, what negentropic procedures have already started to erode the environment. Leading climatologists now agree that global warming is a verifiable condition produced largely by the greenhouse effect, which in turn arises from the buildup of carbon dioxide and fluorocarbon pollution. The design professions have been fully cognizant for more than forty years of the patterns of land settlement that ought to be adopted in response to this condition, along with the mix of transportation systems that should be employed to maintain efficient distribution, and so on, without further exacerbating the current rate of global pollution.

In his remarkable book *Livable Environments* of 1972, the distinguished Austrian architect Roland Rainer outlines what a truly critical practice of environmental design should amount to today in view of the socioecological dilemma that currently confronts the species at every conceivable level. For Rainer, the ecological crisis is a direct consequence of the irresponsible way we have settled urban land since the mid-1920s and particularly since the end of the Second World War. He has in mind, of course, the endless proliferation of the suburban, auto-dependent single family dwelling as a biologically dystopian paradigm of development that will end, at the current rate of its continued expansion, only in the total destruction of the biosphere. Twenty-five years ago, he wrote:

> We must become quite serious about the self-evident principle that motor traffic is not an end in itself but only a means to an end and that a city is never to be adapted to a traffic system but that the traffic system has to be adapted to the city. Instead of kowtowing, at high cost and ridiculously, to the car and to the idea of economic growth, there must grow a feeling of responsibility for life and for the coming generations; we must think in terms of cities built in accordance with biological principles.[29]

Rainer is alluding here to the fundamental principle of sustainable growth, as opposed to the maximization of development and with it developmental waste of every conceivable kind, including the rapid amortization of the building fabric and the generation of building waste (not to mention the profligate ruination of agricultural land) as a factor rarely taken into consideration in the balance sheet of economic maximization. He argues that our present motopian society has to be made acutely conscious of unalterable biological limits and of the already pressing necessity to radically increase vegetation, above all the need to plant trees to compensate for the destruction of the rain forests, accumulation of carbon dioxide, and our escalated consumption of oxygen.

On the other hand, in light of our highly prized privacy, Rainer analyzes the conditions obtaining in the average suburban house in the following terms (figures 15.6 and 15.7):

> . . . the house and its pertinent open areas is exposed to all external influences—from the windows and gardens, from the streets and pavements, one can look into the private spheres of other people in front of, beside and behind their houses. . . . we have neither privacy nor neighbourliness, but we do not get any authentic public realm either, since even the street does not constitute a clearly defined sphere of its own.[30]

To all intents and purposes this is virtually the same critique coupled with a similar advocacy of low-rise, high-density housing previously advanced by Serge Chermayeff and Christopher Alexander in their joint study *Community and Privacy* of 1963, in which they wrote:

> Within a few years human intervention will be sufficiently expanded to affect the whole human species; and man, if he is to survive, will face the inescapable need to design an all embracing ecology of his own; even, perhaps, the need to transform himself. Accelerating population growth, interference with and mastery over the natural, will make man's escape from man wholly impossible, and will force him to ac-

15.6–15.7

Roland Rainer, 1972. Didactic diagrams comparing levels of privacy in low-rise, high-density courtyard housing versus the typical suburban subdivision. Source: Roland Rainer, *Livable Environments*.

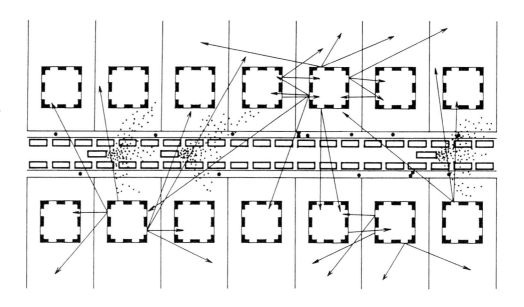

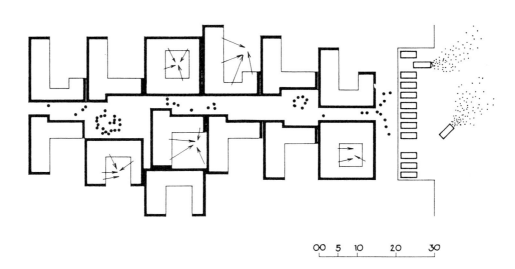

cept responsibility for every phenomenon on the surface of the earth. He will have to design and build his own ecology, his own adaptation to the environment of his own making.

Until he recognizes the more subtle and devastating aspects of his complacent submission to "human nature," he cannot begin to prepare for this responsibility in terms of human engineering. First, he can and must be responsible for the physical form of the environment that is to be the framework for his ecology.

Either he must learn to preserve the existing equilibrium of life or he must introduce a new equilibrium of his own making. If he does neither, his present unplanned conduct may deform human nature beyond all cure, even if it manages to survive the more violent holocausts. The threat of sudden destruction is more dramatic but not more severe.[31]

Morris's unusual claim in "Science, Art and Technology" that a technical a priori is also an ethical a priori returns us to the way our liberal society arrives at consensus and the way this consensus is validated or otherwise in the light of our scientific knowledge. Thus the way consensual values are mediated today by technique has broad implications not only for the practice of architecture at an intimate scale but also for the application of technology at an environmental level.

This inescapable interface between techno-science and politics was convincingly articulated over thirty years ago, and little has transpired since then to indicate that the fundamental terms of the "diagnosis" have changed in any significant way. I have in mind Jürgen Habermas's address, "Technology and Science as Ideology," given in 1968 on the occasion of Herbert Marcuse's seventieth birthday. In distinguishing between two kinds of rationalization, Habermas wrote:

> . . . *two concepts of rationalization must be distinguished.* At the level of the subsystems of purposive-rational action, scientific-technical progress has already compelled the reorganization of social institutions and sectors, and necessitates it on an even larger scale than heretofore. But this process of development of the productive forces can be a potential for liberation if and only if it does not replace rationalization at another level. *Rationalization at the level of the institutional framework* can only occur in the medium of symbolic interaction itself, that is, through removing *restrictions on communication.*[32]

Evidently Habermas had in mind some form of radical democracy in which there would be continual public debate as to the suitability and or desirability of applying certain techniques and norms rather than others. As he puts it later in the same text:

> The question is not whether we completely *utilize* an available or creatable potential, but whether we *choose* what we want for the purpose of the pacification and gratification of existence. But it must be immediately noted that we are only posing this

question and cannot answer it in advance. For the solution demands precisely that unrestricted communication about the goals of life activity and conduct against which advanced capitalism, structurally dependent on a depoliticized public realm, puts up strong resistance.[33]

That this resistance has much of its origin in prevailing forms of global investment, economy, and speculation was soberly addressed in Emma Rothschild's 1973 analysis of the American automobile industry, in which she wrote:

> Social partiality made possible auto domination and the extraordinary profits of the auto industry. Yet because it was supported by quite specific partiality, auto power is comprehensible, contingent, reversible. It required national sustenance which will be reduced as auto ascendancy declines. Auto sales and profits were able to expand not only because of the opportune efficiency of auto companies but also because the costs of auto development were ignored. The 1940's, 1950's, and 1960's in auto transport, whose costs now seem ever more apparent . . . American autos and trucks each year burn 40 percent of all the petroleum in the world. They are designed in such a way as they waste fuel. . . . [and yet] it seems within the grasp of social ingenuity over some decades to preserve the benefits of auto freedom without the present and perceived excesses of automobile waste.[34]

As we have seen, if there is one area where the finer points of advanced technoscience have the capacity to make a critical impact on the culture of building, it is in the field of energy conservation and climate control, as we may judge from recent achievements in the field of so-called high-tech architecture. Needless to say, even here each case must be evaluated on its merits, for that which appears to be technologically advanced may turn out on closer inspection to be regressive in socio-economic terms. Not infrequently one may encounter maximization of one value at the expense of other equally valid considerations from the point of view of user comfort or even with regard to the durability of the building as a whole. With respect to the user, the maximization of air conditioning is a case in point, particularly where this entails fixed glazing with no provision for the manual operation of windows on an individual basis. It so happens, this practice remains forbidden under the provisions of current Dutch building law; such are the variations in cultural norms as one passes from one country to another.

The evolution of the thick, double-glazed thermal wall, as initially developed by Renzo Piano for application to the Daimler-Benz high-rise office building in Potsdamerplatz, Berlin (1992–96), is particularly promising. Here, the application of a thermally active wall in the form of a "thin conservatory," enveloping the entire surface of the building, would have remained open to individual adjustment without compromising the overall economy of the thermal system. The motive in this case was to dispense with air conditioning as it is commonly understood through the installation of a ducted (i.e., fully distributed) heat pump system capable of conserving heat and cold and of balancing out the building's warm and cold sides under changing temporal and seasonal conditions.[35]

Thomas Herzog is another architect who has been particularly concerned of late with the ecology of building form and who has recently demonstrated his capacity in this respect in a number of buildings, above all in his exhibition hall for the Deutsche Messe in Hanover, completed in 1995.[36] This singular clear-span structure demonstrates how the rationalization of building production and engineering form (in this instance, a suspended catenary roof) may be optimally combined with natural lighting and naturally induced ventilation within the hall. In this instance energy efficiency, responsive to climatological changes, is maintained throughout by delivering conditioned air where it is precisely required through the use of long-range nozzles and extracting foul air through natural convection. By a similar token natural light is brought into the main body of the hall largely through rooftop reflectors, and artificial light is provided indirectly by being bounced off the inner soffit of the catenary roof (figures 15.8 and 15.9).

This essay has attempted to address the theme of the reciprocal limits of architecture and science, and in so doing, it has gravitated toward an extensive critique of technoscience as it has been applied or misapplied to building and urban development in the course of this century. I have tried to argue that although the scions of late modern, high-tech architecture have demonstrated their considerable capacity to master the resources of modern technology in an exceedingly complex and humane way, the general domain of the human habitat has not only expanded exponentially in terms of its geographical area but has also engendered forms of ecological and sociological entropy the ramifications of which are only now becoming fully apparent.

In the course of this survey, I have tried to show how various attempts to reconstitute the practice of architecture as an applied science have necessarily focused on the optimization of subsets at the expense of demonstrating any kind of comprehensive synthesis, which nonetheless remains the field's ultimate mandate. Despite the reputable scientific method of isolating variables in order to treat a given domain in terms of its operable components, one would be naive to overlook a certain ideological evasiveness that at times seems complicit with the hegemony of late capitalist production and consumption. In the exchange between science, architecture, and technology, the constraints imposed by commodification return us to the relevance of critical theory as the necessary ground from which to respond with scientific rigor to the project of a rational architecture, irrespective of the degree to which this practice may be informed by advanced technique. Thus whether we like it or not, the interface between architecture and science returns to the political, and as Guess implies, the epistemic criteria of a critical practice aren't just out there waiting to be appropriated and applied. On the contrary, they have to be formulated in the process of constructing a world, just as building comes into being about the convergence of a set of forms and materials that do not as yet exist.

15.8–15.9

Thomas Herzog, Exhibition Hall No. 26, Hanover Fair Grounds, 1995. General view and transverse section showing an ecological ventilation for the catenary roof structure. Courtesy Thomas Herzog, and Heike Seewald, photographer.

| Konzept für das Raumklima | Concept for indoor climate | Concetto per la ventilazione |

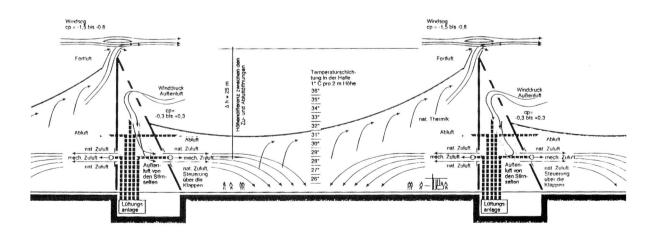

| Funktionsschema (August 1994) | Operational diagram (August 1994) | Schema (Agosto 1994) |

Darstellung der mechanischen Lüftung für den Heizbetrieb: Luftzuführung durch schwenkbare Weitwurfdüsen | Mechanical ventilation for heating operations: warm-air input via long-range nozzles | Principio di ventilazione forzata in regime di riscaldamento: immissione di aria attraverso ugelli orientabili a lunga gittata

Notes

The quote that opens the chapter is taken from Raymond Guess, *The Idea of a Critical Theory* (London: Cambridge University Press, 1951), p. 91.

1. See Rivolta and Rossari, *Alexander Klein*. Alexander Klein was born in Odessa in 1879. He studied architecture at the institute of civil engineering in St. Petersburg. In 1920 he moved to Berlin, where he became involved primarily in the problem of low-cost housing. In 1933 he left Germany and eventually took up residence in Palestine, where he remained almost to the end of his life. He died in New York in 1961.
2. Schnaidt, *Hannes Meyer,* p. 97.
3. Schnaidt, op. cit., p. 95.
4. In his essay "Aufbau/Bauhaus," Peter Galison points out the strong links established between Meyer's Bauhaus and the "unified science" movement of the Vienna Circle, Rudolf Carnap, Otto Neurath et al.
5. Neutra, *Survival Through Design,* p. 86.
6. Aalto, *Synopsis: Painting, Architecture, Sculpture,* pp. 15–16.
7. Olgyay, *Design with Climate*. This "regional" factor is even more noticeable in Aladar and Victor Olgyay's *Solar Control and Shading Devices,* in which many of the examples were drawn from Latin America, particularly from Brazil, where the application of *brise soleil* for sun control seemed to have had particular relevance. Given the latitude of New York and many other North American cities, *brise soleil* should also have been part of standard practice here, in terms of scientific rationality. Indeed if there is one aspect where the "irrationality" of late modern architectural practice may be immediately revealed, it is in this area, for here we may have concrete evidence that although the technical know-how exists, it is not generally followed in a great deal of modern practice. We should also note that the application of adjustable roller blinds and shutters, which were almost standard elements in the 1920s, has become a lost art, as it were. The reason for this climatological regression, so to speak, is largely due to formal preferences; in a word to the dictates of aesthetic fashion.
8. Knowles, *Owens Valley Study.*
9. Hesselgren, *The Language of Architecture.*
10. Jencks and Baird, *Meaning in Architecture.*
11. Eco, *A Theory of Semiotics.*
12. Hesselgren, op. cit., p. 10.
13. Hesselgren, op. cit., p. 11.
14. Vitz and Glimcher, *Modern Art and Modern Science.*
15. March and Steadman, *The Geometry of the Environment.*
16. Alexander, *Notes on a Synthesis of Form.*
17. March and Steadman, op. cit., p. 304.
18. See March and Steadman, op. cit., pp. 315–316. They write:

    > Furthermore, in an extensive plan, we require that the circulation routes of the building, the corridors and staircases, form some coherent and economical system, and do not ramble about chaotically. But using a method which assembles a plan piece by piece and where rooms or activities are treated as relatively independent units which can be added together one by one, then it is inevitable that these overall *systematic* constraints, acting

on the geometry of the building envelope and on the structure of its circulation routes, will not be satisfied. The whole must be more than the sum of the parts. The kind of plan perimeter resulting from an additive or constructive type of method in their simplest applications is often irregular and ragged; there has to be some tidying and reorganizing of the layout done by hand afterwards, before the result is acceptable as a building design. . . .

19. Lindinger, ed., *Ulm Design.* See in particular Herbert Ohl, "Industrialized Building at Ulm."
20. Pike, "Failure of Industrialised Building." He writes: "Comparisons between car production and house construction have always terminated at the financial level. Here the problem seems insurmountable. To build from scratch a new car factory costs between £70 million and £100 million. It is doubtful whether any firms engaged in house construction would have capital of this value available. Even if they had, it is unlikely that they would be prepared to risk investments on an item so untried and unproven . . . it is probably more fruitful to seek answers by separating the organisational procedures from the techniques." (p. 508.)
21. Banham, *Theory and Design in the First Machine Age.* Banham moved toward the conclusion of his study with the words: "It may well be that what we have hitherto understood as architecture and what we are beginning to understand of technology are incompatible disciplines . . ." (p. 329). He had in mind the categoric opposition between Fuller on the one hand and the tradition of modern architecture on the other.
22. See Fuller, "Conceptuality of Fundamental Structures."
23. Thompson, *On Growth and Form.*
24. Price, "Potteries Thinkbelt."
25. Piano, *Logbook,* p. 152. On p. 242, Piano also gives a brief account of the complex global way in which the work was achieved with various components of the airport being made in Britain, France, and Italy and transported to Japan by air and sea.
26. Despite their polemical dimensions, these works still display an authoritative treatment of the period scope, particularly *Mechanization Takes Command,* which in terms of everyday instrumentality remains one of the finest anonymous histories ever written.
27. I am thinking in particular of Piano's single storey Lowara office building, erected near Vicenza in 1985, where if a certain temperature is exceeded, a row of sprinklers on the upper edge of the curved roof switch on automatically so as to wet the roof providing instant evaporation that cools down the roof and, with it, the volume of the building beneath. It should go without saying that the introduction of such sophisticated devices necessitates continuous collaboration with consulting engineers of high calibre.
28. Morris, "Science, Art and Technology."
29. Rainer, *Livable Environments,* p. 23.
30. Rainer, op. cit., p. 49.
31. Chermayeff and Alexander, *Community and Privacy,* p. 47. Rainer (op. cit., p. 22) cites statistics from Hans Reimer's book *Earth, the Refuse Dump Planet* to the effect "The percentage of carbon dioxide in the air in the industrial regions rose in the first twelve years of the 'economic miracle' as much as in the previous hundred years. In 1961 technical combustion processes used up more than 11 million tons of oxygen, which represents the annual requirement of 33 billion people."

32. Habermas, *Toward a Rational Society,* p. 118, italics in original.
33. Habermas, op. cit., pp. 119–120.
34. Rothschild, *Paradise Lost,* pp. 247–249.
35. For details of the initial design for the thermal wall in the Daimler-Benz building see Peter Buchanan, *Renzo Piano Building Workshop,* Vol. 2, London, Phaidon, 1995, p. 219. In the event this design proved to be too expensive to realize.
36. Herzog, *Die Halle 26.*

BIBLIOGRAPHY

Aalto, Alvar. *Synopsis; Painting, Architecture, Sculpture.* (gta vol. 12) Basel: Birkhäuser, 1970.
Alexander, Christopher. *Notes on a Synthesis of Form.* Cambridge: Harvard University Press, 1966.
Banham, Reyner. *Theory and Design in the First Machine Age.* London: Architectural Press, 1960.
Chermayeff, Serge, and Christopher Alexander. *Community and Privacy: Toward a New Architecture of Humanism.* Garden City, NY: Doubleday and Co. Inc., 1963.
Eco, Umberto. *A Theory of Semiotics.* Bloomington: Indiana University Press, 1976.
Fuller, R. Buckminster. "Conceptuality of Fundamental Structures." In *Vision + Value Series/Structure in Art and Science,* 66–88. New York: Braziller, 1965.
Galison, Peter. "Aufbau/Bauhaus: Logical Positivism and Architecture Modernism." *Critical Inquiry* 16 (Summer 1990): 709–752.
Giedion, Sigfried, *Mechanization Takes Command,* New York: Oxford University Press, 1948.
Giedion, Sigfried, *Space, Time and Architecture,* Cambridge: Harvard University Press, 1941.
Habermas, Jürgen. *Toward a Rational Society, Student Protest, Science and Politics.* Boston: Beacon Press, 1970.
Herzog, Thomas. *Die Halle 26.* Munich and New York: Prestel, 1996.
Hesselgren, Sven. *The Language of Architecture.* Lund, Sweden: Student-litteratur, 1967.
Jencks, Charles, and George Baird. *Meaning in Architecture.* New York: Braziller, 1969.
Knowles, Ralph. *Owens Valley Study: A Natural Ecological Framework for Settlement.* Los Angeles: University of Southern California, 1969.
Lindinger, Herbert, ed. *Ulm Design: The Mortality of Objects Hochschule für Gestaltung Ulm 1953–1968.* Berlin: Ernst & Solm, 1990.
March, Lionel, and Philip Steadman. *The Geometry of the Environment: An Introduction to Spatial Organization in Design.* London: Royal Institute of British Architects Publications, Ltd., 1971.
Morris, Charles. "Science, Art and Technology." *Kenyon Review* 1.4 (Autumn 1939): 409–423.
Neutra, Richard. *Survival Through Design.* New York: Oxford University Press, 1954.
Ohl, Herbert. "Industrialized Building at Ulm." In Lindinger, ed., *Ulm Design,* 197–212.
Olgyay, Aladar, and Victor Olgyay. *Solar Control and Shading Devices.* Princeton, NJ: Princeton University Press, 1957.
Olgyay, Victor. *Design with Climate.* Princeton, NJ: Princeton University Press, 1963.
Piano, Renzo. *Logbook.* New York: Monacelli, 1997.
Pike, Alexander. "Failure of Industrialised Building in the Housing Programme." *Architectural Design* 37 (November 1967): 507–509.
Price, Cedric. "Potteries Thinkbelt." *Architectural Design* (October 1966): 484–497.

Rainer, Roland. *Livable Environments.* Zurich: Artemis, 1972.

Rivolta, Matilde Baffa, and Augusto Rossari. *Alexander Klein Lo studio della piante e la progettazione degli spazi negli allogi minimi. Scritti e progetti dal 1906 al 1957.* Milan, Italy: Mazzota, 1975.

Rothschild, Emma. *Paradise Lost: The Decline of the Auto Industrial Age.* New York: Random House, 1973.

Schnaidt, Claus. *Hannes Meyer: Buildings, Projects & Writings.* London: Tiranti, 1965.

Thompson, D'Arcy. *On Growth and Form.* Cambridge: Cambridge University Press, 1917.

Vitz, Paul C., and Arnold B. Glimcher. *Modern Art and Modern Science.* New York: Praeger, 1984.

Wachsmann, Konrad. *The Turning Point of Building.* New York: Reinhold, 1961.

16

The Hounding of the Snark

Denise Scott Brown

While I was considering how and where science and architecture intersect for designers, Lewis Carroll's snark came to my mind. I had thought the "quark" derived from the snark, but Lewis Thomas said the quark came from Joyce. Either origin suggests, metaphorically, a role for artistry and intuition in the sciences, as does Donald Johanson's famous fossil, Lucy, named for a Beatles' song.[1] Taking my cue from these, I free-associated further, from science *and* architecture to science *in* architecture, Science '*n ar*chitecture: snark. Then I thought, "hounding not hunting, some scientists hound us architects." I've had a fair amount of experience with that.

There's also the Hound of Heaven—in our time the sciences *are* God—and I remembered, as well, an article I read for an urban sociology class on the tendency of scholars in disciplines that lack public credibility to assume the mantle of the sciences, hoping it will gain them acceptance. And there I had my theme.

In architecture we are dogged by the model of the sciences. For much of this century, architects have tried to escape what they feel to be their discipline's spurious side by cloaking themselves in what they think are the sciences, but what frequently turns out to be the emperor's clothing: scientism.

The sciences have, as well, become a battleground for territorial warfare within architecture. Would-be invaders—social planners in the 1960s, "computational" designers in the 1990s—have used claims of superior scientific rigor to support raids on our field. As a graduate student and young academic in architecture and urban planning I was embroiled in several such *intra*disciplinary struggles.

At that time (the late 1950s) the New Left was developing in America. It started, for me, in Paul Davidoff's planning classes at Penn's School of Fine Arts. Social scientists were then pouring into planning schools and turning activist. Seeking ways to establish their presence where they found themselves, in schools devoted primarily to architecture, they attacked that target, defining it as unscientific.

This accusation was leveled again in the 1970s, when computers moved into schools of architecture. In the late 1970s, in a computer research program at a school that shall be nameless, I saw an attempt to apply "scientific method" to an architect's design for a community health center. Using the architect's program, that is, the schedule of building spaces and the adjacency and functional relationships required among them, the computer scientists had evolved a plan for the center that achieved, they claimed, a far more rigorous solution to the problem than had the architect using "craft methods," as they called them. One look at the computer-generated plan showed it had been derived solely from the given sizes and relationships of spaces; the resulting plan couldn't be framed, that is, enveloped by the simple modular building structure such a center would need to span and roof its spaces economically. If the first variable architecture must satisfy is adjacency, the next is certainly structure. The computer-generated design had not dealt with the second, let alone the third, fourth, or hundredth variable, nor with those that can't be measured or even ranked.

Architectural designers can and must deal with these. Although we use computers intensively in our office, eventually all requirements of a project must revolve together, and be resolved together, in our minds. As we design, our minds must, like a computer, hold myriad variables in consideration at once; but unlike a computer, they must handle the unmeasurable as well as the measurable, deal in multilayered nuances, make unprogrammable value judgments and take uncharted shortcuts through the material. Anything less is likely to produce some components of a design but not a broad and coherent synthesis.

Waves of scientism seem to roll regularly over architecture. Each decade the latest invaders claim everything before them was irrational and bereft of rigor and assure the profession it will die without the changes they offer. To one who has been in the profession some decades, the new cries sound familiar indeed. There follow the usual attacks on the power base to be occupied: "studio"—the learning-by-doing part of architectural training, purview of the art of architecture, and monopolist of the students' time. The invaders ride into architecture but soon ride out again, finding it too monstrously intuitive for them. Some then head toward the body of the university where, naive souls, they get eaten up in academic wars far older than those in architecture.

The hounds of architecture consider their debate with the field to be between artistic laxity and scientific efficiency. They claim they have method, whereas others have only intuition. Disagreeing passionately, on both scientific and artistic grounds, I was happy to find these passages in *The Lives of a Cell*, by Lewis Thomas:[2]

> The essential wildness of science as a manifestation of human behavior is not generally perceived. . . . The difficulties are more conspicuous when the problems are very hard and complicated and the facts not yet in.

(That is how design in architecture begins.)

> Solutions cannot be arrived at for problems of this sort until the science has been lifted through a preliminary, turbulent zone of outright astonishment. Therefore, what must be planned for, in the laboratories engaged in the work, is the totally unforeseeable. If it is centrally organized, the system must be designed primarily for the elicitation of disbelief and the celebration of surprise. . . .

> Scientists at work have the look of creatures following genetic instructions. They seem to be under the influence of a deeply-placed human instinct.

(See our studio during a "charrette.")

> They are, despite their efforts at dignity, rather like young animals engaged in savage play. When they are near to an answer their hair stands on end, they sweat, they are awash in their own adrenaline. . . . In the midst of what seems a collective derangement of minds in total disorder, with bits of information being scattered about, torn to shreds, disintegrated, deconstituted, engulfed in a kind of activity that seems as random and agitated as that of bees in a disturbed part of the hive, there suddenly emerges, with the purity of a slow phrase of music, a single new piece of truth about nature. . . . It is instinctive behavior, in my view, and I do not understand how it works. It cannot be pre-arranged in any precise way; the minds cannot be lined up in tidy rows and given directions from printed sheets. You cannot get it done by instructing each mind to make this or that piece, for central committees to fit with the pieces made by the other instructed minds. It does not work this way. . . . It is like a primitive running hunt, but there is nothing at the end to be injured. More probably the end is a sigh. But then, if the air is right and the science is going well, the sigh is immediately interrupted. There is a yawping new question and the wild tumbling activity begins once more, out of control, all over again. . . .

> Locally, a good way to tell how the work is going is to listen in the corridors. If you hear the word 'Impossible!' spoken as an expletive, followed by laughter, you will know that someone's orderly research plan is coming along nicely.

(With us, it would be the expletive itself.)

Thomas's description suggests that scientific method and scientific rigor are qualities more complex than is dreamt of by architecture's hounds, and that the orderly disorder of architectural design can make some claim to being scientific.

During the 1960s, the social-scientist planners' views of the roles of analysis and design in our field puzzled me. They called themselves "analysts," but part of their work was to make recommendations for urban policy, economic, social, and physical. This seemed to me to be synthesizing activity, requiring design as well as analysis. In fact, what could be more synthetic than the great urban and regional computer-modeling projects of that time? In claiming to be only "analysts," these planners were able to neglect their responsibility to the whole, synthesizing only those variables they chose to handle. This is to criticize their rigor, not their creativity. Although creativity is frequently linked to synthesis, it was not an issue for these planners, because they believed none was required. They scorned the idea of creativity as being dangerously close to "art," yet I wondered why the architects should have the sole right to creativity in urbanism. Why should there not be creative economic or social visions for cities? Social scientists in urban planning who maintained their work was not creative, were, I felt, a sad-sack lot who crimped our field.

They appeared to believe, as well, that we architects did not work analytically at all. In fact, architecture is both analytic and synthetic. In school and in practice, architecture's subject matter is tackled analytically by its components—inter alia, structures, construction, mechanical and electrical systems, acoustics, theory, history—then reunited, in different ways, in design. Simultaneous or quickly oscillating subroutines of analysis and design occur continually and at many levels throughout the architectural design process.

However it is arrived at, the design must be tested rigorously. Sometimes architects use analogies as heuristics to help them design: "I want that fountain to be like an Edwardian lady's hat." This could be frightening to even a Lewis Thomas, steeped in metaphors, if the implication is that this is as far as architecture goes; but the fountain, once designed, is there for the architect and others to evaluate by measurable techniques. That we use analogies is not the issue: What *is,* is that architectural creativity, however frightening its intuitive aspects, can and should be subject to rationally based critical analysis. This analysis, not the sometimes mystic synthesis, takes most of the designer's time.

Thinking in this way, I was critical of the dicta on planning method social planning theoreticians formulated during the 1960s. First, they said, gather data; second, analyze and synthesize the data; then enunciate goals and survey alternative means to achieve them; then choose democratically between alternatives and develop the one chosen.[3] It was, I felt, a good method, but it should happen over and again, cyclically and at many levels throughout the planning process. At times it could proceed in reverse, going from means to goals. For example, in designing a building, the unique way two activities or spaces are brought together may give rise to the option of a third: A and B, as they are joined, may allow for C, which would not have been possible had A and B been differently arranged. Sometimes these serendipitous activities and their spaces become the most loved in the building. They often occur at junction points in a building's circulation system, where corridors meet, or perhaps opposite a stairway, especially if there is a window and a place for a seat or two. We watch for such opportunities in designing; in our firm's Lewis Thomas Molecular Biology Laboratory at Princeton, for example, at each end of the building, beyond the lab grid, there is a bulge off the corridor. Here a bay window, window seats, a blackboard and a coffee machine offer scientists a rest, a shift of vision, perhaps a conversation with a colleague.

As Kenneth Frampton said elsewhere in this volume, we architects have a holistic responsibility, while designing, to satisfy multiple variables. Some are less measurable than others. To ignore what cannot be measured is, I would have thought, the least scientific approach. Recently I was a member of a design jury for a competition in innovative transportation facilities presided over by a renowned engineering guru. During the discussion I mentioned that a garden of contemplation included in one project would be good for "the soul." He strongly disagreed that "the soul" and its needs could be included in our criteria for judging, and finally said, "If you can measure it, I'll consider it." He may be in a field where he can omit what he cannot measure; I don't have that luxury.

For example, though we can't make scientific statistical projections for the year 2099, urban plans and building designs implemented now could affect the city then and for centuries thereafter, much as Roman and Medieval plans are at the core of many European cities, and some buildings of the world's ancient civilizations have lasted in use a thousand years and more. So we must consider the long-range impacts of our recommendations whether or not we have statistical means of measuring them. And this applies, as well, to shorter periods that are impossible to predict; for example, short-range, area-specific projections of suburban growth and development, which are notoriously untrustworthy.

Faced with this dilemma, some planners choose a figure partway between their highest and their lowest projections, and plan for that. Some architecture and planning scientists bemoan the paucity of information available, then make firm recommendations based on the little information they have, without considering the possible degree and cost of the error that could result. To my way of thinking, these approaches, no matter how many complicated formulae accompany them, are unscientific. Lewis Thomas would not, I believe, be so simplistic; he would face the uncertainty itself.

There are ways of being rational about uncertainty. When unmeasurables form part of the calculus, planning sophisticates may turn to probability theory. More simplemindedly, variables that cannot be measured can often be ranked in some sort of order of importance, or trade-offs can be made between them, or between unmeasurables and measurables. Because architecture is concrete and measurable to that extent, a city can make money trade-offs that reflect broader, more subjective policies or goals, allocating, for example, between quality of building finishes and amount of space to be provided, or between educational plant and teaching salaries.

There are ways of allowing for change you cannot predict. One is through contingency planning: If A happens, then B; for example, a campus plan could suggest sites suitable for new student housing, should a decision to increase student population be made in the future. Another way is through maintaining a level of generality: designing a mitten rather than a glove; for example, a scientific lab building designed as a loft can accept change more easily than one designed specifically to meet individual scientist's present demands. As Bob Venturi shows later in this volume, our Lewis Thomas lab is supported by a series of repetitive structural bays scaled to the dimensions of lab modules. On the facade, windows serving the labs echo the rhythm of the bays. This generic, loftlike building form is not too different from that of Nassau Hall, Princeton's traditional college hall that, over two centuries, has seen several cycles of change in use. A third method is by leaving space for expansion: For his Superconducting Super Collider lab, Moshe Safdie has designed a series of separate and parallel building elements that can in the future be extended along defined channels of growth.

In the city, one way to plan for change is by separating subsystems—for example, public transit from private automobiles—so each can alter or extend its use pattern without disturbing the others. (Such separations are not necessarily desirable on other counts, or at all scale levels in the city: Urbanism is not simple.)

Planners should recommend short-term policies based on acceptably reliable projections and predictions; for long-range decisions, or those that could have long-term implications, they should plan to leave as many options open as possible. The goals of long-range planning should shift in response to new information, some of it derived from monitoring the impact of short-term policies.

Words to do with applying rationality to uncertainty and creativity—"stochastic" and "heuristic," for example—are seldom used by the would-be scientists who hound architecture, but they appear in Lewis Thomas's writing. Thomas had wanted to be a poet and writer, then found what he wanted to write when he became a scientist. In *The Lives of the Cell* he's a poet-scientist, a poetic Andy Rooney for medicine, a literate Tom Wolfe for the sciences.

That's a good way to be for people who have a specialist profession but like overarching concepts, too. They love art, they love science, they have a way with words; they fight to bring the whole together. I think that's what Arnie Levine said he was hoping to do in the Lewis Thomas lab. It sounds like a wonderful enterprise, a focus for both scientific and artistic creativity.

Notes

1. Johanson, Donald, and Maitland Edey. *Lucy, the Beginnings of Humankind.* New York: Warner Books, Inc., 1981, pp. 18, 285–289.
2. Thomas, Lewis. *The Lives of a Cell.* New York: Bantam Books, 1974, pp. 117–140.
3. Davidoff, Paul, and Thomas Reiner. "A Choice Theory of Planning." *American Institute of Planners Journal* (May 1962): 103–115.

V

Princeton after Modernism:
The Lewis Thomas Laboratory
for Molecular Biology

17

Thoughts on the Architecture of the Scientific Workplace: Community, Change, and Continuity

Robert Venturi

Introduction

I should first warn you that I come to this subject of relationships of architecture and science as a practicing architect.

—My generalizations are really pragmatic responses to everyday experience.

—And, at that, to everyday experience squeezed into the life of the fin de siècle architect. A life that allows little time for generalizing, as he or she works—simultaneously—as lawyer, business administrator, salesman, socialite, world traveler, psychiatrist. As referee, among committees of bureaucrats who relish meetings over production; among consultants galore, each requiring his part to be perfect at the expense of the whole; among historical commissions who fear making history in their time; among goody-goody community boards promoting deadening urbanity that very often derives from distorted versions of the architect's own ideas of twenty years before—all this while the architect somehow works into a process the details that God is in and good design entails.

I shall try here to use plain words to make sense out of pragmatic responses-as-generalizations-within-a-context-of-chaos, even though plain words and making sense are unusual—some of you may know—for architects of our time who measure the profundity of their theory in terms of the obscurity of their verbiage.

But if I go from the particular to the general in my everyday thinking, I shall attempt in this essay to go from the general to the particular,

—and focus—at the end—on some work of ours, particularly, the Lewis Thomas Laboratory at Princeton University (figures 17.1 and 17.2).

* * *

17.1

Lewis Thomas Laboratory, Princeton University. Exterior. Photo: Matt Wargo.

17.2

Lewis Thomas Laboratory, Princeton University. Exterior, end. Photo: Matt Wargo. See also color plate 7.

The Research Laboratory

As an architect, I write here about the Lab Building as architecture more than about science in architecture.

The most important thing, in my opinion, about the architecture of the Research Laboratory is that it is *generic*—and I shall focus here on the relevance of GENERIC ARCHITECTURE for such a building type, in particular—but also of its relevance in general for architecture for our time.

And my discussion of a kind of generic architecture concentrates on three of its characteristics:

—The element of FLEXIBILITY—spatial and mechanical—that is promoted inside,

—The imageries of SETTING & PLACE that are accommodated inside,

—The elements of SYMBOLISM & ORNAMENT—permanent and/or changing—that enhance imagery on the outside.

* * *

But before elaborating on these three elements that characterize generic architecture as I am defining it, I should briefly refer to the historic TRADITION that distinguishes generic architecture.

TRADITION—A WIDE-RANGING TRADITION THAT EMBRACES:

1. The *New England Mill:* and other industrial-factory kinds of buildings of the last century—where you find space as loft that flexibly accommodates changing functions and systems inside (figure 17.3). (The Fagus Shoe Works by Walter Gropius, arguably based on American industrial vernacular architecture and arguably considered the seminal building of the Modernist International Style, is significantly generic in terms of its aesthetic-symbolic reference as well as its functional program.)

2. The *Italian Palazzo:* where you find chambers en suite surrounding a cortile—starting as a dynastic residence and evolving into a civic building—as a diplomatic headquarters, museum, etc. (figure 17.4).

3. The *Early American College Building:* housing classrooms and dormitory that evolved into college or university administration headquarters—spatial and symbolic, highly symbolic—as at Harvard, Princeton, Brown, Dartmouth, William and Mary (figure 17.5).

17.3

Historical New England mill building.

17.4

Palazzo Rucellai, Florence.

17.5

Massachusetts Hall, Harvard University.

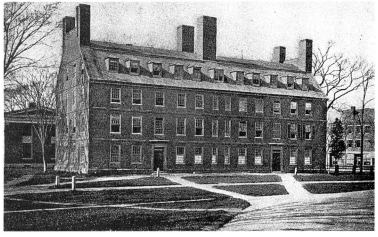

All of these generic prototypes contain adaptable/flexible spaces that evolve over time. This tradition includes as well:

> 4. The *Conventional Laboratory Building:* whose spatial and mechanical flexibility is particularly significant today for accommodating dynamic change inside, change that involves processes and technologies. The Conventional Lab Building whose tradition can embrace Thomas Edison's various sheds and lofts originally in New Jersey as well as Cope and Stewardson's turn-of-the-century Medical School buildings at Penn, whose stylistic Jacobean exteriors symbolize their academic dimension.

> 5. And there is the particular relevance of the *Generic Loft Building for the Architecture of Today in General:* where dynamic functions change over time at a quick pace and when architectural essence is becoming no longer spatial but iconographic. This point I shall come back to.

* * *

To return now to the elements of Flexibility, Setting & Place, and Symbolism, there is:

Flexibility—for today:

> Where no longer does form follow function. But that's not ambiguous enough—it is rather that *form accommodates functionS:* functions that are inherently changing, as they are complex and contradictory.
>
> > —Functions that are accommodated rather than expressed. This is more and more relevant in science buildings in particular and in architecture in general, to accommodate change that is more characteristically revolutionary than evolutionary and that is dynamically wide in its range: spatial, programmatic, perceptual, technical, iconographic. In our time, functional ambiguity rather than functional clarity can accommodate the potential for "things not dreamt of in your philosophy."
> >
> > —Functions that engage spatial and mechanical systems inside, and symbolic and ornamental dimensions outside.
>
> Where, generally, the work space is at the edges near the windows for enjoying the amenity of natural light and the view, and the mechanical space is in the center and at the top for maximum access.
>
> Where the scale of the architecture is physically generous, to create an aura of generosity as well as accommodate the dynamics of flexibility.

* * *

SETTING & PLACE—COUNTERPOINTS:

—For concentration: *Setting*.
—For communication: *Place*.

Setting—as Lab—along windows.
Place—as eddy—off corridor.

Setting—as working labs located along the edge to connect with natural light from windows.
Place—as niche, window seat, eddy—off the circulation route, possibly exploiting natural light as an attraction and an amenity at one end of the corridor (figures 17.6 and 17.7).

Setting—as background for work and focus—alone or with a group of colleagues.
Place—as opportunity for meeting—meeting that is incidental rather than explicit. (Academics are perverse: If your architecture explicitly pronounces a place for interaction, they might not use it; a certain ambiguity concerning the function and nature of the space is perhaps essential here.)

Setting—as more or less local.
Place—as signification of a whole (or perhaps a suggestion of a whole): a Community, an Academic Community within a campus—so you function alone *and* in a community.

Setting—a probably messy space: for the clutter of creative action, analytical, intuitive, physical.
Place—a mostly orderly space: for re-creation, so to speak.

Setting—within the consistent structural order—the generic order—of the loft, accommodating variety within order: spatial, perceptual, functional, mechanical.
Place—as an exception to the rule of the consistent order of structural bays, accommodating a special space.

Setting—accommodating change and dynamics in the lab: Wow, look at this!
Place—accommodating a permanent ambience where one anticipates the comfort of the familiar, but where there can be the surprise of the unfamiliar: "Wow! John, of all people, said something brilliantly relevant this morning as we chatted!"

Setting—neutral, recessive architecturally, to diminish distraction. Artists' studios are in lofts not essentially because artists are poor, but because they feel they can't create a masterpiece in someone else's masterpiece. A setting for inspiration *and* perspiration: artists and scientists are not priests performing rituals.
Place—imageful architecturally, to create amenity and identity.

17.6

Incidental meeting nook at end of corridor, Lewis Thomas Laboratory, Princeton University. Photo: Matt Wargo.

17.7

Sketch, exterior, showing end of building with communal niches. Project for laboratory for the School of Medicine, Yale University.

Setting—for table-distance work focus.
Place—for long-distance eye focus with a view.

A technical IRONY regarding the significance of Place for now: imageful Place that is local might be more essential than ever in our era of electronic communication to and from all over, in our era of networking.

An aesthetic IRONY is that accommodation to Place in this architectural context promotes a rhythmic exception within generic order and thereby creates aesthetic tension.

* * *

Symbolism & Ornament—that engages iconography on the exterior:

Symbolism—as the attendant flourish within Generic Architecture that eloquently breaks the consistent order—that consistent rhythmic order—on the outside (and some places inside), thereby creating aesthetic tension.

Symbolism—involving ornament, sign, iconography—on the outside. Traditionally evident, as in the cupola of Nassau Hall at Princeton, the *portone* with *stemma* of the Italian palazzo, the sign atop the mill, the hieroglyphics all over the Egyptian temple.

The difference between outside and inside—at least in the lab building, and especially for the academic lab building—is very relevant where architectural consistency and neutrality of the workplace inside is counterbalanced by explicit symbolic content on the outside that acknowledges the significance of the institution as a whole.

Another role of exterior *Symbolism* and *Ornament* in the generic Academic Lab Building is that which acknowledges, accommodates, and enhances the context of the architectural campus. And it is true that exterior symbolism on our lab buildings has so far derived more from accommodating particular architectural context than from general scientific sources (figures 17.8 and 17.9).

The significance today of iconography for a generic architecture as a whole, and the significance of electronic technology as the medium for iconography, I touch on below.

* * *

17.8

View of the Clinical Research Laboratory in its urban context from a distance, University of Pennsylvania. Photo: Matt Wargo.

17.9

Exterior ornamental wall pattern, from close up, Clinical Research Laboratory, University of Pennsylvania. Photo: Matt Wargo.

17.10

Exterior of the Richards Medical Laboratory Building at the University of Pennsylvania. Architect Louis Kahn. Photo: Julie Marquart.

WHAT THE GENERIC LABORATORY ACADEMIC BUILDING IS NOT

It is not an architectural vehicle for sculptural articulation that makes for expressionistic architecture that is heroic and original—that constricts flexibility and promotes distraction (figure 17.10).

Science and Technology & the Expression of Technology

This subject is the focus of work by other authors in this volume. I merely mention some issues briefly as I see them:

Expressing function stinks, as it inhibits change and encourages conformity. Accommodating functions is what works.

Modernism employed industrial engineering imagery as an architectural aesthetic, sometimes called the machine aesthetic, along with a Minimalist-Cubist-abstract aesthetic. It did this via an adaptation of the vernacular vocabulary of the industrial loft—essentially the American generic loft of the turn of the century—and this exemplified a wonderful/valid historical-architectural evolution (or revolution) formally acknowledged in this country when Dean Hudnut invited Walter Gropius to Harvard.

The Neo-Modernist movement in the architecture of today involves a revival of engineering expressionism more explicitly ornamental than before. It involves, in the end, an ironical architectural vocabulary based on industrial imagery as industrial *rocaille,* an imagery that is now around 100 years old and in our admittedly post-industrial age no more current or relevant than—and no less historical than—that imagery of the Classical orders of the Renaissance that are 500 years old. Everyone agrees that the Industrial Revolution is dead, but few architects acknowledge that Electronic Technology is what can be fundamentally relevant for architecture today. Electronic Technology, combined with generic order, can enhance, can indeed signify, an iconographic dimension—an iconographic dimension that is for *now* unlike the spatial-structural dimension that was for *then*—but an iconographic dimension nevertheless with a vivid tradition behind it.

So industrial and engineering/structural imagery of space is incidental for now, while an ornamental/symbolic imagery of applique is valid for now.

And perhaps historical precedents for a generic-iconographic architecture that is for now are the Egyptian pylon (figure 17.11) with its hieroglyphics; the American roadside architecture adorned with signs, commercial or evangelical (figure 17.12); Sant'Apollinare Nuovo in Ravenna, whose iconographic murals ornament a generic basilica inside; a Russian Constructivist project (figure 17.13); a Sullivan bank; or an electronic-architectonic feature of Tokyo. Let function, structure, and space take care of themselves efficiently and without fanfare, for scientific workplaces in academic communities and for architecture as a whole. Perhaps the conventional scientific laboratory within this definition is the prototype for a valid and vivid architecture for now as a whole.

* * *

17.11

Wall of ancient Egyptian temple.

17.12

American rustic scene.

17.13

Images of Russian Constructivist architecture.

397 | THOUGHTS ON THE ARCHITECTURE OF THE SCIENTIFIC WORKPLACE

Perhaps the conventional scientific laboratory within this definition is the prototype for a valid and vivid architecture for now as a whole.

* * *

The scientific laboratory buildings designed by our office that are represented here illustrate variously qualities of the generic loft, whose interior flexibility accommodates programmatic, spatial, and mechanical evolution over time and whose exterior ornamentation, within the consistent rhythmic composition of the loft, accommodates symbolic dimensions appropriate for a communal academic building. Exceptions to these forms of order deriving from incidental interactive spaces enrich the composition of the whole inside and out.

The designs of the buildings illustrated here represent work we have done in association with Payette Associates Inc. of Boston who are most significantly responsible for the major interior spatial-mechanical-programmatic elements of the architecture. And it is Jim Collins of that firm with whom we have had the pleasure and honor of working in the last decade.

18

THE DESIGN PROCESS FOR
THE HUMAN WORKPLACE

James Collins Jr.

Early in 1983, Princeton University deliberated its declining position as a forerunner in the field of the biochemical sciences. Its President, William Bowen, recognized that recent advances in the field of molecular biology were mobilizing modern science. Monumental advances like DNA mapping and genetic engineering would have implications not only for science and medicine, but also for the associated areas of health care and health care legislation. Clearly, if Princeton intended to enhance its reputation in the discipline of molecular biology, the university would require a new facility that seamlessly integrated design and technology. To this end, the university committed to the construction of a $29 million, 114,000 square foot facility intended to house a new department that would compete directly with those of the other educational institutions that had anticipated this "intellectual revolution" in science.

After a nationwide search, Dr. Arnold Levine and Dr. Thomas Shenk were recruited from the State University of New York at Stonybrook to head up the new Department of Molecular Biology. In keeping with Levine's goal to create a "community of scientists" at Princeton, a site for the new facility was chosen for its proximity to the other scientific disciplines—biology, geology, physics, and mathematics—in a conspicuous and distinguished locus on the Princeton campus.

One of the primary objectives of Princeton University in constructing the Lewis Thomas Laboratory was to lure a team of the field's most prominent research scientists to the school. Ideally, the university hoped to achieve its goal by offering the potential faculty a state-of-the-art research and teaching laboratory. The new laboratory would challenge Princeton University's conservative construction policy by promising to be one of the largest buildings on campus, and undoubtedly one of the most expensive.

Because of the urgency of the project, the traditional method for hiring architects was abandoned, and the schedule was accelerated. The responsibility for hiring architectural and design firms at Princeton fell to the Office of Physical Facilities. Typically, the office would compose a short list of appropriate design firms and then manage a selection process in which representatives from the particular department to be housed in the new building would evaluate these firms and forward a recommendation through the President's Advisory Committee on Architecture to the University's Board of Trustees, which would ultimately confirm the hiring of the proposed architectural firm. In an effort to accelerate this selection process, President Bowen proposed the marriage of two architectural firms that were already actively working on the campus, Payette Associates Inc. and Venturi, Rauch and Scott Brown (or VRSB, as the firm was known until it became Venturi, Scott Brown and Associates in 1989).

We at Payette Associates were the architects of more than 100,000 square feet of renovation and new construction in the Frick Chemistry Complex. We understood the requirements of designing laboratories at the university. We were also recognized as one of the nation's experts in the design of molecular biology laboratories. We had recently designed a new molecular biology research facility for Harvard University, the Sherman Fairchild Biochemistry Building, as well as a similar building for the Massachusetts General Hospital in Boston.

Princeton University alumnus Robert Venturi and his firm had successfully completed a critically acclaimed building on campus, Gordon Wu Hall, and had recently developed the campus design guidelines for the area of campus in which the proposed building was to be located. They were therefore an obvious choice to develop the new building's exterior.

At President Bowen's urging, Payette and VRSB were invited to submit a proposal and were chosen to lead the project. Payette, because of its expertise in technical buildings, would develop the program and design the interior spaces. VRSB, with a budget of $2.5 million, would work with the facade and site plan, integrating the new building with its traditional, neo-Gothic surroundings. Each organization had a clearly bounded sphere of influence. As Thomas Payette put it, any design on the outside was VRSB's ultimate decision, and design on the inside was Payette's ultimate decision. Payette Associates would be the architects of record and would have overall responsibility for the project management and documentation. By the very nature of this collaborative endeavor, teamwork was stressed over individual inspiration.

It was determined that the building would be developed around the "generic" laboratory vision of the new chairman of the Department, Arnold Levine. To establish a baseline from which to develop a common vision of the building, Payette Associates invited Levine and Robert Venturi to tour the Sherman Fairchild Laboratory at Harvard.

The Sherman Fairchild Building was a good place to start because it physically embodied many of the fundamental planning and design concepts that Payette promoted. All the laboratories were located along the exterior of the building, maximizing natural views and natural light. The building was organized around a simple rhythm of generic lab modules, expanses of glass within the building brought exterior light into interior spaces and visually connected spaces and people, and a large amount of natural wood humanized the scientific environment.

Levine summed up his overall acceptance of this design approach when he told Tom Payette, "Why don't you just give me a building just like this!" The visit was also useful in addressing some of Bob Venturi's preliminary concerns. In his early study of the College Walk site, Venturi had recommended that the building's shorter end be located along the walk, on the assumption that windows would exist in the few office spaces and the lab itself would be a big windowless box. A study of many of the research laboratories in existence at that time might lead anyone to that assumption. With the realization that the preferred planning approach, which the department chair supported, allowed for extensive exterior windows, Venturi realized that a whole world of proportion and rhythm could be explored. The site visit was a success, as everyone now appeared to have a strong basis of common values from which to develop the design.

As the majority of the building's occupants were yet to be recruited, the program and the concept of the building organization had to be developed on a generic basis by a small team. Arnold Levine, his associate Tom Shenk, and I, as Payette project architect, formed the core of the programming effort. On a weekly basis I would fly to Arnie's laboratory at Stony Brook to discuss the project. Arnie had truly enjoyed the Fairchild visit and wanted to take many of its planning approaches one step further. Arnie, Tom, and I agreed that the Sherman Fairchild lab module was a good model with which to start, but that two major adjustments would be introduced to the module at Princeton. Since the completion of the Sherman Fairchild Building, Payette Associates had recently completed the construction of the Welman Research Building at the Massachusetts General Hospital. Although it embraced many of the same planning concepts as the Harvard building, it introduced others into the model. It did not employ a race track corridor system; rather, it made use of an implied corridor system, and the labs employed a large "open lab" concept.

This open lab concept was not new. The large open lab had been successfully used as early as 1965 in Louis Kahn's design of the Salk Laboratories in La Jolla, California, but Salk and a few other notable laboratories were the exception. At the time the Lewis Thomas Lab was being planned, the dominant approach was to design discrete, small laboratories reflecting the hierarchical nature of "senior scientists" and "junior assistants." Most scientific laboratories did little to encourage interaction among scientists, either through their organization or through their architecture: Offices were tucked away in labs, making them inaccessible and inconvenient, and inflexible laboratory modules reflected the philosophy of "server" and "served."

During the 1980s, Payette Associates and other laboratory design firms challenged this archaic model. As corporate and private funding for basic research in the life sciences dissipated, governmental sources of funding became increasingly common. Pressure to reexamine the marketability of scientific research increased, and governmental organizations like the National Science Foundation liked to see their dollars funneled into potentially lucrative research projects, often carried out at new, multidisciplinary science and technology centers. As a result, scientists were interacting more than ever with their peers, with scientists from other disciplines, and with the business community, in an effort to "cross-pollinate." Impeding this interaction, however, was the architecture of the standard laboratory design: Protected areas and containment facilities kept researchers physically isolated; hard, sterile work surfaces discouraged sociability; and a highly controlled mechanical environment similarly set up barriers.

Arnold Levine emphasized the importance of discussion and interaction to the successful realization of molecular biology. In the formulation of new strategies, molecular research scientists tend to rely on the input and free exchange of ideas from their colleagues and students. Essentially, they act and react as an interdisciplinary and interdependent organism. Given the framework of this communal relationship, a design approach was established that demanded wide open expanses, or the semblance of such, to facilitate interactive behavior. We believed that, through our manipulation and demarcation of space in the laboratories, an environment that encouraged growth and scientific activity could be created.

We have maintained that the success of the Lewis Thomas Laboratory is largely grounded in the free communication that evolved among the university, Arnold Levine, Payette and VRSB—a veritable symbiotic relationship that prevailed throughout the length of the project. The team functioned in a manner strikingly similar to the behavior we hoped the design would support. We acted and reacted as an interdisciplinary and interdependent organism. Discussion and interaction became essential for the successful realization of our concepts. In formulating new strategies, we relied on the input and free exchange of ideas from our colleagues.

The project, proposed in March 1983, was slated for completion and occupancy by October 1985. Because of the abbreviated schedule, project meetings with the owner, scheduled every two weeks, helped to refine the building's functions and goals. Although there were technical constraints, we continued to pursue the goal of a structure that provided ample light, windows, open spaces, and spaces for reflective study. By May 1983, the building had developed a very straightforward and unassumingly rectangular shape. The rationale was simple: The laboratory had evolved as a type of building whose exterior appearance is derived from the nature of its interior functions. In essence, specific requirements of program and interior layout delineated its facades in terms of scale, rhythm, and proportion. Arnold Levine had suggested that, in his experience, a three-story building seemed to allow for the most interaction among groups. Although two or four floors were also deemed acceptable, three floors also seemed ideal for the number of required shared facilities. Levine envisioned five research divisions, distinguished by the biological material with which they worked: higher eukaryotes, prokaryotes, viruses, lower eukaryotes, and plants. (See figure 18.1.) Lower eukaryotes and prokaryotes were

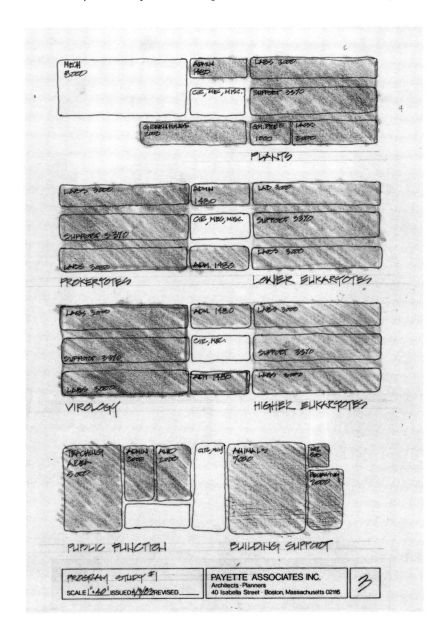

18.1

Color-coded early program study for the organization of space in the Lewis Thomas Laboratory for Molecular Biology at Princeton University. Five research divisions were defined, distinguished by the biological material with which they worked. See also color plate 8.

not closely related to any specific support facility and thus could be on their own. The smallest of the groups, plants, could also be located in a more remote location, since it had no requirement for being in close proximity to either the animal facility or to the animal research groups. In spite of this segregation, we wanted to create an environment in which no one would feel isolated, even though his or her biological concentration dictated a separately controlled environment. This goal was accomplished through the use of glass instead of solid walls, creating a visual connection that promotes awareness of fellow researchers. In addition to the sense of connected space produced by moveable fixtures and glass windows between labs and internal support areas, this technique also provided maximal diffusion of natural light throughout the structure. To enhance further the human element in a highly technical laboratory, natural materials such as oak were chosen for all wainscoting, doors, furniture, and casework.

By July 1983, we had a firm breakdown of space, furniture, and equipment needs. Clarifying these internal relationships finally freed VRSB to plan the exterior and approach the siting constraints for a rectangular building. In short, the character of the exterior was approached as a response to the simple rhythms and proportions of the floor plans. The highly ornamental patterned surfaces of the long facades evoke a sense of history. Robert Venturi is notably influenced by Elizabethan manor houses and New England mills, and both served as prototypes for the overall appearance of the building. As Venturi has remarked, "Architects have traditionally used symbolism in architecture to enrich its content and to include other dimensions, some almost literary, which make architecture a not purely spatial medium."[1] The distinctive patterning, this time in brick and stone to reflect Princeton's traditional building materials, would come later.

The building was planned to have three stories in the front and four stories in the back, reflecting the slope of the site down to the south. The two long elevations would indicate a consistent arrangement of identical bays, both inside and out, via repetitive window openings. These openings had a deep reveal, or wall thickness, to accommodate a mechanical zone on the exterior wall as well as to provide natural sun shading, and to underscore the overall aesthetic tone, which complemented that of the surrounding buildings. These facades were determined to utilize varying orders of scale; large to reflect the building's institutional quality, resulting from the height needed to accommodate the mechanical systems, and small to relate to the users' perception of the building. In its overall design, the building was highly repetitive and orderly. However, inconsistencies were intentionally introduced into the design to add visual contrast and a sense of character. Following some deliberation, it was decided that the entrance and lobby would be asymmetrically placed in the front facade in reaction to the placement of shared core support facilities and to relate to the landscape design of the neighboring Guyot Hall. VRSB chose to manipulate the building's rectangular shape and reduce its blockiness by softening the edges of the end elevations. VRSB's desire for an exception in each of the end elevations provided the perfect opportunity to develop small informal lounge spaces that would foster and support a variety of activities. In turn, the positioning of the lounge areas at these localities would have further implications for VRSB's ornamental masonry and was ultimately to assume a far greater sense of importance when it was realized that the building is typically approached obliquely.

The building was to be situated parallel to College Walk, the university's main pedestrian path, to reinforce the walk's spatial identity. This created an entrance from the facility to the main campus on Washington Road from the south and visual identification with the Guyot Hall complex to the north. VRSB's intent was physically and symbolically to connect the Guyot complex and the molecular biology building through both the new building's siting and its landscape design. Here, we attempted to create a sense of amenity in the space between the two buildings through landscaping, which involved terracing, paving, planting, and modifying the configuration of College Walk at this end. Although a low retaining wall was created, the resulting formalization was tempered through a design that utilized a maximum of lawn and a minimum of paving in the tradition of the Gothic residential courts at Princeton University. In effect, the resulting eastern end of College Walk would contrast with the nearly fully paved Butler plaza at the western end.

By February 1984, it was understood that each level of the new building would measure approximately 278 × 85 feet. A 1,000 square foot central meeting room, numerous lounge areas, and an outdoor patio, as well as laboratories, classrooms, and offices would house the 220 occupants of the building. The ground floor would possess a large and a small seminar room, teaching laboratories, and mechanical and electrical equipment areas. Additionally, a gallery would occupy the western section, and animal cage rooms, workrooms, offices, and general storage rooms would be located in the eastern section. The penthouse area, encompassing nearly one half the width of the roof and running its full length, would consist of five greenhouses, mechanical equipment rooms, and small laboratory rooms. A loading dock that offered storage rooms for gas cylinders, chemical waste, and solvents would be located outside along the building's east wall, adjoining the ground-floor level.

Because this building would involve research in a rapidly progressing field, the goal of planning for unpredictable change continually challenged our design approach. Indeed, in any laboratory design, planning for the future is nearly as important as is planning for the present. Since molecular biology is evolving almost daily, there is constant pressure to adjust to ever changing standards and trends. Understandably, because a great deal of flexibility was expected in order to meet these unforeseen future challenges, these attributes were fundamental to the project's viability. For example, Princeton expected at some point to convert some perimeter laboratories into high-containment, high-equipment-density laboratories. To accommodate this need in the particular laboratories identified for conversion, dual supply and exhaust boxes were specified to handle increased cooling load requirements, even though the possibility remained that the univer-

18.2

Typical open lab space in the Lewis Thomas Laboratory for Molecular Biology at Princeton University. The ceiling equipment was left exposed to expedite access to and maintenance of pipes, ducts, and cables.

sity might decide against this conversion, if the space became earmarked for another future use.

The laboratory module was also designed to accommodate additional piped and electrical services, enabling easy conversion of an instrument space, for example, to a biochemistry laboratory, or a seminar room to a physics laboratory. Planning with such a strong degree of forethought usually proves invaluable, both to the client and the designer. Typically, during the programming phase, however, certain palpable factors become apparent that do restrict a design's flexibility.

There were initial problems associated with the placement of certain specialized mechanical rooms, such as the biological containment facility, where infectious agents would be isolated. We placed it at the core of the facility, with all user spaces on the periphery; thus, its placement delineated the layout of the lobby and ground floor. Our reasoning for this position was initially derived from the simple premise that humans react favorably to their connection with the outdoors through window exposure and associated views. Hoods would be on the corridor side of the laboratory to allow for desk and work space placement at the perimeter. Ceilings were conspicuously absent from these laboratories and main corridors. By this time, our goal—an overwhelming sense of light and space, enveloped by technical precision—was nearly achieved.

The facility with which the mechanical systems would be maintained also became a priority. Because many mechanical systems in laboratories precisely control often hazardous environments, these systems are quite complex, and they require expert maintenance. The people who customarily service these systems face subsequent difficulties when attempting to approach these systems as they would any other. To avoid dangerous situations involving either themselves or the laboratories' inhabitants, either these technicians need to be highly specialized, or the systems' maintenance needs to be simplified. As David Rowan, the principal in charge of the project, related, "I think that maintenance is a big problem. Very few institutions have knowledgeable maintenance departments that really know how to adequately maintain the equipment. You try to keep things as simple as possible."[2] One of many ways we attempted to adhere to this strategy was by designing the hot lab without a hung ceiling to facilitate access to the charcoal filters that purify the air. The requirement to make filters easily accessible was not based solely on the frequency with which they would be changed. In fact, charcoal filters can be quite heavy, and attempting to manipulate them while one is perched on top of a ladder can be difficult and dangerous. As a result, longer intervals may elapse than are appropriate between changes. For low maintenance and ease of reconfiguration, an open ceiling with exposed pipes, ductwork, and a data network cabling tray was chosen. (See figure 18.2.)

Because a full 25 percent of the Lewis Thomas Laboratory was dedicated to the associated mechanical systems, the placement of these systems in numerous instances had to be scrupulously negotiated. Educated guesses informed these negotiations as much as did

factual realities. There were unforeseeable equipment requirements: Were lasers to be used? If so, could our proposed cooling system handle this future load? Other mechanical issues, such as this building's high ventilation rate requirements, also presented challenges. In the animal area, ventilation systems were designed to operate at full capacity twenty-four hours per day. When the space requirements for this facility were in the planning stages, the animal care committee members believed that the present animal population would be exceeded in the future. If the population were to be expanded by, say, 50 percent, we needed to insure that the proposed cooling system could handle this additional strain. In another of many examples, the degree of acidity in proposed calculations of chemical waste was unknown. R. G. Vanderweil Inc., the mechanical engineering contractor, understood that there would be some acid waste that would require treatment, but the quantity remained uncertain. We ceaselessly searched for the answers that eluded us. What type of allowances needed to be made? Would an increase in the acidity create additional problems in the storage of this hazardous waste? Had these storage facilities in the loading dock area been designed to handle an increase in materials? At the same time, we needed to conform to the requirements of regulatory agencies such as the Environmental Protection Agency and the National Fire Protection Agency in planning for the disposal and storage of these hazardous and potentially flammable materials. As I earlier stated, we perpetually planned for the unknown future and direction of molecular biology. Judgments were usually made by tailoring our previous design philosophies to our current client's needs and special situations. Exhaust stacks were designed to accommodate a ten-foot increase in height, to provide an effective means for correcting unanticipated problems once the building was completed. Also, laboratories were exhausted individually through separate ducts to the penthouse before connecting to the central exhaust system. These ducts would provide for flexibility in accommodating future hoods.

Throughout the entire design process, allowances needed to be made, predictions ventured, and safety issues assiduously addressed. Whereas some elements of the program appeared to have been fixed through program space requirements, in actuality, the planning was structurally tight, while allowing for a great deal of flexibility. Space above each lab was planned to allow the mechanical and electrical systems to be moved easily and installed elsewhere without disrupting the lab below. Large, open lab spaces could be reconfigured and subdivided in the event that existing laboratory needs changed. Because much of the electrical and utility space was run up through shafts at either end of the loftlike laboratories, rather than up through each individual laboratory station, a great deal of flexibility was incorporated.

Admittedly, although the creation of a viable and effective laboratory involves numerous technological constraints, at Payette Associates, we know that research is ultimately about people. People like choice, thus we sought to provide a variety of different kinds of spaces within the building: closed, quiet spaces for contemplation and individual work; open public spaces for spontaneous activity and discussion outside the laboratories; and research space that also encourages the continuous exchange of information between investigators. A generous staircase, for instance, generally invites exchanges between and among floors, as people constantly pass each other and relate their laboratory experiences. (See figure 18.3.) Traffic patterns can be tightly controlled when a single corridor functions as a main thoroughfare. At the Lewis Thomas Laboratory, offices were grouped in clusters rather than in separate laboratories to reinforce the strong sense of community felt among the members of this interdisciplinary group. Blackboards were strategically placed to invite spontaneous interactions and impromptu gatherings. In essence, by manipulating the frequency with which researchers exchange information, architects can effectively promote the sharing of knowledge through the sharing of space, resources, and facilities.

18.3

The central staircase, connecting the ground and first floors of the Lewis Thomas Laboratory for Molecular Biology at Princeton University. Students, faculty, and staff all come together and mix within this space throughout the day. See also color plate 9.

18.4

Central foyer and stairwell on the first floor of the Lewis Thomas Laboratory for Molecular Biology at Princeton University. The extensive use of glass brings "borrowed light" well into the building's interior. Here, light from the foyer and the distant conference room is brought into the enclosed stairwell, encouraging passers-by to stop and converse on the lower landing. See also color plate 10.

Researchers and students at the Lewis Thomas Laboratory appreciate the building for simple reasons: its flexible laboratory spaces and its sense of openness, with ample natural light and generous spatial opportunities that encourage commingling and casual conversation. (See figure 18.4.) They like the lab because it feels like an academic building, not a spaceship. They like the oak trim. They like it because they have to spend most of their lives there and it is a nice place in which to live. On this level we architects and these scientists are very much the same. We are interested in discovering, understanding, and influencing life.

In recent years, architectural theorists have disagreed over whether or not the social behavior of a building's users is influenced, even determined, by the physical environment in which that behavior occurs. Proponents of this influence—architectural determinists—believe that designers can direct social behavior through their work. Using the Lewis Thomas Laboratory at Princeton University as a case study, we can positively attest that there is a direct correlation between the work environment and the workers' intellectual and physical activity. Furthermore, a sense of order, continuity, and cohesive structure are all expected to have a positive impact on the way scientists relate to their surroundings.

We can, as architects, through our definition and manipulation of space, create a positive and nurturing environment for research scientists. Spatial planning can foster the paradoxical factors inherent to the research laboratory: innovation and replication, discussion and reflection, teamwork and competition.

Through our design of the Lewis Thomas Laboratory, we at Payette have demonstrated the significant influence that human behavior must have on the environmental planning and design professions. Many in the field continue to view research labs as highly controlled environments supported by intense, space-consuming mechanical systems. They cite examples where major science has been accomplished in the most inhospitable of places. Given our own experience in building for the scientific community, we are inclined to believe that the very opposite is true. Perhaps Dr. Jonas Salk, who discovered the polio vaccine and founded the research institute that bears his name, summed up our theory most accurately when he discussed Louis Kahn's design of his institute:

> My ambition was to optimize the functioning of the human mind, to deal with the issues and questions with which the human mind is concerned. I wanted to create something that would influence the realm of the mind—the minds of those who would gather here to carry on this kind of work. I was seeking a retreat atmosphere for reflection and work, away from the business and noise of the world. . . . Architecture is used here. Some people pursue science for human use, in contrast to science for the sake of science. This architecture is for human use, to serve a purpose.[3]

Notes

1. Robert Venturi, "Diversity, relevance and representation in historicism, or plus ça change . . ." *Architectural Record* 170 (June 1982): 114–119, quote on p. 115.
2. Conversation with the author.
3. Salk is quoted in Michael J. Crosbie, "The Salk Institute," *Progressive Architecture* (October 1993): 47.

19

Life in the Lewis Thomas Laboratory

Arnold J. Levine

The Lewis Thomas Laboratory, home of the Department of Molecular Biology at Princeton University, took shape and came about by a series of extraordinary coincidences, much like the evolutionary origins of any life form. The time (1983–1986) was right for Princeton University to raise the funding for a new department and a new building. The major donor, Laurance Rockefeller, class of 1932, wished to name the building for his good friend Lewis Thomas, class of 1933. Rockefeller appreciated the contributions of molecular biology to medicine because of his position on the Board of Trustees of Memorial Sloan-Kettering Cancer Center. Princeton's president, William Bowen, wanted to complete an ongoing and successful fund-raising campaign by correcting an old set of problems in the life sciences. Bowen realized he had to do something special to turn around a slowly wilting biochemistry department at Princeton. One tangible way to accomplish that was to build a new, state-of-the-art teaching and research facility to be designed by Robert Venturi, class of 1947. By 1984, it was clear that the intellectual revolution in the biological sciences, led by molecular biology, was going to have a major impact not only on the way medicine would be practiced in the twenty-first century, but on society as a whole. Any university that wished to remain in the forefront of intellectual endeavors had to have real excellence in the newest areas of biology. Bowen sought the marriage between the architectural firms of Payette Associates Inc. in Boston, with a good deal of experience in laboratory design, and Venturi, Rauch and Scott Brown of Philadelphia (or VRSB, as the firm was known before it became Venturi, Scott Brown and Associates in 1989), a successful union that has since resulted in seven additional science buildings. The Molecular Biology faculty design team worked closely with these two sets of architects, a Mr. Inside (James Collins, Jr., of Payette) and a Mr. Outside (Robert Venturi). But before all this could come together, William Bowen, Neil Rudenstine (then Provost), Aaron Lemonick (then Dean of the Faculty) and Walter Kautzman (Acting Chair of Biochemistry) recruited Tom Shenk and myself to head the new Department of Molecular Biology, help design the building, raise funds, recruit new faculty, and create a new department in the life sciences at Princeton. The economy, the science, the people, the architectural teams, and the donor, who gave such a special name to the building, all came together to create a remarkable research and teaching center, named for a scientist-humanist and placed at a university that has been called "the most unspecialized of American universities."[1]

This paper has two goals: first, to tell the story, in a personal and reflective mode, of the events that led to the Lewis Thomas Laboratory and its design requirements; second, to answer some questions about our years of research and teaching experience in that laboratory. How does the physical space, the building, affect the way we do science? Can a building actually change the science we do? Would we have been different scientists in another building?

In the spring of 1983, I had already accepted a position at Princeton University to begin in September 1984. I was then chair of the Department of Microbiology at the Medical School of the State University of New York at Stony Brook, and I received a phone call from Robert Venturi, who informed me that he was the architect for our new building and he wanted to visit with us at Stony Brook. He wanted to spend a few hours in a laboratory and my office to observe and talk about design. The phone call triggered in my memory the last time I had met a well-known architect and discussed his design of a building.

I was a first-year graduate student in January 1961 at the University of Pennsylvania, and we had just moved into a new laboratory, the Richards Building. (See figure 17.10.) I was assigned a bench at a corner laboratory on the second floor. The room was composed of bench-to-ceiling windows on two sides (a southern and western exposure) and sinks and cabinets on the other two sides. There was no place for laboratory equipment, refrigerators, or even desks for the students. But worst of all, from 8 A.M. until dusk, the sun intruded into the laboratory, melted the ice in our ice buckets, heated up the room, and in the bad times, impaired our vision. The graduate students responded swiftly and efficiently. We covered the higher portions of the windows with newspapers. As time went by, these spaces became our bulletin boards, with comics, headlines, and articles of interest protecting us from the sun. One morning when we arrived at the lab, all our newsprint and protection was removed. Building engineers warned us, in the sternest terms, that the architect did not want newspapers covering the windows. After a short meeting, we responded by using laboratory silver foil to cover the windows, on the premise that it was more artful and hopefully acceptable. Our timing was extraordinary, a whole year before Andy Warhol would cover the windows of "The Factory" with silver foil. The next time I arrived in the laboratory it was to see and meet a short man, standing on my bench, removing the foil with a large ruler. He explained that he was Louis Kahn. He had designed the building and was not amused that we chose to alter its appearance, even for our comfort. Kahn, of course, was to go on and design the laboratories at the Salk Institute, which are clearly among the best research spaces in the United States. But for me, I carried the memory of an architect standing on my bench and designing a work space—which was not to be altered by the workers. Fortunately we came to an accomodation, when fine screens to shade the sun were installed on the windows of the Richards Building. Over the years, though I cherished the memory of meeting and receiving a lecture from Louis Kahn, I was not sure what to expect from my next encounter with an architect.

When Venturi came to Stony Brook, he and I talked about the building I was working in there. A monster, eight floors of solid brick walls on the outside, rising like an impenetrable object to dwarf a person approaching it. The brick wall allowed no windows in the laboratory. The only windows were at the end of each hallway and in the offices for the faculty (not the students). It seems I had gone from too much sun to no sun, and I was ready for a compromise. Venturi spent a couple of hours in the laboratory, watching students and postdoctoral fellows carry out repetitive tasks, moving incredibly small volumes of liquid from one tube to the next, heating it for four minutes, cooling it in ice baths, placing it on filters, drying the filters, and then placing hundreds of filters in glass vials. He noticed on each wall there were pictures of windows that looked out on soft landscapes. He noted that there was often a great deal of time between experimental procedures, a time to discuss ideas, results, problems, and sports. Venturi spent two hours in my office, observing the endless meetings of academics and the excitement of a student discussing her results and what they must mean.

Out of this came some observations that were to have a major impact on our program for design. Molecular biology is a blue-collar science. It is labor intensive and involves repetitive sets of action. It employs people from all socioeconomic groups, equal numbers of males and females. It is time consuming, with students working long hours, seven days a week. Experiments often fail, and it takes persistence to get them to work. The science is, by and large, done by young graduate students and postdoctoral fellows. They have the stamina, aspirations, and drive to be productive and contribute a fresh perspective. The long hours are punctuated by time waiting for something to incubate or develop. It is during those free hours when communication, criticism, and new insights are developed and the future is planned. There is a need for spaces to hold each of these activities. We asked for the following in our building: lots of light—windows, views, an open, warm feeling, and an atmosphere where people would enjoy working twelve to fourteen hours a day; spaces to communicate and talk about our science; a horizontal building—long and low—where people could communicate with others on the same floor (six or seven research groups per floor); combined teaching and research facilities in one building, reflecting the philosophy of Princeton University; faculty offices adjacent to their laboratories, and blackboards everywhere—in the halls, labs, lounges—to promote communication; the creation of spaces to enhance communication between groups, between floors (vertical communication); and a central office and lobby, a gathering place for everyone, for cookies and coffee, at 3 P.M.

The building's shoe box shape and its repetitive modular format (both outside and inside) remind one of the New England factories where long hours of repetitive work were carried out. The windows, outside and inside the building, bring borrowed light into all the building spaces. From a hallway deep inside the building, one can look in any of four directions and see light coming from outside windows. There are no ceilings in the laboratories or hallways, giving work spaces an open feeling. There are two lounges per floor with two walls of blackboards and built-in seating. A kitchenette is hidden in each lounge. The building has one central staircase, used by all for vertical traffic. (See figure 18.3.) This staircase has large open landings decorated with art, where people can stop to talk, and it is encased in glass and oak, extending the open spaces and light to this vertical communicator. Hallways, lounges, stairways, all with blackboards and places to talk outside the laboratories, are bright and open spaces. The ground floor houses teaching laboratories, a large and a small lecture hall, and a wall of glass, bringing light in from a pleasant patio with a sitting wall. The use of oak throughout the building for all wood trim and to frame all the inside glass windows provides the warmth and open feeling that we requested.

Payette and VRSB designed the Lewis Thomas Laboratory with the input of the scientists who would inhabit it, and it works. (See figures 19.1 and 19.2.) It now houses twenty research groups, some led by junior faculty with two to four people in the lab, and other groups as large as twenty to twenty-five researchers in a team. There are six or seven research groups (faculty) per floor, organized so that individuals who work with similar organisms and use similar technologies (have similar needs) are clustered on the same floor. On the third floor are the microbial genetics groups using microorganisms (bacteria and yeast) to probe the problems of cell biology and developmental biology. On the second floor are research groups studying the control of cell division, cancer, and viruses that cause cancer. On the first floor are several groups that study developmental biology and the genes that control development, utilizing the mouse as their experimental organism. On this same floor are groups that study how proteins are assembled into structures, forming the three dimensions of a cell. The building's physical organization reflects an experimental or functional organization, maximizing horizontal communication between groups that share experimental approaches and similar resources, thus encouraging collaborations. The lounges at the ends of each floor have become centers for the exchange of ideas within and between groups. They provide a space for lunch and coffee breaks, as well as a place for informal lab meetings. (See figure 19.3.) The building maximizes communication, and that fact alone surely affects the way we do science and the science we choose to do.

19.1

Researchers in the lab of Arnold Levine, on the second floor of the Lewis Thomas Laboratory for Molecular Biology at Princeton University. The biologists here are carrying out research on the p53 protein and the role that viruses play in causing cancer.

19.2

Students at work in one of the ground-floor teaching labs of the Lewis Thomas Laboratory for Molecular Biology at Princeton University. The proximity of student teaching labs and faculty research labs in one building reflects Princeton's larger commitment to maintaining an equal balance between teaching and research at the university.

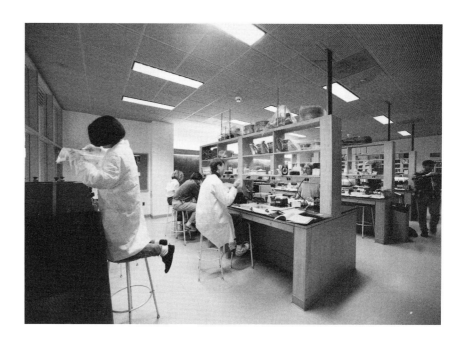

19.3

Members of Arnold Levine's laboratory research group informally discuss ongoing work on the p53 project in one of the second-floor lounges of the Lewis Thomas Laboratory for Molecular Biology at Princeton University.

From the beginning, when the Trustees of Princeton University agreed to commit $42 million to create the Department of Molecular Biology with twenty new faculty in a new research and teaching facility, they made it clear that this department was to have a broader mandate than just teaching and research in the sciences. The discipline of molecular biology will surely have a considerable impact on health care and related economic issues, such as biotechnology and health insurance. It will additionally address social and ethical questions concerning our genetic endowment and our right to change it. As we learn about life at the molecular level, we approach the answers to some of those questions that have been asked before by philosophers: Who are we? How did we get here? Why are we here? The name given to the department's teaching and research facility, the Lewis Thomas Laboratory, reinforced our ties to the humanities and social sciences. Few have explored the relationship between the study of life and the values we hold better than Lewis Thomas:

> Nature is a very strange affair, and the strangeness already encountered by our friends the physicists are banalities compared to the queer things being glimpsed in biology, and the much queerer things that lie ahead.
>
> As these turn up . . . they will inevitably change the way the world looks. And when this happens, the view of life itself will also shift; old ideas will be set aside; the look of a tree will be a different look; the connectedness of all the parts of nature will become a reality for everyone, not just the mystics, to think about; painters will begin to paint differently; music will change from what it is to something new and unguessed at; poets will write stranger poems; and the culture will begin a new cycle of change.
>
> What I hope for, and expect to happen, is that biology will fit neatly into the humanities, and that all the human studies—perhaps with luck even including linguistics which is after all the most biological puzzle of all, the species-specific mark of our species, along with music and the human tendency to ask incessant questions that have nothing to do with the matter at hand—will one day become the closest intellectual neighbors of biology, with the life of the mind at the center of the life sciences.[2]

The department that began in 1984, as a zero-based entity benefiting from the coincidences of its origins, is now an adolescent. To be sure, it has grown: by 1994, to a faculty of 33, in three buildings, with 114 graduate students, 122 postdoctoral fellows, 168 undergraduate majors, 64 technical staff, and 31 administrative staff. It has all the problems of adolescence, a too rapid growth rate and lots of changes to accommodate to, but it has the advantages of youth as well. No description of the Lewis Thomas Laboratory would be complete without imparting a sense of the science being accomplished in that building. The research, carried out by undergraduates, graduate students, and postdoctoral fellows, is the heart of this building and over the years, important things have been accomplished. As an example, I have chosen one story to tell. It happened in my own research group, and so I know it well.

In 1979, my research group at Stony Brook was working with a virus that could cause cancer in rodents. Viruses, the smallest and simplest of living things, were excellent model systems to understand the origins of cancer. A virus, with as few as five or six bits of genetic information or genes, should permit us to dissect which of the viral genes encodes the information to cause a cell to continue to divide in an uncontrolled fashion, that is, cancer. By 1979, we knew that a single gene of that virus produced a protein, called a tumor antigen, that caused the cancer. The question then became, how can a single protein, the tumor antigen, redirect a cell to divide and form a tumorous growth? One of my graduate students, Dan Linzer, found that a cellular protein, whose size he measured at 53,000 molecular weight units, tightly bound to the tumor antigen and communicated with it in some way. This led to the hypothesis that the tumor antigen, which causes the cancers, acts through the protein of 53,000 units, called p53, to cause the cancer. To test this idea, it was useful to isolate the gene from mouse cells that contained the genetic information for the p53 protein. In the parlance of the day, we needed to clone (isolate in pure form) the gene for p53. This task fell to a postdoctoral fellow in my group, Moshe Oren (newly arrived from Israel after Dan Linzer had graduated). Now the repetitive work began. It was necessary to search through thousands of gene clones to find one that produced p53 protein. Although we could enrich the pool of genes that contained p53 information from one part per 1,000,000 to one part per 1,000 or 10,000, the search took years. In 1983, we found part of the gene, and in 1984, we had a complete copy of the genetic sequence of information for this p53 protein and the gene. When we parted in 1983, Moshe Oren returning to start his own laboratory in a new position at the Weizmann Institute for Science in Israel, and I moving to Princeton from Stony Brook, Moshe and I agreed to become friendly competitors in the race to find out how the p53 gene and protein caused cancer.

By 1984, Moshe Oren had shown that the information or gene copy for p53 derived from a mouse in Israel could be introduced into rat cells in a culture (petri) dish and change these cells from normal to cancerous. Here was direct proof that the p53 gene copy could contribute to a cell's transformation from normal to cancerous. My own research group at Princeton had isolated a copy of the p53 gene from a different mouse and also tested it for the ability to cause transformation of normal cells into cancerous ones. A postdoctoral fellow, Cathy Finlay, and a graduate student, Phil Hinds, in my group had teamed up to compete with Moshe Oren and get the answer. The problem was, we got a different answer. Our clone of p53 genetic information failed to change normal cells to cancer cells. Moshe Oren's group, with the positive result, published his work; our lab, with a negative result, had little to publish and much to talk about—why we failed to find this property. Being friendly competitors, we shared our copies (clones) of p53 genes with each other and two things rapidly became clear: (1) We had no trouble repeating Moshe Oren's observation: His clone transformed cells, our genetic clone failed to transform cells; and (2) There was one chemical difference between his gene and our gene copy, one genetic change. One of us had a mutation, a genetic change, while the other had a normal copy of mouse DNA. For Cathy Finlay and Phil Hinds, that meant the need to discriminate between two hypotheses:

1. We had the normal gene that doesn't cause cancer, and Moshe had a mutant gene that does cause cancer, or
2. We had a mutant gene that had lost the ability to cause cancer, and Moshe had the normal gene which, when overexpressed in abnormally large quantities, causes cancer.

Examples of both types of oncogenes were well known in 1987–88 as Cathy and Phil debated ways to distinguish between these ideas. In 1989, Cathy and Phil published two papers that changed everyone's ideas about the p53 gene. The normal p53 gene actually prevents cancer. It stops other oncogenes from causing cancer in cells and even in animals. Moshe Oren's gene was a mutation that changed p53 from a gene that prevents cancer, a tumor suppressor gene, to a gene that causes cancer, an oncogene. This entirely new concept, born in the lounges and laboratories of the Lewis Thomas Laboratory, did not have to wait long for validation, nor did it wait to become central in our understanding of the origins of human cancers. In the same year, 1989, Bert Vogelstein and his collaborators at Johns Hopkins Medical School published a paper that showed mutations in the p53 gene in human colon cancers. The only way to interpret his results was to conclude that p53 was a cancer-preventing gene or a tumor suppressor gene, in complete agreement with Cathy Finlay and Phil Hinds. A year or two later, several families that inherit predispositions to some types of cancer were shown to harbor mutations in their p53 gene. The inherited basis for some of the cancers in humans (a small number of families) is due to this gene. By 1994, it had become clear that mutations in the p53 gene (in tumor cells and not the inherited type) are the single most common genetic change to be observed in cancer cells of humans. Almost 55 percent of all human cancers have genetic faults in the p53 gene of their cancerous cells.

Starting with a model virus that causes cancer in rodents, a protein was identified, a gene was cloned, a conflict in results and interpretation was resolved, a new concept emerged and its generalization to an important human disease was established. And that is just one of the detective stories that can be multiplied by the twenty research groups in the Lewis Thomas Laboratory. The postdoctoral fellows and students who have been trained in that environment have already gone out and populated medical schools, research institutes, biotechnology companies, and much more. The students of Lewis Thomas Laboratory are the future of our science.

Has this physical space changed how we do science, the way we do science? Surely! We collaborate and interact between research groups more than in the past. Part of that derives from the unity of life Lewis Thomas spoke about. We now find the same genes in bacteria, yeast, worms, flies, frogs, mice, and humans. That brings the first, second, and third floors together for joint meetings, shared graduate students and common research themes. Part of our collaborative spirit is clearly fostered by our surroundings where pleasant meeting places welcome the investigator to stop and talk. Has the physical space of Lewis Thomas Laboratory changed the science we might have done elsewhere? Most scientists are more reluctant to accept that premise. Because no experiment can be done to test this hypothesis, perhaps it is not a good question. But I suspect the answer is yes, for subtle reasons. I notice the students respond to the spaces—filled with oak, and glass, and light, and openness—in Lewis Thomas Laboratory with more respect and a standard or expectation of excellence and achievement. From 1986 through 1990, Lewis Thomas loved to return to his laboratory and talk with the students. He would sit in the lounges and see who wandered in to talk about the issues of the day, the problems in the lab, and the experiments that failed. And he would say:

> In my view, (that) is the kind of useful work a young biologist can now look forward to if he or she is willing to take a long, long look ahead and run all the risks in order to have all that fun.[3]

NOTES

1. Lewis Thomas, 1986. "A High Hope for Princeton." Speech given at the dedication of the Lewis Thomas Laboratory. Reprinted by and available from the Department of Molecular Biology, Princeton University, Princeton, N.J., 08544.
2. Ibid.
3. Ibid.

20

Two Faces on Science: Building Identities for Molecular Biology and Biotechnology

Thomas F. Gieryn

What is sociologically interesting about buildings for science? Possibly this: Built places materialize identities for the people, organizations, and practices they house. Through their very existence, outward appearances, and internal arrangements of space, research buildings give meanings to science, scientists, disciplines, and universities—for those who work inside and for those who just pass by. Emphatically, different buildings announce different things about the stuff inside, which implies that the design process is at once a negotiation of architecture and identities and an eventual settlement of distinctive physical spaces and social faces.

    This chapter compares the Lewis Thomas Laboratory (LTL) at Princeton to another facility for molecular biology seventeen miles up the road in Piscataway: the Center for Advanced Biotechnology and Medicine (CABM). Their architectural differences, I suggest, measure the distinctive identities each was designed to enact. In choosing which functions or people to put where or how to decorate the shed, the design teams at LTL and CABM first had to decide the constituents of each building, the tendencies and idiosyncracies of its occupants (and of those who would be excluded), and its institutional or organizational ambitions. In effect, they designed people along with walls and windows, and gave them identities. Full-time users, visitors, audiences, consumers, and outsiders were typified and fleshed out in and through decisions about net usable square footage to be located here or there. The two finished research buildings say different things about *who* science is.

I begin with a brief review of the sociological literature on identity and its associations with physical and material spaces. The comparison of LTL to CABM moves through three steps. First I look at their entrances, lobbies, and public spaces: Who were these science buildings built to welcome? Who are their constituencies and audiences? Then I move inside the laboratories themselves: What qualities are attributed to "molecular biologists" such that these differently designed spaces each become suitable to house them? Finally, I consider those left outside: What principles were used to exclude some people and activities from the new buildings, and what does this imply about the identities of those who would become its residents? The analysis is based on documents left behind from the design process: sketches, early floorplans, meeting minutes, correspondences (supplemented by retrospective interviews with architects, scientists, and facilities managers). A methodological subtext runs through the chapter: Assumptions about the architectural materialization of identities based only on post hoc readings of finished buildings are thin and may be misleading without attention to the design decisions argued over during the process leading up to physical construction.[1] My goal is to recover from the design processes at LTL and CABM a sense of the individual and collective identities the design team hoped to build in, which enabled the designers to decide just what kind of spaces and places would be most functionally efficient for *these* people and *these* organizations.

IDENTITIES IN PLACE

The connection between buildings and identities has not gone completely unnoticed by sociologists. Research divides loosely along micro and macro lines: Some look at how individuals create and sustain a sense of self by dressing up their personal spaces; others look at public monuments and other places as tangible translations or expressions of collective values. For social psychologists, *identity* is the meanings attached to oneself, chosen by that person and/or attributed by others.[2] As an answer to the question "Who am I?" identity is an irreducibly social phenomenon (rather than merely personal or psychological) in ways that implicate buildings and other material objects in its construction. Identities give substance to social roles, as they translate structural positions into meaningful features of self: Just what is it for me to be male or female, Black or White, scientist or plumber? Identities also link individuals to groups and organizations: Membership in such collectivities can become part of one's identity as qualities of the aggregate are claimed for (or attributed to) oneself. Moreover, identities exist in social interaction: They are in effect a co-construction by oneself and others involving endless processes of feedback, revision, and negotiation. Finally, constructions of identity vary historically, culturally, and even situationally—but not infinitely so. Some elements of identity may endure throughout an individual's lifetime, just as the meaning of a social role or group affiliation may remain the same through the passage of time and even across distinctive contexts of beliefs and values.

This conceptualization of identity has spawned a line of inquiry on domestic and workplace architecture centered on this question: How are individual identities wrapped up in the places people live or work? For example, the role transition from high school student to college undergraduate involves (for most new undergraduates) a move from one's bedroom at home to a dormitory room. Ira Silver finds that the objects freshmen use to decorate their new living quarters fall into two categories: Some (like a well-worn teddy bear) become "anchors" establishing continuity through the transition, whereas others (recently purchased posters of alluring embodiments of the opposite sex) become "markers" of the new identities students display for others.[3] Other material mementos of childhood are left behind at home because they do not coincide with the new identity a student hopes to fashion: a collection of dolls, for example, or of baseball cards. Because objects (and, by extension, rooms) can be invested with personal meanings, but also because they carry conventional meanings more or less consistently discerned by dorm room visitors, Silver concludes that the strategic decoration of physical environments is crucial for identity formations especially during unsettled periods of transition. Erving Goffman has suggested that architecture is equally important as a means to hide "backstage" activities so as to preserve the "front region" identity a person hopes to convey, a process he calls "impression management." His classic example is a Shetland Isles' restaurant whose walls and doors separate kitchen from dining room, insulating the crofting culture of Scottish workers ("the scullery boys would use the coal bucket as a target for the well-aimed expulsion of mucus; and the women on the staff would rest sitting with their legs up in unladylike positions") from the proper decorum and politesse demanded of workers and diners alike in the eating area.[4] The design of buildings enables the segregation of behavior, so that different identities may be selectively performed for different audiences. Even a casual comparisons of front yards to back yards in American suburbia would bear out Goffman's point.

David Halle's explorations into the homes of working-class and elite residents of New York (and environs) lead him to conclude that the identities materialized in the art they chose to hang on their walls are not tightly coupled to their social structural locations. A taste for "calm landscapes" and "depopulated" scenes of nature is expressed as often in working-class as in upper-middle-class living rooms and thus cannot be easily interpreted as a "reflection" of class structure onto a status hierarchy of low vs. high culture. Instead, Halle suggests that decorations of domestic spaces gather up and give tangibility to personal values and current life experiences; that is, they create and sustain an identity not determined by class or other structural attributes. Still, the manipulation of domestic spaces to construct individual identities may not be a cultural universal.[5]

Amos Rapoport suggests that in most traditional societies—tightly knit into small and dense social networks, unified by shared values and common understandings—dwellings display the group's identity rather than differentiate individuals.[6] It seems, for example, that the Kabyle houses of Algeria analyzed by Pierre Bourdieu vary hardly at all one to the next. Indeed, his discussion of how the internal arrangements of spaces and functions objectify and reproduce symbolic oppositions and societal cosmogony—female spaces are low, dark, and wet; male spaces are high, light, and dry—refers consistently to *the* Berber house.[7]

Cultural sociologists working on a macro level, joined by those interested in social and political movements, have shifted their attention from individual to collective identity: Who are *we?* In the wake of a rejection of any essentialism that would fix the characteristics of a collective in either biology or social structure, the focus is on the social construction of we-ness: The molding of a common identity for diverse gender, racial, ethnic, class, religious, nationalist, or corporate groups, useful for mobilization and solidarity but also for members' self-definitions.[8] To accomplish these ends, a collective identity must at minimum achieve a stability of contents and continuity through time, sufficiently discriminate us from them, and allow for the mutual recognition of members.[9] As with individual identities, collective identities vary in terms of their comprehensiveness (that is, the degree to which they organize members' social lives and in terms of the consistency between insiders' asserted identity and those assigned by outsiders.[10]

How buildings shape collective identity is transparent in the design and construction of corporate headquarters. In September 1997, IBM moved into new headquarters vastly different in design and purpose from the digs they had occupied since 1964—reflecting, in part, recent reversals in which the corporation lost billions of dollars and downsized away more than 100,000 jobs. The new building is much smaller than its predecessor, "quietly nestled in a woodsy ravine," in stark contrast to the former "fortress": a gray, faceless, imposing, modernist behemoth no longer in keeping with IBM's sprightly new image of innovation, creativity, and spunk. "The function of a corporate headquarters has always been as much about the identity of the corporation as it has been about supporting the workers in that building." But the new headquarters are also designed for a different IBM worker: Rather than endless hallways of closed doors and private offices, almost all managers and executives work at cubicles in large wall-less spaces, an architectural reproduction of the growing corporate emphasis on increased flexibility and teamwork. Perhaps because of IBM's more modest position in the computer industry, the new headquarters are not "a monument," but rather a "living example of the company's values."[11]

Monuments, on the other hand, are exactly what nation-states have so often used to express their identities, values, traditions—and power. Benedict Anderson mentions the efforts of British and Dutch colonial regimes to unearth and refurbish historic monuments or shrines built by the native peoples of South and East Asia—ironically, not to kindle traditional loyalties of the indigenes but (because these were British or Dutch restorations) to show that "contemporary natives were no longer capable of their putative ancestors' achievements."[12] Different kinds of buildings are also important for fragile *post*-colonial nation-states to secure legitimacy, as Lawrence Vale's study of the design of national capitols suggests so well. The 1984 Parliament House for newly independent Papua New Guinea was an "impressive and monumental" attempt to fuse 1,000 tribes speaking 700 languages and dialects into a unified nation-state committed to representative democracy, in a place where neither nation nor parliamentary democracy had long roots. Designers faced the following tensions: A building that recovered indigenous architecture and displayed tradition-laden iconography would establish cultural continuities but risk activation of deep hostilities among the different tribes and peoples now somewhat accidentally pulled together; a building that abandoned tradition for one or another international style might seem appropriate for the inauguration of an utterly new political entity and for giving symbolic weight to its authority but would risk criticism that it was an elite's extravagance remote (symbolically and literally) from the ordinary lives of too many citizens living in poverty. Perhaps no single plan could succeed on all fronts: The prominent roofline for the as-built Parliament House features a sweeping, rising arrow that evokes the traditional pattern of spirit houses among the Sepik culture, though such a pattern is totally alien everywhere else in the archipelago. It is decorated throughout by diverse carvings and mosaics done "in the style of" Papua New Guinea's many distinctive indigenous peoples. But because these artifacts were created anew for the Parliament House and used modern media (for durability) largely unknown among traditional practices, they come off more as tourist art than authentic expression. Still, says Vale, "as an advertisement for independence and as an idealization of . . . representative democracy, the building inspires confidence" and gives shape to a "nationalist assertiveness."[13]

Sometimes place itself—its built and natural features—becomes the basis for collective identity, often reinforcing bonds of ethnicity, race, or tribe. Robeson County, North Carolina, is the "home place" for the Lumbee Indians, whose collective identity "depends heavily" on this geographic spot. Anthropologist Karen Blu writes that the place is "pivotal for their current sense of peoplehood, their ethnicity" despite the fact that the Lumbee share Robeson County with both Whites and African Americans.[14] Perhaps place is salient for Lumbee identity precisely because the Lumbee have not been gathered up and moved to a reservation: Many live in houses clustered around crossroads' general stores scattered amid the flat, piney swamps often drained for farmland. For a culture whose official Indian-ness political authorities have only grudgingly recognized, the Lumbee's historic concentration in Robeson County, where they have built specifically Lumbee churches and schools, is evidence for their very existence as a collective.

Pulling together these scattered hints and thinking ahead to the design of laboratories for molecular biologists, it might be said that buildings solve three problems faced by any identity, personal or collective. First, buildings give tangibility and demonstrability to what is intrinsically elusive and implicit: The research centers at Princeton and Piscataway display the meanings of "scientist" or "biotechnologist," and so provide prima facie answers to the question "Who are we/they? Second, buildings give stability and durability to the potentially transient: The two labs fix (in different ways) the goals and purposes of the scientific people and practices they contain, through walls, utilities, and finishes not easily or often changed. Third, buildings discriminate identities that are always at some risk of blending: Architecture does a masterful job of spatially segregating insiders from outsiders, us from them. If buildings do all this (and more) for identities, then the design of buildings must also be the design of a self (or a collective). LTL and CABM are distinctive materializations, stabilizations, and discriminations of "science" or "molecular biology." How so?

In some respects, the two laboratory buildings are not so different. Floor plans (see figures 20.1, 20.2, and 20.3) show much in common: a rectangular footprint; multiple

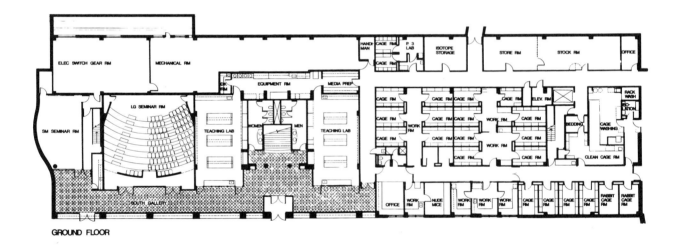

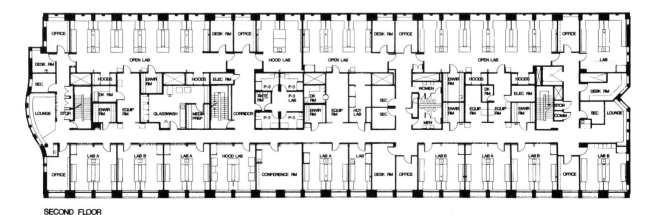

CENTER FOR ADVANCED BIOTECHNOLOGY AND MEDICINE
Payette Associates • Architects • Planners

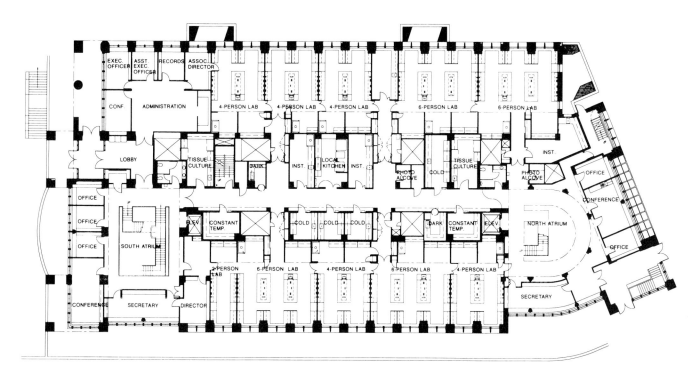

First Floor

20.1

Lewis Thomas Laboratory, Princeton, New Jersey. Ground floor plan. The ground floor of the Lewis Thomas Laboratory is divided in half: To the left are spaces mainly for students (teaching labs, a seminar room, a lecture hall); to the right are the animal facilities (mice cages), along with the receiving dock. Between the two teaching labs is the central staircase, where undergraduates, graduate students, faculty, and staff collide on their way *down* to classes or *up* to research labs and office.

20.2

Lewis Thomas Laboratory, Princeton, New Jersey. Second floor plan. The corridors on the second floor of the LTL form a kind of racetrack: Research labs, offices, and lounges are on the outside perimeter of the building; communal equipment rooms and special facilities (glass wash room, media storage, and darkrooms) are on the interior. The labs vary considerably in size: Big and open labs are shown at the top of the plan, segmented and smaller spaces on the bottom.

20.3

Center for Advanced Biotechnology and Medicine, Piscataway, New Jersey. First floor plan. The first floor of the New Jersey Center for Advanced Biotechnology and Medicine is a classic "double-loaded" corridor design. A single hall runs down the middle, lined on both sides with communal cold rooms, constant-temperature rooms, darkrooms, tissue culture rooms, and facilities for glassware sterilization. One enters the labs via a series of cross-corridors, passing through some noncommunal heavy equipment space before reaching the wet-lab benches and desks. Faculty offices are clustered around expansive atria stairwells at each end of the building.

entrances from the outside; a tunnel; a double-loaded, single-corridor plan, with perpendicular cross-corridors linking work spaces on either side; elevators and stairwells; repetitions of laboratories and researchers' offices around the perimeter; interior core space taken up by heavy equipment (refrigerators, centrifuges, freezers, incubators), cold rooms, constant-temperature rooms, glass washing facilities, media prep rooms, darkrooms, and tissue culture rooms; administrative areas with spaces for secretaries, mailboxes, and photocopying machines; conference/reading rooms; storage; mechanical rooms; building services; and rest rooms. Laboratory benches are similar as well, whether attached to a wall (single) or peninsular (two work spaces back to back): Ordinarily, interior benches have a sink at the end closest to the middle of the building; a drying rack; a run of wet-bench spaces with piped services (gas) and electrical receptacles, ending with a dry desk at the end by the window; shelves above the bench; and drawers, cupboards, more shelves, or kneeholes below. "By the numbers," LTL and CABM echo each other again (see table 20.1): CABM has 2.2 percent more gross square feet than LTL, but LTL has 5.7 percent more net square feet (usable, programmable space—labs, offices—as opposed to what is needed for mechanical systems). Estimated project costs differed by only $1 million, though actual project costs for CABM came in about $5 million more than LTL. (In constant 1987 dollars, the estimated costs per gross square foot are almost identical.) CABM was built to house 180 eventual occupants, LTL was built for 164.

TABLE 20.1
Two science buildings, by the numbers

	Lewis Thomas Laboratory, Princeton, NJ	Center for Advanced Biotechnology and Medicine, Piscataway, NJ
Project inception	1982	1984
Completion date	1986	1989
Gross square feet	109,600	112,167
Floors	4	4
Perimeter	270' × 83'	208' × 112'
Net square feet	65,800	62,543
Efficiency (NSF/GSF)	.60	.55
Project cost (actual dollars)	$29 million	$35 million
Cost/GSF	$210	$243
Number of occupants	164	180
Number of scientists	20 faculty	18–20 principal investigators

Not everything looks the same, of course. CABM is the fatter rectangle (with a length-to-width ratio of about 2:1, but about 3:1 at LTL) and has several features absent from LTL: an aisle between the wet lab bench and the dry desk by the window, facilities for nuclear magnetic resonance and protein microchemistry, a boardroom and an executive conference room, two atria, a room for computer workstations, and house vacuum piped to each benchtop. There are also features at LTL not found at CABM: informal lounges tucked into ends of the main corridor, kitchenettes, two teaching labs, a lecture hall (capacity 150), animal cage room and prep facilities, and greenhouses on the roof. Although the same architectural firm, Payette Associates Inc., of Boston, designed both buildings, different principals were assigned to the two projects: Jim Collins, Jr., to LTL, where he collaborated with Venturi, Rauch and Scott Brown (Venturi, Scott Brown and Associates since 1989) and William Wilson to CABM. As this list of differences between LTL and CABM gets longer, it becomes clear that although each facility is functionally efficient and aesthetically pleasing, neither "functional" nor "pretty" may have a common meaning.

Entering

Recklessly ignoring advice about judging books from covers, I believe that doors say much about a building. How one gets into a research facility—and, once inside, how one moves through lobbies, atria, and corridors on one's way to laboratories or offices—has a sociological dimension quite apart from fire code definitions of safe egress or architects' models of circulation and flow. Entrances and public spaces measure the different constituencies of these two science buildings; doors and corridors give a reading of those for whom they were built.

Each building has two major entrances from the outside, but similarities end there. At the Lewis Thomas Laboratory in Princeton, the exterior doorways are on the rectangle's long sides. The building sits on a slope: the front entrance is on the north side, facing College Walk and the campus, and opens into the first floor. The other entrance is on the south side, facing a small (privileged) parking lot and—across the fields—the larger Lot 20, and opens into the ground floor. Bikes surround the campus-side entrance, and there is a steady flow of pedestrians in and out of LTL all day long. (See figure 20.4.) Parked cars surround the south-side entrance, and pedestrian flow is negligible except around 8 A.M. and 5 P.M. From either doorway, one enters a modest lobby with a grand staircase of warm oak: Those who enter from campus go down those stairs to the lecture hall and teaching labs; those who enter from the parking lot go up to department offices on the first floor or to labs and faculty offices beyond.

The two doors open for two discrete groups of users: Undergraduates coming from their campus residences, the library, or other classroom buildings enter the first floor from the north; those who drive to work—graduate students, postdocs, technical and

20.4

Lewis Thomas Laboratory, Princeton, New Jersey. Campus entrance with bicycles. The campus-side entrance to the LTL, adjacent to dormitories, libraries, and other classroom buildings, is almost always surrounded by bicycles. Very few bikes are left by the entrance on the other side of the building, adjacent to a large parking lot where faculty and staff leave their cars. Both doorways enter into a central staircase, the agora of the LTL.

secretarial staff, faculty scientists—enter at ground level from the south. They are forced architecturally to mix, in the lobbies and on the grand staircase: Most undergraduates visit LTL to attend lectures or to do experiments in the teaching labs, both downstairs; faculty, staff, postdocs, and graduate students first go up the grand staircase to check mail at the departmental office, and then continue up to their labs. "Faculty and staff would park to the south of the building, whereas the student traffic would typically come on foot or bicycle from the upper campus, and come into the main lobby area where everyone would interact."[15] Such mixing was by design. LTL was built for faculty, and for the staff they needed to do cutting-edge research in molecular biology. But Princeton undergraduates were designed into this building as well: The front door faces campus, where these students live, and they arrive here in prodigious numbers, for lectures or lab sessions downstairs.

The bi-level lobby has become the heart and soul of LTL, and of Princeton's Department of Molecular Biology. Here, the building's two constituencies get stirred up—slurping coffee, having parties, welcoming visiting speakers, hanging out, and finding information (permanent and temporary bulletin boards were festooned on one visit with photographs of molecular biology majors in the classes of '95 and '96, about seventy or so in each; examination scores; answers to recent tests; Department of Molecular Biology Undergraduate Committees, 1994–95; lists of fall courses, including a freshmen seminar on viruses by Department Chair Arnold Levine; the hourly payroll schedule; photographs of the department's twenty-one new graduate students, about half female; announcements of upcoming colloquia and seminars, here and at nearby places of research such as CABM; precepts for Molecular Biology 342; a notice of a computer for sale; "Molecular Biology Majors: Great Job Opportunities in Asia!"). Backpacks, briefcases, and white lab coats merge to the sound of clinking clean glassware on carts heading back to the labs.

Add up the overall square footage in LTL built for students: Two teaching laboratories (plus preparation space), a large lecture hall for 150 students, and a smaller seminar room outfitted for undergraduate discussion sections together eat up about half of the usable space on the ground floor. Again, this was by design: "Princeton's reputation for excellence in teaching attracts exceptionally able applicants. The new department will benefit from—and educate—the very brightest students at both the undergraduate and graduate level, providing a stream of outstanding people who will assume leadership roles in the pharmaceutical industry and academia."[16] Levine's willingness to shell (i.e., leave unfinished) research lab space in the event that the budget would not allow a large lecture hall indicates the design priority given to teaching. Administrative space at LTL is described in meeting minutes as the "Departmental Office Facility" (14 April 1983), and though it houses grants administration and equipment purchasing, it has the feel of a place where undergraduate majors come to check their progress toward degree requirements. The lounges at the ends of the main corridors on each floor have a welcoming grunginess (soft cushions on built-in benches are frayed from overuse) that lures students out from the labs to eat, talk, and sleep. Levine told me: "The need for facilities to socialize, have coffee, not eat in the lab, be away from the lab yet be so close that you can walk right to it, really makes the lounges as a concept a higher and higher priority in all our minds."[17] A concern for making space for undergraduates in LTL reached the highest administrative levels, as seen in a letter from Robert Venturi to (then) Princeton President William Bowen (17 May 1983): "This is to assure you that I think all of us—Arnie Levine, Payette and ourselves—are aware of the importance of the teaching spaces functionally and symbolically, and they will not be slighted in their expression despite the greater amount of purely research space in the building."

Why was it so important at LTL for this new research building also to take on the identity of a home for Princeton undergraduates? Donors, whether individual or institutional, drive design in subtle ways: "We knew what donors thought were important issues, and we could convince donors that what we wanted was an exciting plan for them." Although $5 million for the construction of LTL came from New Jersey's Commission on Science and Technology, the rest of the tab was to be covered by donations—from whom? "The teaching facility became a half-floor, and could be so-called marketed as such by the development office. . . . The reason that is a very good marketing tool, in the parlance of Princeton, is that that is in the teeth of where the finances are for the university—undergraduate education. The alumni are undergraduates by and large with money, and they understand that issue [teaching] much better than say, anybody who would have enough money to understand research."[18] The building is named for Lewis Thomas at the request of major donor Laurance S. Rockefeller, both Princeton men.

The main entrances to the Center for Advanced Biotechnology and Medicine at Piscataway are not immediately apparent as one approaches the long side of the rectangular building from parking lots across Hoes Lane. The doors are tucked into the short ends, facing (to the right) the Waksman Institute of Rutgers' Department of Biology and (to the left) the research tower and other facilities of the University of Medicine and Dentistry of New Jersey (UMDNJ). (Architecture critic Ada Louise Huxtable once described the 1960s UMDNJ tower as a "hovering mass of dour concrete" that stands in sharp contrast to the homey, low-slung brick of CABM.) Positioning is everything: The entrances point to two constituencies of CABM, New Jersey's two cooperative but also competitive state universities, and the building between the doors is an attempt to link them, physically and functionally, in the pursuit of scientific excellence. CABM sits on contested terrain, a space between two universities, each coveting the plum research center it was funded to become (see figure 20.5). Think of CABM as a kind of architectural meiosis, pulled by its entrances toward UMDNJ to the left and Rutgers to right, stretched along the axis between them. The doors on each end address one or the other university, and because they have been made less pronounced, they signal CABM's uneasy dependence on its two parent universities.

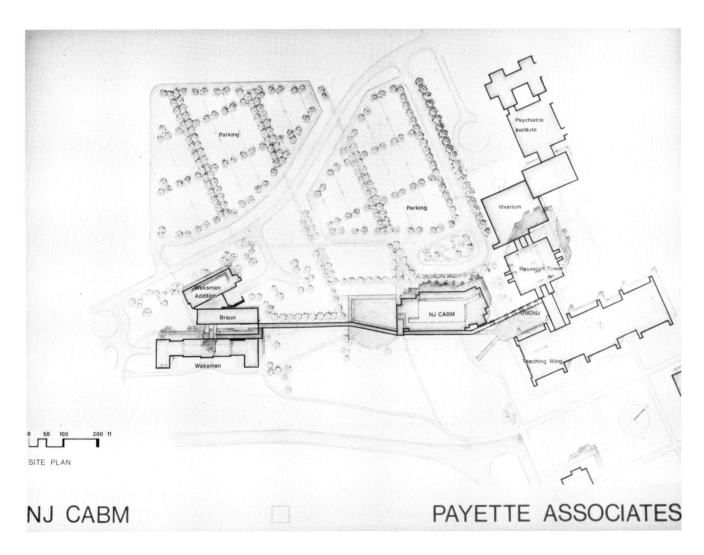

20.5

Center for Advanced Biotechnology and Medicine, Piscataway, New Jersey. Site plan. To the right is the complex of buildings making up the Piscataway campus of the University of Medicine and Dentistry of New Jersey. To the left is the Waksman Institute, housing the biology department of Rutgers University. In the middle sits the CABM—on terrain symbolically contested by the two public universities—tethered to UMDNJ by an underground tunnel (dashed lines) and to the Waksman by an above-ground corridor.

The process of designing entrances for CABM was contentious, unlike that at LTL, where all immediately hailed the bi-level and grand staircase arrangement as perfect. CABM was destined to have no front entrance, but only two side ones: "The entry . . . from the parking lot side, is not well articulated. The building doesn't have a strong sense of entry. . . . There's no front door. . . . It's a building that in fact is open in every direction, in a lot of ways, but it gives the impression of not having any way in, at first."[19] This bothered Aaron Shatkin, who (like Arnold Levine at Princeton) was hired early to shape and staff the new facility. The building committee meeting minutes for 5 December 1986 read: "A. Shatkin stated that the CABM Main Entryway should be highly visible and maintain its own identity from that of the neighboring UMDNJ complex." But that identity would not come at the expense of either of the two universities it sits between. "The configuration of construction proposed for the site must preserve a strong sense of identity for the UMDNJ-Rutgers Medical School Buildings and must preserve a strong sense of arrival."[20] William Wilson's notes of the Rutgers Trustees' Meeting on 3 October 1985 read: "The Trustees are concerned that the building will have a separate identity. They do not want 'an addition onto UMDNJ.' They asked what the building would be made of: I said that there would be brown stone bands, which is a Rutgers motif." CABM could belong neither to UMDNJ nor to Rutgers, and everything architectural—entrances, skin—had to indicate that it was part of both of them (and only a little bit on its own).

Architectural meiosis continues inside CABM. Whether from the Rutgers end or the UMDNJ end, one enters a multilevel atrium, filled with natural light from above, almost glitzy with mahogany woodwork and terrazzo floors. (See figure 20.6.) Public space in the building has a barbell shape, pulled apart by the UMDNJ atrium on one end and the Rutgers atrium on the other. The atria were designed to achieve vertical communication among scientists, who could (and do) have impromptu chats hanging over the railings or on the stairs. But why two atria, which segregates occupants who then must choose to hang over one or the other set of railings? Two constituent universities means architectural meiosis: "I tried very hard to eliminate one of them. And it didn't go over, quite frankly because . . . I think I'm remembering this correctly, the Rutgers end of the building had an atrium, therefore the UMDNJ end of the building had to have an atrium too. . . . That's the type of relationships that go on when you have a joint building such as this. . . . But not only did we have a new center, which had its own entity involved, but you also had the two universities involved. So you had three competing groups who were involved not only in usage but also in image, and their relationship to the shared building."[21]

The two entrances and the two atria are linked by a single central corridor on each floor—but how different a space than at LTL in Princeton (where corridors are crammed with "beer coolers," freezers, and incubators, and a rainbow of cables droops down from exposed ceilings packed with pipes and ducts). No heavy equipment or drooping wires

20.6

Center for Advanced Biotechnology and Medicine, Piscataway, New Jersey. Interior view, atrium. A sumptuous atrium stairwell at CABM, trimmed in mahogany woodwork and terrazzo floors, exuding an upscale corporate feel. The space is also designed to enhance spontaneous faculty meetings and conversations: Their offices line the atrium. There is another atrium very much like this one at the other end of CABM, needed to keep Rutgers and UMDNJ ever in balance.

in CABM's halls, but rather mahogany-framed cabinets holding clean glassware out of the way on either side of a shiny, blank floor. (See figure 20.7.) The dark woodwork contrasts with the light maple in the labs themselves, a demarcation-by-color of public spaces from work spaces. Concern for "the public" ran throughout the design process at CABM, leading (for example) to a suggestion that more toilets be built because "there will be significant numbers of visitors."[22] CABM was designed not just for Rutgers and UMDNJ, but for a third constituency, who might only visit the place: the citizens of New Jersey, who voted for a bond that provided most of the capital for building CABM, and who would benefit (in the form of high-tech jobs and tax revenues) from its becoming the catalyst for the state's biotechnology industry. Managers and scientists from the huge pharmaceutical companies that line Route 1 (Johnson & Johnson has headquarters across the Raritan in New Brunswick) were expected to enter the doors of CABM to hear the latest news from the research frontier, and the glitzy tidy public spaces reminiscent of a "bank"[23] were built to welcome them. Is it an accident that the setting for CABM was described as looking "just like a corporate park"?[24]

Hardly. Funding for CABM came from the State of New Jersey, via its two public universities and also the Commission on Science and Technology (which drew $7.6 million from the Jobs, Science and Technology Bond fund).[25] Educational goals were certainly not overlooked: Planning documents routinely mention training—"The Center will serve the educational needs of the state by providing opportunities for graduate and postdoctoral students to train in laboratories of excellent quality using state of the art techniques presented in formal educational programs"[26]—along with "continuing education" for the state's biotechnological workers.[27] However, undergraduate education was not on CABM's agenda, so special labs for teaching and large lecture halls never got on its drawing board.

Instead, the Center is pictured as a kind of high-tech industrial extension service, modeled after the old agricultural extension networks through which university experts helped farmers grow better vegetables. "The development of advanced technical training and research programs at New Jersey universities is a central component of a larger effort to restore and maintain the economic prosperity of the state. This effort, which constitutes, essentially, a state industrial development policy, also includes provisions for establishing a grant program for applied research, the construction of 'business incubation facilities,' and the creation of a 'technology extension service.' [CABM] will be a physical and organizational focal point for reciprocal interaction between industry and academia in the biotechnology fields. This interaction will foster advanced research and development in some of the State's most important industries, such as pharmaceuticals and health care."[28] These anticipated outputs are inscribed throughout the building. Administrative space has none of the clutter of LTL's "departmental office"; instead, the reception room is outfitted with upscale couches and tables (no fraying edges) and leads to offices for an executive director and an assistant executive director and to an executive confer-

20.7

Center for Advanced Biotechnology and Medicine, Piscataway, New Jersey. Interior view of laboratory space. A long look through one of the upscale labs at CABM, where bench work of rock maple contrasts with the dark mahogany woodwork of public spaces.

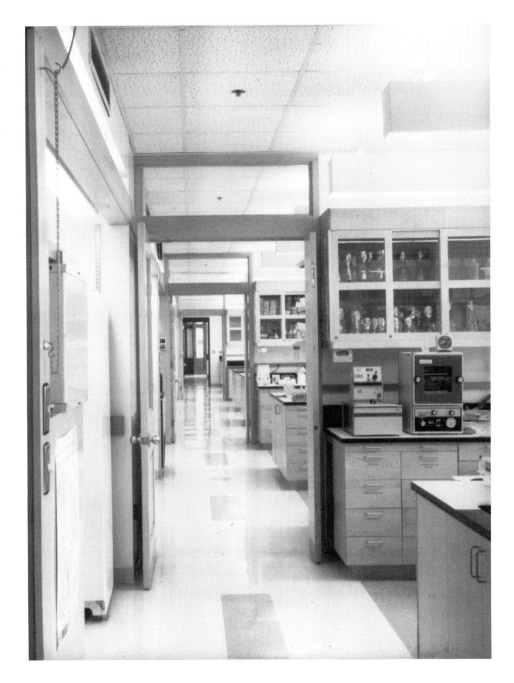

ence room. McGraw-Hill's Biotechnology Watch writes glowingly of CABM (3 June 1985): "Corporate carrots include matching grants for industry-supported basic research, low-cost rental space, and 'network and satellite laboratories' providing equipment, technical staff, pilot plant and clinical drug testing unit. . . . New Jersey's future CABM will buy a highly leveraged window on technology. . . ." CABM was designed to become an incubator in the "technology transfer" sense, an architectural "bootstrapping for the economy."[29]

Typifying

Both LTL and CABM were "zero-based," designed and built almost completely for occupants who were unknown until after construction was completed. Neither Princeton nor Rutgers nor UMDNJ had yet achieved a stellar reputation in molecular biology, and the new facilities on each campus were intended to lure the necessary stars. No extant members of biology faculties were scheduled to move into LTL or CABM: Space there was reserved almost exclusively for new hires.[30] Such a fresh start is both the best and worst design situation. With no flesh-and-blood scientists on board, there is no one around to demand customized space that suits them uniquely (but is more expensive to build and costly to renovate for the next occupant). Turf wars over square footage are precluded. On the other hand, it is difficult to assign space without some understanding of the specific equipment needs of the individuals who will eventually inhabit the space. The solution is to typify[31] eventual occupants of the new building by creating hypothetical scientists who are given just enough of an identity to facilitate their easy and rational arrangement in space—but no voice to complicate matters or complain.

Typification of hypothetical scientists enabled the design teams to build up LTL and CABM from a generic lab module in a remarkably formulaic process. A dollar budget fixed a maximum square footage, and the game was to fit in as many standardized lab modules as possible, with just enough space left over for equipment rooms, offices, administrative facilities, mechanical systems, and other nonresearch requirements. However, the LTL and CABM design teams chose to work with slightly different lab modules, and then the different modules were cloned and arrayed in even more distinctive ways.

The LTL team devised a working "four-person" module of about 500 square feet. In the end, fifty-five such modules were fit into the building, every one just like the next: four wet benches, two against each side wall, two back to back to form a peninsula in the middle (some are low for sitting, but most are high for standing); piped and wired services at the back of each bench space (gas, air, electrical receptacles, cup sink, water); four dry benches (or desks) at the window end of each wet bench; open shelves above the wet benches; a combination of drawers, cupboards and kneeholes underneath (filled with small refrigerators if under a high bench); a big sink in the hallway end of the peninsular benches; and a fume hood at the end of the wet bench along the side wall opposite the doorway. Though described as a four-person module, it was assumed that each could hold up to six people (one or two might be rotating undergraduate majors, learning skills from postdocs and graduate students). Decisions about how many modules would be given to each faculty member were settled by academic rank: For planning purposes, assistant professors were given 1–1.5 modules (500–750 square feet) for 6–8 people; associate professors got 2.5 modules (1,250 square feet) for 12–14 people; and full professors got 3–4 modules (1,500–2,000 square feet) for 15–22 people. These numbers were working averages, for it was assumed that there would be variations within rank in the size of each lab "group" (i.e., a faculty member and those who work in his or her lab).

Lab modules can be stacked up vertically from floor to floor or spread out horizontally on each floor, and choices here determine the building's footprint and height. At LTL, such decisions took account of two factors: the critical mass of scientists per floor needed for good communication (assumed to be primarily horizontal and deterred by stairs or elevators), and the equipment or special research facilities needed by a particular group. Early plans show three or four groups each on four laboratory floors, later revised to six or seven groups each on only three laboratory floors. But who goes where? So far, the hypothetical occupants of LTL have been typified only by academic rank. To distribute them throughout the building required that they be typified by research specialty as well: Those researching higher eukaryotes (mice) and prokaryotes (bacteria) were located on the first floor; virology (tumors, cancer research) and more prokaryotes on the second; lower eukaryotes (yeast, slime mold, worms, and fruit flies) and plants on the third. "What we did was to actually plan out a faculty, in a broad sense. We didn't pick individuals, but we picked areas, we picked directions."[32]

Assignment of each research specialty to a certain floor was determined by the special facilities each needed—equipment and instruments not housed in the generic lab modules. Otherwise nameless and faceless scientists were given minimalist identities as they were characterized by the big machines on which they depended: "Higher Eukaryotes to be near animal facilities and could form one floor with virology. Virology with a P-3 lab with small autoclave. Plant group and greenhouse could form one floor. Tissue culture (monoclonal antibody) to be near Higher Eukaryotes and Virology. Lower Eukaryotes and Prokaryotes can be on their own and not closely related to any specific support facility."[33] Levine is a virologist, and as chair of the department-in-the-making, would probably have been located with other virologists on the first floor near the central administrative space. But because this research uses infectious agents and requires a P-3 containment and "hot" (radioisotope) room nearby, it was decided to move these researchers up to the less-public second floor. Animal cage rooms were hidden away on the ground floor, so it made sense to locate mice researchers doing developmental biology nearby on the first floor, just as it made sense to locate plant biologists on the third floor, near the rooftop greenhouses. Prokaryotic researchers required no special facilities beyond those to be located on each research floor (environmental chambers, warm and cold; darkrooms; glass-washing and kitchen facilities; instrument rooms for counters and balances; instrument alcoves—in the cross-corridors—for centrifuges, freezers, and other heavy bulky equipment); and so they were split between the first and second floors.

The arrangement of lab modules, equipment rooms, special research facilities, and faculty offices on each floor was determined by the following principle: Humans are heliophilic, machines are heliophobic (or maybe just indifferent). Lab modules and faculty offices ring the perimeter of LTL, taking advantage of natural light pouring in from ribbons of windows. Levine recalls: "I was five years in a building that had no windows, except in the offices, at Stony Brook. . . . When Venturi came to visit me, I brought him in the lab, and . . . it was just walls. There were no windows, and I told him that was an impossible situation."[34] The location of instruments and equipment was determined via the following equation: Distance from the wet lab bench (and its windows) increases if the equipment is infrequently used, requires only brief tending (start it and leave), or kicks out heat, dirt, or noise. Most centrifuges end up in the interior cross-corridors; most electrophoresis setups for gel running sit on the wet lab benches. But both humans and machines often cross the threshold of windowless core/sunlit perimeter, a circumstance anticipated in design: "[G]lass walls between labs and internal support spaces bring natural light and orienting views through the width of the building."[35] But in actual use, bulky refrigerators or freezers block those interior windows, not because these machines want more sun, but because researchers must have them especially close at hand.

Two final decisions were debated at length. First, lab modules could each be walled off from those adjacent ("closed lab") or strung together without walls between to form a capacious biological atelier ("open lab"). The decision fell to Levine, who chose both: Labs on LTL's north side are open (stringing together as many as three modules, to house as many as eighteen people), whereas labs on the south side are discrete lab modules with a wall between each pair. Levine assumed that molecular biologists come in two species, those who prefer open labs (for increased communication, sociability, and accessibility) and those who prefer closed (for safety, security, noise suppression, less commotion, or privacy). As Levine put it later, why jeopardize Princeton's opportunity to hire the best scientists by not being able to offer both possibilities? Second, faculty offices could either be clustered together or distributed singly throughout the lab spaces on each floor. Levine opted for the latter, placing a premium on proximity and easy interaction among members of a research group (sociologically, maximizing vertical communication between a faculty member and those who work in his or her lab). Such a plan puts distance between faculty colleagues, who would then be more likely to meet at mailboxes, over coffee, in lounges, or in the halls, and also makes faculty more difficult to find for a casual visitor to the building. But "the big advantage is that you are close to your lab. It's easy for people to come into your office from your lab; it's easy for you to spend your free time in your lab. The highest priority is your dynamic with your laboratory. In a place like Princeton, where undergraduates visit faculty members, it would have been more convenient to cluster offices, but we chose once again to stay near our labs."[36]

All these decisions at LTL seem rational, almost inevitable, until one sees from a comparison to CABM that they could have been made in very different ways. There is nothing natural, inherently more efficient, or universally more cost effective about any of the design decisions made at LTL. Such decisions make sense only in terms of the working identities attributed to typified scientists during the design process, and the identities created for scientists that CABM hoped to lure are different in architecturally consequential ways. Begin at the wet benches (which are just as generic as they are at LTL): At CABM, black epoxy bench tops are covered by a replaceable, white adhesive film (easier to keep clean); a "house vacuum" is available at the back of each bench; shelves above each bench have sliding, wood-framed glass doors (though they are easily removed); and overhead lighting is direct. (The indirect lighting at LTL was judged inadequate after members of the CABM design team visited the Princeton lab and saw additional task lighting at almost every workstation.) Even more obvious is the aisle between the end of the wet lab bench and the writing desk by the window, in effect making LTL's peninsula benches into island benches at CABM. Gains in safety and in sociability balance the implied loss of bench space (however, CABM benches are wider than LTL's). The cut-through allows a second egress should an accident occur in a lab bay, just as it allows for easier communication and interaction between the bays in each module.

The importance placed on safety in the design of CABM cannot be underestimated. The Concept Document for CABM (26 July 1984) reads: "The building will be a model of safety." Alba LaFiandra recalls that the "number one issue, as far as I was concerned, since this was sort of placed in my lap, was safety, because we knew that we would potentially have pathogenic organisms in here."[37] The primordial typification of CABM scientists as "those who work with hazardous substances" guided another crucial design decision that further distinguishes the two buildings. LTL opted for a mix of open and closed labs; CABM designed only closed labs, though of different sizes. Two standard modules are replicated throughout the building: The four-person lab has one back-to-back island bench between two benches on the walls, and (like the corresponding model at LTL) was really built for six people; the six-person lab has two island benches to go with the two wall units and is built for eight researchers. The larger modules were designed for "senior principal investigators (PI)," the smaller ones for "junior principal investigators," with five PIs slated for the first floor, six each on the second and third floors. But whatever the size, each lab module is a closed lab, walled off from its neighbors (each does have a connecting door, usually shut), each with autonomous access to the main corridor, and each with a lockable mess of doors at the lab end of the cross-corridors. The architects would have preferred open labs: "University requirements [state that] . . . all labs have an entrance and can be locked. . . . Right there [points to plan]. Zillions of doors, God, there's a lot of doors right here."[38] But the coupling of university safety requirements dictating that all labs have corridor access and lockable doors with a strong preference for closed labs by the two identified users (Shatkin and LaFiandra) ruled that out: "Now we, in our experience, have found that large labs like that don't work very well, because of the interaction of people. We're going to be reasonable here, practical and very realistic. You can have a problem in the lab. If you have smaller labs like this, you can isolate them, so that they don't cause problems for everyone else. When you have everything that's 12's [big labs for twelve people] . . . you're just dead in the water. . . . There are safety issues, as far as I'm concerned, when you have too crowded a lab."[39]

A concern for safety was not alone in steering CABM toward closed labs. At least from the standpoint of facilities managers at Rutgers and UMDNJ (who were very active in the design process, especially before Shatkin and LaFiandra were brought in), the typification of scientists as "territorial" and "security-conscious" also militated against the large open labs, where bench space might be shared and access to another's equipment unimpeded. "We talk about the open lab concept. But my feeling is . . . when you get down to it, they still say: Well, we're bringing a PI on, and he has a research team, and what are the physical constraints? In other words, this is his kingdom: Where do you put the wall? The total open lab is not something that the researchers still seem to be comfortable with." "In this institution, and I believe in Rutgers as well, the scientists are territorial. From Day One until the end. . . . Most of them want their own territory. They just don't want a big open space shared with everybody else."[40]

The presumed territoriality of hypothetical CABM scientists may also have justified an allocation of space for heavy equipment and special research facilities noticeably different from their placement at LTL. The equipment alcoves that line the interior walls of each lab module (usually two alcoves per module) offer more square footage and are better defined than the equivalent run of free wall space at LTL. Besides making CABM a fatter rectangle than LTL, these roomy alcoves also mean that each PI can house (and use) more "off bench" equipment in space that he or she "owns," in their own, lockable closed lab ($-70°$ freezers, refrigerators, sorvalls, counters, microwaves, speed-vac lyophilizers, high-speed spincos). At LTL, more heavy equipment was stored in "communal" spaces: Cross-corridors there are a kind of no-person's land. True, CABM scientists must also leave their labs and go out to the main corridor to gain access to special equipment rooms (tissue culture, cold rooms, constant-temperature rooms, darkrooms, glassware and sterilization facilities). But simply put, at CABM, there is more off-bench equipment space within the extended lab module—on a scientist's own turf—than at LTL. "When you really push, most people say, yeah, we can share equipment. But when it really comes down to it: I can't be inconvenienced in terms of when I'm doing something. . . . So if you've got a wall, you've got your own space, your equipment is in there, that's where you live. If somebody wants to share it, yes, they can share it, but on your terms. They come to you. It's very obvious who owns it. . . ."[41]

Interestingly, the typification of eventual occupants of CABM by their research specialties seemed to play little role in the design of the building, with one notable exception. In early programming, the Center was loosely described as a place for molecular genetics, cell and developmental biology, and molecular pharmacology. But these distinctions had no appreciable impact on the design of the three laboratory floors, which offer carbon copies of the special research facilities that line the main corridor (except for a unique radioisotope room on the first floor), and made it easier to locate arriving scientists just about anywhere in the building. The one exception, which made the ground floor at CABM profoundly different from those above it, illustrates how dramatically the replacement of a typified occupant by flesh and blood alters the design process. Stephen Stein was hired in 1986 to provide a cornerstone to CABM's evolving focus on structural biology (protein structures, biomolecular crystallography). "His space was specifically designed for protein microchemistry work":[42] Rooms were created for nuclear magnetic resonance equipment, X-ray diffraction, lasers, and lots of souped-up computers for modeling protein structure.

Faculty offices at CABM are clustered: Four line the side of each atrium (about eight offices per floor). Several factors shaped the choice not to follow the LTL model of offices distributed throughout the labs. There is little question that the clustered arrangement encourages peer communication among principal investigators, though this may come at the expense of immediate access to those working in one's lab. "We've got the double helix stairway; it's kind of a joke, but it works. . . . We know the people on the next floor. They're not just separated by a floor with narrow stairs somewhere—look up, it's like an Italian Renaissance village or something. You can stand in the square and call up!"[43] Moreover, by locating offices visibly on the atria, they are easier for visitors to find, if not necessarily the undergraduates that throng to LTL, then perhaps representatives of New Jersey's hoped-for biotech industry. Finally, since this is CABM after all, architectural meiosis requires that offices at the Rutgers end be matched by an identical number at the UMDNJ end. "There are two organizational entities in the building. Rutgers is down here, and UMDNJ is up here, and so the building has two entrances. . . . This building could be . . . a quantum leap better if these were all pulled in the middle, both [sets of] offices were put together. . . . It's a bi-nuclear building, no question."[44]

In planning these two science buildings, designers gave different identities to the scientists who would, with time, move in. These differences in hypothetical occupants in turn steered important design decisions in distinctive directions. Certainly it is not the case that those who designed LTL were unconcerned about lab safety or never suspected some scientists of being territorial, just as it is certainly not the case that those who designed CABM believed that communication between principal investigators and those working in their labs is a trivial matter. A concern for safety in dealing with potentially hazardous materials led, at LTL, to the enclosure of upper-level stairwells so that corridors could be defined as research space, and at CABM to a generic island bench and closed lab modules. My point is, rather, that there is no single obviously perfect building for contemporary molecular biology or biotechnology, and to resolve myriad design dilemmas (most with several reasonable and effective solutions), the design teams typified eventual users and endowed them with different traits, wants, and needs, which then served as evidence supporting one or another choice. Scientists had to be designed before their workplaces could take shape.

Outjobbing and Tethering

Design of any science building begins with a "primordial squeeze": The new building simply will never have enough space. Scientists/users always want more labs, more space for equipment, larger offices; university architects and facilities managers temper the scientists' big eyes with reminders of budgetary limits and finite land; architects try to squeeze everything the scientists want into a building that the owner/client can afford, and sometimes they succeed. The competing pressures of scientists' wants and budgetary constraints need a relief valve, without which the design process could become an unproductive and inefficient war. One source of relief is to act as if the building now on the drawing board is just the first of additional new facilities yet to come; that is, the design team collectively sustains the reality of "future expansion" as a place for everything that cannot fit. At LTL, the possibility of "future expansion" was tied to growth and change in the field of molecular biology and governed the siting of what would become LTL: "Locate the building to extend the full length of the Guyot complex in order spatially to contain and complement that complex from the south. This allows for appropriate expansion of Moffett to the south and of MBB [LTL] to the east."[45] Such plans were realized with the post-LTL construction of Schultz Lab immediately across College Walk. In Piscataway, talk of future expansion became a kind of manifest destiny: Fill up the rest of the space along the axis between UMDNJ and Rutgers' Waksman Institute. (So far, it hasn't happened.)

Another relief valve might be called "outjobbing," in which certain people, facilities, and functions are assigned to space in other extant buildings nearby (either taking advantage of facilities already there or remodeling old space as part of the new building project). If some research activities and equipment can be outjobbed, the new building's finite square footage can be used for other, possibly more important purposes, so long as the outjobbed facilities are reasonably nearby and accessible. Arguments for outjobbing are usually couched in the language of "avoiding wasteful duplication" or "promoting shared use" (by residents of different buildings). Still, the decisions about who or what to outjob must be justified not just architecturally but sociologically: Identities must be assigned to people or collective units to justify their exclusion, which, in turn, requires the design teams to add still more detail to the identities of those worthy enough to move into the new facilities.

The particular lists of outjobbed facilities at LTL and CABM are instructively different. At LTL, the following functions were outjobbed to the Guyot-Moffett biology complex just to the north: central stockrooms, the main biology library, and central research facilities, such as oligonucleotide synthesis, peptide synthesis, DNA sequencing, confocal microscopy, electron microscopy, fluorescence-activated cell sorting, monoclonal antibody production, media production, additional animal facilities, and computer facilities. Several of these research facilities were initially located in standard lab modules at LTL, and then moved across College Walk as the full complement of faculty members in molecular biology were hired. The decision to move these service facilities to Moffett-Guyot was based in part on the fact that LTL was never specifically designed to house them in the first instance, and its modular labs were less than accommodating to the sometimes bulky equipment. It is also a tacit indicator of the status hierarchy among those contributing to molecular biology at Princeton: Moffett is not just old space, but less appealing space than LTL, even after a series of renovations. Who ends up over there? The service facilities are run by "a group of twenty technicians, technical staff, each of those facilities is run by a senior technical staff person." Levine puts it this way: "We centralized all of those facilities in a single floor of Moffett; we choose Moffett because it was better to scavenge bad lab space for that than good lab space." This outjobbing suggests that some of the seventeenth-century "invisible technicians" described by Steven Shapin are still not there three centuries later.[46]

Outjobbing at CABM was more extensive and encouraged by concerns other than mere efficiency. Some facilities were excluded from CABM on grounds that they were already easily available in the research tower of UMDNJ or at Waksman, both next door: the vivarium (animal facilities), media resources, computer mainframes, library, cafeteria, and shops at UMDNJ; greenhouses at Rutgers; and lecture halls and teaching labs at both other buildings were deemed sufficient to meet the needs of those who would move into CABM. But desires to avoid costly redundancies only begin to explain outjobbing in Piscataway: CABM was conceived as a physical and organizational means to knit together molecular biologists at Rutgers and UMDNJ into webs of instrumentational interdependence that would, with the infusion of new blood into CABM, elevate the performance and reputation of them all.

Even before construction of a new building for CABM began, two of its component parts were already established at other sites. First, Network Laboratories, initially for bioimaging, biomaterials development, DNA synthesis, flow cytometry/cell sorting, molecular biological computing, and monoclonal antibodies (not all survived for long), were distributed throughout UMDNJ and the Waksman. The goal of the Network Labs is to provide research services (equipment and expertise) for academic and industrial users. Second, two Satellite Facilities (the name itself highlights outjobbing), one for fermentation in Rutgers' Waksman, another for clinical research in UMDNJ's Robert Wood Johnson University Hospital in New Brunswick, focus on practical application of the latest tools of molecular biology (for example, producing cells for clinical research or carrying out drug trials). Network Laboratories and Satellite Facilities came on-line before construction began on CABM's building, mainly because of the availability of funding that, in effect, needed to be spent immediately.

These facilities and personnel were spread around various sites of UMDNJ and Rutgers to insinuate the hoped-for excellence of CABM into extant scientific mediocrity. But the *insulation* of CABM from UMDNJ and Rutgers was just as important: Evidently, the staff who tended machines in the Network Labs and Satellite Facilities were not up to the standard demanded of people to be hired into CABM, and so they took their organizational place inside but their physical place outside. "Aaron Shatkin had a vision of a very well defined, high quality basic research organization. . . . He looked at the people who were running these network labs and said: 'No way I want any of them part of my organization. They're not high enough quality. Okay to continue them as a service organization, they can certainly chunk out some DNA syntheses or run some cell sorts. . . . But I'm not giving any of my space to them.'"[47] Status distinctions and judgments of "quality," as at LTL, played a part in deciding who got outjobbed from the newest and fanciest space.

Outjobbing solves one problem as it creates another: Needed facilities located outside the new building must be accessible, for scientists and their equipment, in sometimes inclement weather. The solution is "tethering," by which the new building is physically linked to others nearby and thus functionally linked by the distribution of needed facilities in more than one place. Tethers are not cheap. At LTL, "the cost of the service tunnel between Molecular Biology [LTL] and Moffett is included in the $22 million. It is presently estimated to cost $500,000."[48] Tethers are not just physical and functional linkages between science buildings; they may also become symbolic or organizational linkages creating a shared collective identity for those who are architecturally segregated—unwanted connections, in the case of Princeton. "Although a connection between Biology animal facilities and M.B. animal facilities is desirable, the two should be separate and autonomous facilities."[49] The tethering of LTL to Moffett and Guyot should be functional but unannounced: "Make the physical connection between the MBB [LTL] and Moffett underground rather than above-ground for varying reasons involving aesthetics, economy, circulation and varying floor levels." "The Committee felt very strongly that there should not be a bridge connection to Moffett."[50] Tethering was a threat to the distinctively special and separate place that LTL would become, averted by putting the link underground.

Tethering at CABM approaches the absurd. The new building is physically linked to the UMDNJ complex via a short underground tunnel, and to Rutgers' Waksman Institute by a long (800 feet), aboveground, enclosed "connector" (cost: also $500,000; See figure 20.8). With facilities needed by CABM-housed researchers located in both UMDNJ and the Waksman, these tethers are in principle justifiable on functional

20.8

Center for Advanced Biotechnology and Medicine, Piscataway, New Jersey. Exterior view, connector to Rutgers/Waksman. The facade of CABM, lacking the distinctive decoration of the Lewis Thomas Laboratory, but crisp and handsome nevertheless. The stacks of UMDNJ's "hovering mass of dour concrete" loom behind, while the aboveground corridor begins to reach for Rutgers' Waksman Institute to the left.

grounds (moving equipment, shielded from weather), but "function" is only a small part of this story. Unlike Princeton, where the tether is effaced (by going underground) to display LTL's autonomy, the tether from CABM to the Waksman prominently and inescapably announces that the new building is as much a part of the relatively distant Rutgers as of the much closer UMDNJ. "So they built a tunnel [to UMDNJ]. This starts to get very suspicious. . . . Does that mean that someday UMDNJ is going to take over the program, because it's actually then physically a part of the UMDNJ complex? So, then we got the word to build what we call 'the leash'. . . . It looks really stupid. If that building [CABM] was going to be connected to UMDNJ with a tunnel, then it was going to be connected to our [Waksman Institute] above ground. Nobody ever uses it. In fact, there's no traffic between these two organizations. It's strictly a matter of physically linking the two facilities, so that . . . the joint nature of the enterprise would be preserved."[51] "A [half] million dollars worth of symbolism; that's the covered all-weather connector to the Waksman."[52] Tethers in Piscataway announce a collective identity that, at the organizational level, draws together CABM, UMDNJ, and Rutgers.

LTL and CABM have each succeeded fabulously as new buildings for science, in the eyes of their users and their designers. Arnold Levine says of LTL: "This is the best lab I've ever seen or ever had the pleasure of working in. It's just a very, very good place to work. . . . I'm appalled by most labs after I've worked in this one."[53] The praise for CABM is as loud: "It's an absolutely marvelous building." "I personally think that it's probably the best laboratory building that's been designed in this country."[54] But people in Princeton and Piscataway draw on different metaphors to help them describe how each building succeeds so well. LTL is likened to an Elizabethan manor house and an old schoolhouse; CABM is likened to a temple, a showplace, and a Lord & Taylor among Kmarts.[55]

The different metaphors are an apt synthesis of the different identities designed and built into the two places for biological research. It is too simple to cast the obviously academic LTL as producing basic research and the corporate-tinged CABM as producing applied research: Each building, during design, played with the basic/applied dichotomy in distinctive ways. The "case statement" for LTL (2 March 1983) suggests that "[w]hile Princeton's scientists will focus on basic research in plant genetics, the methods they will use and develop will be important to agricultural scientists developing new crop strains in the battle against worldwide hunger." The program document (18 September 1985) suggests that an identical hope inspired CABM: "To become a world-class center for research and basic sciences in biotechnology. . . . To facilitate the transfer of basic advances in biotechnology and biomedical science to practical application by the creation of a core of scientists who understand the basic research needs of New Jersey's biomedical industry." Both facilities were built to nurture excellent basic and applied research alike; their identities would straddle that amorphous, if consequential, boundary.

LTL and CABM were designed for different users, constituencies, audiences, and consumers, and so our sense of the identity (or meaning) of science and scientists changes as we move from one to the other. Princeton's unsurpassed commitment to its undergraduates made teaching labs and large lecture halls a necessary feature of LTL; the State of New Jersey's desire to use CABM to spur its biotechnology industry made it just as necessary for that building to include CEO-friendly reception rooms and to make faculty offices easily reachable from mahogany-and-terrazzo atria. We learn something about the identities of the scientists who work in the two buildings by seeing how they were typified as design decisions were haggled over and settled: Biologists conceived in terms of academic rank, research specialty, and a dependency on this or that equipment (LTL) yield labs of a certain size, shape, and location different from those based on scientists conceived as safety conscious, security conscious, and territorial (CABM). Finally, architectural segregations (or connections) of people and practices reified and identified the collective identity of "molecular biology of Princeton" or "biotechnology at Piscataway." Establishment of a top-flight molecular biology department at Princeton University depended in part on distance and visual separation from other, less-distinguished programs on campus. (The physical link is only underground.) CABM was not allowed to become an autonomous center of excellence: It is prominently attached to both of its jealous co-owners, and its guts are pulled first toward UMDNJ and then toward Rutgers.

Buildings stabilize identities simply through their durability: It would cost considerably more than a haircut and new wardrobe to give LTL or CABM a fresh look. But stabilize how much? A decade or so after their opening, the buildings look much the same. Yet they welcome different constituents, as Princeton students graduate and fragile biotech start-up companies come and go. They house different names and faces, as faculty get lured away to even newer facilities on campuses elsewhere. With age, they blend in with their surrounds, almost seamlessly becoming a part of the local landscape. Do these new constituents, residents and neighbors also take on the identities that LTL and CABM were designed to achieve more than ten years ago—or different ones?

Notes

1. Goodman, "How Buildings Mean." See also Shoshkes, *Design Process*, which includes a discussion of LTL.
2. Gecas and Burke, "Self and Identity"; and Stryker, "Identity Competition."
3. Silver, "Role Transitions." Hormuth, *The Ecology of the Self*, also considers the move from home to dorm room in terms of identity. On the general subject of identity and "things": Csikszentmihalyi and Rochberg-Halton, *The Meaning of Things;* and Douglas and Isherwood, *The World of Goods*.
4. Goffman, *Presentation of Self*, p. 117.
5. Halle, *Inside Culture*.

6. Rapoport, "Identity and Environment."
7. Bourdieu, *Outline,* pp. 114 ff.
8. Cerulo, "Identity Construction." On buildings and gender identity: Spain, *Gendered Spaces.* On buildings and social class identity: Harvey, "From Space to Place."
9. Melucci, "Process of Collective Identity."
10. Cornell and Hartmann, *Ethnicity and Race.*
11. Zuckerman, "IBM's New Headquarters."
12. Anderson, *Imagined Community,* p. 181. Other works on the architecture of colonialism include Metcalf, *An Imperial Vision* and Wright, *The Politics of Design.*
13. Vale, *Architecture,* p. 185. Other works on buildings and political identity include Goodsell, *Social Meaning;* Keith and Pile, eds., *Place and the Politics of Identity;* and Sennett, *Flesh and Stone.*
14. Blu, "'Where Do You Stay At?'" pp. 207, 218.
15. Interview, Gary Ireland, Princeton University staff architect, LTL 10 November 1994.
16. Case statement for LTL, 2 March 1983.
17. Interview, Arnold Levine, 11 November 1994.
18. Both extracts in this paragraph are from an interview with Arnold Levine, 11 November 1994.
19. Interview, Matt Leslie and Jon Romig, Payette architects, 13 November 1991.
20. Letter, Norman Smith (UMDNJ member of the programming committee) to William Wilson, 2 August 1985. Until the late 1960s, the Piscataway branch of UMDNJ was known as the Rutgers School of Medicine, to distinguish it from the main medical campus in Newark.
21. Interview, David Rosenblatt, Rutgers University, Facilities and Construction, 11 June 1991.
22. Letter, Robert Namovicz, CABM administrator, to David Rosenblatt, 10 September 1986.
23. Interview, Matt Leslie, 13 November 1991.
24. Interview, Bob Schaeffner, Payette architect, 14 November 1991.
25. Memorandum, New Jersey Commission on Science and Technology, 21 January 1987.
26. CABM Program document, 18 September 1985.
27. CABM Concept document, 26 July 1984.
28. CABM Program document, 18 September 1985.
29. Interview, Robert Namovicz, 5 December 1990.
30. Princeton created a Department of Biochemical Sciences in 1968 that in the end failed partly for geographic reasons: Its faculty (which included Arnold Levine) was split between buildings for chemistry and for biology and crowded into each. The university let the hybrid department dwindle to almost nothing (Levine left for Stony Brook in the mid-1970s) so that it could start fresh in 1983 with a freestanding Department of Molecular Biology. Levine was hired back to Princeton in 1984 with a mandate to build a department that would eventually fill twenty new faculty lines. Levine began when the design of LTL was literally on the ground floor. The remainder of the university's biologists formed the Department of Ecology and Evolutionary Biology, located across College Walk in Moffett and Guyot Halls.

    The story at CABM is much the same: The building was intended to recruit molecular biologists who were better than those already employed at UMDNJ and Rutgers. Recruitment started with Aaron Shatkin and Alba LaFiandra, both from Roche Laboratories, who entered the process after much of the building had been programmed and even designed (by facilities managers and scientists who would never move in at the two constituent universities, along with Payette architects). Eventual CABM scientists would also have an academic affiliation in either Rutgers or UMDNJ (in roughly equal numbers). "The rationale for this approach is that it provides a mechanism for bringing scientific 'new blood' into the state" (CABM Program Document, 18 August 1985).

31. Social phenomenologist Alfred Schutz developed the concept of "typification" (*Collected Papers,* Vol. I, pp. 15 ff.).
32. Interview, Arnold Levine, 11 November 1994.
33. Meeting minutes, 31 March 1983.
34. Interview, Arnold Levine, 11 November 1994.
35. LTL Progress report, 13 May 1983.
36. Interview, Arnold Levine, 11 November 1994.
37. Interview, Alba LaFiandra, CABM Lab Manager, 5 December 1990.
38. Interview, William Wilson, CABM Project Manager and Payette architect, 14 November 1991.
39. Interview, Alba LaFiandra, 5 December 1990.
40. Interview, Harry Borbe, UMDNJ, Facilities and Planning, and David Rosenblatt, 11 June 1991.
41. Interview, Harry Borbe, 11 June 1991.
42. Interview, Robert Namovicz, CABM Executive Officer, 5 December 1990.
43. Interview, Aaron Shatkin, 11 June 1991.
44. Interview, William Wilson, 14 November 1991.
45. LTL Progress report, 13 May 1983.
46. Interview, Arnold Levine, 11 November 1994; and Shapin, "The Invisible Technician." Shapin's article describes transparent assistants who contributed to the experimental prowess of Robert Boyle and other prominent natural philosophers, but who remained faceless in graphic depictions of laboratories, out of view during real experiments, and whose names were typically effaced from reports of findings.
47. Interview, Robert Namovicz, 5 December 1990.
48. Meeting minutes, 15 September 1983.
49. Meeting minutes, 14 April 1983.
50. LTL Progress report, 13 May 1983; meeting minutes, 28 April 1983.
51. Interview, Mark Rozewski, Rutgers Planning Office, 12 June 1991.
52. Interview, Norman Edelman, UMDNJ Dean, 12 June 1991.
53. Interview, Arnold Levine, 11 November 1994.
54. Interview, Norman Edelman, 12 June 1991; Interview, Jon Romig, 13 November 1991.
55. In sequence: Letter, Robert Venturi to Jon Hlafter, Director of Physical Planning, Princeton University, 4 August 1983; Interview, Gary Ireland, 10 November 1994; Interview, name withheld ["I mean to use words which I hope you never attribute to me . . ."]; Interview, Harry Borbe (twice), 11 June 1991.

Bibliography

Anderson, Benedict. *Imagined Communities: Reflections on the Origin and Spread of Nationalism.* Rev. ed. London: Verso, 1991.

Blu, Karen I. "'Where Do You Stay At?': Homeplace and Community Among the Lumbee." In Steven Feld and Keith H. Basso, eds., *Senses of Place,* 197–227. Sante Fe, NM: School of American Research Press, 1996.

Bourdieu, Pierre. *Outline of a Theory of Practice.* Cambridge: Cambridge University Press, 1977.

Cerulo, Karen A. "Identity Construction: New Issues, New Directions." *Annual Review of Sociology* 23(1997): 385–409.

Cornell, Stephen E., and Douglass Hartmann. *Ethnicity and Race.* Thousand Oaks, CA: Sage, 1998.

Csikszentmihalyi, Mihalyi, and Eugene Rochberg-Halton. *The Meaning of Things: Domestic Symbols and the Self.* New York: Cambridge University Press, 1981.

Douglas, Mary, and Baron Isherwood. *The World of Goods: Toward an Anthropology of Consumption.* New York: Norton, 1979.

Gecas, Victor, and Peter J. Burke. "Self and Identity." In Karen S. Cook, Gary Alan Fine and James S. House, eds., *Sociological Perspectives on Social Psychology,* 41–67. Boston: Allyn and Bacon, 1995.

Goffman, Erving. *The Presentation of Self in Everyday Life* (1959). Woodstock, NY: Overlook, 1973.

Goodman, Nelson. "How Buildings Mean." In Nelson Goodman and Catherine Z. Elgin, *Reconceptions in Philosophy and Other Arts and Sciences,* 9–48. Indianapolis, IN: Hackett, 1988.

Goodsell, Charles T. *The Social Meaning of Civic Space: Studying Political Authority Through Architecture.* Lawrence: University Press of Kansas, 1988.

Halle, David. *Inside Culture: Art and Class in the American Home.* Chicago: The University of Chicago Press, 1993.

Harvey, David. "From Place to Space and Back Again." In *Justice, Nature and the Geography of Difference,* 291–326. Oxford: Blackwell, 1996.

Hormuth, Stefan E. *The Ecology of the Self: Relocation and Self-Concept Change.* New York: Cambridge University Press, 1990.

Keith, Michael, and Steven Pile, eds. *Place and the Politics of Identity.* New York: Routledge, 1993.

Melucci, Alberto. "The Process of Collective Identity." In Hank Johnston and Bert Klandermans, eds., *Social Movements and Culture,* 41–63. Minneapolis: University of Minnesota Press, 1995.

Metcalf, Thomas. *An Imperial Vision: Indian Architecture and Britain's Raj.* Berkeley: University of California Press, 1989.

Rapoport, Amos. "Identity and Environment: A Cross-Cultural Perspective." In James S. Duncan, ed., *Housing and Identity,* 6–35. London: Croon Helm, 1981.

Schutz, Alfred. *Collected Papers.* Vol. 1, *The Problem of Social Reality.* The Hague, the Netherlands: Martinus Nijhoff, 1971.

Sennett, Richard. *Flesh and Stone: The Body and the City in Western Civilization.* New York: Norton, 1994.

Shapin, Steven. "The Invisible Technician." *American Scientist* 77(1989): 554–563.

Shoshkes, Ellen. *The Design Process: Case Studies in Project Development.* New York: Whitney Library of Design, 1989.

Silver, Ira. "Role Transitions, Objects, and Identity." *Symbolic Interaction* 19(1996): 1–20.

Spain, Daphne. *Gendered Spaces.* Chapel Hill: University of North Carolina Press, 1992.

Stryker, Sheldon. "Identity Competition: Key to Differential Social Movement Participation?" In Sheldon Stryker, Timothy Owens and Robert White, eds., *Self, Identity and Social Movements.* Minneapolis: University of Minnesota Press, forthcoming.

Vale, Lawrence J. *Architecture, Power and National Identity.* New Haven, CT: Yale University Press, 1992.

Wright, Gwendolyn. *The Politics of Design in French Colonial Urbanism.* Chicago: The University of Chicago Press, 1991.

Zuckerman, Laurence. "IBM's New Headquarters Reflects a Change in Corporate Style." *The New York Times* (17 September 1997).

VI

Centers, Cities, and Colliders

21

Architecture at Fermilab

Robert R. Wilson

In this essay, I describe a scientific project in which architecture was of crucial importance. In 1967, a 6,700-acre site in Illinois was chosen on which to construct a large laboratory for the study of the deepest depths of the atom. The proposed laboratory was to be built under the aegis of the Universities Research Association (URA), an association of college presidents.[1] The URA was responsible to the Atomic Energy Commission (AEC) of the federal government for the laboratory's construction and operation. The lab, named after the physicist Enrico Fermi, is now called Fermilab, short for Fermi National Accelerator Laboratory. (See figure 21.1.)

The research was to be conducted by building a large synchrotron to accelerate protons to energies never before attained.[2] It is curious that to examine the smallest details of nature, the largest instruments must be used. Thus the synchrotron would be comprised of 1,000 magnets arranged in a ring-shaped tunnel more than one mile in diameter. The design for this 200-billion-electron-volt[3] (eV) monster had been executed at the venerable Lawrence Berkeley Laboratory in the firm expectation that it would be built there in sunny California.

However, a battle royal had erupted between physicists in the West and those in the East about the location of the accelerator. As a result, when Congress authorized the construction of the project, the location of the site was moved, almost by force of arms, to Illinois. The URA then asked the offended and bitter Berkeley designers to come to Illinois to oversee the construction of their machine. Being human, and perhaps being accustomed to the pleasant life of California, the Berkeley people refused as a group to move to Illinois. After that hullabaloo, I was chosen to be the Director in early 1967.

21.1

Fermi National Accelerator Laboratory (Fermilab) site, Batavia, Illinois, looking north. The large ring is 1.25 miles in diameter. The Central Lab building, named Robert Rathbun Wilson Hall in 1980, is visible at the left side of the ring. Photograph courtesy of Fermilab Visual Media Services.

Although I was the only employee when we started, the staff had grown to more than 1,000 when I left eleven years later. Thus there are many tales to be told. I shall tell this particular tale as though I were the only employee, but in truth I am but reporting on the work done by many.

At the time I was offered the job, I was a happy professor of physics at Cornell University. As I accepted the offer, I was intrigued by the idea of helping create new knowledge about the atom. But I was also attracted by the opportunity to participate in an exciting architectural adventure. There would, however, be more than just romantic adventure involved in the architecture of the lab: money, for one thing.

The Berkeley group had estimated that the proton accelerator they had designed would cost about $350 million to build. That included everything necessary to turn a vacant site into a functioning laboratory. About half of the costs were for the synchrotron and the concomitant laboratories and scientific equipment; the remaining half would go for architectural and engineering constructions, that is, buildings, utilities, roads, site improvement, and so forth.

It was bad enough news for the Berkeley physicists to learn of the changed site for the accelerator, but at the same time the total funding for the project was reduced quite arbitrarily from $350 to $250 million. This blow contributed further to the Berkeley group's unhappiness and probably contributed to their decision not to move to Illinois. I had been a fierce critic of the Berkeley design and had claimed, perhaps too vociferously, that it cost too much and did not give enough energy. So when I became the director, it was up to me to make good on my claims, even though the money had been seriously reduced. Clearly I had a problem. I needed architects to perform miracles by building larger and more beautiful structures but at half the cost. Could they, would they, do that?

There were many reasons for desiring good architecture. For one thing, we were competing with other laboratories for a select group of expert physicists to join us in building our laboratory. Their expertise could well make the difference between our success or failure, between excellence or mediocrity. Physicists, for the same aesthetic reason that they appreciate the physical world, also enjoy theater, music, museums, and good architecture. The competition in this respect from the other national laboratories was formidable. Berkeley has its low hills overlooking the San Francisco Bay; the European Center for Nuclear Research (CERN) in Geneva has its magnificent view of the Alps; Brookhaven on Long Island is close to sailing on the Sound. They were centers of beauty and culture from which it would be hard for us to move anyone to the rigors of the prairie.

Make no mistake: The laboratory's success would be judged entirely on the quality of the physics research done there, on the discoveries made, not on the quality of the architecture. Nevertheless, many intangible ingredients go into making a successful creative laboratory, and architecture is one of them.

If there were reasons to desire good architecture, there were also reasons to be apprehensive about the likelihood of achieving it. Most other AEC laboratories were just plain ugly. Perhaps it is apocryphal, but at that time it was said about AEC builders that their buildings did not have to be cheap, they just had to look cheap! This fear of AEC standards was reinforced when some of the AEC administrators stated that, because there was no experience in design or construction at our nonexistent laboratory, they wanted the AEC Construction Division to be in charge of all of the nonscientific construction. In fact the congressional committee in charge of our funding[4] had suggested that the AEC should require exactly such a scheme. It made my blood run cold.

For those reasons, before accepting the position of Director, I wrote a letter to the Chairman of the AEC (Glenn Seaborg, an old friend from our student days at Berkeley) to the effect that I would accept the position only if I were responsible for both technical and nontechnical designs and construction. I added that "although I did not believe that good architecture would always require extra costs, there would be some occasions when the architecture would require extra expenditures." In agreeing with me, Seaborg also expressed his desire to see some good buildings go up, but even so I still had to stay within the $250 million limit . . . or I would roast in Hell. I happily accepted the stricture.

Our site might not exactly be one's first choice for distinctive architecture. It is not only dead flat but is set on a plain that is flat all the way from Buffalo to Denver. It has a long, brisk winter and a withering summer. Still, I became fond of it. I liked the idea of a clean slate on which to create a new laboratory. I liked the remnant of the "Gran Bois" on the site with its memories of the French voyageurs and of the economic adventurer John Law, whose "Mississippi Bubble" collapsed in 1720, a symbol, perhaps, to physicists of the opportunity and danger of flying too close to the sun.

The site did have great possibilities, if only we did not spoil them. We treasured the trees, and our deer, and our grouse, and our owls, and even our bugs as we awaited the coming of the brave architects. It was not a question of finding the architects: They found us. As soon as I had been named Director of the lab, I received a call from a Colonel William D. Alexander, the project director of DUSAF, which was a consortium or joint venture of four firms that had previously been qualified by the AEC.[5] They had been chosen by the Berkeley study group for their design of the original accelerator project. Col. Alexander asked for the opportunity to make a presentation to me in hope that his consortium would be chosen for the new Illinois project.

I just hated the idea of choosing them. I was not impressed by their plans. Everything seemed overdesigned, and "no birds sang," at least not to me. Their structures seemed to accord with the AEC dictum of "looking cheap and being expensive." If I was unfairly prejudiced, perhaps it was because the DUSAF people had been so loyal to their clients in defending their designs and the project during the arguments about moving the site from California. I wanted a "new broom." Nevertheless, I did give them a hearing, largely because the AEC procedure for qualifying a new architecture/engineering firm was so complex that it would have required about a year to qualify a new group.

An important part of my strategy for saving money was to move as quickly as possible. I had announced that, instead of the seven years required by the Berkeley plans, I would do it in five years. Hence one full year of marking time without architects loomed large as a delay in our plans. It was a good reason to choose DUSAF if at all possible.

A team of half a dozen DUSAF people along with Col. Alexander arrived at my Cornell office for the presentation. Unfortunately, it did not go well. The DUSAF experts patronized me (as well they might). They could not understand how, in the wilderness of Illinois, it would be possible for one professor to build from scratch a fully staffed new laboratory. They knew that it had taken the great E. O. Lawrence more than thirty years to build the Lawrence Radiation Lab at Berkeley. Accordingly, they had concluded that DUSAF would have to do the whole job, including the accelerator and the experimental facilities, as a "turn key" operation.

Presumably, this was how they had previously executed military projects, never mind the cost nor where they would find physicists with the requisite skills. They brought a huge organization chart with many little boxes showing how they would get the job done. A small box at the bottom right of the chart indicated the Director and a few colleagues receiving a key which would turn the whole thing on!

I listened quietly to their presentation, and when they had finished, I went to the blackboard and drew on it my own chart, also with a small box at the bottom right. It was marked DUSAF; I remarked that we might want to have some badly needed outhouses built by them. Of course they were just as offended as I had been, and they angrily departed. No, it had not gone well. Science and architecture were off to a bad start.

A few days later, though, I received a call from Col. Alexander, a Virginia gentleman if there ever was one. He explained to me that they had misjudged the situation, that they wanted the work, that they would please me, that they could make things quickly, beautifully, and economically, too. There would be no personality conflicts because anyone who even appeared to be in conflict would be removed at once to another location. DUSAF, he said, wanted to and would do a great job for us. I was impressed by his flexibility and perseverance, and so was Norman Ramsey, president of the URA, who had sat in on their presentation. Norman made a careful study of their competence, and eventually we agreed that, despite our fears, if we could negotiate a mutually acceptable fee, then we would accept them for the job. It was a decision I have never regretted.

At first, we did not get along at all well. I attended their weekly meetings because so much was at stake. They did not know quite what to make of me. I was a physicist, so I could calculate the strength of support beams or the flow of liquids or gases through pipes. To make matters worse, I had also attended the Academia Belli Arte in Rome and studied sculpture (see figures 21.2 and 21.3). Thus I could, and did, criticize the curves and volumes of the buildings they were creating. The architects continued to repeat the kind of designs that they had made at Berkeley. My occasional reaction to a too expen-

21.2

Robert R. Wilson at work in 1978 on his sculpture *Aqua alla Funi,* which now sits in the cooling pond adjacent to Wilson Hall at Fermilab. Photograph courtesy of Fermilab Visual Media Services.

21.3

Broken Symmetry, sculpture by Robert R. Wilson at the main entrance to Fermilab, with Wilson Hall visible in the distance. Depending on the point of view, one sees either perfect or broken symmetry, reflecting an important principle of the particle physics being studied at the lab. Thick iron plates from the USS Princeton were used in neutrino experiments at Fermilab, and deck plates from the same decommissioned ship were used in the sculpture—"swords into plowshares." Photograph courtesy of Fermilab Visual Media Services.

21.4

The sixteen-story Robert R. Wilson Hall at Fermilab, exterior. To decide how high this central lab building ought to be, Wilson went up in a helicopter and had the pilot hover at various altitudes as he plotted an "aesthetic factor" as a function of height. The curve rose sharply to about 75 feet, where it began to flatten as the Fox River Valley came into view. He concluded that the building should be at least 200 feet tall, and taller if possible. Wilson Hall is 250 feet tall. Photograph courtesy of Fermilab Visual Media Services.

sive or too awkward design was to have a mild fit. This got my message across that we just did not have that kind of money to spend, but it was no way to foster a good relationship.

This difficulty was resolved when I reminded Col. Alexander of his pledge. He immediately put in a new project manager, E. Parke Rohrer, who solved all my administrative and cultural problems with the architects. Parke threw himself into the project with such enthusiasm that I took a chance and appointed him as Associate Director in charge of all architectural construction. Usually in a project of this size, a whole department goes over plans and costs before going to the AEC for approval. My scheme short-circuited all that and saved us all kinds of money and time, both of which were in short supply. Of course everything depended on having such a reliable and talented person as Parke Rohrer.

The first task was to locate the synchrotron and its experimental areas on the site plan.[6] This was done mostly to maximize the length and number of experimental areas, but ecological considerations for the site, such as trees, were given due regard. My own sentiment then was to have just one large building located right on a tangent of the accelerator and to have as much as possible of the critical parts of the accelerator and the utilities located in the lower floors. The offices and laboratories would then be above. I hated the clutter and bad communication that results from having a multitude of small buildings. To decide how high this central lab building ought to be, I went up in a helicopter and had the pilot hover at various altitudes as I plotted an "aesthetic factor" as a function of height. The curve rose sharply to about 75 feet where it began to flatten as the Fox River Valley came into view. The sky, the sunsets, the Illinois landscape, all looked better at the higher levels. I concluded that the building should be at least 200 feet tall and taller if possible (it turned out to be 250 feet). (See figure 21.4.)

Years earlier, in France, I had been delightfully involved with the question of height while driving from Paris to see the Chartres Cathedral. As you go along, at first you see it, then you don't, then it seems to flirt with you, and finally bursts out in all its radiant splendor. Perhaps it was hubris to hope for a similar effect on approaching Fermilab. Ultimately, it was not Chartres, but Beauvais Cathedral—with its soaring interior height—that was to have a close resemblance to the Central Lab. (See figures 21.5 and 21.6.)

When the architects asked if I had any predilections concerning the style of the building, I responded that the Ford Foundation Building in New York City, with its handsome atrium, appealed to me, except that it had been one of the most expensive buildings ever built. Could they not do a similar building, but inexpensively? They took up the challenge.

Competition developed in which each of the firms comprising DUSAF (except for the Fuller Co.) designated a group to make a design. We announced a date when all the designs would be presented. I would then select one for further development. We all gathered on the designated day for the dramatic moment. There was a leaning building

21.5–21.6

Robert R. Wilson Hall at Fermilab, interior. Models for the soaring interior space included the great gothic cathedrals of France, particularly Beauvais, and also the Ford Foundation Building in New York City. Photographs courtesy of Fermilab Visual Media Services.

of triangular cross-section, a narrow but high circular building (a Prell shampoo bottle had been used for the model), a far-out, upside-down pyramidal structure presented by Max Urbahn and his associates, our official beaux arts firm. Finally, there was a darkhorse entry by a young architect named George Adams, whom Parke Rohrer had hired "off the street." His building was a truncated cone with a domed atrium. A cost estimate as well as a model had been submitted for each entry. After discussing the merits of each of the buildings with our staff, to Urbahn's disgust (his building had been architecturally most significant, except for being upside down) I chose the Adams entry as coming closest to satisfying our technical and aesthetic needs, as well as being substantially lower in cost. By that time, I had realized how slowly our money would be appropriated, so I announced that the accelerator would be built first and the Central Lab building would come later. Still, we had learned a lot from the exercise; we had also enjoyed it tremendously, and Adams' building did eventually metamorphose into the Central Lab, which was principally designed by Alan Rider.

To get our buildings done on time and within budget—and this was crucial for keeping our fast schedule—we followed an effective procedure. First, the buildings were designed for efficient, economical construction. We then solicited equally economical construction bids. Once the lowest bidder had been chosen, Parke Rohrer would sit them down for a heart-to-heart talk at which he offered a financial bonus if the contractor would finish his building ahead of schedule. It became a game—and a point of pride—for the contractor to get the "prize" money, and usually they succeeded. Yet the total of the premium was but a few thousand dollars out of millions, and it was worth much more to us to get things done on time. Cost penalties often do not work because the builder's lawyers can usually find an excuse for delay without penalty.

I had many dreams for the project as it started, dreams of physicists coming to do superb physics from all over the country, indeed from all over the world, dreams of ecological delights, and dreams of an architectural paradise. How much those dreams became reality is for others to judge. As an idiosyncratic director I was clearly bound to come into conflict with a new micromanaging Department of Energy in Washington. But when I left in 1978, I was pleased by the experiments that had been made and the architecture that had materialized (see figures 21.7 and 21.8). The next Director, Leon Lederman, led Fermilab to new heights of physics during his term of ten years. The subsequent and current director, John Peoples, and his staff have attained prodigies of discovery. So a healthy Fermilab goes on! Change is a mark of health in physics, but the stones of architectural construction have remained the same, a mark of good health too.

21.7

In the Fermilab pump house at Casey's pond, a logarithmic spiral housing was constructed to enclose some water pumps and to evoke the work of Leonardo da Vinci. Photograph courtesy of Fermilab Visual Media Services.

471 | ARCHITECTURE AT FERMILAB

21.8

Utility Poles, Fermilab. Wilson considered the standard utility pole to be ugly, so he went to Commonwealth Edison with his own design for power poles for Fermilab. They were originally "outraged," but after a lot of fighting, the electric company gave in and used Wilson's design. Photograph courtesy of Fermilab Visual Media Services.

Notes

Thanks to Adrienne Kolb, Archivist at Fermilab, for assistance with the images and captions that accompany this essay.

1. The URA was an organization set up specifically to serve as contractor to the government for the construction and operation of the new laboratory. It consisted of 46 member universities, the presidents of which formed a Council of Presidents, who elected a Board of Trustees, which chose a President of the URA and a Director of the Laboratory. I was the first to be chosen as Laboratory Director.
2. A synchrotron is a device for accelerating protons to high energy. It is a circular chain of many magnets, 1.25 miles in diameter in this case. (Although small compared to the late, lamented Superconducting Supercollider in Texas, the Fermilab accelerator remains the most powerful in the world.) After acceleration through the magnet ring, the protons emerge from the synchrotron in a narrow beam that can be directed at a target of other particles at rest. Collisions occur, and the measurement of the properties of the particles resulting from the collisions gives clues that allow the nature of the particles' inner structure to be deduced.
3. The electron volt is a unit of energy.
4. The Joint Committee on Atomic Energy. U.S. Congress. The Chairman in 1968 was Senator John Pastore of Rhode Island and the Vice Chairman was Congressman Chet Holifield of California. John T. Conway was the Executive Director for the Committee.
5. The four joint-venture firms that made up DUSAF were Daniel, Mann, Johnson & Mendenhall; the Office of Max O. Urbahn; Seelye Stevenson Value & Knecht, Inc.; and the George A. Fuller Company (the same venerable firm that built the Lincoln Memorial).
6. I am quoting freely for the next few pages from pages 173 to 177 of the 1987 *Annual Report of the Fermi National Accelerator Laboratory, Batavia, Illinois.*

22

THE ARCHITECTURE OF SCIENCE: FROM D'ARCY THOMPSON TO THE SSC

Moshe Safdie

Science has touched my work through a series of inspirations. The first of these came after my graduation from McGill University in 1962, when I went to Philadelphia to work for Louis Kahn. Kahn, I felt, was the architect from whom I could learn the most, and this intuition was quickly confirmed. Through Kahn and his collaborator, Ann Ting, I was introduced to the book *On Growth and Form,* by D'Arcy Thompson.

The book, written at the turn of this century, describes the science of morphology. Reading it, I felt I had discovered a profound connection between the processes that generate natural objects and their ultimate physical forms. Although obvious in one sense, to me this connection seemed to transcend commonplace experience. I became fascinated with the causality of form.

Perhaps the key sentence is in Thompson's introduction, when he writes about "forms which are so concomitant with life that they are seemingly controlled by life."[1] The analogy to architecture is to recognize that the form of buildings is also generated by a response to a multitude of requirements and physical constraints. "Concomitant with life" translates in architecture to the life intended in a building, the manner in which an architect shapes spaces and light in anticipation of the way a place will be used and experienced. That is how the program of requirements translates into a particular organization of spaces. But as in morphology, there is also a connection between structure and form. The properties of building materials, such as stiffness, surface tension, compression, and tensile capacity, for example, generate particular forms in architecture as well as nature. Even the manner in which natural organisms accomodate growth (as in gnomonic growth) and provide for distribution of data and sustenance in their nervous and arterial systems seems to have significant connections to building. In architecture I began to see that irreducible physical conditions like gravity, the properties of building materials, or the natural features of a site might contribute fluidly and seamlessly to the production of built form.

Thompson himself often made an explicit connection between science and architecture. In writing about the logarithmic formal structure of the nautilus shell (see figure 22.1), for example, he quotes the Reverend H. Moseley: "But God hath bestowed upon this humble architect the practical skill of a learned geometrician." Thompson argues, "the same architecture which builds the house constructs the door. Moreover, not only are house and door governed by the same law of growth, but, growing together, door and doorway adapt themselves to one another."[2] Thus a sort of spiritual reciprocity between form and material seems not only possible, but inevitable. Later, Thompson's connection between the vulture's wing and architecture—that the wing is efficient like the Warren truss—describes how a delicate, lattice-like, three-dimensional structure evolves to achieve maximum strength with minimal weight.

For decades, these observations have continued to haunt my own architectural imagination. The seasonal transformation of plants, for example, suggests that architecture should also be able to transform from season to season, that it should be convertible or changeable as it responds to different climactic conditions. The spiral patterns of leaves maximizing exposure to sunlight and the continuous movement of cactus leaves avoiding overexposure both offer beautiful metaphors for architecture that could respond to *specific* climates and topography.[3]

I later found a concept of architecture that most closely embodied Thompson's ideas in Bernard Rudofsky's book and great exhibition, *Architecture Without Architects*.[4] Rudofsky hypothesized that vernacular buildings, those not designed by architects, were exquisitely beautiful, elegant architecture. In Rudofsky's words, they merited the term "architecture" without being conceived by architects. Throughout the world, houses, granaries, villages, and towns have utilized traditional modes of construction such as mud-brick, stone, and wood. Roofs have been made of thatch, logs, and beams, as well as mud-bricks and small stones that form vaults and domes. All these techniques and forms have evolved over time.

Rudofsky's most profound message was that these buildings were beautiful, and that their beauty derived from their fitness to the purpose for which they had been constructed. Here, beauty was not something to be sought as an objective in itself or to be applied according to the artfulness of the maker, but—like the beauty of a flower, a beehive, or a nautilus shell—the beauty of architecture would emerge when its forms grew directly out of its purpose.

22.1

The nautilus shell. The close relationship between human-made designs and design in nature is apparent in the field of morphology. Thompson's analysis of the causality of the form of the nautilus shell is provocative: "The same architecture which builds the house constructs the door. Moreover, not only are the house and door governed by the same laws of growth, but, growing together, door and doorway adapt themselves to one another." D'Arcy W. Thompson, *On Growth and Form,* p. 777. Photo © Andreas Feininger.

22.2–22.3

The Pigeon House. The pigeon house is a cylinder within which a series of vaults create little cells for pigeons to leave their dung. The form appears as a direct response to the process of construction, the functional need, and a certain unselfconscious attitude to design. This vernacular architecture is as beautiful as any church or mosque, deriving its form directly from its purpose. Bernard Rudofsky has used similar examples in his seminal book *Architecture Without Architects.* Photos © Moshe Safdie.

On a trip to Iran some years ago, I discovered the architecture of which Rudofsky wrote. In Iranian villages, vaults and domes grow directly from their environment and the most efficient methods of construction. Where wood is available for construction, distances are spanned with straight beams that create steeply sloping and flat roofs. But in much of the Middle East, the scarcity of wood is visible in the architecture as well as the vegetation. Across skylines of most Middle Eastern cities, small stones and mud-bricks shaped or found locally are crafted into graceful, arched vaulted and domed roofs. An expressive example of an exquisitely beautiful series of structures is the "fertilizer factory," or pigeon house (see figures 22.2 and 22.3). As beautiful as any cathedral or mosque, the building is a cylinder within which a series of vaults create little cells for the pigeons to leave their dung. The form appears as a direct response to a process of construction, a functional need and a certain unselfconscious attitude to design.

The question of architectural form derived from tectonic principles and natural precedents is particularly relevant today, as stage-set architecture crowds out work that we recognize as "authentic." I have always been more inspired by and more able to identify

with a Medieval cathedral than with a Baroque church. In a Baroque church, the dominant effects are painted decoration on plaster walls, independent of the architectural tectonics. In the Medieval cathedral, visual richness results directly from the modes of construction: What you see is what you get. Stone piers, flying buttresses, and walls pierced with giant windows visually represent the physical forces that travel through them. The spatial experience is a play between structure and the manipulation of natural light. The whole interior glows and fades as the sun and stained glass windows jointly decorate the space.

In the age of Disney, we wonder what is real and what is not. We visit Venice to experience a certain tradition of urbanism, of buildings in relation to water, of the fusion of classicism and the architecture of geometries. We then experience the same thing at Disney, but made with papier-mâché and plaster held up by steel studs and other slap-dash techniques. We know it is not real. We know it is created for effect, just as a stage set in the theater is created for short-term use. But can life be sustained by continuous theater? How do we satisfy a need for the real and the authentic when so much of what surrounds us is there purely for effect? Is there a significant difference between the real and the virtual? Can daylight and sunlight be simulated so perfectly that we would not mind living in an interior world forever?

The study of vernacular architecture impresses on us one aspect of architecture analogous to morphology: that the evolutionary ongoing refinements and expressiveness of modes of construction and the needs of the building inhabitants will result in a spatially and formally satisfying aesthetic experience that we associate with the pleasures of observing organisms in nature.

Yet vernacular buildings also reveal the fundamental difference. In natural forms, we assume a direct causality among form, material, and survival. In man-made constructions, the decision to decorate, celebrate, assert power, ritualize, symbolize, or worship introduces a complexity that transcends causality. As Kenneth Frampton said, nature is not culture. In *Man and His Symbols*,[5] Carl Jung describes certain themes, forms and patterns that recur in cultures vastly separated in time and space. The connection between these forms and collective memory is one of the main sources of meaning in architecture. Architectural experience has the power to be at once rational and spiritual. Beyond providing shelter, architecture must attend to our emotions.

The idea that design combines the rational and the intuitive, causality and the arbitrary, is by no means accepted universally. The often quoted dictum "form follows function" asserts that, like the evolution of natural forms in morphology, a design with a perfect fitness to purpose shall in itself yield greatly satisfying aesthetic experience. This was suggested in the early work of Christopher Alexander in *Notes on the Synthesis of Form* (1970) and in my own writings in *Beyond Habitat* (1970).[6] For myself, I became increasingly uneasy with the qualities associated with the term "function," which increasingly was debased to mean the most utilitarian and simplistically defined aspects of program. Hence by 1982, I titled my book *Form and Purpose,* making the case that "purpose" embraced a much broader array of human needs than "function."

Some years ago, I came to discuss these concepts with General Israel Tal, the designer of the main Israeli battle tank, the Merkava. Tal, who had then just embarked on developing a concept for the Merkava, was intrigued by the notion that the most functional design would yield a beautiful design. To test the theory, somewhat tongue in cheek, he suggested I join the team that was busy designing the tank in its most embryonic phase.

Unlike other tanks, the engine would be at the front, so that the troops could enter and exit from the rear protected, not through the turret. The profile would be streamlined to minimize exposure to missiles and bullets. A plywood mock-up had been constructed.

I vividly remember my first visit to the top-secret warehouse where the plywood mock-up was constructed. There were all sorts of objects protruding from the main shape of the tank: ventilation outlets, equipment—it looked more like a crusty rhinoceros than a streamlined shark. The turret, too, while lying low, was awkwardly shaped, its geometry unrelated to the tank's body. My main objective was to internalize all the protrusions to minimize the tank's vulnerability, accommodating its functions within a streamlined envelope.

In time, an integrated geometry developed to generate the form of the body and turret. The first cast-metal model was somewhat reminiscent of the radical Citroën DS design. Smooth and streamlined, it was, many said, the most elegant tank on the road. Tal reported later that internalizing all protrusions made the tank less vulnerable in combat by deflecting incoming missiles and bullets. At least for Tal, the theory held. (See figure 22.4.)

Could we conclude from this experiment that form follows function, and that beauty would be achieved by a single-minded commitment to resolving "functional" issues? The case could probably be made that it is a valid theory for single-purpose objects (a tank, an airplane, a submarine) that are so governed by physical performance criteria that one could apply a direct morphological analogy. Most of what we design and build does not fit into such singular physical criteria. Myth, the human psyche, and our emotional response come to play a major role, hence the replacement of the term "function" with that of "purpose."

Recent debate among architectural theoreticians has attempted to set up the aspects of ritual and iconography as polar opposites of utility. The dichotomy seems to fall between basic shelter and high culture: A higher level of creativity and inventiveness, like that of the artist, comes to bear only in architecture produced for high culture. These assertions echo the prevailing tendency elsewhere in academia to separate and polarize the sciences and the humanities. But to recognize that each realm occupies a different side of the human brain ignores the fact that they occur always and inextricably in the same

22.4

Merkava Battle Tank. Challenged by the designer of this battle tank, General Israel Tal, to test the theories of "form and purpose," I assisted in developing an integrated geometry for the form of the tank's body and turret. Tal's innovation was to place the engine in the front so that troops could enter and exit from the rear protected. The tank form was streamlined by internalizing many protruding vents and projections. This transformed its appearance, making it reminiscent of the classic Citroën DS design. Although this made the tank elegant indeed, it also improved its performance by making it less susceptible to shell damage. Photo © Moshe Safdie and Associates.

head. Even the most cursory examination of vernacular architecture demonstrates that the tendency to build efficiently and the tendency to ornament and decorate are inseparable.

In recent decades, this debate over architecture has become increasingly polarized. Architecture is seen either as a high art form, exactly similar to all other visual arts as a means of personal expression, or, in the tradition of Vitruvius, as a creative process intended to satisfy certain needs like "commodity" or "delight" and to draw heavily on patterns that have evolved over time.

The more avant-garde view of architecture is to push toward architecture as art, to imply that it is governed by the immeasurable, that the personal takes precedence over the communal. In the words of Philip Johnson, "There are no rights or wrongs in any of the arts and architecture today, only the world of wonderful freedom."[7]

Certainly the second view, which today is most often labeled conservative, draws vernacular architecture firmly within the scope of architecture as a discipline. For one might argue vernacular buildings are our most pure artifacts of technological exigencies, programmatic necessities, and a desire for beauty. But this debate also raises very provocatively the whole question of the relationship between science and art.

By convention, art is intuitive, personal, and expressionistic, whereas science is rational and analytical. Vitruvius's values of "commodity, firmness and delight" touch upon the spiritual as well as the rational, upon the intuitive as well as the measurable. Thus, in the "traditional" interpretation, architecture bridges art and science.

My own view remains that a comprehensive understanding of the "life intended in a building" involves both the science and the art of design. To me, the articulation and expression of the tectonics of construction are fundamental if architecture is to possess meaning, authenticity, and a measure of timelessness. In this sense, the emerging interest in integrating scientific and artistic thought is bound to enrich the world of architectural theory.

Theory into Practice

One of the best opportunities to experiment with these ideas in practice came with my involvement in the Superconducting Super Collider (SSC) and my related visit to the Fermi National Accelerator Laboratory (Fermilab). As I walked about Fermilab with Melissa Franklin, a friend and physicist, I learned of the dreams and objectives of Fermilab's director from 1967 to 1978, Professor Robert Wilson. Franklin said that Wilson was interested in getting people to interact with each other, in creating a setting for the sort of spontaneous exchanges that rarely occurred otherwise in a place like a specialized lab. Wilson is a physicist genuinely interested in the environment and in the reciprocal relationship between people and places.

Wilson's Fermilab emerged as a high-rise tower of laboratories and offices with common facilities like the library, cafeteria, and auditorium at its base (see figure 22.5). But walking through the Fermilab with Professor Franklin raised many questions about this realization of Wilson's objectives. The Fermilab site stretched flatly into the horizon, its tower lonely against a vast sky. Industrial-type warehouses popped up along the length of the four-mile-long underground particle accelerator tunnel attached to the tower. Did the Fermilab, in the final analysis, live up to Wilson's dreams?

The laboratory itself became fairly decentralized over time, draining out of the original tower across the campus and surely dissipating some of the intensity Wilson had envisioned. A high-rise building did not seem the most effective way to achieve the kind of interaction that normally occurs horizontally, in what we might call the "street." I questioned also whether a high-rise spatial organization could ever be flexible enough to provide the wide range of spaces that experimental physics requires today, from the little cubes of space for computers to the larger workshops needed to assemble major machinery.

Another serious question that arose from talking with Franklin concerned the relationship between the architect and the client. In the *Kindergarten Chats*,[8] Louis Sullivan writes that great buildings occur only when there is a great client working with a great architect. Or put another way, he says that clients usually get the architects that they deserve. This resonates as well with Filarete's discourse on the role of the patron: The client should love his architect as his spouse, says Filarete, and the architect should try his utmost to serve his client's needs and wishes.[9]

I remember my experience at Louis Kahn's office, when I worked briefly on the Salk Institute of Biological Studies (see figure 22.6). There in the office, as I worked out some construction details, I witnessed a by-now famous dialogue between Kahn and Salk. Kahn, probably the greatest American architect after Wright, had a passion for program, for understanding the life activity of a building and how that life generates architectural

22.5

Fermi National Accelerator Laboratory (Fermilab), Batavia, Illinois. Sitting in flat prairie land, Fermilab consists of a single high-rise tower of laboratories and offices with common facilities such as a library, cafeteria, and auditorium at its base. Industrial-type warehouses are interspersed along the length of the four-mile particle accelerator tunnel. Professor Robert Wilson, Fermilab's director from 1967 to 1978, originally conceived of this lab as a setting that would be conducive to spontaneous exchanges among the scientists. Whether the architectural form of the complex achieved Wilson's goal is open to debate. Photo © Fermilab Visual Media Services. See also color plate 11.

22.6

Salk Institute of Biological Studies, La Jolla, California. The success of this building is a direct result of the close working relationship between the client, Jonas Salk, and the architect, Louis Kahn. The basic scheme integrates places for solitude that reach forward into a long courtyard from the places for collective work, the great, flexible laboratories. The program of "life intended in the building" becomes an inspiration for the architect. Photo © Moshe Safdie and Associates.

22.7

Biocenter, J. W. Goethe University, Frankfurt, Germany. Peter Eisenman Architects. The forms and location of different components of the program were inspired, according to Peter Eisenman, by Mandelbrot's fractal mathematics. As an architect, Eisenman seeks inspiration from disciplines outside architecture, an approach not unlike that of many contemporaries who look to linguistics or philosophy for their inspiration. The "life intended in the building" does not seem to be a central area of inspiration or concern. Photo © Dick Frank Studio.

form. Jonas Salk was at the height of his career, ambitious and passionate about creating the greatest place possible for exploring biological studies. Salk was an unusual client. He had strong feelings about what should constitute a center for biological studies, about how scientists work both alone and in groups. It seems inevitable that a great and intense dialogue should unfold between those two strong-willed and passionate men. And that dialogue is undoubtedly played out in the form of Kahn's building.

I thought that Wilson was also a great client. Even more intensely than Salk, he had a vision of the kind of interactions, spaces, and landscapes that would be ideal settings for research in high-energy physics. But a great physicist and client, he fashioned himself architect as well, hiring architects who were submissive to "draw up his vision." He failed to realize that the great physicist-client should have sought a great architect who could have joined him in the design of Fermilab.

Two weeks after my visit to Fermilab, the director of the Superconducting Super Collider project in Waxahachie, Texas, showed up at Harvard, having heard from Franklin of my critical comments about the Fermilab. Roy Schwitters, who would become my client, had come to invite me to meet with a think tank of sixteen physicists responsible for coordinating the events in Waxahachie. Later, there would be visits to several high-energy labs to study the relationship between scientific workplaces and experimental installations, but as I prepared to go to Waxahachie, I reflected on the different attitudes architects have expressed toward similar problems. On the one hand, I thought of the Salk Institute; on the other, the Biocenter Complex in Frankfurt designed by Peter Eisenman, a competition-winning proposal for a major biological scientific research center (see figure 22.7). These two buildings obviously have very similar programs. Although a generation separates their architects, it seemed meaningful to me to reflect: What motivated and generated the thoughts of Kahn and, in comparison, Eisenman?

Kahn was obsessed with how he might create a space that would enhance the creative activity of scientists. He was impressed with the fact that scientific activity today requires solitude as well as collaboration. This led him to develop the basic scheme for Salk: places for solitude reaching forward into a long courtyard from the places for collective work, the great, flexible laboratories (see figure 22.6). For Kahn, likewise, understanding the life "intended" in a building is key to generating its organization and form. This goes back to D'Arcy Thompson's "form that is so concomitant with life that it is seemingly controlled by life."

By contrast, Eisenman's inspiration for the Biocenter Complex clearly comes from outside architecture (see figure 22.7). In the text that accompanied his proposal for the complex, he writes about his interest at that time: Mandelbrot's fractal mathematics. Through this text, he explains that his forms and the location of different components of the program are generated by fractal mathematics. Not only does the idea for this project come from outside architecture, but it comes from outside the science for which the

building was designed. Eisenman's approach is common to many contemporary architects who have gone out of the medium of architecture to seek inspiration. These are somewhat like "tea leaf explorations": looking into philosophy or linguistics for concepts or inspiration. By applying the most improbable ideas, an original and provocative design would emerge.

Eisenman has often presented such a strategy as subversive, a method by which the status quo and traditional solutions are shattered. But it is difficult to see how unexpected formal solutions can be effectively subversive in relationship to the subject on hand, in this case a scientific research laboratory. It is true that any deviation from the norm is bound to make us think and reconsider, a kind of subversion by shock treatment. However, new inventions and reinterpretations of a building's program would seem to me to have a greater potential for challenging the norms and values that have generated traditional solutions.

My own attitude, as I approached the assignment of the SSC, was much more similar to that of Kahn. For me, the central question was, What would be the most effective, creative, and wonderful place for two to three thousand physicists and technical support people to undertake one of the greatest scientific experiments in history? Every agenda external to that seemed secondary. I believed that focusing on social interaction, access, privacy, orientation, flexibility, sense of place, and the most effective tectonic means of creating such places would lead to appropriate and inspiring designs and meaningful formal inventions.

Designing the SSC

The existing SSC that I first visited was far different from the futuristic experiment I had envisioned. The project was housed in three buildings purchased by the Department of Energy: two buildings a few miles to the north of a central facility, which had been a Sears warehouse. In the central facility, 750 physicists and support staff were going about their work in ten-by-ten-foot cubicles, with not a single window, not even a skylight to light their way.

I wandered through the internal alleys formed by rows of cubicles. Maps indicated how to find each person. I asked how a community of physicists—enlightened, creative people—was prepared to put up with such an oppressive, inhuman environment. I personally got a headache after about ten minutes. I was told that everyone was so into physics that they did not really care. I was told that to the extent that funds are available, they should go into the experiment, not into architecture. Among the committee of sixteen physicists with whom I worked, however, there was no controversy on this subject: The new campus was a necessity.

In fact, I discovered there were officials at the Department of Energy who had decided that the current location was a perfectly fitting environment for the project. A memo issued by the Department asserted that there was no need to build a campus or laboratories. Perhaps in the future another warehouse would be purchased to accommodate some of the required staff expansion. But in general, the memo claimed, the existing buildings would serve adequately for the next twenty-five years.

The prospect of building an SSC campus raised a debate that recurs almost daily among policy makers. What portion of limited resources should be invested in building schools versus teacher's salaries and educational programs, for example? The real question in all of its different manifestations is, How much do we value the physical environment in which we work, study, or live?

Clearly the "temporary" SSC facilities near Waxahachie, Texas, represented an extreme view: that the environment is of little or no significance. The most minimal shelter was provided so that all available resources could be invested in physics. This translated into large warehouses subdivided into cubicles, totally dependent on artificial climate and lighting. There were those who contended that such minimalist space was essential under the critical eye of Congress, lest anyone think public funds were being wasted to provide daylight or other luxuries. What seemed to be missing was a constituency, vocal and articulate, to demand minimum work space standards. In that sense, too, the SSC reflected the world outside. As the great corporations of America increasingly resort to constructing megaspace for their white-collar megastaff, the most minimal, standardized environment prevails. There, too, one has yet to hear the voices of unions and employee associations directed to the issue of the quality of the work space.

Despite the highly charged political climate surrounding the SSC, I began by focusing on the facility itself: where it should be and how it should be built to serve the experiment. The assumption was that within the next four or five years the temporary building would be abandoned and sold, and the experiment (or those working on it) would move into permanent, custom-built facilities.

The central concern was clearly similar to that posed by Fermilab. The landscape is very flat, almost featureless. How do you create a sense of place in this burnt-out Texas prairie? There was also a new question of scale. The Fermilab accelerator tunnel is four miles in length, whereas the SSC tunnel extended fifty-four miles, with the town of Waxahachie in the middle. The distance between the accelerator to the west and the area where the detectors would be to the east is about twenty-five miles. With a size more than ten times that of Fermilab and facilities extending on both sides of the tunnel, the idea of one central campus location could not, in itself, be taken for granted. This scale was new; it meant thousands of people and communications problems that were difficult to comprehend.

As a way of understanding this location and learning, as well, from other places, I superimposed the plans of several existing laboratories—SLAC at Stanford, Brookhaven, Fermilab, and CERN in Geneva—over the high-energy beam of the SSC. SLAC and Fermilab are centralized. Brookhaven is campus-like in a rolling landscape. CERN seems almost like a medieval town, having begun to grow outward from one end. In apparent disorder, its workshops, laboratories, and central meeting spaces spread out over the landscape.

It was CERN from which I learned the most. Though it had grown incrementally and in a seemingly unplanned way, certain things about the campus seemed to work. For example, a network of passages connects many of its buildings, some of which are weather protected. All lead to a central building that houses a cafeteria, an auditorium, some of the control facilities, and the library. When I talked to scientists about the different laboratories, they said that when they traveled to collaborate on an experiment, the only place where they ever casually met people was CERN. One reason might be the central cafeteria's good coffee, but the architectural reason has to do with the strong horizontal connections within the facility and the fact that it weaves a variety of different kinds of spaces together.

At the SSC, the great challenge of dispersed components was the result of a decision made late in the process. A geological report concluded that although the accelerator and its high-, medium-, and low-energy beams were on the complex's west side, the two main detectors (the observation points for the experiment) that were initially adjacent to the accelerator had to be moved to the east. This meant that the accelerators were twenty-five miles away from the two main detectors and the magnet building, and that the physicists would need to travel vast distances.

From these spatial requirements several questions arose. First, was the original desire for a centralized campus possible or even necessary? With computers, teleconferencing, and other communications technology facilitating the exchange of information, why shouldn't the scientists be physically separate, with the experimenters near the detector where they want to be and where they are often at Fermilab, for example? Why shouldn't the accelerator people be near the accelerators and the magnet people be in their own building? We might run buses back and forth, set up with computers so that people could jump on the bus and do some work while commuting. Why not recognize that the idea of a single, strong center might be simply romantic, even outdated?

In the group of sixteen physicists with whom I worked, there was strong opposition to this conclusion. Led by Schwitters, most believed that it was essential to have a strong intellectual and social center for the community of scientists engaged in this experiment. No matter what communications facilities were available, they believed, the need for interaction brought us continually back to CERN as a model. Many physicists believed that certain facilities, such as a library, auditorium, and hotel, would be indispensable and should form part of a single space where all could interact.

Hence the decision to design a centralized campus. Satellite centers would be used when needed, on a cyclical basis, depending on whether an experiment was underway or work was being done on the detectors. But there would be only one center. The question then arose: Where should this center be located? Because I had several friends who were on the experimental side of things, I was convinced, briefly, that those doing the experiment would not journey far from their detectors, and thus it had to be located in the east. After all, if the center was located in the west, they might rarely use it. But as soon as the east was suggested as the location of the center, the accelerator people and theoretical physicists insisted that it had to be in the west. Eventually, after much debate, it was decided that the center would be built on the complex's west side.

A Sense of Place

After locating the center, I became convinced that any scheme that would cost one cent more than the approved budget would have no chance of getting anywhere with the Department of Energy. The idea evolved that we should try to use an element already funded by the scientific budget as the campus's focal point.

A cooling pond was already approved and underway at the medium-energy beam. One way to achieve a sense of place might be to make use of that great expanse of water necessary for the cryogenics system cooling the accelerators. The campus could face south, toward the cooling pond, in the middle of the medium-energy accelerator ring. Putting the campus within the ring, I thought, would define it symbolically. Although the ring eventually proved too restrictive as a perimeter, another larger cooling pond to the south became the feature around which the whole campus was organized (see figure 22.8).

At this point, the plan evolved a step further: We recognized that the program for the campus divided into public and private sectors. On the west side of the pond, we proposed offices and laboratories: a private side that would grow and expand in layers, not unlike a mill town along a river. The public face would be on the east side: an education center, hotels and villas for visiting scientists. Joining the two sections, in the form of a bridge or a dam, would be the "street" (see figures 22.9 and 22.10).

Approached from the east, the street consisted of an upper-level public entry through which the public could visit without interfering with the experiment. At the other end was a lower-level entry for the working community. Both bridge entrances would lead to the facilities. The bridge building would be open, shaded, and transparent toward the south and almost like a dam toward the north, creating a strong backbone for the space. The street was lined with cafeterias, meeting rooms, a large auditorium, a library, and two control rooms—one for the accelerator and one for the detector—served by entrances at either end. Platforms would extend outward from the bridge into the water, with terraces oriented toward the labs and offices serving the cafeteria and meeting rooms. The street itself, running between the facilities, was to be the campus's place of interaction.

In the labs and offices, in acknowledgment to Kahn, we attempted to differentiate between places of solitude in the form of individual offices and places of collaboration in

22.8

Superconducting Super Collider Laboratories, Waxahachie, Texas. In my proposed design, a large cooling pond needed to cool the detector also served as the principal organizing element for the entire campus, creating a sense of place and orientation. Photo © Michal Ronnen Safdie. See also color plate 12.

22.9

Superconducting Super Collider Laboratories, Waxahachie, Texas. The program divided the campus into public and private sections. On the west side of the cooling pond were private offices and laboratories. The public face would be on the east side and would include an educational center, hotels, and villas for visiting scientists. Joining the two sections would be the "street" in the form of a bridge or dam. This is the great interactive space, the "downtown" of the complex. Photo © Michal Ronnen Safdie.

22.10

Superconducting Super Collider Laboratories, Waxahachie, Texas. Approached from the east, the "street" as seen here consisted of an upper-level entry through which the public could visit without interfering with the scientists' daily life. At the other end was a lower-level entry for the working community. The bridge would be open, shaded, and transparent to the south and damlike to the north, creating a strong backbone for the space. The street, lined with cafeterias, meeting rooms, an auditorium, a library, and control rooms, was to be the campus's central place of interaction. Photo © Michal Ronnen Safdie.

the form of larger laboratories and working areas. Toward the water's edge, almost like a fractalized extension of its surface, are the places of solitude backed by flexible laboratories. Extending almost like islands into the water, the offices are interwoven by a system of streets that culminate in the center: the bridge over water.

The facility would be able to expand, layer by layer away from the water's edge, over the next twenty or thirty years. Parking lots could be consumed by new buildings whenever they were required, to be replaced by structured parking. To the east of the pond, the education center and the hotel would project into the pond, their rooms looking across it toward the meeting rooms, the control rooms, and the auditorium.

Elsewhere on the grounds, many purely utilitarian industrial structures were needed. The SSC had completed building the first large warehouse for the magnet assembly at the time of my arrival. I believed that each of these buildings and landscape interventions should, indeed must, contribute to the SSC's overall environment. To this end, we proposed using corrugated sheet metal to span the warehouse-like buildings. This material could be brought to the site in large sheets and bent into shape as the basic method of construction, forming silvery reflecting vaults of large and small spans, soft and gentle elements in the prairie landscape (see figure 22.11).

Another environmental concern was handling the million cubic meters of spoils produced by the tunnel excavation. When I arrived, an elaborate scheme for spreading the spoils over the landscape was already in place. Another group was studying how much topsoil would be required to cover the waste thoroughly. We suggested an alternative: using the spoils "artistically." By forming the excess into spiral mounds on the land, we could collect drainage water and generate natural plantings to take the place of the topsoil. The forms could then become sculptural, save the search for topsoil, and turn what would otherwise have been waste into a positive feature for the campus.

Robert Smithson, Christo, and other artists have created "earthworks" and other megalandscape schemes similar to this elsewhere. But their projects cost millions of dollars in soil and bulldozing. At the SSC, millions of cubic yards of earth were about to be dumped on the landscape without much thought. Solutions that unite art with necessity are so rare in our society. One must ask how much richer we might make both our art and our everyday life if, every so often, we were to unite them. The artificial split we have established between the so-called scientific (i.e., analytical, rational) method and the artistic (i.e., intuitive, expressionistic) in the process of design has been detrimental to our understanding of the complex process of the creative search for appropriate designs for our physical environment, as well as to the development of a lasting architectural theory. As in morphology, as designers we must achieve a deep understanding of a building's physical and behavioral aspects. But it is simplistic to assume that even if a perfect fit could be achieved, that in itself would answer other more subtle needs. The human psyche, both individual and collective, seeks more. We are more likely to discover our full potential when artistic and scientific modes of thinking merge into one.

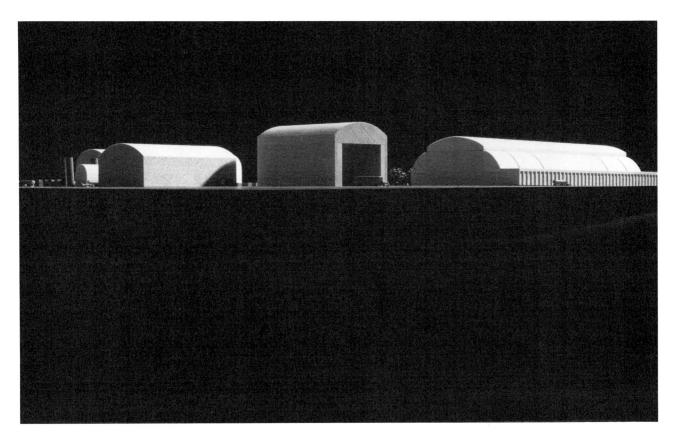

22.11

Superconducting Super Collider Laboratories, Waxahachie, Texas. All of the elements of the campus, including the utilitarian buildings shown here, were intended to contribute to the SSC's overall environment. To this end, the use of corrugated sheet metal to span the warehouse-like buildings was proposed. This material would form silvery reflecting vaults of large and small spans, soft and gentle elements in the prairie landscape. Land excavated from the tunnel was to be used to form sculptural counterpoints to the structures. Photo © Michal Ronnen Safdie.

Notes

1. Thompson, D'Arcy W. *On Growth and Form.* (1917) 2 vols. Cambridge: Cambridge University Press, 1959, quote is from Vol. 1, p. 15.
2. Ibid., vol. 2, p. 777.
3. Safdie, Moshe. *Form and Purpose.* Boston: Houghton Mifflin, 1982, p. 13.
4. Rudofsky, Bernard. *Architecture Without Architects: A Short Introduction to Non-Pedigreed Architecture.* Albuquerque: University of New Mexico Press, 1987. Original publisher: New York: Museum of Modern Art, 1964.
5. Jung, Carl Gustav. *Man and His Symbols.* Garden City, NY: Doubleday, 1964.
6. Alexander, Christopher W. *Notes on the Synthesis of Form.* Cambridge: Harvard University Press, 1970. Safdie, Moshe. John Kettle, ed. *Beyond Habitat.* Montreal: Tundra Books, 1970.
7. Personal letter from Philip Johnson to the author.
8. Sullivan, Louis H. *Kindergarten Chats and Other Writings.* (1918) New York: Dover, 1974.
9. For more on Filarete, see the essay by Pamela O. Long in this volume.

23

Factory, Laboratory, Studio: Dispersing Sites of Production

Peter Galison and Caroline A. Jones

Productive Spaces

In 1963, the same year that he closed his studio and opened a new work space dubbed "The Factory," Pop artist Andy Warhol announced: "I think everybody should be a machine. . . . I think somebody should be able to do all my paintings for me."[1] Across the North American continent, at the Stanford Linear Accelerator Center (SLAC) in 1966, physicist Luis Alvarez told his audience of high-energy physicists that individual laboratory work and the building of tabletop instruments might well be a "delightful activity" seemingly without the constraints of lucre, but he warned that their science needed engineering, a bottom-line calculus of producing the most events per dollar.[2] Now, Alvarez asserted, "we operate a very large business," plants replete with day shifts, night shifts, and swing shifts; Warhol was soon to identify himself in similar terms, as being in "The Business Art Business."[3] At this mid-1960s moment, both men set out to transform the architectural and discursive spaces thought to be peculiar to their disciplines, namely, the laboratory and the studio. Both sought to be seen as architects of new kinds of spaces that aimed to evoke, if not replicate, the postwar economy's astonishing success. Both were strikingly successful late modernists, seeing themselves as steersmen of a centralized production. And yet in retrospect, we can see how both were poised at the threshold of modernism's end in their respective disciplines, their type of late-modernist centralization beginning to give way to a dispersal of production among multiple authors at multiple sites.

This paper focuses on historical sites of production in literal and figurative terms: actual places and architectures in which science and art were produced, as well as discursive sites where they were made meaningful to a wider culture. Despite their location in different disciplines in different parts of the country, American postwar artists and scientists occupied common ground and experienced shifts in that common ground at particular synchronic moments from the 1940s to the 1980s. The timeworn two-culture debate polarizes the *products* of such spaces, but multiple links still bind science and art together as *practices*. We argue that a dominant but highly contested model for both experimental physics and advanced painting during the "Pax Americana" of the post–World War II period was the wartime factory, and demonstrate that the choice of this factory production model had crucial implications for work in both domains. As our paper aims to articulate, the development of studio and laboratory sites and production methods takes place in three broad phases that implicate different kinds of subjects. From an initial epoch of humanistic individuality, we argue that the experience of World War II precipitated the development of artistic and scientific sites that were enlarged and centralized to adopt industrial modes of production, modes that in turn engendered conflicts in the construction of a centered humanistic subject. From this triumphant but conflicted period of centralization, we trace, after around 1970, an era of dispersed production, peripheral sites, and a decentered subject—a subject that would prove difficult to describe in traditional humanistic (or modernist) terms.

Through its critique of prior modes of production that had been characterized by gendered tropes of genius and solitary creation, the mid-twentieth-century exemplar of the factory transformed both laboratory and studio, first spatially, then experientially. "Big science," "executive art," and "the business art business" were at their peak by 1965. But we argue that the changed practices evident at this late-modernist moment shifted again after 1965, as previously centered "executive" authorship begins to be dispersed among multiple sites and multiple authors. As the single-site factory itself became less prevalent after 1970, the studios and laboratories that once participated in such centralized models of production also evolved. The architecture of modernist science becomes a postmodern architecture of data flows and ethernets; the postmodern production of art takes place not in a factory-modeled studio but in the spectacular and discursive realms of print, film, and photographic media.

Centralization: Factory Site and Executive Subject

From early in World War II until the mid-1960s, the understanding of productive sites was fundamentally one of centralization, whether that centralization was constructed in rhetoric or built into laboratory, studio, and factory design. Industrial architecture had emerged from the war transformed, as the American Institute of Architects conceded. Older professional standards forbidding professional architects from taking part in construction were now judged to be obsolete. Indeed, long before the armistice, hybrid architect-engineer-building companies were producing massive new structures that contained bigger open areas, higher-capacity limits for floor loading, more flexible space, better climate control, brighter and more even lighting, higher bays, and more powerful new mechanical means for moving heavy equipment. Exemplary of these wartime builder-architects was The Austin Company, founded early in the century. By the beginning of 1945, Austin had completed more than 6,000 design-engineering-construction projects worth more than $900 million (of which two-thirds had been expended during the war). Having introduced standardized trusses in 1915, Austin perfected welded trusses without gusset plates by 1925, and in 1935 the company assembled a diesel locomotive assembly shop in La Grange, Illinois, that ran a 200-ton traveling crane over an 1,100-foot aisle on a 104-foot span. In World War II, this type of economical, welded-truss construction became ubiquitous for Austin. Among their many projects was one with a 300-foot clear span, making available nearly a million square feet for the production of B-29 superfortresses. These vast, cavernous, climate-controlled spaces had "one common feature—flexibility—which should render the plants adaptable to a variety of uses when war contracts have been fulfilled."[4] After the war, Austin (like many other war factory builders) turned its skills to the building of large-scale laboratories, such as one for Sinclair Oil in Harvey, Illinois. In postwar labs as in factories, total flexibility remained a key goal—in striking contrast to prewar laboratories' many dedicated spaces built with thick, immovable walls.[5]

Other architecture-engineering firms recognized the enormous pent-up demand for laboratories, and again wartime factories provided the template.[6] Until his death in 1942, Albert Kahn, the founder and leader of Albert Kahn Associates, Architects and Engineers, was putting out a staggering 20 percent of American large factories. Like Austin, Kahn participated in what *Fortune* (May 1945) called the "revolution" of full climate control that had been fully realized when government money, no-window blackout demands, and twenty-four-hour production converged in 1940.[7] New laboratories at Los Alamos, Hanford, and Oak Ridge shared many of these same characteristics and often were built by the same firms: Stone and Webster, for example, took on both defense plants and laboratory work in the Chicago branch of the Manhattan Project.[8] (See figure 23.1.) Among the lessons learned during the war was a commitment to keeping research and manufacture close together; this was as true of the wartime airplane plants as it was of the nuclear weapons facilities and microwave electronic development centers.[9] Laboratory and industrial plant designers also learned during the war to routinize the design process through elementary "modules," repetitive spatial units that were provided with appropriate heat, illumination, and ventilation along with gas, air, and electricity. Finally, postwar laboratory builders made extensive use of "advanced building techniques" developed by companies like Austin during the war. For example, Giffels & Vallet's wartime Aircraft Building for the River Rouge Plant of the Ford Motor Company generated solu-

23.1

Hanford Nuclear Site. Laboratory and factory merged in the Manhattan Project as physicists, engineers, metallurgists, and mathematicians joined industrialists in producing some of the largest factories ever constructed. The Hanford Nuclear Site was built to house reactors that turned uranium into plutonium and then extracted the plutonium in vast structures known as "canyons." Source: United States Department of Energy, Hanford Photography, C26706-1CN.

23.2

Ford Motor Company, Aircraft Engines Building Exterior, River Rouge Plant, Dearborn, Michigan. Built near the famous complex of Ford's automobile plant, Giffels & Vallet, Inc. and L. Rossetti Architect-Engineers produced both interior and exterior of this factory. Aircraft training took place in a three-story division with hermetically sealed windows; manufacturing and assembly stood in a two-story division over the building's 1,000 feet of depth; engine testing was located in a two-story windowless division. The building's low, streamlined design finds clear repetition in Giffels & Vallet's 1949 research and manufacturing facility for General Electric, Electronic Park, in Syracuse, New York. See also Giffels & Vallet's postwar Federal Telecommunication Laboratory, figure 23.4. Source: *Pencil Points* 22 (1941): 656–657.

tions adopted in their postwar Federal Communication Laboratory and Microwave Tower in Nutley, New Jersey (see figures 23.2, 23.3, and 23.4). In this laboratory design, the architect-engineers employed such wartime-proven methods as metal exterior skins insulated with fiberglass and service piping run under the floor to keep space clear and to allow movable partitions. No permanent pipe shafts or vents would "fix" laboratory benches.[10]

Architects of postwar laboratories considered centralization to be a precondition of flexibility at the periphery. In a "Discussion on the Design of Physics Laboratories" held in London in 1947, several participants insisted on the importance of maintaining a hub, preventing the edge from drifting away: "If you have all your scientific staffs near one another in a central position, it presumably follows that services and other activities which do not call for the direct attention of scientific staffs should be peripheral." Architecturally, this meant "a core consisting of a multi-story structure with single-story buildings going out from it as needed." Working against what they considered to be such laudable centralization was a contrary and damaging tendency for each of these smaller peripheral units to develop parallel systems. Such duplication had to be stopped.[11] At the Berkeley Radiation Laboratory, the multistory Central Laboratory conceived just after the war represented such a nodal point where researchers would have their home base. From this center, it was hoped, physicists and engineers would then walk to the host of specialized buildings planned for the periphery. The center would secure participation in the group identity; at the periphery a subsidiary individuation could be sustained. This "center of thought and research" would combine office space, small laboratories, a library, and the theoretical group. Explicitly, it would be a center that "differs little from the conventional industrial research laboratory."[12] The architectures of centralized industrial research, scaled-up production, and academic nuclear studies were united in the Berkeley Radiation Lab.[13]

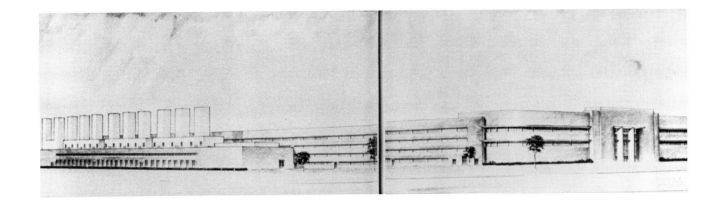

23.3

Ford Motor Company, Aircraft Engines Building Interior, River Rouge Plant, Dearborn, Michigan. The flexible use of floor space, high bays, advanced lighting, environmental controls, and traveling cranes characteristic of wartime factories like this one served as a model for many postwar nuclear physics laboratories. So too did the work structure of wartime laboratory-factories, as the postwar laboratory carried over the day, night, and swing shifts, with its mixture of engineers, scientists, and technicians. Source: *Architectural Forum* 75 (1941): 105.

23.4

Federal Telecommunication Laboratory. Like many factory designers after the war, Giffels & Vallet brought their new techniques to the peacetime task of designing large industrial, military, and national laboratories. In the Federal Telecommunication Laboratory and Microwave Tower in Nutley, New Jersey, the architects held to aluminum facing skin, fiberglass insulation, paper backing, and a cellular steel panel. To ensure flexibility of interior space, all interior walls and partitions were of Transite, so they would require no finishing and would be easy to move. As in many of the war factories, service lines were made easily extendable and located so they would not constrain the useful work space; again as in many of the war plants, modular interior space was employed. Source: *Architectural Digest* 101 (1947): 61.

Even before the United States joined the Allied forces, managers began to view centralization as essential for increased industrial productivity. Gearing up for wartime production, huge new plants integrating previously diverse functions were pulled together in the great airframe factories of Los Angeles, the tank centers of Detroit, and the shipyards of San Francisco and Newport News. Whereas mural artists of the 1930s had depicted the precursors to these booming industrial spectacles, wartime science immediately adopted (and adapted) industrial centralization itself. Sometimes indirectly (as in the formation of Los Alamos) and sometimes through direct collaboration with industrial giants such as DuPont (Metallurgical Laboratory in Chicago), Raytheon (MIT Radiation Laboratory), and Tennessee-Eastman Co. (Oak Ridge Laboratory), science was quick to respond to the perceived need for centralized and scaled-up facilities. Contrary to scientists' often dismissive retrospective accounts, it was by no means arbitrary that General Leslie Groves was assigned the overall control over Los Alamos and the Manhattan Project. Not long before, his assignment had been the building of the Pentagon, a monumental attempt to centralize architecturally the U.S. military directorate in the largest building on earth.[14] By the war's end, credit for victory in the war was attributed to a countless array of cultural and economic forces. But for the physicists, chemists, metallurgists, and electrical engineers drawn into these large wartime laboratories, the war's principal lesson was that a centralized, hierarchical, collaborative, and mission-directed production was the key to both power and discovery in the postwar era.[15] Artists would draw that same lesson, but not for some time. Scientists had learned already to think big.

True, the roots of "big physics" can be tracked back to the 1930s, in the room-sized cyclotrons built by Ernest O. Lawrence and the great telescopes perched on Palomar and Kitt Peak.[16] But the massive drive toward expansion and centralization, and the hierarchical subject that they entailed, came with World War II. At laboratories and departments of physics across the country, the Japanese attack on Pearl Harbor precipitated a fundamental reorganization of research. As late as 1941, MIT's Radiation Laboratory (where radar was being designed) was still a fairly small, quasi-academic entity employing but a score of physicists and their graduate students. Under the enormous pressure of war-driven research, that intimacy had to change. A wide debate ensued about how to think about the structures and processes of the "Rad Lab," as scientists transformed a few rooms at MIT into a multibillion-dollar military-academic-industrial complex.[17] Those at Los Alamos had similarly seen the Manhattan Project grow from a handful of men in a few rooms to a $2 billion enterprise that altered war, physics, industry, and what it meant to be a scientist, all at once. And finally, the wartime nuclear weapons facility at Oak Ridge was not just the largest physics plant in the world, it was the largest factory of any type, anywhere.

The new industrial-style management of such structures had deep implications for the architecture of science and for the inhabitants of that architecture. Just a few weeks after D day, Princeton's Henry Smyth let his colleagues know that the war experience would not recede like the tide. It was not just the funding that had been altered; the practice, meaning, and sites of physics production had also changed forever. Here is Smyth:

> Forty years ago the physicist working on a research problem usually was largely self-sufficient. He had available a certain number of relatively cheap instruments and materials which he was able to assemble himself into an apparatus which he could operate alone. He then accumulated data and interpreted and published them by himself. Most of the special apparatus that he needed he himself constructed with his own hands. He was at once machinist, glassblower, electrician, theoretical physicist, and author. He instructed his students in the various techniques of mind and hand that were required, suggested a problem, and then let the student work in the same fashion under his general supervision.[18]

No longer. The 1930s had seen the development of teamwork within a university setting, but now (Smyth proclaimed) even the universities were too small. Unmentioned (because of wartime secrecy) but always understood was the huge complex of laboratories that marked the model for the future conduct of physics. Smyth dreamed of a new laboratory with 300 feet per physicist, employing about 100 physicists. Above their heads, 15-foot ceilings; below their feet, 5,000 square feet of floor space for large installations. Industrial design offered architectural guidance: "The laboratory should be essentially of factory-type construction, capable of expansion and alteration. Partitions should be nonstructural." Emblematic of the new age of science was this detail: "panelled offices for the director or any one else should be avoided."[19]

Like Smyth, the physicist Luis Alvarez spent his time during the war in massive scientific efforts, first at the Radiation Laboratory and then at Los Alamos. It was a transformative moment, one in which he was jarred quite suddenly from bench-top tinkering to "line work" with a military-industrial order of command. From careful, individual projects with Geiger counters before the war, Alvarez found himself plunged into radar research. He then shifted to the atomic bomb work at Los Alamos, and ended the war flying in the chase plane over Hiroshima, minutes after the city was incinerated. Scaling up became a matter of course. After V-J day, Alvarez took on new construction projects, commensurate in size with the new large-scale physics: first a massive linear proton accelerator, and then a staggeringly large classified effort designed to produce Cold War nuclear materials for weapons at the newly founded Livermore Laboratory.[20] But there were counter-currents to this enthusiastic growth.

When Donald Glaser, a young cosmic ray physicist, proposed a new device he dubbed "the bubble chamber" in 1952, he did so with the hope of "saving" small-scale cosmic ray physics. Above all, he wanted to avoid working in what he called the "factory-like" environment of big machines that had emerged during the war. Nervously presenting his instrument at the American Physical Society on the last day of its conference, he faced only a handful of auditors. Alvarez was among them—and the senior physicist had not one iota of interest in Glaser's dream to preserve the life-world of the individual investigator. Within months of adopting Glaser's device, Alvarez was bringing in engineers, cryogenic experts, and calling on his Atomic Energy Commission (AEC) connections to expand it. He even had the hydrogen liquefier for his bubble chamber hauled up from Eniwetok, the Pacific atoll where the world's first hydrogen bomb had erased the nearby island of Elugelab from the face of the earth. The resulting device resembled nothing so much as the "factory" that Glaser had most feared. (See Figure 23.5.)

Decades after the war, it is hard to recreate the effect such industrial structures had on those who had not directly participated in the expansion: admiration tinged with horror. To the architectural community, the first sign of the coming postwar expansion emerged during the war itself. As the British journal *Architect and Building News* reported in November 1944, the Douglas Aircraft Assembly Plant in Oklahoma City was "rather terrifyingly prophetic." Constructed by the Austin Company in just five months, "the whole conception of this great plant, with its endless artificially-lit interiors, breathes an impression of mechanical precision in which the human element seems to become merely a cog in a process." Such scruples would have to be set aside: the "latent uneasiness as to how far man should allow himself to become involved in

23.5

Luis Alvarez on the 72-inch Bubble Chamber of the Lawrence Berkeley Laboratory. Luis Alvarez transformed the daily lives of experimental physicists as he orchestrated the building of the world's first large-scale particle detectors. Here engineers, rather than technicians, began building and operating instruments—and here formal, factory-like twenty-four-hour-a-day shift work was introduced both for the experimenters and for the scores of women "scanners" who converted the images into computer-readable information. For scientists emerging from the Radiation Laboratory and the Manhattan Project, such industrial-scientific spaces had become the norm. Courtesy Lawrence Berkeley Laboratory Archives.

machine-like organization cannot blind us to the drama of great plants like these."[21] Just two months later, in January 1945, another plant, and another stunned reaction from the same journal: "Just as we are beginning to recover a little self-respect from reading the White Paper on Britain's War Effort, bang! wallop! comes another set of pictures of an American war plant," as another Douglas Aircraft Plant came on line in Chicago.[22] If the scale of wartime factories restructured the experience of self and startled European architects, the new industrial-scale laboratories built on their model truly shocked continental physicists.

European physicists, who obviously had not shared their American colleagues' experience of war work in factory-type laboratories, recoiled in dread at the scale of the new architecture of physics. The dean of French cosmic-ray physics (a branch of physics characterized by individual or small-scale work in the mountains or in more traditional laboratory space), Louis Leprince-Ringuet, warned his colleagues at a conference: "We must go quickly, we must run without slowing our rhythm: we are being chased . . . we are being chased by machines." He continued:

> . . . we are . . . in the position of a group of alpinists who scale a mountain, a mountain of extraordinary height, maybe almost indefinitely high, and we are clambering in ever more difficult conditions. But we cannot stop to sleep because coming from below, from beneath us, is a rising tide, an inundation, a flood which builds progressively and forces us to climb ever higher.[23]

Contained in this breathless cry of despair are many things: a fearful nostalgia for a lost way of doing physics, a forlorn assessment of the triumph of American mechanization (rising like an inundation of machines from below), and an implicit desire to transcend that materialism through an ascent to more rarified, even sacral domains. Climbing ever higher into the Alps was at once a metaphor and a reality: the cosmic-ray stations at Pic-du-Midi, Jungfraujoch, and elsewhere stood as actual isolated laboratories and as instantiations of one set of meanings of what counted as postwar physics (see figure 23.6). Such meanings could not be further architecturally, conceptually, or socially from the emerging physics factories in the Berkeley hills.

Even some established American physicists shared these concerns, and like Glaser and his alpinist colleagues, found reasons to resist the factory model. One laboratory, the Harvard-MIT Cambridge Electron Accelerator, prided itself on its informal, nonindustrial structure. After a devastating explosion at this laboratory in 1965, however, resistance to the factory model crumbled. For warranted or not, the lesson drawn by the physics community was clear: Centralized authority, organization charts, and industrial production procedures were no longer negotiable. Universities came away with the conviction, in the face of disaster, that an older, decentralized mode of scientific work could not function in the age of the laboratory factory.[24]

23.6

Cosmic-Ray Refuge: The Sonnblick Observatory, Austria, 1954. In the late 1940s and early 1950s many European particle physicists conducted their experiments at high-altitude stations like this one. In the aftermath of World War II, such small-scale, romantic mountaintop sites served as a dual symbol: both as a refuge from American accelerator prowess and, provisionally, as a way of maintaining an older, less-industrialized form of experimental life. Source: S. Korff, ed., *The World's High Altitude Research Stations*. New York: New York University, 1954.

For artists, the factory model was much slower to make itself felt, and resistance to centralization was much more programmatic within the individualizing ethos of the avant-garde. Already in the 1930s the Works Progress Administration (WPA) had planted the seeds for a new collectivization in its Federal Art Project, that nearly utopian system under which artists drew salaries as apprentices to muralists and sculptors producing the massive public art projects so characteristic of the Depression era. Paradoxically, these government-sponsored collectives evolved after the war's end into something quite different, as the artists who had benefited from the WPA's forced centralization drew back from a public, representational art. They reformulated their previous social and political groupings into splintered enclaves of fierce individualism and alienated existential angst. The failure of a more material collectivity to develop seems overdetermined. Most obviously, the isolated artistic subject merely continued an equation of author and site that had appeared in science with the alchemist's cave and in art with Michelangelo's studio.

23.7

Nikola Tesla, Colorado Springs Laboratory, 1899. Tesla did everything he could to ensure a romantic isolation for his research. Lodging the laboratory against a mountainous backdrop, he eschewed corporate investment and even the routine of engineer assistants. Physical laboratories of 1900 typically had thick, immoveable walls to block noise and vibrations; working space was essentially individual. On Tesla, see Hughes, *American Genesis,* pp. 37ff. Photo courtesy Burndy Library, Dibner Institute for the History of Science and Technology.

Both Michelangelo and alchemist had been updated in the nineteenth century, when Romanticism cemented both artist and scientist to a vision of the solitary creator. The artist's studio became the site of creation most privileged in narratives of humanist individualism, paralleled occasionally by images of the modern laboratory such as the 1899 photograph of Nicola Tesla in his Colorado Springs electrical lab, supremely isolated as a sublime rain of electricity showers from above (see figure 23.7). Although public witnessing would become an important component of the epistemology of early modern science,[25] the image of the solitary artist remained dominant in art until well after the end of World War II. The representation of Tesla, Zeus-like in his isolated laboratory, can be compared with Romantic constructions of the isolated genius such as Eugène Delacroix's painting of an isolated *Michelangelo in His Studio* from fifty years before: a view of the artist sculpting "with his mind"[26] in a studio populated only by his work, seemingly but concretized emanations of his genius. Even as late as 1950, these same

tropes still dominated American art, appearing most notably in photographs by Hans Namuth portraying the New York School painter Jackson Pollock at work in his Easthampton studio (see figure 23.8).

As in representations of Tesla's turn-of-the-century laboratory, such constructions of genius (here Dionysus rather than Zeus) were predicated on a site that was centralized, but entirely individual. Assistants, wives, and the entire outside world were figured as inhabitants of a periphery that bears no functional relationship to the laboratory/studio domain. The site of the modern studio had long been set apart—linguistically, socially, and economically—from its market-linked analogues (the French *atelier,* Italian *bottéghe,* and British workshop). The studio functioned as a site of purely individual contemplation and creation, closer to its monkish and medieval sources in the Latin *studium,* for zealous learning, than to the bustling scriptorium or guild workshop. In this it resembled the premodern alchemist's lair, with its secret inner chambers for the individual alchemist's private use, bearing a structural similarity to the genial space established for the Romantics' Michelangelo and later adapted for the Abstract Expressionist generation in New York. This proximity is inscribed even in the art: a 1947 Pollock painting titled *Alchemy,* for example, pledges its allegiance to those sanctuaries of laboratory life from an earlier age, rather than the centralized and collaborative production facilities more characteristic of science and industry at the time of its making.

These artists' desires to be linked metaphorically with alchemists and religious isolates were deeply willful and anachronistic, their way of resisting the gigantism and massification evident in their culture at the time. Like Glaser, the artists of the emerging New York School were laboring to preserve space for a particular kind of subjectivity in a world that seemed increasingly to threaten that subject position. They worked to preserve their roles as individual creators, but in fact they were already working in specific kinds of industrial studios: gigantic, non-domestic spaces that had been carved out of cast-iron urban factory sites. Rather than literal garrets, figurative ivory towers, or constructed alchemical retreats, these Abstract Expressionist artists worked in urban lofts redolent of their former identities as "daylight factories," high-ceilinged industrial buildings from earlier economic booms.[27] At least in spatial terms, the factory had already replaced the studio room as the site of postwar production. And that transfer had already begun to make a mark on the canvases that these artists produced (whatever their titles). Robert Motherwell, when asked why the Abstract Expressionists' paintings were so large compared to contemporaneous European canvases, commented, "The scale of America is different. . . . I would say that most American painters live in what were once small factories, whereas European artists work either in apartments or studios that were designed in terms of the scale of easel painting."[28] The characteristic centralization we have identified in the immediate postwar period may have initially expressed itself in art as a heroic, humanistic individualism, but the expanded sites of postwar production were also marked in these enormous canvases, symbolized by the non-domestic scale of a wall-sized canvas installed within the cavernous space of a daylight factory loft.

23.8

Hans Namuth, *Jackson Pollock in His Studio*, 1950. In a famous film and series of photographs of this preeminent Abstract Expressionist, German immigrant Hans Namuth drew on Romantic and Modernist tropes of the isolated studio to show the artist making his paintings in a trancelike state of possessed, creative energy. Courtesy Hans Namuth.

23.9

Frank Stella painting in his West Broadway studio, New York City, 1959. Photograph by Hollis Frampton. Stella's friend Frampton pictures the artist as a common laborer, patiently layering bands of black enamel on a stretched canvas. The studio here is an abandoned commercial loft in downtown Manhattan; the artistic subject is split between executive ideation (in which Stella conceives the diagram for each painting) and working production (in which he gets the job done). Courtesy Frank Stella, A.R.S.; by permission of Hollis Frampton estate.

The Abstract Expressionists, then, joined Glaser and the cosmic ray alpinists in resisting the factory model's larger implications, even though (like Glaser) they were already working within a scaled-up architecture of industrial dimensions. Their anxious existentialist rhetoric can be seen as a defense that had to be mounted with increasing energy and complexity as the postwar economy continued to boom. The painter Barnett Newman, for example, presents a compelling example of such resistance, constructing a centered individual subject for the vast, absorptive expanses of his loft-sized canvas fields:

> . . . the painting should give man a sense of place: that he knows he's there, so he's aware of himself. In that sense he relates to me when I made the painting because in that sense I was there. . . . I hope that my painting has the impact of giving someone as it did me, the feeling of his own totality, of his own separateness, of his own individuality. . . .[29]

Most scholars of postwar American art have viewed such refrains of "separateness" and "individuality" in the discourse around these paintings as a by-product of the political and ideological emphasis on a "vital center"[30] of democratic freedom, separated and individuated from totalitarian (whether fascist or communist) systems of control. That vital center and the isolated studio loft constructed as its spatial equivalent also represented the particular kind of centralization felt to be appropriate for the first phase of the Pax Americana, in which the ideological emphasis on the free individual in a democratic society made the group work projects of the 1930s and 1940s seem "communistic." But the factory lofts in which the postwar works were made and the large museum and gallery institutions in which they demanded to be seen began to reveal other, more social implications. By the end of the 1950s, the culture of the factory's aggregative and social mode of production exercised an ever stronger appeal. As we have seen, most of experimental physics had already adopted a more dynamic centralization modeled on the wartime industrial collaborations; even if the alpinists at Pic-du-Midi or the lab workers of the Cambridge Electron Accelerator yearned for the days of table-top experiments, others like Alvarez rushed to marshall wartime resources and materiel to incorporate ever more diverse functions under the new banners of big science. Such duplications of the functions and structural layout of postwar industry did not prove appealing to American artists until the late 1950s and early 1960s, when movements as diverse as Minimalism and Pop began to express an expanding fascination with "being a machine" and producing art in a way that seemed analogous to producing cars—or neutrinos.

So Frank Stella, a young painter only two years out of college and already showing at The Museum of Modern Art, galvanized the audience of a panel discussion on "Art [19]60" when he provocatively claimed that an artist who had been roundly criticized for having his paintings industrially fabricated was "a good painter when he doesn't paint." Stella then praised the delegating painter as "an executive artist" and added that he himself "would welcome mechanical means" to paint his own banded canvases.[31] (See figure 23.9.)

Stella's paintings were unusual, but still potentially within the frame established by the Abstract Expressionist canvas; his painting *method* was more problematic. Dark, brooding bands of black enamel were layered without inflection across the unsized canvas, lined up in a regular pattern with only the barest of "breathing lines" between them. Nihilistic, but perhaps not yet incomprehensible. The electrical charge of Stella's remarks, however, proved scorching, creating a context for the black paintings that changed their possible perception from absorptive, individuating fields à la Newman to the rigidly geometric "pinstripes" perceived by Stella's critics in 1960. Stella's viewers had immediately begun to link these works to the prevailing uniform of bankers and industrial tycoons.

The audience and panel at "Art 60" were apparently united in their hostility to Stella's suggestion that fabricated enamel-on-steel tondos were good *as paintings,* resisting the possibility that something made in an industrial process could even be art, much less "painting." The anxiety Stella's posture produced can be measured finally by a single remark attributed to one participant as he left the event: "That man isn't an artist, he's a juvenile delinquent."[32] But far from the juvenile delinquent, Stella had aspired, as he said, to be the *executive* artist (a perception reinforced by the misprision of the canvases as "pinstriped"). The inversion of these social categories is indicative of the confusion generated by what Stella represented in the culture of the time, the new kind of subjectivity his works, and his rhetoric, seemed to offer. The executive was "delinquent" because he presented a threat to the old life-world of the artist, that isolated creator whom Glaser, too, had yearned to protect.

In part responding polemically to his critics, Stella produced an even more industrial-looking series of aluminum canvases in 1960, painted with a specific commercial space in mind (see figure 23.10). No longer painted as individual canvases in an evolutionary/chronological order, the works in the Aluminum series were mapped out in advance and completed with as little tolerance for variation as manual production would allow. Their seemingly die-cut shapes were planned for a spectacular inaugural array at the Leo Castelli Gallery's midtown loft. The point of this industrial aesthetic seems to have been well taken. Castelli himself mused suggestively that these canvases evoked "old-fashioned cash registers"; later, reviewers looked for Stella's annual series with the kind of expectation they brought to the Ford Motor Company's latest lines: "What's That, the '68 Stella? Wow!"[33] Stella's paintings had iconically provoked a reference to industrial production, if not yet achieved its corresponding line assembly. That move would be made by his contemporary Andy Warhol, who achieved the notoriety for Pop art that Stella had won for Minimalism.

While Stella was announcing his affinity with "executive artists," Warhol was determining to leave his successful career as a commercial illustrator and make a bid for Fine Art. His commercial art career had been characterized by the delegated production of ostentatiously "hand-drawn" illustrations that obscured the actual assistance of other

23.10

Installation of Frank Stella's "Aluminum Series" at Leo Castelli Gallery, New York City, 1960. A shimmering row of larger than life-sized canvases were conceived as a series and made with this commercial gallery space in mind. Although painted by hand, their hard-edged metallic perfection, seemingly die-cut shapes, and strict geometric variation prompted associations with commercial objects and industrial manufacturing. Photograph courtesy Castelli Gallery, A.R.S.

23.11

Andy Warhol silk-screening a *Campbell's Soup* painting with his assistant Gerard Malanga, as a visitor looks on, 1965. Here the factory model reaches its apogee in the studio site. The canvases (or boxes) to be silk-screened are lined up along the floor, and a primitive line production ensues, with no pretense of maintaining the solitary production and mysterious creative forces of the early modern studio. Photograph by Ugo Mulas.

commercial artists. Significantly, as he positioned himself as a fine artist in the early 1960s, Warhol moved toward a more obviously reproductive technology, the photographic silk screen, and announced with pride whenever interviewed that "Gerard does all my paintings for me." In photographs publicizing this process, we see documentation of Warhol's crudely aggregative assembly line, modernized only by the photographic silk-screen technology he employed (see figure 23.11). These silk screens were produced off-site by suppliers to the burgeoning advertising industry then defining the identity of Madison Avenue, just blocks away from the old brick structure that would become Warhol's notorious "Factory" in 1964. Just before moving into the new work space, Warhol hired Gerard Malanga as his talented assistant. In this he looked to benefit from Malanga's prior experience as a silk screen assistant in a custom textile shop for the garment industry, aligning himself with factory production techniques just as aggressively as Alvarez had (although the rag trade and the munitions industry were leagues apart in their manufacturing modes).

23.12

View of Warhol's 47th Street studio building, dubbed "The Factory" in 1964 when Warhol and his assistants moved into the fourth floor. Because this nineteenth-century industrial building type was already a misfit among the glass-curtain-wall skyscrapers of midtown Manhattan (and no longer maximized the revenue-generating potential of the real estate on which it sat), it was demolished in 1968; the arrow indicates the buildings to be razed in this developer's photograph. Warhol moved to the Union Building on Union Square, where "The Factory's" organizational structure became more hierarchical, organizing itself vertically on two floors rather than sprawling horizontally as in this earlier commercial site. Copy photograph by Caroline A. Jones from New York Department of Buildings files.

The brick-faced structure in which Warhol and Malanga set up shop was located in Manhattan's Midtown, the boom neighborhood of the fifties, but it had been constructed much earlier as a commercial building based on nineteenth-century mill architecture. The photograph shown in figure 23.12 was taken shortly before its demolition (which accounts for the arrow drawn across its face). The facade's rows of narrow windows were aimed at bringing daylight into the entire work space, but the building type had little in common with the glass-faced, reinforced-concrete "daylight factories" long identified as the heart and soul of modernist architecture.[34] The "daylight factory" tradition had come to dominate postwar industrial architecture, and glass-curtain-wall buildings (just visible looming over the Factory in figure 23.12) were home to the advertising industry Warhol knew so well. The neighborhood signified modern business and contemporary industry so strongly that Malanga argued against Warhol's taking the space, because the neighborhood "wasn't Romantic enough. It was too business-like."[35] That seemed to be the point. Although the surrounding buildings housed managers, corporate executives, and advertising mills rather than heavy industry, the East 47th Street building in which Warhol rented a floor for his workshop had, in fact, previously housed a hat manufactory. Despite Malanga's evident objections, Warhol and his other associates took this salient prior tenant as their model and source for the "Factory" name.[36]

The space of the Warhol Factory was left open and unstructured, but at Warhol's request it was entirely covered with silver foil, including the windows and pipes.[37] The name was clearly intended to displace the time-honored trope of the isolated studio with a term that would make room for the whole motley crew of hangers-on and assistants in its collective embrace, yet this "Factory" also signaled its countercultural intentions with its silvered interior, part decadence and part "space age" in its glamorous sheen. "Factory" suggested wider operations of the modern world, rather than the narrow (if romantic) ivory tower. The functional archaism of its assembly line and the willful occlusion of the daylight factory aesthetic (most obviously through the silver covering of the windows themselves) reflects the ambivalence of the moment, much as the collectivization in the Factory bore within it the seeds of tensions over leadership and control (below we will indicate how the Alvarez group would also splinter and dissaggregate, becoming the anonymous "group A"). Certainly the stated intention of Factory as symbol was quite contemporary, as was the intention behind its medium, the silk screen. As Warhol explained at the time: "In my art work, hand painting would take much too long and anyway that's not the age we're living in. Mechanical means are today."[38]

Thus for Warhol, there were complex architectural and productive meanings signified by the Factory and coded in its serial objects. This cluster of meanings resonated with the forces of postwar production: a hip reference to postwar industrial modularity was amalgamated with a nineteenth-century collective's less organized sites of production. There was a new, space-age "look" of silver and a democratic deployment of unstructured space (no paneled walls here, either!) Most crucially, the crude assembly line worked to reinforce the sense of centralized collectivity itself; at the same time, Warhol's status as "boss" would be continually undercut by his own insistence that "Gerard" and "Brigid" made the art. The Warholian subject, still tenuously centered as the "boss" of the Factory, was already eroding under the decentered implications of his own Factory assembly line. As in Alvarez's group of scientists, Warhol's Factory workers began by performing centralized, quasi-industrial tasks. But by 1968, both space and subject in these sites seemed to register definitively that the modernist center would not hold.

Decentralization: Dispersal and the Postmodern Subject

At the very apogee of the Pax Americana, the charmed spaces of industry and centralized production began to disperse. In both science and art, the ideal of the modernist factory began to give way to other modes of production characterized by decentralization and dispersion; the boss and the executive were exchanged for other, more nebulous roles. In the work and rhetoric of artist-theorists such as Donald Judd, Robert Morris, and Robert Smithson appeared challenges to the centralized picture of the studio, even in its post-Warhol configuration of studio-as-factory. First, the authorial function of the individual "genius" was undercut through an emphasis on the industrially fabricated or deliberately unauthored nature of the artwork, and, in a related move, the site of production was itself dispersed from a single studio to multiple peripheries (territories that would be seen as postmodern rather than modernist sanctuaries or factory fiefdoms). Scientists, too, began challenging centralization in favor of physical dispersal and new modes of authorship. The "Alvarez group" became "group A," and group A became a moveable element in the multi-institutional (indeed multinational) collaborations of the 1970s and beyond. At the same time, the physical laboratory itself began a conceptual diffusion. As data tape and then computer links bound work sites together through an expanding electronic infrastructure that would become known as "the net," it became increasingly incoherent to speak of the experiment as taking place in any single location. More than the passing of leadership from one generation to another, these developments, we argue, signaled the emergence of a new episteme only partially captured by the contested term "postmodernism."

It becomes possible to argue in this context that neither physics nor art simply "appropriated" or "mimicked" a preexisting industrial model for their productive activities. Their dispersive activities of the 1970s in some ways preceded and informed the design of industrial and architectural practices that were not realized widely until the 1990s. In only the most dramatic example, the software innovator Tim Berners-Lee first proposed the ideas behind the World Wide Web in 1989 at the European accelerator laboratory (CERN) to manage and distribute hypertext information regarding the massive new high-energy physics experiments then in the planning stages. Berners-Lee made it clear that the laboratory was not a mere reflection of the wider industrial world. On the contrary, in his 1989 proposal, he envisioned the laboratory as a prototype for everywhere else: "The problems of information loss may be particularly acute at CERN, but in this case . . . CERN is a model in miniature of the rest of the world in a few years' time."[39] And indeed, through internet links and the World Wide Web, high-energy physics laboratories became a new dispersed architecture, producing industrial and governmental visions of the "information superhighway." Software (designed by computer scientists) became the vehicle for the new "digital factory" intended to propel the United States back into a position of global leadership, diminished since the mid-1960s.[40] Even as the moon launch had promised a new burst of momentum for "the American century," artists' efforts to expose the image-based, post-industrial nature of the U.S. economy had largely preceded business and public policy agencies' acknowledgment of a new economic phase.[41]

Artist Robert Smithson positioned his own voice—dominating the late 1960s in mordant essays published monthly in art and general interest magazines—as defiantly alternate to a particular form of late modernism still supported by important critics such as Clement Greenberg and Michael Fried. Smithson's works (both objects and essays) defined an explicitly "post-studio" role that built on Warhol's performative factory-studio, but articulated a cunning critique. Smithson saw the remnants of centralization that characterize Warhol as late modernist—the single Factory, and the arguably unique museum object that it produced—as unnecessary centers that had to be dispersed in favor of a more resonant periphery that could never be fully pinpointed or definitively mapped.

Smithson's site/nonsite works participated in the general conceptual thrust of the late-1960s art world and began to define the decentered subject we have discerned in both science and art toward the end of the century. In these dialectical and discursive "sculptures," a "site" was designated in the world outside the gallery system, for example, a slate quarry in Bangor-Pen Argyl, Pennsylvania. Photographs and samples were taken of and from the site and brought into the museum or gallery with appropriate documentation of their source. Such samples were presented as heaps of rough-hewn shards, pebbles, or monochrome rocks, bracketed, bounded, or bisected by mirrors; alternatively, larger-sized chunks were installed in slatted metal bins or trapezoidal wooden boxes. In each nonsite work, descriptive labels, directions, and cryptic maps of the site of origin were mounted on a wall adjacent to the sample material.

Smithson's own discursive and spatial positioning generated the primary theoretical work done by these pieces. The bins and mounds of rock in the gallery were the only "objects" on view, yet they were presented resolutely as "non-sites," absences from the site of origin, their presence intended merely as indexical signs pointing the viewer out of the gallery to what Smithson privileged as the *real* site, always located in what he celebrated as the periphery or fringes of the urban center. Smithson constructed the site/nonsites specifically as rejections of the modernist unitary object. In a 1970 symposium on Earth Art held in conjunction with an exhibition where one of his mirrored nonsites was presented, Smithson commented, "Most sculptors just think about the object, but for me there is no focus on one object so it is the back and forth thing."[42] Smithson demanded of viewers, and of himself, explicitly durational processes: at minimum, a kind of mental scanning that would go "back and forth" between an object and its origin, between center and periphery, between city nonsite and noncity site. In this scanning process, the art's meaning would be made in its viewers' minds. The studio becomes moot as a site for production or a guarantor of authenticity; the object itself is dispersed, and it is discourse that reigns.

Of course, the very construction of periphery implies and requires a center, and this ambiguity is what marks Smithson's position at the crux of the modern/postmodern turn. Smithson's importance within the art world as an engineer of this "postmodern turn" depends heavily on the notion of dispersal and on Smithson's own construction of himself as a "post-studio" artist. This is the inverse and conclusion of our observation that the artist's studio is uniquely valorized within the discourse of modernism. By destroying the studio as a site of production, Smithson effectively emptied the core of modernism and began the attempt to construct a new and peripheral architecture of postmodern dispersal. Yet through the term "post-studio," he reaffirmed the terms of a binary (in much the same way that post-modernism continues to be parasitic upon modernism's remains).

As we have argued, the modernist canvas (such as Jackson Pollock's *Alchemy* or one of Barnett Newman's punctuated fields of saturated color) is meant to "stand for" or reconstruct the solitary artist in his cavernous studio loft, much as the "golden event" of centralized big science can still be connected (if with increasing difficulty) to the individual physicist credited as its "discoverer." But where the modernist studio served to legitimate the still-dominant canvas, which remained the object most capable of reconstituting the individual, Smithson's scattered sites are at least theoretically dominant over his nonsite objects; moreover, they explicitly "fail to cure" or make whole the fragment implied by the gallery nonsite. The subject is left in her scanning, de-differentiated state. As we argue, the remnants of centralization in the dispersed collaborations of late-twentieth-century physics also fail to reconstruct any coherent author or dominant subjectivity for the members of the experimental group. The spaces of production—the architectures of postmodern science and art—instantiate and enforce this reading.

Spiral Jetty is the earthwork seen to be the quintessential post-studio achievement of Smithson's career (see figure 23.13). For a very brief moment in its existence, *Spiral Jetty* could be described as a 1,500-foot, inward curving jetty made from rock and compacted earth in the shallow, microbially polluted water of the Great Salt Lake. Seen here in a photograph taken shortly after it was completed in 1970, the *Spiral Jetty* was known physically to only a small group of people comprising the artist and his colleagues: his dealers who funded the work; his wife, artist Nancy Holt; filmmakers and photographers; and the bulldozer and dump truck crews and their families who came to picnic during the virtuosic earth-moving activity. We know the limited size of this audience with some certainty, for the rising waters of the Great Salt Lake inundated the rock-and-dirt jetty shortly after it was finished, and it remained invisible for two decades. It was only during the summer of 1994 that the lake receded briefly to reveal an erratic string of salt-encrusted rocks marking fragments of the *Spiral Jetty*'s path.

23.13

Robert Smithson, *Spiral Jetty*, 1970. The site of Smithson's most famous work was an industrial outback in the Great Salt Lake of Utah. Fragments of abandoned machinery can be seen along the shore; Smithson filmed the jetty's construction, and through film, photographs, essays, and interviews made it clear that this earthwork was posed against the modernist studio. Proud of the substantial crew of bulldozer and backhoe operators that it took to build the jetty, Smithson opened his work to evocations of mining and civil engineering as he worked on the peripheries of discourse and art world alike. Photograph courtesy The Robert Smithson Estate and the John Weber Gallery.

Luckily for us, Smithson did not rely on the jetty's material existence to establish its cultural value. His film *Spiral Jetty* was coeval with the earthwork; even Smithson's preliminary staking out of the spiral was filmed as work began on the site. Smithson and others took still photos throughout the construction process while Robert Fiore served as movie cameraman and sound man; a professional still photographer, Gianfranco Gorgoni, produced more sumptuous documentation of the completed earthwork for Smithson's dealer. The film, completed with the earthwork in 1970, was then shown with photographs and texts at the New York gallery of *Spiral Jetty*'s primary backer, Virginia Dwan, shortly before both gallery and jetty closed down; Smithson published his essay on the jetty shortly thereafter. The essay, the exhibition, and the vigorous continuing life of the film and photographs constructed *Spiral Jetty*'s widespread cultural meaning. The physical jetty had been rendered for all intents and purposes a "profilmic event" as irrelevant to visual culture as an abandoned back lot of Metro-Goldwyn-Mayer. The site's former primacy was now definitively upended. The unitary modernist object was hopelessly dispersed among the mirrors of its reproductions. The various "nonsite" devices of text, photographs, film, and so forth now constituted all there was to know of *Spiral Jetty*, with film privileged as the main purveyor of cultural meaning. And the film recapitulates the decentered quality of both the jetty and its subject, entities constructed by discourse and cultural exchange.

After footage of dinosaurs, dumptrucks, and exhilarating aerial takes of the jetty's expanse, the film suddenly falls silent, and the camera cuts to an interior. In this last scene of the film, we see a filmmaker's editing table, spiraling reels of celluloid centered under a large black-and-white photograph of the completed *Spiral Jetty*. The constricted architectural space we see is *not* an artist's studio, nor is it a factory. It is not even Smithson's, since, as we learn immediately in the credits that follow, he was not the editor of this film. The final image of the editing table surfaces the film's status as a constructed artifact and reminds us that *Spiral Jetty*-the-film is a collaborative endeavor. This final sequence witnesses the critique of the isolated modernist studio and the postmodern dispersion of production. To the extent that we can ever know it, we must acknowledge *Spiral Jetty* as a discursive entity, still linked to industrial sites and processes, but ones that become, by 1970, conflicted, peripheral, and dispersed.

By his own antimodern, "post-studio" route, then, Smithson had arrived at a decentered architecture of dispersion not unlike what physicists were constructing on a similarly gigantic scale. Just as it makes little sense to try to localize an experiment conducted in computerized data flows over the ethernet, so it remains largely irrelevant whether the rock and dirt component of the *Spiral Jetty* is, or is not, above its inland sea. Artists and scientists engaged in such projects are equally unfixed as subjects, experiencing themselves as moveable links on an endless chain of relays that only together count as productive of the art or science involved.

While the Great Salt Lake was gently rising in the early 1970s, completing the *Spiral Jetty*'s trajectory into an explicitly discursive form, experimental physics was also undergoing expansion and dispersal. Dwarfing even the large-scale operations of Luis Alvarez, in which ten to twenty collaborators had worked on a million-dollar bubble chamber, experiments in the 1980s involved 200–700 physicists from ten to twenty institutions working on 500 million dollars' worth of electronic equipment. In the canceled SSC (projected for the millennium), each detector was to have cost about $1 billion and encompassed the work of about 800 physicists. Inheriting the SSC's momentum in 1996, CERN's Large Hadron Collider (LHC) has two detectors, each with some 150 participating institutions from 30 countries. By the end of the twentieth century, the LHC embraced between 1,600 and 2,000 physicists and engineers (see figure 23.14), a significant percentage of *all* living high-energy experimentalists.

If Alvarez (like Warhol) could be thought of as running a business, or more specifically a bubble-chamber picture-producing factory, post-1970 colliding-beam experiments like the TPC, the SSC detectors, or the LHC's detector "ATLAS" would be more akin to Smithson's enterprise in their dispersed structure, and more like multinational corporations in their productive decentralization. Alvarez had borrowed his managerial structure, his physical plant, his safety procedures, and his scanning assembly lines from wartime laboratory-factories; the newer, post-1970 generation of experiment leaders would attend management seminars with high-level representatives of global corporations. The first incorporation of experimental physicists had brought physicists together; this second incorporation was a corporation of corporations. With this shift, control, coordination, and ultimately authorship itself were dispersed, conditions that physicists were forced to accept, but Smithson anticipated and enacted.

We do not see the decentering of the laboratory or the dispersal of the studio as merely "mirroring" some independent, underlying change in the economy. Instead, we find large consortia of laboratories participating in the same dynamical process of coordination that one finds among consortia of companies; equally, we see the production of artistic meaning since the 1970s as honing the very discursive systems that enable a service-and-consumer economy to flourish. When they expend hundreds of millions of dollars, these laboratories *are* multinational enterprises of consequence. Large, heterogeneous physics collaborations borrow management techniques from the multinationals, but these same large corporations have also used techniques (including microwave data links and the World Wide Web) pioneered by physicists to solve their own problems of interlaboratory coordination. By the same token, when postmodern artists such as Barbara Kruger employ the authorless, authoritative visual language of advertising to convey cultural messages, it can no longer be seen as surprising when conservative media organizations appropriate those same strategies.[43] Positing an industrial "base" that either is autonomous, or univalently determinative, of the cultural sphere of scientific-artistic work fails to capture these worlds' fluidity, permeability, and coextensiveness.[44]

23.14

ATLAS-1 (1996). The Large Hadron Collider at the European Center for Particle Physics will house two detectors, each staffed by some 2,000 physicists from 150 or so institutions. Dwarfing the Alvarez-era "big science" by a factor of nearly 100 in personnel, budget, and tonnage, collaborations of this size create a different kind of sociological, physical, and electronic architecture. No one person heads up the collaboration, and even the category of scientific authorship is contested. Source: CERN, *Experiments at CERN in 1996* (Geneva, 1996), p. 322.

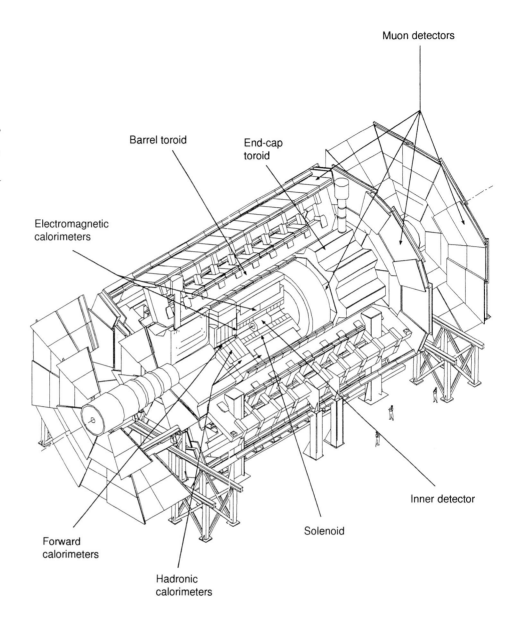

The architectural implications of this late twentieth-century transformation can be seen in the profession's recent literature. A 1993 article in *Architecture,* for example, reported that we are now in a phase of laboratory design marked by closer relations between business and research. This tighter affiliation, so the article argues, is driving a decloistering of science as the formerly separate worlds of applied and pure research merge. A more corporate mode of work is becoming widespread: "Today, scientists spend more time in an office environment, simulating experiments on computers. Hypotheses are often tested electronically before experiments are conducted in real life. . . . 'You don't even have to go to the lab to get your results,'" according to Janet Brown of CUH2A, a consortium of Princeton-based lab design specialists, "'they appear on your computer screen.'" This movement of researchers away from the centralized facilities allows even greater mobility in the design of work modules, well suited to the new style of organizational volatility. "Teams are reconstructed with remarkable rapidity during projects," noted another lab architect. "They need to be able to create dedicated support or additional office space . . . We have come to expect this in general office environments. The same pressures exist in labs."[45] Alternations in the flow of information reshape the way the physical spaces of scientific work are configured.

Understood in this way, the architecture of software is not merely homonymic; software architecture *alters the positioning of workers and walls.* For illustrative purposes, consider the Time Projection Chamber (TPC) facility at the Stanford Linear Accelerator Center (SLAC). The TPC is a particle physics facility, novel in its extraordinary ability to use electronics to produce striking "images" of the subvisible world, yet typical of a wide class of particle physics laboratory units from the late 1970s through the 1990s. Not only in scale, but in its social, electronic, and mechanical architectures, facilities like the TPC differ radically from Alvarez's centralized factory bubble chamber. Where the bubble chamber is a single-principle device (essentially a homogeneous vat of liquid heated beyond the boiling point but kept under pressure until a particle passes through), the architecture of the post-1970s "facility" is in every respect hybrid. Known as TPC/PEP-4/9, the specific detector installed at SLAC represents a facility that was actually two different experiments. Each "experiment" (PEP-4 and PEP-9) was directed by a different consortium of university laboratories, and each participating laboratory had special responsibility for an individual detector component within the device. The University of California at Riverside, for example, controlled the outer drift chambers; Tokyo held proprietary interest in the cylindrical calorimeter. The days of an absolute leader like Alvarez had, for the most part, passed. Riverside, Tokyo, and SLAC were hardly going to submit to the authority of a single individual from the Lawrence Berkeley Laboratory (LBL). It was just this lateral affiliation of components and university groups (rather than hierarchical structuring) that marked these big colliding-beam facilities as very different from the prior generation of bubble chamber experiments. The division of labor necessitated by these post-1970s hybrids had powerful consequences. Each group had to guarantee not only the physical construction of its particular component, but ultimately the physics results that came out. For the device to function at all, a process of coordination among equals had to take place on every level.

The architectural coordination required for functioning raises a series of questions that are simultaneously physical, sociological, spatial, and epistemic. How do the subsystems mesh with one another mechanically, how do they interact electronically, and how are data synchronized? Who counts as an experimenter, and what complex modifications of the subject must be articulated before scientific authorship can be ascribed? A post-1970s demonstration in physics is a multiply coordinated assemblage of architectures conducted at each of these levels (sociological, hardware, and software architectures). Like Smithson's site/nonsite works and Barbara Kruger's image interventions, these experiments are dispersed social-technical-spatial entities in which meaning is constructed at several peripheries, and no single center can hold.

As might be imagined, the problems such experimental collaborations faced in the 1980s were very different from those facing experimentalists of the early 1960s. Alvarez's concern in that early postwar period had been to compete successfully with other groups—the Thorndike-Shutt Group at the Brookhaven National Laboratory, for example. The gigantic teams of the 1980s and 1990s, by contrast, are not so much competing with others as they are simply trying to hold together against the forces of a dispersed production and a decentered site.

By contrast with the ever present threat of dispersal at laboratories in collaborations like the TPC/PEP-4/9 or the planned SSC, the various bubble chamber groups of the 1950s and 1960s had remained secure in their centralized, well-structured hierarchy (more secure than Warhol's ambivalent "Factory" would be). The bubble-chamber laboratories were systems,[46] with well-defined centers powerfully endorsed and reinforced architecturally and socially by the scientific culture as a whole. By the time one gets to the TPC of the 1970s and 1980s, however, the hierarchy had come to have a multitude of focal points. Each of the constituent groups entered into the collaboration with its own hierarchy: professors, associate professors, assistant professors, postdocs, and graduate students were replicated at each dispersed locale. Inside the collaboration, however, the political economy of the whole was such that no single group could dominate absolutely. Although LBL could lay claim more than most to the title of "center," even it had to work alongside the other university groups from Riverside, Yale, Tokyo, Johns Hopkins, and UCLA. *Intra*university structures were perfectly adapted to hierarchy; *inter*university collaborations resisted such rank dominance at every turn.

Given the network structure of such collaborations, new modes of governance were necessary. The Alvarez organization had borrowed, as Warhol did, from 1940s industry, but it also modeled itself on the military command structure of the war; it had a clear hierarchy with a head, a chief of staff, and a host of other deputies directing various "departments." The TPC, by contrast (on a scale that Smithson could only dream of), absorbed many of its structural features from the huge corporate consortia and computer-coordinated projects such as the submarine-launched, nuclear-tipped missile, Polaris. The Critical Path Method (CPM) was one of many computer-based attempts adopted by the TPC to distribute resources in a way that would rationalize priorities and guide various roughly equal subgroups toward a common goal. Computer flow charts began to replace industrial organization charts as maps of experimental practice.

The TPC participated in a second institutional restructuring that reflected the shift from centralized hierarchy to dispersed network. Where laboratories had earlier relied almost entirely on the guidance of individuals (e.g. the "Alvarez group"), members of the TPC collaboration militated for an organization revolving around committee-based tasks, rather than individual authors or an absolute center. As committees came to decide what would count as a result and who would be an author, the physicist-as-subject changed. Indeed, the very notion of "experimenter" as a fixed, centered identity began to come apart.

Without formalized trading zones,[47] the dispersing TPC threatened to tumble into chaos when, for example, physicists reversed engineers' decisions without any sense of the economic or time consequences. The simple director-follower relation that had existed between physicists and engineers during the days of centralization had to be retooled, and both spaces and subjects were modified. "Next time, build a circus tent to house everyone," joked one of the TPC's physicists, invoking both convivial and temporary architectures in his nomadic metaphor.[48] For the first time in LBL's history, physicists began to plead with engineers to stand up to them, to argue, to present separate areas of control and become, as it were, separate authors. As the TPC's leader put it: engineers needed to inject "reality into the physicists' . . . dreams and then . . . translate [those dreams] into a 'working piece of hardware.'"[49]

The hardware and software the TPC collaboration and others like it introduced had profoundly dispersive effects. In the first stage of dispersion, it became possible for data to be analyzed at different laboratories. But this transport of magnetic tape was not different in any important way from the export of bubble chamber film in the 1950s and 1960s. Delocalization of the experiment progressed a significant step further in the late 1970s, when the TPC collaboration established a microwave computer link between LBL and SLAC (an early form of email), followed by a full-scale network among the participating institutions.

By the 1980s, delocalized networks had gone much further, as in the L3 experiment at the European particle physics laboratory CERN. In that collaboration, communication was conducted through LEP3NET, mapped in figure 23.15, linking the 500 or so collaborators together by purely electronic means, and making possible the joint composition of massive programs for the design, planning, and analysis of experiments. With such high-volume networks it was possible for physicists to transfer data instantaneously around the world to the myriad of participating laboratory sites. It becomes obvious that

23.15

LEP3NET (1990). In a high-energy physics experiment of the 1990s, computer architecture is not merely a metaphor for physical architecture. The computer not only serves to design detectors (such as the Atlas), it also gathers information and controls the detector. Furthermore, by distributing the data to the widely dispersed collaborators, the computer linkages effectively make distant laboratory sites part of the same experiment. With commands ordered from one site, the events occurring in another, and on-line analysis elsewhere, it is impossible to say unambiguously where the experiment actually *is*. In this sense, electronic architecture fundamentally reshapes physical architecture. Source: B. Adeva et al., "L3 Experiment."

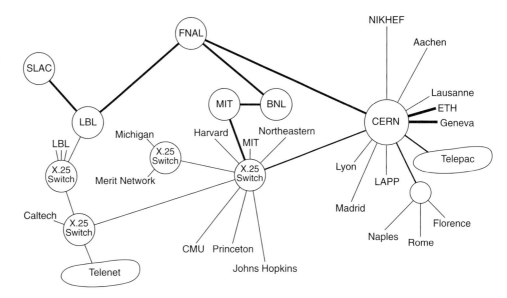

looking at beam characteristics or particle production on a monitor located at CERN is no more "direct," no "closer" to the phenomena than those same data, packet-switched by microwave and fiber optic cable between collaborators in Stanford, Madrid, Florence, Princeton, or Michigan.[50] Even more dramatically, before the cancellation of the multibillion dollar Superconducting Super Collider (SSC), control rooms were being planned at sites on several continents. As we have argued, if the control rooms are multiple and the data are in play across the world, it is no longer clear *where* the experiment can be said to be. The experiment cannot be localized, for the circus tent spans the globe.

When the SSC was in its planning stage, with Moshe Safdie as its architect, there was a great deal of discussion about how a "central campus" was needed to bring together an otherwise dispersed community of theorists, experimentalists, and detector engineers (the "circus"). Safdie's design (see his essay in this volume) included a "main street" within the structure of the laboratory complex that would serve to bring diverse sectors of the lab together, at least transiently. But at the same time there was a forceful move to establish "regional centers" away from the laboratory—from Europe through the United States to Japan—from which the SSC could be run. By this point, the "center" is frankly acknowledged to be a strategic construct (like the New York and Los Angeles galleries in which Smithson's *Spiral Jetty* film was first shown), socioscientific creations necessary to counterbalance the forces of physical as well as institutional disaggregation.[51] An experiment's spatio-temporal bounds, the notion of a scientific author, and the methods for arriving at a scientific demonstration were all in flux.

The effects of dispersion are, perhaps, less complete in the art world than in the physics world, because the art market works continuously to recuperate authorship and define a locus of production that can guarantee uniqueness in an economy structured in those terms. Even here, however, the effects that were only emergent in the 1970s have spread much farther, with an unprecedented surge in collaborative authorship.[52] The emphasis on discourse and dispersal is also unprecedented; multi-lingual word and image are codependent, and often coextensive;[53] new artists' Web sites and "virtual studios" jostle with other marketing and information systems on the net.[54] Verbal discourse stamps itself on the image to ensure an ironic, postmodern reading and to suggest a "re-productivity" and transferability that aims to cancel the studio romance. These works of art (rarely can one call them "paintings") announce that they become meaningful only in the dispersed spaces of print and verbal exchange, rather than harkening back to a privileged center, an isolated studio, or a genial subject. Artists manage their avatars as the physicists manage data, calibrating, cutting, sorting, and analyzing. No one person "owns" the resulting data in either case.[55]

Conclusion

Despite their material concreteness, architectural spaces always allude elsewhere; laboratories and studios are no exception. They can be modeled on great public institutions: The seventeenth-century Gresham House, predecessor to the Royal Society, was symbolically structured to mimic the British Parliament in order that disputes could be resolved with witnesses and public disputation.[56] Or they can be sites of introspective and solitary meditation patterned on religious retreats: Mendel's literally monastic settings of the late nineteenth century, Delacroix's Michelangelo in his cerebral *studium,* the mid-twentieth-century European cosmic-ray physicists clambering ever higher into their alpine huts, or the Abstract Expressionist painters' veiling of their industrial lofts as spaces staged for confrontations with the existential void. In these last, those allusive spaces of humanist individualism were then further mapped onto the mural-sized, shimmering, individuating canvases themselves. Denying the daylight factories in which they worked, New York School artists labored to link the painter metaphorically with alchemists and theologians, mystical scientists and philosophers whose search was thought to be equally dependent on secrecy and isolation.

We have argued that World War II brought enormous pressures to bear on these sites of scientific and artistic production, transforming the allusions they made. Domestically, the war economy created icons of productive prowess visible in every newspaper, newsreel, and industrial district: ships, planes, and tanks rolling off the assembly lines in proud formation. Coming out of the Depression, participants in the new wartime economy witnessed a postwar surge in production and consumption that sent the Pax Americana into global circulation. No thinking American could remain unaware of the postwar economy's changed scope and scale; soon that scope and scale were expressed in dramatic physical terms as superhighways, housing developments, and consumer electronics brought industrial power into everyday view.

At one level, it seems pointless to deny that physicists' conceptions of the laboratory, or artists' views on the studio, were in some sense responsive to changes in this broader economy: from individual handwork, through factory work, to the work of multinational or multiauthor coordinations. But this "responsiveness" no longer serves to describe fully the history of the period's art or science. Such a blindered historiographical approach leaves both realms of knowledge (science/art) mistakenly *outside* the means of production, reducing the laboratory or studio to an aspect of a superstructure that can only be explained elsewhere by some more extensive basis. As we have argued, this reductive explanation is manifestly inadequate in light of the coproduction engaged in by physicists and industrialists collaborating in the laboratories of World War II. Microwave instrumentation, plutonium separation, electronic computation—these activities oscillated back and forth between new physics, breadboard mock-ups, and scaled up, mass-produced weapons for the battlefield. This lived experience of creating physics and physical devices on an industrial scale, with the ever present necessity of expansion and multiplication, was transformative of the physicists' conception of their work space and, indeed, of themselves. Oak Ridge and Hanford were not laboratories "like" factories, they were themselves factories; the MIT Rad Lab did not mimic production, it was production. When postwar laboratories like Brookhaven were designed, their space was conceived within an industrial world that had already been produced in the plutonium "canyons" of Hanford, the laboratories of Los Alamos, and the isotope separation facilities of Oak Ridge.

Given the postwar laboratories' full-fledged participation in—even leadership of—the trend toward massive, flexible, round-the-clock factory production, the Abstract Expressionist artists of the immediate postwar period appear ever more isolated in their resistance to this powerful industrial movement. Contrasting their creations with the expanding production lines setting up shop in the suburbs (from corporations that had vacated the smaller-scale factory spaces now converted into these artists' Manhattan lofts), the New York School painters constructed their isolated studios as places of refuge and alienation from the crowd. It was a classic modernist gesture of separation, if a final one, and it echoed many of the older tropes of individual humanism that were still in circulation: heroic genius, solitary manhood, even alchemical retreat. But if such tropes were still in circulation, they were also being challenged at every turn. The futility of the Abstract Expressionists' resistance seemed immediately evident by the end of the 1950s. Stella and Warhol abandoned any interest in the romantic tradition of the artist-hero. Centralization in the studio became focused not on the individual, but on production and reproduction. The talk of "process" waned, and it became possible for the first time to speak of sites of artistic "production." Objects, some of them intended to be identical, were produced on assembly lines by an aggregate of assistants and hangers-on. Warhol, Stella, and others no longer saw themselves as critics of society, perched in some "view from nowhere," some isolated studio that gave moral force to their critiques. Their complex works may not always have endorsed the industrial aesthetic they purveyed, but there was no question that they wanted to locate themselves squarely within that order: ". . . I geared myself," as Pop artist James Rosenquist said, "like an advertiser or a large company. . . ." His colleague Roy Lichtenstein concurred, expressing his desire to make "industrial painting" because ". . . it's what all the world will soon become."[57]

So many theorists (some in lamentation, others in celebration) have characterized just this refusal to stand outside large-scale modes of production as the beginning of the postmodern turn. Rather than merely "responding" to larger developments in an economic base, these 1960s artists can also be seen as historical agents of change in their own sectors of social, visual, and economic culture, leaders in signaling "what all the world will soon become." Just as Alvarez built his laboratory as a physics production plant, so Warhol reconfigured his Factory to become a silvered production line for serial art. These architectures were meant to allude elsewhere, but inevitably they became themselves models for physicists, artists, and even industrialists to come.

These first moves were tentative, and still centralized, as we have seen. Alvarez inaugurated management charts, shift supervisors, safety officers, and production quotas, but he maintained a central site: from the Bevatron and the 72-inch bubble chamber came the blueprints and spools of film that were physically distributed around the world. Similarly, Warhol delegated production, but he was still "the boss" of the business art business, and there was only one Factory at its source. And if the architecture of the laboratory and studio was still centered, so too was the architecture of authority. The architecture of authorship at this midcentury moment was still individual, if tenuously so. Although both Warhol and Alvarez did much to vex the concept of the unitary creative author—both vaunted the collaborative, industrial modes of their production—there was little doubt among "the Alvarez Group" or the Factory denizens that the site of authority was East 47th Street or Berkeley's Cyclotron Road. And there was, in the end, little doubt that "image manager" Warhol would accept the market's designation of him as an author-function, still needed to market the work, or that it was Alvarez, as the head of the laboratory—and not the accelerator technicians or even his dozens of collaborators—who would win the Nobel Prize. Despite the increasing fragmentation of production, both physicist and artist could still be embedded in the discourse of an individual, sometimes genial creator.

More challenging to traditional modernist notions of authorship were the spatial and authorial complexities presented by Robert Smithson's assorted productions, and correspondingly, by the dispersal figured in the authorless title for the post-Alvarez collaboration, "group A." In Smithson's work, the isolated studio was abandoned for an ambiguous "post-studio" site, emblematized by the *Spiral Jetty,* which never had a studio to contain it, outside of the editing room in which it was manipulated under someone else's name. But more than merely opening the construction of artistic meaning to other authors, Smithson abandoned the unitary object itself. Neither shrine nor "outdoor sculpture," the submerged jetty is no Mount Rushmore. It is an image circulated in still and movie photographs, a collaborative construction described in essays and reviews, a concept traded in discourse. In its polyvalency and post-studio status, the discursive *Spiral Jetty* is explicitly leveled against modernist faiths in a centralized, unitary work of art. Once the earthwork *Spiral Jetty* is, through its submersion, present only in its other visual and textual manifestations, authorship itself comes into question. Where is the artwork produced? In the editing of the film? In the earthmoving? In the planning of the jetty? In the publishing of the essay? And more vexing still to the market: Where is the object that can be bought or sold?

So it is with the transition from the late modernist age of Alvarez to the post-1970s TPC. Alvarez's group rather symbolically loses its preeminence, and does not become the titular group. It becomes, simply, group A, and group A loses no time in joining other groups in the construction of a project that *ab initio* has no single, unchanging central authority. The corporation of corporations exists, from the start, as a dispersed and decentered entity. Again, the disposition of space reveals a great deal. For as data from the experiment are exchanged over the computer links, the importance of the original trace in the bubble chamber has vanished: there *is* no "golden photographic event," instead even individual events are laboriously assembled through a network of software and hardware. There is no unitary object, just as there is no single author. Most aspects of the experiment can be run from any terminal, and the data can be imported, processed, and sent from any number of sites stretching from Europe to Japan. When the trigger for the experiment is set by computer from someone fifty miles from Fermilab, and data are evaluated in Tokyo; when a teleconferenced meeting about the experiment is shared among Japan, Italy, Chicago, and Cambridge, we can well ask: Where is the experiment? As with the art market, much of the institutional apparatus of physics cuts against the grain of these very-large-scale projects: Authorial recognition must still be constructed, anachronistically, around the individual—in awards like the Nobel Prize and in the academic processes that raise individuals from graduate student up through the ranks to full professor and principal investigator. But even here one sees the suggestions of a change. In 1984 (for the first time ever) the award of the Nobel Prize for physics was given jointly to a physicist and an engineer. Symbolic and limited as this gesture was, it nonetheless spoke volumes about the reconfiguration of the production of scientific knowledge and the simultaneously dispersed and collective identities of its contemporary subjects.

The dispersing of artistic sites of production into discourse, and of scientific sites of production into the ethernet, presents us with entirely new architectures to consider. Architectures of software and managerial structures gain an unruly complexity; dispersed architectures of "post-studio" nonsites and elusive architectures of images and texts are distributed over networks of film and type. To paraphrase Foucault's essay on the author-function[58] that appeared around the same time as the later developments we have outlined here: Where do these architectures appear? Who moves through them, and to what uses are they put? Where do these structures join, and where do they give way? What subjects do they presuppose? To what "elsewheres" do they allude? Laboratories, factories, and studios disperse and recombine in ways unimaginable half a century ago.

Notes

> Much of our thinking began with an attempt to understand the relation between our ideas in Peter Galison, *Image and Logic* (1997), and Caroline Jones, *Machine in the Studio* (1996); as we worked through the common themes and histories in these books, we were given the opportunity to present an early version of this essay as "Centripetal and Centrifugal Architectures" at the *Anyplace* conference, May 1995, in Montreal. We are grateful to Sanford Kwinter for including us in that forum, and to our participating colleagues for useful suggestions. At the conference on "The Architecture of Science," Neil Levine provided helpful comments that helped clarify the nature of the architectural spaces in which the New York School painted. For research assistance and help with permissions we thank Debbie Coen, Jaimey Hamilton, Kirk Hanson, and Kristen Peterson.

1. Swenson, "What is Pop Art?" Part I, p. 26.
2. Alvarez, "Round Table Discussion on Bubble Chambers," p. 272. For this reference and a detailed discussion of the Alvarez bubble chamber, see Galison, *Image and Logic,* chap. 5.
3. Alvarez, "Round Table," ibid. Warhol and Hackett, *The Philosophy of Andy Warhol,* p. 92.
4. "Building in One Package," p. 95.
5. Haines, "Planning the Scientific Laboratory," p. 107.
6. Haines (of Voorhees, Walker, Foley & Smith, designers of the original Bell Labs in Murray Hill) began an article on the scientific laboratory with the words, "the end of the war brought into sharp focus industry's need for research facilities." Haines, "Planning the Scientific Laboratory," p. 3.
7. "Last Word in Factories," p. 129. For more on Kahn, see Rappaport, "Albert Kahn and the Modern Factory"; Albert Kahn, "Kahn's Views on Factory Design"; Reid, compiler, *Industrial Buildings;* and Hildebrand, *Designing for Industry.*
8. Galison, *Image and Logic,* p. 252.
9. On warplane construction, see Robert Ferguson's excellent dissertation, "Technology and Cooperation in American Aircraft Manufacture During World War II." On typical plant layout, see *Warplane Production* 2:3–4 (1944): 15–16, cited in Ferguson, ibid., pp. 22–23.
10. Giffels & Vallet Inc., L. Rossetti, Engineers and Architects, "Federal Telecommunication Laboratory."
11. "Discussion on the Design of Physics Laboratories," pp. 163 and 165.
12. Hertzka and Knowles, "A-E Research Buildings": Central Laboratory on pp. 105–106, quotation about "industrial research" on p. 106. For planning documents, see, e.g., D. Cooksey files, Lawrence Berkeley Laboratory Archives, Berkeley, California, box 5, folder 38, "The General Laboratory Building, A Study," 9 January 1948.

13. For Gerald McCue, architectural design began with industrial laboratories. As a 26-year-old, he found himself in 1952 on a Standard Oil commission to build an industrial laboratory, and that was followed by many others, including a host of buildings at Chevron Research Company's Richmond laboratories beginning in 1954. Design of Berkeley Radiation Laboratory's 88-inch cyclotron (1958) fell into the "pure" as opposed to "industrial" domains, but many of the architectural design features carried over smoothly, as McCue had to build a high-bay staging area with a thirty-ton crane, while providing the facility with the capacity to remove some two megawatts of heat, deionize and demineralize the cooling water, and offer stable temperature and humidity for the electronic systems. Interview with Peter Galison; see also "Laboratory as Machine"; McCue, Boone, Tomsick, no title.
14. On Groves, see Goldberg, *Fighting to Build the Bomb.*
15. See Galison, "Physics Between War and Peace."
16. See, e.g., Seidel, "The Origins of Lawrence Berkeley Laboratory."
17. Galison, *Image and Logic,* esp. chaps. 4 and 9.
18. Smyth, "Cooperative Laboratory," 25 July 1944, file "Postwar Research 1945–46," box "Physics Department Departmental Records, Chairman, 1934–35, 1945–46, no. 1," Princeton University Archives, Princeton, NJ. Cited in Galison, *Image and Logic,* chap. 4.
19. Smyth, "Cooperative Laboratory," revised version, 7 February 1945, cited in Galison, ibid.
20. Galison, *Image and Logic,* chap. 5.
21. "A Douglas Aircraft Assembly Plant."
22. "Douglas Aircraft Plant Chicago."
23. Leprince-Ringuet, "Discours de Clòture," p. 288.
24. See Galison, *Image and Logic,* chap. 5.
25. Shapin and Schaffer, *Leviathan and the Air-Pump.*
26. Michelangelo had been quoted as advising artists to "paint with the mind, not with the hand." Michelangelo Buonarotti, quoted in Wackernagel, *The World of the Florentine Renaissance Artist,* p. 312.
27. See Banham, *A Concrete Atlantis,* for a full discussion of the "daylight factory" and its significance for modernism. For another take on factory style and structure, see Amy Slaton, "Style/Type/Standard."
28. Robert Motherwell in de Antonio, *Painters Painting,* p. 65. For more on the New York School painters, see Jones, *Machine in the Studio,* chaps. 1 and 2.
29. Barnett Newman, interviewed by David Sylvester, Easter 1965; transcript in Auping et al., *Abstract Expressionism,* p. 140.
30. Schlesinger, *The Vital Center.*
31. The panel "Art 60" was held at New York University on 21 April 1960; Robert Goldwater moderated. Quotes and paraphrases are from Sandler, *The New York School,* p. 284. For more on Stella see Jones, *Machine in the Studio,* chap. 3.
32. Recalled to C. Jones in this context by Irving Sandler in a phone conversation of 15 May 1990. See also Sandler, *American Art of the 1960s,* p. 2, where Sandler generalizes this comment well beyond the terms of the panel discussion.
33. Castelli, quoted in Stuart Preston's review, *The New York Times,* Sunday, 2 October 1960; clipping in Castelli Gallery archives, New York. Castle, "What's That, the '68 Stella?"
34. Banham, op. cit., 23ff.

35. Caroline Jones, interview with Gerard Malanga, 25 October 1989. For more on Warhol, see Jones, *Machine in the Studio,* chap. 4.
36. Ibid.
37. Warhol admired the tinfoil-covered apartment of his friend and assistant Billy Linich (a.k.a. Billy Name) and asked Linich to do the same to the Factory. Like other Warhol assistants and associates, Linich contributed a great deal to the "Warhol aesthetic."
38. Arango, "Underground Films"; Castelli Gallery archives, Warhol file.
39. Berners-Lee, "Information Management: A Proposal," March 1989, May 1990 (http://www.w3org/History/1989/proposal.html).
40. Bylinsky, "The Digital Factory."
41. The phrase "postindustrial" goes back at least as far as Arthur J. Penty, who published *Post-Industrialism* in 1922. Daniel Bell was one of the first American academics to make the notion of a "post-industrial society" widely known, although David Reisman and Alain Touraine had earlier used the term. See Bell, "The Post-Industrial Society," and *The Coming of the Post-Industrial Society.* Paolo Portoghesi connects the post-industrial economy with post-modernism in *Postmodern: The Architecture of the Postindustrial Society.* For an overview, see Margaret A. Rose, *The Post-Modern and the Post-Industrial.*
42. Smithson, "A Sedimentation of the Mind," p. 87.
43. See Jones, "La politique de Greenberg et le discours postmoderniste."
44. David Harvey, in his *Condition of Postmodernity,* very persuasively argues against attributing a free-floating autonomy to the cultural sphere. Instead, he sees the increasing dispersion and decentralization of cultural forms as reflecting a broad shift from Fordist-Taylorist forms of accumulation to what he calls "flexible accumulation": the stage of economic development propelled in large measure by the rise and globalization of finance and credit. Harvey does recognize what he calls an "aesthetic" back reaction into the industrial sphere from the artistic domain. Where we differ, perhaps, is that we see (for example) the scientific "cultural" shift from central to dispersed not as separate from the broader technological-economic world, but constitutive of it. For an important critique based on the political economy of city planning and architecture, see Davis, *City of Quartz.*
45. Solomon, "Laboratory Innovations," on pp. 124 and 126. We agree with the author that the twentieth-century lab divides into three stages, but we periodize and explain this division rather differently. She argues that individual, isolated laboratories appeared only with Bell Laboratories in the late 1930s; we see Bell Labs as quite continuous with a long chain of fixed space laboratories stretching back to German labs of the late nineteenth century. She dates a second wave of lab design to Louis Kahn's 1965 Salk Institute, where interstitial space beneath the floor allowed the passage of heat, electricity, gas, and water and therefore permitted flexible walls. (On Louis Kahn, see the interesting and critical view of the Salk Laboratory in Anderson, "Louis I. Kahn in the 1960s" and "Louis I. Kahn"; and the classic text by Scully, *Louis Kahn.*) As we argue in the text, we see the trend toward flexible use of laboratory space as ultimately having its roots in factory design during World War II.
46. The literature on systems and systems engineering is, of course, vast. Of particular interest is the work of Thomas Hughes, whose *Networks of Power* and related articles offer ways of thinking about innovation within complex technological systems.

47. This draws from the theory of the "trading zone" that (locally) knits these subcultures together; see Galison, *Image and Logic,* chap. 9.
48. Quoted in Galison, *Image and Logic,* p. 619.
49. Jay Marx to David Nygren, "Endemic Problems with PEP-4 Mechanical Engineering," 17 July 1979, file "Project Management," box 1, David Nygren Papers, II, Lawrence Berkeley Laboratory Archives, Berkeley, California.
50. Adeva et al., "The Construction of the L3 Experiment," esp. pp. 98ff.
51. On the architecture of the SSC, see Galison, *Image and Logic,* pp. 673–678.
52. Functioning artist pairs who gained prominence in the 1980s and 1990s included Gilbert and George, Clegg and Guttman, the Starn Twins, Komar and Melamid, Gerlovina and Gerlovin; other groups of artists and artist-activists include Silence=Death, Gran Fury, Group Material, Guerrilla Girls, Tim Rollins and K.O.S., and Women's Art Caucus.
53. Here we are thinking of the work of artists such as Barbara Kruger, Christopher Wool, and Jenny Holzer, among others.
54. Media artist and virtual reality designer Perry Hoberman has commented to Caroline Jones that he considers his home page a "virtual studio" where he posts work-in-progress. An artists' Web site "ada.web" is also notable in this regard.
55. For "avatars," see the work of artist Matthew Ritchie from New York. Ritchie generates numerous avatars that he then manipulates in paintings, murals, and reliefs; the suggestion here of a decentered artist-subject is, we think, compelling.
56. On the relation of early modern experimentation to post-restoration politics, see Shapin and Schaffer, *Leviathan and the Air-Pump;* see also Shapin, *Social History of Truth.*
57. James Rosenquist and Roy Lichtenstein, interviewed by Swenson, "What is Pop Art?," Part II and Part I respectively, each p. 63.
58. Foucault, "What Is an Author?"

Bibliography

Adeva, B., et al. "The Construction of the L3 Experiment." *Nuclear Instruments and Methods in Physics Research* A289 (1990): 35–102.

Alvarez, Louis. "Round Table Discussion on Bubble Chambers." In *Proceedings of the 1966 International Conference on Instrumentation for High Energy Physics.* Stanford: Stanford Linear Accelerator Center, 1966.

Anderson, Stan, "Louis I. Kahn in the 1960s." *Boston Society of Architects Journal* 1 (June 1967): 21–30. Also printed as "Louis I. Kahn," *a&u* (Tokyo) (1975): 300–308.

Antonio, Emile de, and Mitch Tuchman. *Painters Painting: A Candid History of the Modern Art Scene, 1940–1970.* New York: Abbeville Press, 1984.

Arango, Douglas. "Underground Films, Art or Naughty Movies." *Movie TV Secrets* (June 1967): n.p.

Auping, Michael, et al. *Abstract Expressionism: The Critical Developments.* New York: H. N. Abrams, and Buffalo, NY: Albright Knox Art Gallery, 1987.

Banham, Reyner. *A Concrete Atlantis: U.S. Industrial Building and European Modern Architecture, 1900–1925.* Cambridge: MIT Press, 1986.

Bell, Daniel. "The Post-Industrial Society: the Evolution of an Idea." *Survey* 17:2 (Spring 1971): 102–168.

Bell, Daniel. *The Coming of Post-Industrial Society: A Venture in Social Forecasting.* New York: Basic Books, 1973.

"Building in One Package." *Architectural Forum* 82 (January 1945): 95–110.

Bylinsky, Gene. "The Digital Factory." *Fortune* (14 November 1994): 92–110.

Castle, Frederick. "What's That, the '68 Stella? Wow!" *Art News* 66:9 (January 1968): 46–47, 68–71.

Davis, Mike. *City of Quartz: Excavating the Future in Los Angeles.* London and New York: Verso, 1990.

"Discussion on the Design of Physics Laboratories." *Journal of Scientific Instruments* 25 (1948): 157–166.

"A Douglas Aircraft Assembly Plant, U.S.A., Architect-Engineer-Manager The Austin Company." *The Architect and Building News* 180 (3 November 1944): 78–79.

"Douglas Aircraft Plant, Chicago, The Austin Company Engineers and Builders." *The Architect and Building News* 181 (5 January 1945): 4–5.

Ferguson, Robert. "Technology and Cooperation in American Aircraft Manufacture During World War II." Ann Arbor, Michigan: UMI Dissertation Services, 1996.

Foucault, Michel. "What Is an Author?" In Paul Rabinow, ed., *The Foucault Reader,* 101–120. New York: Pantheon, 1984.

Galison, Peter. "Bubble Chambers and the Experimental Workplace." In P. Achinstein and O. Hannaway, eds., *Observation, Experiment, and Hypothesis in Modern Physical Science.* Cambridge: MIT Press, 1985.

Galison, Peter. *Image and Logic: The Material Culture of Twentieth Century Physics.* Chicago: University of Chicago Press, 1997.

Galison, Peter. "Physics Between War and Peace." In E. Mendelsohn, P. Weingart, and M. Roe Smith, eds., *Science, Technology and the Military,* 213–251. Cambridge: Cambridge University Press, 1989.

Galison, Peter, and Caroline A. Jones. "Centripetal and Centrifugal Architectures: Laboratory and Studio." In Cynthia Davidson, ed., *Anyplace,* 78–87. New York: Anyone Corporation, 1995.

Giffels & Vallet Inc., L. Rossetti, Engineers and Architects. "Federal Communication Laboratory, Electronics Laboratory and Microwave Tower, Nutley N.J." *Architectural Record* 101 (January 1947): 58–65.

Goldberg, Stanley. *Fighting to Build the Bomb: The Private Wars of Leslie R. Groves.* South Royalton, VT: Steerforth Press, forthcoming.

Haines, Charles. "Planning the Scientific Laboratory." *Architectural Record* 108 (July 1950): 107–127. Reprinted in *Building for Research,* 2–19. New York: F. W. Dodge Corporation, 1950.

Harvey, David. *The Condition of Postmodernity: An Enquiry into the Origins of Cultural Change.* London: Basil Blackwell, 1989.

Hertzka and Knowles, Architects. "A-E Research Buildings, U.S. Atomic Energy Commission. University of California." *Architectural Record* (February 1952): 105–113.

Hildebrand, Grant. *Designing for Industry: The Architecture of Albert Kahn.* Cambridge, Mass: MIT Press, 1974.

Hughes, Thomas. *American Genesis: A Century of Invention and Technological Enthusiasm.* New York: Penguin Books, 1989.

Hughes, Thomas. *Networks of Power: Electrification in Western Society, 1880–1930.* Baltimore: Johns Hopkins University Press, 1983.

Jones, Caroline A. "Andy Warhol's 'Factory': The Production Site, Its Context, and Its Impact on the Work of Art." *Science in Context* 4 (Spring 1991): 101–131.

Jones, Caroline A. *Machine in the Studio: Constructing the Postwar American Artist.* Chicago: University of Chicago Press, 1996.

Jones, Caroline A. "La politique de Greenberg et le discours postmoderniste." *Les cahiers du Musée national d'art moderne* 45/46 (1993): 105–135.

Jones, Caroline A., and Peter Galison, eds. *Picturing Science, Producing Art.* New York: Routledge, 1998.

Kahn, Albert. "Kahn's Views on Factory Design." *Architect and Engineer* 148:2 (1942): 33–34.

"The Laboratory as a Machine." *Architectural Forum* 124:3 (1966): 40–47.

"Last Word in Factories: The Architect-Engineers Create a New and Efficient Industrial Design." *Fortune* 31 (May 1945): 126–132.

Leprince-Ringuet, Louis. "Discours de clôture." *Bagnères de Bigorre.* 1953. Mimeographed typescript.

Lipke, William C., ed. "Fragments of a Conversation (February 1969)." In Robert Smithson, *The Writings of Robert Smithson.* New York: New York University Press, 1979: 168–170.

McCue, Boone, Tomsick. [Descriptive Brochure]. San Francisco, 1974.

Penty, Arthur J. *Post-Industrialism.* London: G. Allen and Unwin, 1922.

Portoghesi, Paolo. *Postmodern: The Architecture of the Postindustrial Society.* Translated by Ellen Shapiro. New York: Rizzoli, 1983.

Preston, Stuart. "Review." *The New York Times* (Sunday, 2 October 1960).

Rappaport, Nina. "Albert Kahn and the Modern Factory." *Metropolis* (1996): 25–31.

Reid, Kenneth, compiler. *Industrial Buildings.* New York: F. W. Dodge Corporation, 1951.

Rose, Margaret A. *The Post-Modern and the Post-Industrial: A Critical Analysis.* Cambridge: Cambridge University Press, 1991.

Sandler, Irving. *American Art of the 1960s.* New York: Harper and Row, 1988.

Sandler, Irving. *The New York School: The Painters and Sculptors of the Fifties.* New York: Harper and Row, 1978.

Sandler, Irving. "New York Letter." *Art International* 4:9 (1 December 1960): 25.

Schlesinger, Arthur. *The Vital Center: Our Purposes and Perils on the Tightrope of American Liberalism.* Boston: Houghton-Mifflin, 1949.

Scully, Vincent. *Louis Kahn.* New York: Braziller, 1962.

Seidel, Robert. "The Origins of the Lawrence Berkeley Laboratory." In Peter Galison and Bruce Hevly, eds., *Big Science: The Growth of Large Scale Research,* 21–45. Stanford, CA: Stanford University Press, 1992.

Shapin, Steven. *A Social History of Truth: Civility and Science in Seventeenth-Century England.* Chicago: University of Chicago Press, 1994.

Shapin, Steven, and Simon Schaffer. *Leviathan and the Air-Pump: Hobbes, Boyle, and the Experimental Life.* Princeton, NJ: Princeton University Press, 1985.

Slaton, Amy. "Style/Type/Standard: The Production of Technological Resemblance." In Jones and Galison, eds., *Picturing Science, Producing Art.*

Smith, Patrick. *Warhol: Conversations About the Artist.* Ann Arbor, MI: UMI Research Press, 1988.

Smithson, Robert. "Frederick Law Olmsted and the Dialectical Landscape." *Artforum* (February 1973). Reprinted in Smithson, *The Writings of Robert Smithson,* 117–128. New York: New York University Press, 1979.

Smithson, Robert. "A Sedimentation of the Mind: Earth Projects." *Artforum* (September 1968). Reprinted in Smithson, *The Writings of Robert Smithson,* 82–91. New York: New York University Press, 1979.

Smithson, Robert. *The Writings of Robert Smithson.* New York: New York University Press, 1979.

Solomon, Nancy. "Laboratory Innovations." *Architecture* 82:3 (1993): 123–127.

Swenson, Gene. "What is Pop Art? Answers From Eight Painters: Part I." *Art News* 62:7 (November 1963): 25–27, 61–64.

Swenson, Gene. "What is Pop Art? Part II." *Art News* 62:10 (February 1964): 40–43, 62–67.

Wackernagel, Martin. *The World of the Florentine Renaissance Artist* (1938). Translated by Alison Luchs. Princeton, NJ: Princeton University Press, 1981.

Warhol, Andy, and Pat Hackett. *The Philosophy of Andy Warhol: From A to B and Back Again.* San Diego, CA: Harcourt Brace Jovanovich, 1975.

Index

Italicized page numbers refer to illustrations and captions.

Aalto, Alvar, 13, 355
"ABC fordert die Diktatur der Maschine" (Schmidt and Stam), 240, 250n8
"Abstract Architecture II" (Meyer), *239*
Abstract Expressionism, 510, 512, 514, 530, 531
Abstract machine, 248, 249
Academic medical centers, 294
Accademia dei Lincei, 46, 47, 54n71
Accademia del Cimento, 79, 94n1
Achromatic lenses, 142–143, 152, 154, 155, 156, 158n1, 159n23
Ackerman, James, 53n51, 74n19
Acoustical Corporation of America, 276n59
Acousti-Celotex, 263
Acoustics. *See* Architectural acoustics
Adams, George, 470
Adjacency, 7, 14, 376
Adler, Dankmar, 186, 204n16
Aesthetic View from Pegasus, The (Schinkel), *115*
Air conditioning, 367
Akoustolith, 11, 262, 263, 264, 274n37

Alberti, Leon Battista. *See also De re aedificatoria*
  Aldrovandi as disciple of, 40
  as architect, 87
  on the architect, 40, 87
  architectural influence of, 38
  audience for architectural writings of, 52n32
  Books on the Family, 37, 43
  on empiricism, 88
  gender in architectural plans of, 4, 36
  as humanist, 86
  mechanical inventions of, 87–88
  Navis, 87
  on openness, 6, 86, 88–89
  and the patronage system, 92
  on restricting women within the home, 43, 44
  on rulership and architectural style, 88
  on the study, 33–40, *37*
  techne and *praxis* in writings of, 85–89
  on villas, 41, 42, 52n36
  on the visual arts, 87
  Vitruvius criticized by, 86, 89

Albert Kahn Associates, 18, 499
Albrecht, Prince, 120, *121,* 124
Alchemia (Libavius), 60, 67, 71, 72
Alchemy. *See also* Hieroglyphic monad
    artist's studio compared with alchemist's lair, 510
    "dispersion of knowledge" technique, 71
    and distilling and brewing, 160n57
    donum Dei, 67
    Geber, 67, 71, 75n25
    Libavius on the ideal alchemist, 67
    Libavius's defense of, 67
    Libavius's laboratory, 4–5, 59–77
    Paracelsians, 62, 63, 67, 68, 70, 72–73
    philosophers' stone, 5, 66, 68, 71
    princely patronage in, 80
    secrecy of, 59, 60, 67
Alchemy (Pollock), 510, 521
Aldrovandi, Ulisse
    as amateur architect, 40–41
    as disciple of Alberti, 40
    humanism of, 41, 42
    as ideal reader of Alberti, 52n32
    Lexicon of Inanimate Things, 44
    library of, 41
    the museum as conceived by, 32
    museum occupying more than one room, 40
    Pandechion Epistemonicon, 44
    On the Remains of Bloodless Animals, 45
    Catalina Sforza visits museum of, 29–30
    Studio Aldrovandi, 29, 48, *49*
    villa of, 41–43
    women excluded from museum of, 4, 30–31, 43
Alembert, Jean Le Rond d', 318, 319
Alexander, Christopher, 357, 364, 366, 480
Alexander, William D., 462, 463, 466
Alexandra (steamboat), 108, *108,* 118, Plate 2
Algarotti, Francèsco, 257
Alpers, Svetlana, 329

Altenstein, Karl Freiherr von, 133, 134
Altes Museum (Berlin), 131, *131*
Alvarez, Luis
    authority of, 532
    bubble chamber work of, 505, 506, *506,* 524, 526
    as competing with other groups, 527
    group A, 518, 519, 533
    at Los Alamos, 505
    Nobel Prize for, 532
    on physics as business, 497
American Institute of Architects, 292, 498
American Museum of Natural History, 172
American roadside architecture, 396, *397*
Analysis
    and design, 378
    in eighteenth-century philosophy, 319
    in engineering, 319
    in modern thinking, 12
Anamorphosis, 344
Anderson, Benedict, 426
Ansicht der neuen Dampfmaschine bei Sanssouci (Freydanck), *126*
Ansicht des Königlichen Schlosses Sanssouci mit den Terrassen (Freydanck), *124*
Anthropological museums, 165–180
    Boas-Mason debate on display in, 7, 167–172
    culture area displays, 172
    Pitt Rivers Museum Project, 8, 172–176
Anthropology. *See also* Anthropological museums
    architecture of the discipline, 166
    fieldwork, 165–166
    museum arrangement and theory in, 165–180
    museum period of, 166, 167, 191
    paradigmatic traditions of, 167
    spaces of anthropological inquiry, 165–166
Archaeology, 166
Architect and Building News (journal), 506–507

Architectural acoustics, 253–280. *See also*
	Sound-absorbing materials
	as case study of role of science in modern architecture, 254
	geometrical approach to, 257–258
	history of, 257–261
	nineteenth-century attempts at applying, 10–11
	Sabine's study of sound-absorbing materials, 258–259
	soundproof construction, 264, *265*
Architectural determinism, 410
Architectural form. *See* Form
Architecture. *See also* Buildings
	Alberti's influence on, 38
	as applied ergonomics, 354
	architect-engineer-building companies, 499
	artistic aspect of, 322, 483
	as art of mediation, 347
	authority of architects, 41, 89
	avant-garde view of, 483
	Beaux-Arts architecture, 218, 339, 343
	Benedictine architecture, 142, 153–154
	building around the technical, 7–9
	circulation as metaphor in, 9, 213–220
	collaboration with other disciplines, 181, 202
	computer-generated design, 375, 376
	as craft-based, 87, 363
	and the crisis of modern technology, 324–330
	and engineering, 309, 362
	as expression, 322
	Filarete on the architect, 89–90
	generic architecture, 388–390
	and governance, 6, 93
	industrialized building, 357–358, 371n20
	as instrumental after seventeenth century, 13, 339
	is architecture science, 12–14
	as liberal art, 86–87, 363
	mechanical metaphors in, 9, 220–226
	Meyer's scientization of, 10, 233–252
	modular construction, 499
	mutual limits of science and, 353–373
	openness and empiricism in early writings on, 79–103
	Papworths' guide for museums and libraries, 191, 192
	physicians compared with architects, 40
	Plato's *Timaeus* as prototype for, 13, 338
	polarized views of, 483
	professionalization of, 181, 182–190
	in public sphere, 322
	pupilage in education for, 182–183
	"revolutionary architecture," 316
	as science, 337–351
	and science and technology, 309–335
	and science in early modern Europe, 27–103
	and science in Victorian Britain, 181–208
	scientific metaphors in, 9, 213–231, 337–338
	scientism in, 2, 9–10, 14, 375–376
	as scientized, 1
	server and served philosophy, 402
	specialization in sixteenth century, 92
	subject matter of classical and Renaissance, 81
	Vitruvian view of, 483
	World War II transforming, 498
Architecture (journal), 526
Architecture Without Architects (Rudofsky), 476
Aristotle, 97n25, 324
Art. *See also* Earthwork art; Studios
	Abstract Expressionism, 510, 512, 514, 530, 531
	Alberti on visual arts, 87
	architecture seen as, 322, 483
	artist pairs, 537n52
	artistry in science, 375
	artists as solitary creators, 509–510
	conceptual art, 520
	dispersion in, 529–530
	Minimalism, 512, 514

New York School, 510, 530, 531
Pop art, 512, 514
science separated from, 494, 498
WPA Federal Art Project, 508
Artifacts, 12, 324–325, 347
"Art 60" (panel discussion), 512, 514, 535n31
Atomic Energy Commission (AEC), 17, 459, 462, 506
Auditorium Building (Chicago), 186, 204n16
Augusta, Princess, 118
Augustus (emperor), 5–6, 84
Austin Company, 18, 499
Authority
　of Alvarez, 532
　of architects, 41, 89
　of buildings of science, 202
　medical, 299
　of museums, 191
　patriarchal authority controlling space, 37
　of physicians, 40
　Plato as, 64
　of Warhol, 532
Authorship
　the art market and, 529
　by committees, 528
　dispersal of, 498, 519, 524, 532–533
　humanist, 86
　as unstable and contested, 19
　Vitruvius on, 83–84
Averlino, Antonio. *See* Filarete

Babelsberg (estate), 108, 118, *119,* 120
Bacon, Francis, 248
Bad air (miasmatic) theory of disease, 11, 284, *287*
Baird, George, 356
Banham, Reyner, 358, 371n21
Barbaro, Francesco, 45
Barry, Charles, 216, *217,* 218
Bassi, Laura, 48
Bastian, Adolf, 167, 168

Bauhaus
　Carnap's speech to, 19
　on expression of function, 15
　Mandelbrot's criticism of, 343
　Meyer as the "other" director of, 233
　scientized conception of architecture of, 10, 21n14
　unaesthetic aesthetics of, 10
　and unified science movement, 370n4
Bauschule (Berlin), 113
Beaux-Arts architecture, 218, 339, 343
Beccadelli, Ludovico, 38
Becker, Henry, 194, 195
Behne, Adolf, 238
Bell, Daniel, 536n41
Bell Laboratories, 536n45
Belopoeica (Philo of Byzantium), 81–82
Benedictine order
　architecture of, 142, 153–154
　and capitalism, 160n36
　glassmaking tradition of, 151
　labor in Benedictine culture, 142, 152–153
　libraries and scientific instruments of, 152
　secrecy as characteristic of, 142, 153
　silence as characteristic of, 142, 153
　skilled artisans of, 152
Benediktbeuern (Germany)
　distilling and brewing at, 160n57
　forest scientists of, 152
　Fraunhofer's laboratory, 156, *157*
　glass house in monks' washroom, 156, *157*
　Optical Institute located at, 142, 151–152
　secrecy of work at, 154–155
　suitability for optical research, 142
Benjamin, Walter, 348
Bentmann, Reinhard, 41
Berkeley Radiation Laboratory. *See* Lawrence Berkeley Laboratory
Berlin Physikalische Gesellschaft, 132, 134
Berlin-Potsdamer Bahn, Die (Menzel), *111*
Berman, Marshall, 272n3

Berners-Lee, Tim, 519
Bestimmung des Brechungs- und Farbenzerstreuungs-Vermögens verschiedener Glasarten (Fraunhofer), 143, *146,* 158n5
"Beuth" (locomotive), 114, *115,* 128
Beuth, Peter, 112–114, *115,* 124, 126, 133, 135
Beyond Habitat (Safdie), 480
Biagioli, Mario, 54n71, 80
Big science
    high energy physics centers as exemplars of, 17, 18
    at its peak by 1965, 498
    post-1970 physics, 524
    roots of big physics, 504
Bill, Max, 360
Billings, John Shaw, 284, 285, *287*
Biocenter (Frankfurt, Germany), 486–488, *487*
Bioclimatic design methodology, 355
Biological (physical) anthropology, 166, 172
Biology. *See also* Center for Advanced Biotechnology and Medicine; Molecular biology; Natural history
    morphology, 475–476
    physiology, 134
    Salk Institute of Biological Studies, 401, 410, 415, 485–486, *485,* 536n45
    secrecy and rivalry in twentieth-century, 80
Blomfield, Reginald, 183
Blondel, François, 311
Blu, Karen, 427
Boas, Franz
    as cultural relativist, 170
    debate with Mason on museum display, 7, 167–172
    on goal of museum exhibition, 171
    Northwest Coast Hall, 172
    and Pitt Rivers Museum Project, *175,* 176
Bolzoni, Lina, 42

Books on the Family (Alberti), 37, 43
Borel, Pierre, 36
Borsig, August
    "Beuth" locomotive, 114, *115,* 128
    and Egells, 114, 124
    at *Gewerbeschule,* 114, 124
    horticultural pursuits of, 129
    locomotive production, *111,* 128
    Sanssouci engine, 124, *126,* 128
    villa at Moabit, 129, *129*
Borsigsche Etablissement zu Moabit, Das (Kolb), *129*
Boston Lying-In Hospital, 289, 300n26
Boston Symphony Hall, 11, 258, 259–260, *260,* 270, 274n30, 276n64
Boullée, Etienne-Louis
    and artistic dimension of architecture, 322
    Cenotaph for Newton, 314, 320
    on circulation at Versailles, 216
    dynamism as preoccupation of, 320
    engineers using principles of, 317, 320
    geometric compositional technique of, 316
    sensationalism as influence on, 318
    symbolism in, 322
Boundary, 157, 310, 324
Bourdieu, Pierre, 426
Bowen, William, 399, 400, 413, 414
Boyle, Robert, 4, 157–158, 453n46
Brandt, Allan M., 11–12, 15
Bricks, testing of, 188–190, *189,* 205n28
Bridge design, 319
Brise soleil, 370n7
British Association for the Advancement of Science, 134
Broken Symmetry (Wilson), *465*
Brookhaven National Laboratory, 489, 527, 531
Brücke, Ernst, 132
Brunelleschi, Filippo, 87
Bubble chambers, 505–506, *506,* 524, 526, 527
Buffon, Georges-Louis Leclerc de, 331n27
Builder, The (journal), 186

Building materials. *See* Materials
Buildings. *See also* Buildings of science; Dwellings; Libraries; Loft buildings
    circulation in, 213–220
    as determining user behavior, 410
    and identity, 424–431
    as machines for high modernism, 338
    new types in nineteenth century, 181
    Philo of Byzantium on proportions for, 81–82
    as sealed systems, 220, 229n17
    segregation of behavior in, 425
    and the subject of science, 1–25
    vernacular buildings, 476, 478, 480, 484
Building science
    Meyer on basing design on, 248
    in professionalization of architecture, 181, 182–190
Buildings of science. *See also* Laboratories; Museums
    as never having enough space, 446
    scientists affected by, 2–3, 414, 422, 423
    sociological interest of, 423
    variety of types of, 1–2
    in Victorian Britain, 190–202
Bulfinch, Charles, 283, *283*
Bülow, Hans von, 112
Bureau of Topography (Bavaria), 150
Burnham, Daniel, 272n11

CABM. *See* Center for Advanced Biotechnology and Medicine
Calatrava, Santiago, 326, *327*
Calculus, 319
Cambridge Electron Accelerator (Massachusetts), 507
Cambridge University Center for Built Form and Land Use Studies, 354, 357
Candelabra, 122–123
Carl, Prince, 108, 110, 116, 118, 120, 124
Carnap, Rudolf, 10, 19, 370n40
Carnot, Sadi, 134
Casa Giuliani Frigerio (Como, Italy), *225, 226, 227*
Castiglione, Baldassare, 42
Catastrophe theory, 340
Celotex Corporation, 11, 263, 264
Cenotaph for Newton (Boullée), 314, 320
Center for Advanced Biotechnology and Medicine (Piscataway, New Jersey)
    administrative space in, 438
    architectural meiosis in, 434, 436, 445
    atria of, 436, *437*
    and basic/applied science dichotomy, 450
    central corridor of, 436, 438
    closed labs of, 443–444
    educational goals of, 438
    entrances to, 434–436
    exterior view of, *449*
    floor plans, 428–431, *429*
    funding of, 438
    future expansion planned for, 446
    goal of, 452n30
    heavy equipment storage in, 444
    identities made tangible by, 428
    laboratory interiors of, *439*
    laboratory modules of, 440, 443–446
    Lewis Thomas Laboratory compared with, 16–17, 423–455
    lighting of, 443
    metaphors for, 450
    and Network Laboratories, 447–448
    numerical data on, 430, *430*
    offices in, 445
    outjobbing at, 447–448
    praise for, 450
    public space in, 436, 438
    safety in design of, 443–444
    and Satellite Facilities, 447–448
    site plan of, *435*
    structural biology facilities of, 445
    tethering at, 448–450
    two constituencies of, 434
    typifying, 440, 443–446

visitors accommodated in, 438, 445, 451
as zero-based, 440
Center for Built Form and Land Use Studies (Cambridge University), 354, 357
CERN (Geneva)
    horizontalism of, 18, 490
    Large Hadron Collider, 524, *525*
    L3 experiment, 528–529
    Safdie on, 489–490
    site of, 461
    World Wide Web development at, 519
Cesi, Federico, 46
Chaos theory, 340, 342, 343
Chaptal, Jean Antoine, 319
Charlottenhof (palace), 114, 120, 122–123, *122, 123,* 132
Chemistry. *See also* Alchemy
    analysis in, 319
    elements, 318, 319
    Libavius's laboratory, 4–5, 59–77
    Mitscherlich, 133
    openness of, 59, 60
Chermayeff, Serge, 364, 366
Chestnut Hill Mall (Newton, Massachusetts), 296, *297*
Chicago School, 272n11
Christina of Sweden, 47, 54n82
Christine de Pisan, 50
Chromatic aberration, 142–143, 149
Cicero, 62
Circulation
    as architectural metaphor, 9, 213–220
    Durand on, 214, 216
    in economics, 228
    in Fermilab, 17
    Guadet on, 218
    Harvey on, 9, 214, 219, 229n16
    human circulatory system, *215*
    Le Corbusier on, 213–214
    March and Steadman on, 370n18
    as modernist category, 220, 228
    as vehicular traffic, 214

Viollet-le-Duc on, 9, 214, 216, 218, 220, 337
City and Guilds Central Institution (London), 183–184, 188
Civic humanism, 60
Civil Conversation (Guazzo), 46
Clapeyron, Emile, 134
Clarke, Max, 205n28
Classical evolutionism, 168
Climate control, 367–368, 499
Clinical Research Laboratory (University of Pennsylvania), *394*
"Closed lab" concept, 442, 443–444
Coefficients of absorption, 259
Cole, Henry, 185, 200, 204n13
Collections. *See also* Museums
    scientific disciplines emerging from, 191
    secrecy and openness regarding, 38
Collective identity, 426–427
Collins, James, Jr., 15, 398, 401, 414, 431
Commentationes (Libavius), 67, 71
Complex systems, 343
Computer-generated design, 375, 376
Computer images, 329
Conceptual art, 520
Condillac, Etienne Bonnot de, 318, 319, 322
Congress on Hygiene and Demography (1891), 185
Constructivism, 396, *397*
Contingency planning, 380
Control
    architecture as vehicle of, 347
    as dispersed in post-1970 science, 524
    the Panopticon, 248, 300n13, 322, 339
    patriarchal authority controlling space, 37
    technology creating dreams of, 347
    in work of "revolutionary architects," 322
Conway, John T., 473n4
Coolidge and Shattuck, 300n26
"Co-op Construction" (Meyer), 249–250, *249*

"Co-op Zimmer" (Meyer), 240, *241*, 243–244
Corrugli, Benedetto, 38
Cosmic-ray physics, 505, 507, *508*, 530
Cosmology, 341
Cram, Ralph Adams, 254, 261, 270, 274n35, 276n65
Cram, Goodhue & Ferguson, *256*, *257*, 260, 261, *262*, 270
Credit, 84, 154
Crell, Ostwald, 159n22
Critical Path Method, 528
Critical theory, 353, 368
Cugnot, Joseph, 325
Cultural (social) anthropology, 166
Culture area displays, 172
Curtis, William J. R., 272n11
Cuvier, Georges, 193, 315
Cyborg, 21n17

Daimler-Benz building (Berlin), 367
Dal Co, Francesco, 2
Daly, César, 216, *217*, 218, 220, 309
Damisch, Hubert, 20n5
Dampfmaschine zum Springbrunnen im Lustgarten, Der (Leithold), 131
Dardano, Luigi, 44
Dartmouth-Hitchcock Medical Center (Lebanon, New Hampshire), 296, *297*
Davidoff, Paul, 376
Davies, Richard Llewelyn, 354
De architectura (Vitruvius)
    Libavius citing, 64
    on openness and empiricism, 80, 82–85
    in the Renaissance, 311
    sixteenth-century architects drawing on, 38
De azotho (Libavius), 67
Deconstruction, 326, 343, 363
Decoration, 42, 425, 483
Dee, John, 60, 62, *62*, 63
De la Beche, Henry, 194, 195
Delacroix, Eugène, 509, 530

De lapide philosophorum (Libavius), 67
Deleuze, Gilles, 248, 249
De mercurio philosophorum (Libavius), 67, 74n22
De re aedificatoria (Alberti)
    the architect as defined in, 87
    Libavius applying ideas of, 40
    on openness and empiricism, 80, 86
    on restricting women, 43
Desargues, Girard, 338
Descartes, René, 312, *313*
De secretis naturae (pseudo-Ramon Lull), 75n25
Determinism, architectural, 410
Devonian Controversy, 195
"Discussion on the Design of Physics Laboratories" (1947), 501
Disney Celebration Health development, 298
Dollard, John, 142
Domestic architecture. *See* Dwellings
Donnelley, Sir John, 200
Donum Dei, 67
Douglas Aircraft Assembly Plant (Oklahoma City), 506
Dove, H. W., 113, 133
D-sodium couplet, 144
Du Bois-Reymond, Emil, 132, 134
DuPont, 504
Durand, Jean-Nicolas-Louis, 183, 214, 216, 322, 339, 340
DUSAF, 462–463, 466, 473n5
Dwan, Virginia, 523
Dwellings. *See also* Studies; Villas
    decoration of, 425
    early museums located in, 36, 47, 50
    General Panel Systems house, 359, *359*
    Italian palazzo, 388, *389*
    Klein's ergonomic housing, 13, 354, 355, 370n1
    mass production of, 358
    medical metaphor for, 21n17
    and place, 2
    Rainer on privacy in, 364, *365*

Terragni's Casa Giuliani Frigerio, *225, 226, 227*
Törten Housing, 355
in traditional societies, 426
Dymaxion principle, 359–360
Dynamic processes, 319, 320

Eamon, William, 73n2, 155–156
Early American college buildings, 388, *389*
Earthwork art
   as dispersing object, artist, and studio, 19, 520–521
   Smithson's *Spiral Jetty,* 521–523, *522,* 532–533
   Superconducting Super Collider landscaping compared with, 494
Echenique, Marcial, 354
Eco, Umberto, 356
Ecole des Ponts et Chaussées, 317, 319–320
Ecological crisis, 363–366
Education. *See also* Technical education
   medical education in hospitals, 290
   museums in, 191
Egells, F. A. J., 112, 114, 116, 118, 124, 131, 135
Egyptian pylon, 396, *397*
Eisenman, Peter
   Biocenter in Frankfurt, 486–488, *487*
   on the end of classical architecture, 339
   mechanical metaphors of, 9
   on Terragni, 224, *225, 226, 227*
Electronic revolution, 325
Electronic technology, 396
Elements
   in Boullée's and Ledoux's designs, 318
   in eighteenth-century accounts of science, 318–319
   in modern science and architecture, 12
Eléments et Théories d'Architecture (Guadet), 218
Eliot, Charles, 258, 259
Empiricism
   Alberti on, 88

   in fifteenth-century trades, 85
   Filarete on, 89, 90
   logical empiricism, 10
   Philo of Byzantium on, 81–82
   and rhetoric of openness, 5, 79–103
   various meanings of, 80
   Vitruvius on, 82–85
Energy conservation, 367–368
Engineering
   Alberti grounding architecture in, 87
   analysis in, 319
   architect-engineer-building companies, 499
   architecture as divorced from, 309
   Boullée's and Ledoux's principles used by, 317, 320
   engineer sharing Nobel Prize for physics, 533
   in high-energy physics, 497
   Le Corbusier on listening to American engineers, 255, 257
   modernism conflating architecture with, 362
   Philo on experimentation in, 82, 96n15
   in Time Projection Chamber project, 528
English gardens
   disguised steam engines in, 6
   in Germany, 107, 130
   Helmholtz on British, 134–135
   Lenné and Schinkel's tour of, 112
   at *Sanssouci,* 124
Enlightenment, 344, 363
Episteme, 97n25
Ethnology, 166
European Center for Nuclear Research. *See* CERN
Events, 329–330
Evolution
   in Boas-Mason debate, 167–168, 171
   in display at Pitt Rivers museum, 172, 176
Excellent and Learned Defense of Women (Dardano), 44

Exhibition Hall No. 26 (Hanover), 368, *369*
Experimental philosophy
   openness and empiricism in seventeenth-century, 79–103
   practitioner devaluation in, 93
   separating itself from practice and power, 94
   trust in evaluation of experimental knowledge, 158
   women in seventeenth-century experimental culture, 32, 51n10
Experimentation
   in architecture and engineering for Philo, 82, 96n15
   in architecture for Vitruvius, 83
   dispersal of the experiment, 533
   in fifteenth-century trades, 85
   openness in, 79
   and theory as interrelated, 80

Factories
   artists' studios adopting factory model, 18, 498, 508–519
   centralization of, 498
   "daylight factories," 518
   Douglas Aircraft Assembly Plant, 506
   Fagus Shoe Works, 388
   flexibility in, 499
   Ford Motor Company Aircraft Building, 501, *501, 502*
   Kahn in construction of, 499
   laboratories adopting factory model, 498–507, 531
   Ledoux's salt factory at Arc-et-Senans, 320, 322, *323*
   New England mills, 388, *389,* 416
Factory, The (Warhol studio), 497, 516–519, *517,* 520, 532
Fagus Shoe Works (Alfeld, Germany), 388
Fairchild, Sherman, Biochemistry Building (Harvard University), 15, 400, 401
Faraday, Michael, 184
Federal Art Project (WPA), 508
Federal Communication Laboratory and Microwave Tower (Nutley, New Jersey), 501, *503*
Fermi, Enrico, 459
Fermilab (Batavia, Illinois), 17–18, 459–473. *See also* Wilson, Robert Rathbun, Hall
   aerial views of, *460, 484, Plate 11*
   funding for, 461
   location debate, 459, 461
   pump house of, *471*
   Safdie's visit to, 483–485
   site of, 462, 484
   Superconducting Super Collider compared with, 489
   utility poles at, *472*
   Wilson's *Broken Symmetry,* 465
Fieldwork, 165–166
Filarete. *See also Tratatto di architettura*
   on the architect, 89–90
   as artisan, 86, 89, 92
   on empiricism, 89, 90
   on openness, 6, 92
   and the patronage system, 92, 485
   on a school for the impoverished, 90–91
   Sforzinda, 6, 89
   techne and *praxis* in writings of, 89–92
Findlen, Paula, 4, 20n8, 80, 93, 191
Finlay, Cathy, 420, 421
Finsbury College, 184
Fintelman, C. J., 110
Fintelman, C. J. T., 110
Fintelman, G. A., 110
Fiore, Robert, 523
Flexibility, 388, 390, 499
Flexible accumulation, 536n44
Fogg Art Museum (Harvard University), 258, 259
Fontana, Francesca, 30, 42–43, 44–45, 47
Fontana, Lavinia, 31, *31,* 50n8, 51n9
Fonte, Moderata, 4, 48
Forbat, Fred, 355
Force, 134

Ford Motor Company Aircraft Building (Dearborn, Michigan), 501, *501, 502*
Forestier de Villeneuve, F., *317*
Forgan, Sophie, 8, 9, 12
Form
    form-content disjunction, 344, 346
    "form follows function," 480–481
    Kahn on, 485–486
    materials and site contributing to, 476, 478
    Meyer on, 233
    natural and man-made, 480
    psychology and, 356–357
    Thompson on, 475–476
    Venturi on form accommodating functions, 390, 396
Form and Purpose (Safdie), 481
Formula-world, 13
Forsyth, Michael, 273nn21, 23, 275n53
Fortschritte der Physik, Die (journal), 132
Forty, Adrian, 9, 17, 203n2, 337, 338
Foster, Norman, 360
Foucault, Michel, 20n4, 248, 300n13, 533
Foundation Cartier (Paris), *328, 329*
Four Books on Architecture (Palladio), 38
Fowkes, Francis, 185, 204n13
Fractals, 343, 486
Frampton, Kenneth, 13–14, 310, 379, 480
Frank, Philip, 10
Frankl, Paul, 218–220, 222, 229n15
Franklin, Melissa, 483, 484, 485, 486
Fraunhofer, Joseph von, 141–163
    achromatic lens technique, 143, 152
    Bestimmung des Brechungs- und Farbenzerstreuungs-Vermögens verschiedener Glasarten, 143, *146*, 158n5
    as epitomizing German *Handwerkerkultur*, 141
    Herschel's visit to, 155
    monastic laboratory of, 7, 142, 156, *157*
    at the Optical Institute, 151
    publishing his work, 154
    refractive index computation, 148–149
    secrecy about work of, 154–155
    Six Lamps Experiment, 143–147, *146, 147*
    Solar Light Experiment, 147–148
Fraunhofer lines, 141, 143, 148, *149, Plate 5*
Freidorf Cooperative Estate, 238, 240
Freydanck, Carl Daniel, *124, 126, 131*
Frick Chemistry Complex (Princeton University), 400
Fried, Michael, 520
Friedrich Wilhelm III, 124
Friedrich Wilhelm IV, 114, 116, 118, 120, 122, 124, 132
Fuller, Buckminster
    deterministic theory of, 339
    on dymaxion principle, 359–360
    on geodesic domes, 358, *358*, 359
    Viollet-le-Duc as precursor of, 316
Function
    as architectural metaphor, 226, 228
    differentiated spaces of PSFS tower, 268
    form accommodates functions, 390, 396
    "form follows function," 480–481
    in Lewis Thomas Laboratory, 15
    Safdie contrasting purpose with, 481
    Viollet-le-Duc on, 216, 218
Functionalism
    in hospitals, 290, 292
    as metaphor, 337
    of Meyer, 233, 246, 248
    Venturi on abandoning, 17
    of Weimar German architects, 355
Futurism, 254–255

Gadamer, Hans-Georg, 81, 345
Galen, 40
Galileo
    in Accademia dei Lincei, 46, 47
    in fundamental epistemic break of modernity, 13
    and strength of materials, 312

Galison, Peter, 18–19, 80, 310, 370n4
Gallo, Agostino, 41, 53n51
Gardens. *See also* English gardens
   in Fonte's *Women's Worth,* 48
   ideal territories compared with, 320
   as setting for philosophizing, 54n80
Gärtnerschule (Berlin), 113, 116
Gauthey, E.-M., *320*
Geber, 67, 71, 75n25
Geertz, Clifford, 167
Gender. *See also* Women
   in Alberti's architectural plans, 4, 36
   Renaissance moralists on mixing, 46
   in Scamozzi's domestic architecture, 40
   solitary genius as gendered trope, 498
General Panel Systems house, 359, *359*
Generic architecture, 388–390
Genius, 498, 510, 519, 531
Gentlemen, 4, 41, 50, 80, 181, 182
Geodesic domes, 358, *358,* 359
Geoffroy Saint-Hilaire, Etienne, 193, 315
Geological Museum (London), 202
Geological Survey (United Kingdom), 194, 195, 200, 201
Geology
   Geological Museum (London), 202
   theories expressed in Museum of Practical Geology displays, 8, 195–200
Geometry of the Environment, The (March and Steadman), 357
Germ theory of disease, 11–12, 288
Gewerbeschule (Berlin), 113, 114, 133, 134
Giedion, Sigfried, 255, 362, 371n26
Gieryn, Thomas, 16–17, 21n23
Giffels & Vallet
   Federal Communication Laboratory and Microwave Tower, 501, *503*
   Ford Motor Company Aircraft Building, 500, 501, *501, 502*
Giganti, Antonio, 38, *39*
Gilbert, Cass, 264, *265,* 266, 267, 270, 276n63
Gille, Bertrand, 325, 332n43
Giorgio, Francesco di, 66, *66,* 89

Girardin, René Louis de, 320
Glaser, Donald, 505–506
Glassmaking, 155–156
Glienicke (estate), 108–110
   Hahn's panorama of, *108, Plates 1–4*
   steam engines at, 108, *108,* 109, 110, 116, *117,* 119, *Plate 3*
Glimcher, Arnold B., 357
Global warming, 363
Goffman, Erving, 425
Golden events, 521, 533
Goldin, Grace, 299n2
Goldthwaite, Richard A., 40, 41, 52n32, 85
Goodhue, Bertram Grosvenor, 253–254, 271, 272n2
Goodman, Nelson, 226
Gorgoni, Gianfranco, 523
Greenberg, Clement, 520
Greenhouse effect, 363
Gresham House, 4, 530
Gropius, Martin, 273n23
Gropius, Walter, 233, 355, 359, *359,* 388
Groves, Leslie, 504
Guadet, Julien, 218
Guastavino, Raphael
   Rumford tile, 262, *262,* 263, *263,* 264
   St. Thomas's Church, 262, *262,* 270
   thin-shelled vaulting system, 261, 274n36
Guazzo, Stefano, 42
Gubler, Jacques, 250
Guess, Raymond, 353, 368
Guild practices, 6, 7, 85, 155
Guinand, Pierre Louis, 151, 154
Guyot Hall (Princeton University), 404, 405, 446, 447, 448

Habermas, Jürgen, 85, 340, 366–367
Haddon, Alfred, 165
Hahn, August C., 108, *108, Plates 1–4*
Haines, Charles, 534n6
"Hairy woman," 31, *31,* 48, 51n9
Hall, Chester Moor, 142
Halle, David, 425

Hanford Nuclear Site, 18, 499, *500,* 531
Hannaway, Owen
   on early laboratories as domestic sites, 51n20
   on laboratory as site of knowledge production, 161n69
   on Libavius, 4–5, 60, 62, 67, 68, 70, 71, 72
Haraway, Donna, 21n17
Hardenberg, Prince, 110, 112, 120
Harkness, Deborah, 51n10
Harth, Erica, 46
Harvard University
   Fairchild Biochemistry Building, 15, 400, 401
   Fogg Art Museum, 258, 259
   Massachusetts Hall, *389*
Harvey, David, 536n44
Harvey, Sir William, 9, 214, 219, 229n16
Häsler, Otto, 355
Hays, K. Michael, 10, 12
Hayward, C. Forster, 204n21
Health care. *See* Hospitals
Hegel, Georg Wilhelm Friedrich, 222
Heidegger, Martin, 2, 349
Helmholtz, Hermann von
   in *Berlin Physikalische Gesellschaft,* 132, 134
   at British Association for the Advancement of Science meeting, 134
   doctorate under Müller, 130
   English steam gardens influencing, 6, 107
   fermentation experiments, 110, 134
   at *Glienicke,* 110
   in *Königliches Garde-Husaren-Regiment,* 130
   and Leithold, 130
   marriage to Olga von Velten, 130
   On the Conservation of Force (as Work or Energy), 132, 134, 135
   physiology research, 134
   on Potsdam-Berlin train, 110, *111*
   and Puhlmann, 130
   and Schinkel-Lenné landscapes, 132
   and Selo, 132
   and Siemens, 130
   as striving scion of the middle class, 130
   teachers of, 132–134
Herder, Johann Gottfried von, 170
Hermeneutics, 345, 347
Herringbone Rigid Metal Lath, 264, *265*
Herschel, John, 155
Hertzberger, Herman, 362
Herzog, Thomas, 368, *369*
Hesselgren, Sven, 13, 356
Hieroglyphic monad, 62–66, *62*
   Libavius's version of, *63*
   in plan for Libavius's laboratory, 5, 60, 64, *65,* 67, 72, 74n19
   and squaring the circle, 68, *69*
   as symbol of mercury, 62, 66
Higbie, Henry Harold, 246
Higginson, Henry, 258, 259, 264
High culture, 481–482
High-energy physics centers. *See also* CERN; Fermilab; Lawrence Berkeley Laboratory; Superconducting Super Collider
   bubble chambers, 505–506, *506,* 524, 526, 527
   dispersed architecture in, 519, 524–530
   as exemplars of big science, 17, 18
   as multinational enterprises, 524
   Stanford Linear Accelerator Center, 489, 497, 526
   synchrotrons, 459, 461, 473n2
Hill Burton Act (1946), 289
Hinds, Phil, 420, 421
History, 339–340, 341, 344
Hitchcock, Henry-Russell, 255, 270, 272n11, 276n63
Hoberman, Perry, 537n54
Hochschule für Gestaltung (Ulm), 357, 360
Holifield, Chet, 473n4
Holzer, Jenny, 537n53
Hospital, The (Thompson and Goldin), 299n2

"Hospitalism," 284
Hospitals, 281–305. *See also* Massachusetts General Hospital
  academic medical centers, 294
  admissions policies, 282
  and biomedical paradigm, 288–289, 294–295, 299
  Boston Lying-In Hospital, 289, 300n26
  as charitable enterprises, 282, 300n12
  childbirth in, 289
  communal wards, 283, 285
  competitive environment of, 296, 298
  as core institutions, 288
  costs increasing, 295
  criticisms of, 12, 294–295
  as customer-oriented, 295–298
  dangers of, 284, 294
  Dartmouth-Hitchcock Medical Center, 296, *297*
  Disney "hospital of the future," 298
  diversity of care in, 292
  dominant values of medicine expressed in, 281, 284, 296
  and germ theory of disease, 11–12, 288
  grand facades of, 282
  Hill Burton Act of 1946, 289
  increase in number of, 289, 292
  as institutions of last resort, 281–282
  Johns Hopkins Hospital, 285, *286, 287*
  laboratories in, 294
  mediative position between science and public culture, 11, 281
  medical education, practice, and research in, 290, 294
  medical technology in, 290
  and miasmatic theory of disease, 284
  moral regeneration in, 283
  nineteenth-century hospitals, 11, 282–285
  as patient-centered, 295–298
  pavilions, 285, 288
  paying patients, 288
  postmodern hospitals, 12, 295–299
  Pru-Care medical mall, 298
  as scientific institutions, 285, 288
  shopping malls compared with, 2, 295, 296, *297,* 298
  skyscraper hospitals, 290–292
  specialized hospitals, 296
  surgical suites, 288
  two-class system of care, 288
Housing. *See* Dwellings
Howe & Lascaze, *256,* 257, 268, *269,* 270, 271
Hughes, Thomas, 310, 332n43, 536n46
Hull, David, 80
Humanism
  Alberti as humanist, 86
  alchemist as model of, 67
  of Aldrovandi, 41, 42
  civic humanism, 60
  individualism of, 509, 512, 530, 531
  Libavius as humanist, 62, 72
  private contemplation in, 46
Humboldt, Alexander von, 114, 120, 132
Humphrey, John Barnett, 112, 116, 124
Hunt, Robert, 194, 195
Husserl, Edmund, 13, 21n19, 219, 341–342
Huxley, T. H., 184, 185, 195, 200, 203n10
Huxtable, Ada Louise, 434

IBM, 426
Idea of Universal Architecture (Scamozzi), 40
Identity
  architects, 1–3, 9, 12–14, 19
  buildings in, 1–3, 424–431
  collective identity, 426–427
  scientists, 1–3, 9, 19
  as social phenomenon, 9, 424
Illich, Ivan, 294
Image and Logic (Galison), 534n
Imagination, 349
Imperato, Ferrante, 34, *35*
Imperial College (London), 183–184, 200
Individualism, 509, 512, 530, 531
Industrialized building, 357–358, 371n20

International Style, 268, 270, 355, 388
Internet, 519, 530
Italian palazzo, 388, *389*

Jackson, Myles W., 6–7
Jackson Pollock in His Studio (Namuth), 510, *511*
Jarry, Alfred, 346–347
Jed, Stephanie, 37, 46
Jencks, Charles, 356
Jermyn Street Museum (London). *See* Museum of Practical Geology
Jevons, W. Stanley, 200
Johns Hopkins Hospital (Baltimore), 285, *286, 287*
Johnson, Philip, 270, 483
Jones, Caroline, 18–19
Jordy, William, 272n11, 276n59
Judd, Donald, 519
Judd, J. W., 201
Jung, Carl, 480

Kabyle Berbers, 426
Kahn, Albert, 18, 499
Kahn, Louis
 Richards Medical Laboratory Building, 16, *395,* 414–415
 and Safdie, 475
 and Salk, 410, 485–486
 Salk Institute of Biological Studies, 401, 410, 415, 485–486, *485,* 536n45
Kansai Airport (Osaka), 360–361, *361,* 371n25
Kant, Immanuel, 341
Kauffmann, Emil, 316–317
Kautzman, Walter, 414
Kerll, A. L., 134
Kern, Stephen, 255
Kircher, Athanasius, 273n15
Klein, Alexander, 13, 354, 355, 370n1
Knowledge. *See also* Science
 Filarete on artisanal and learned, 6
 Merleau-Ponty on origins of, 219
 museums as privileged site of, 32, 190–191
 and space as intertwined, 3
 women segregated from, 32, 51n15
Knowles, Ralph, 355
Knudsen, Vern, 276n58
Koolhaas, Rem, 13, 326, 329
Korn, Arthur, 355
Kremsmünster (Austria), 152
Kreuzberg monument (Schinkel), 120, *121*
Kriegschule (Berlin), 113, 133
Kruger, Barbara, 524, 527, 537n53
Kuhn, Thomas, 95n2
Kunckel, Johannes, 159n22

Laboratories. *See also* Center for Advanced Biotechnology and Medicine; High-energy physics centers; Lewis Thomas Laboratory
 "big science" laboratories, 17, 18
 centralization of, 498–507, 530–531
 city-sized laboratories, 18
 client's expertise superior to the architect's in, 190
 "closed lab" concept, 442, 443–444
 conventional laboratories as generic, 390
 cosmic-ray stations, 507, *508*
 decentralization of, 519, 524–530
 factory model of, 498–507, 531
 Federal Communication Laboratory and Microwave Tower, 501, *503*
 flexibility in, 499
 Fraunhofer's monastic laboratory, 7, 142, 156, *157*
 in hospitals, 294
 of Libavius, 4–5, 59–77
 "open lab" concept, 401, *407,* 408, 442, 444
 Robins specializing in, 184
 Salk Institute of Biological Studies, 401, 410, 415, 485–486, *485,* 536n45
 Smyth's industrial vision of, 505

as social spaces, 157–158
of Tesla, 509, *509*
for University College London, 186–188, *187*
variety of, 1
LaFiandra, Alba, 443, 444, 452n30
LaFollette, Marcel, 80
Lagrange, Joseph Louis de, 341
Laird, Donald, 266
Land Registry (Bavaria), 150
Landscape. *See also* Gardens
   of Lewis Thomas Laboratory, 405
   Schinkel-Lenné landscapes, 132
   of Superconducting Super Collider, 494
   technology as, 326
   and texture, 329
Laplace, Pierre Simon, 341
Large Hadron Collider (CERN), 524, *525*
La Tourette monastery (Evreux-sur-l'Arbresle, France), 222–224, *223*
Latrobe, Benjamin, 257–258
Lavoisier, Antoine-Laurent, 318, 319
Lawrence, Ernest O., 463, 504
Lawrence Berkeley Laboratory
   bubble chamber, 506, *506*
   Central Laboratory, 501, 534n13
   Fermilab synchrotron design, 459, 461
   as taking more than thirty years to build, 463
   in Time Projection Chamber project, 526, 527
Learned societies, 182
Lears, Jackson, 261
Leavitt, Judith, 289
Le Camus de Mézières, Nicolas, 318, 322
Le Corbusier
   on acoustical aspects of architecture, 273n13
   on architecture as circulation, 213–214
   La Tourette monastery, 222–224, *223*
   on listening to American engineers, 255, 257
   and Lyon, 275n53

   Olivetti project, 220, *221*
Lederman, Leon, 470
Ledoux, Claude-Nicolas
   and artistic dimension of architecture, 322
   dynamism as preoccupation of, 320
   engineers using principles of, 317, 320
   geometric compositional technique of, 316
   as no radical, 317
   salt factory at Arc-et-Senans, 320, 322, *323*
   sensationalism as influence on, 318
   symbolism in, 322
Lefebvre, Henri, 158n2
Leibniz, Gottfried Wilhelm, 343, 344, 345
Leithold, H. R. C. von, 130–131
Lemonick, Aaron, 414
Lenné, Peter Josef
   Albrecht's palace, 120, *121*
   Babelsberg, 118
   Borsig villa, 129
   Charlottenhof, 120
   English gardens built by, 107
   Glienicke, 110, 116
   institutions for gardening culture, 116, 133
   Lustgarten for royal palace, 131, *131*
   Moabit suburb, 129
   Sanssouci, 124
   tour of English gardens, 112
Leprince-Ringuet, Louis, 507
LEP3NET, 528, *529*
Levinas, Emmanuel, 2
Levine, Arnold
   on Fairchild Laboratory, 401
   on generic laboratories, 400
   and goal of Lewis Thomas Laboratory, 380
   on interaction in molecular biology, 402
   on laboratory arrangement, 441, 442
   on Lewis Thomas Laboratory, 450

on Lewis Thomas Laboratory design
team, 401
meeting with Venturi, 415, 442
on outjobbing, 447
p53 gene research, 420–421
in Princeton biochemical sciences department, 452n30
in Princeton molecular biology department, 399, 414, 452n30
"shelling" laboratory space, 433
on socializing facilities, 434
on three-story laboratories, 403
working in Louis Kahn laboratory, 16, 414–415
Lewis Thomas Laboratory (Princeton University), 14–17, 383–455
administrative space in, 433, 438
animal area of, 408, 441
and basic/applied science dichotomy, 450
bi-level lobby of, 404, 433
biological containment facility of, 406
Center for Advanced Biotechnology and Medicine compared with, 16–17, 423–455
central staircase of, 409, *409*, 416, *Plate 9*
chemical waste requirements for, 408
communication as maximized in, 416, 442
design process for, 399–412
dimensions of, 405
as Elizabethan manor house, 404, 450
entrances to, 404, 431–434, *432*
exterior architecture of, 44, *386, 387*, 404, *Plate 7*
Fermilab contrasted with, 17
five research divisions in, 403, *403, Plate 8*
floor plans, 428–431, *428*
funding for, 434
future change planned for, 405–406, 446
as generic laboratory, 380, 388, 400
glass walls of, 404, *411*, 416, *Plate 10*
ground-floor teaching laboratory, *417*
heavy equipment storage in, 442, 444
hiring the architects, 400
identities made tangible by, 428
instruments binding activities together, 16
interior architecture of, 15, 401–404
laboratory modules of, 440–443
landscaping of, 405
life in, 413–422
lighting of, 443
lounges in, 379, *392*, 416, *418*, 433–434
maintenance requirements of, 406
mechanical systems of, 405, 406, 408
natural materials in, 404, 410
number of stories of, 403, 404
numerical data on, 430, *430*
offices in, 409, 442–443
open lab concept in, 401, *407*, 408, 442
outjobbing at, 446–447
p53 gene research, 420–421
praise for, 450
public spaces in, 409
rectangular shape of, 402–403, 404, 416
as repetitive and orderly, 404, 416
requirements for, 415–416
second-floor laboratory, *417*
site of, 405
spending constraints on, 15, 419
tethering at, 448
typifying, 440–443
undergraduates accommodated in, 433–434, 451
variety of spaces in, 408–409
Venturi on exterior architecture of, 15
as zero-based, 419, 440
Lexicon of Inanimate Things (Aldrovandi), 44
Libavius, Andreas, 59–77
Alchemia, 60, 67, 71, 72
alchemy defended by, 67
athannor of, 71, *72*

Commentationes, 67, 71
De azotho, 67
De lapide philosophorum, 67
De mercurio philosophorum, 67, 74n22
on "dispersion of knowledge" technique, 71
division of space in laboratory of, 59–60, 158
on the hieroglyphic monad, 62–64, *63*
hieroglyphic monad in plan for laboratory of, 5, 60, 64, *65,* 67, 72, 74n19
as humanist, 62, 72
on the ideal alchemist, 67
on the Paracelsians, 62, 63, 67, 68, 70, 72–73
plan and elevation of laboratory of, *61*
secret room in laboratory of, 70–73
transmutation practiced by, 71
Tycho's laboratory contrasted with that of, 4–5, 60, 64, 70

Libraries
of Aldrovandi, 41
of Benedictine monasteries, 152
museums compared with, 192
nineteenth-century evolution of, 192
Papworths' guide for, 191, 192
Ptolemaic Museum and Library, 81

Lichtenstein, Roy, 532
Life-world, 13, 342
Linguistic anthropology, 166
Linich, Billy, 536n37
Link, H. F., 113, 133
Linzer, Dan, 420
Lissitzky, El, 234
Livable Environments (Rainer), 363
Lives of a Cell (Thomas), 14
Loft buildings
artists' studios in, 510, 512, 530, 531
as generic architecture, 390, 391, 398
Logical empiricism, 10
Lollio, Alberto, 41
Long, Pamela O., 5–6, 13, 73n2
Los Alamos
Alvarez at, 505
Groves as controlling, 504

growth of, 504
as laboratory-industry hybrid, 18, 531
wartime factory as template for, 499
Lotto, Lorenzo, 34, *34*
Louvre Colonnade (Paris), 312
Loyang Bridge (China), *320*
LTL. *See* Lewis Thomas Laboratory
Lucretius, 167
Ludwig, Carl, 132
Lumbee Indians, 427
Lyon, Gustav, 275n53
Lyotard, Jean-François, 339, 363

McCue, Gerald, 534n13
Machiavelli, Niccolò, 33, *33*
Machine aesthetic, 396
Machine in the Studio (Jones), 534n
McKim, Mead & White
Boston Symphony Hall, 259–260, *260,* 270
revivalism of, 272n11
Madrid School of Architecture, 354
Magnus, Eduard, 134
Magnus, Gustav, 113, 133–134
Malanga, Gerard, 516, *516,* 518
Malinowski, Bronislaw, 165
Managed care, 298
Mandelbrot, Benoit, 343, 486
Manhattan Project, 499, *500,* 504
March, Lionel, 354, 357, 370n18
Marx, Jay, 528
Marxism, 10, 233–234
Mason, Otis
debate with Boas on museum display, 7, 167–172
on goal of museum exhibition, 171
and Pitt Rivers Museum Project, *175,* 176
as practical American democrat, 169
Massachusetts General Hospital (Boston)
Bulfinch building of 1821, 283, *283*
growth and renovation of, 292, *293*
Robert White Memorial Building, 290, *291*
as voluntary hospital, 282

Welman Research Building, 401
Massachusetts Hall (Harvard University), *389*
Materials. *See also* Sound-absorbing materials
  form generated by, 475–476
  natural materials in Lewis Thomas Laboratory, 404, 410
  new materials of the twentieth century, 261
  scientific investigation of, 312
  testing bricks, 188–190, *189,* 205n28
Maternity hospitals, 289
Mathematico-Mechanical Institute (MMI), 150
Mathematics
  in Alberti's conception of architecture, 87
  calculus, 319
  in design, 357
  fractals, 343, 486
  geometrical approach to acoustics, 257–258
  in University of London architectural school, 354
May, Ernst, 355
Mead, Margaret, 166
Mechanization Takes Command (Giedion), 362, 371n26
Medical technology, 290
Medici, Piero de, 89
Medici family, 41
Menzel, Adolf, *111,* 130, 134
Merkava battle tank, 481, *482*
Merleau-Ponty, Maurice, 219, 342, 348
Merton, Robert, 80, 95n2
Metallurgical Laboratory (University of Chicago), 504
Meyenburg, Konrad von, 233, 251n12
Meyer, Hannes, 233–252. *See also* Petersschule
  "Abstract Architecture II," *239*
  Bill as influenced by, 360
  on building as a biological event, 233, 249
  on "building" rather than "architecture," 354
  on composition, 354–355
  "Co-op Construction," 249–250, *249*
  "Co-op Zimmer," 240, *241,* 243–244
  denigrated for his architecture and politics, 233
  on form, 233
  Freidorf Cooperative Estate, 238, 240
  as Marxist, 10, 233–234
  on modern subjectivity, 234–238
  "Die neue Welt," 234, *235,* 240, *241*
  office of, *237*
  scientized conception of architecture of, 10, 21n14
  Scott Brown on, 14
  "Self-Portrait," *242*
  Trades Union School in Bernau, 355
Miasmatic (bad air) theory of disease, 11, 284, *287*
Michelangelo in His Studio (Delacroix), 509, 530
Mies van der Rohe, Ludwig, 233, 268
Mills, Robert, 257–258
Mimesis, 348
Minimalism, 512, 514
MIT Radiation Laboratory, 504, 505, 531
Mitscherlich, Eilhard, 113, 133
Modern Architecture: Romanticism and Reintegration (Hitchcock), 272n11
Modernism. *See also* Bauhaus
  in America, 268, 271
  as antithesis of Victorianism, 254
  on buildings as machines, 338
  circulation as category of, 220, 228
  continuity of traditional American architecture and, 257
  "daylight factories" of, 518
  of Fermilab, 17
  Futurism, 254–255
  historical context of, 272n11
  International Style, 268, 270, 355, 388
  late-modernist sites of production, 498
  machine aesthetic of, 396
  of Meyer, 233–252

neomodernism, 15, 396
  of Philadelphia Savings Fund Society tower, *256,* 257, 271
  Russian Constructivism, 396, *397*
  in St. Bartholomew's Church, 253–254, 272n2
  Smithson's challenge to, 520
  on symbolism, 15
  technology as justification of, 254–255, 361–362
  Venturi on abandoning functionalism of, 17
Modernity
  fundamental epistemic break of, 13
  German modernization, 114, 124
  hospitals, 281–305
  Meyer on subjective effects of, 234–238
  modern space, 9–12
  noise abatement for cacophony of, 11, 264, 266
  on progress, 344
  science and architecture in early modern Europe, 27–103
  technology in modernization, 254
Modular construction, 499
Moffett building (Princeton University), 446, 447, 448
Molecular biology. *See also* Center for Advanced Biotechnology and Medicine; Lewis Thomas Laboratory
  as blue-collar science, 16, 415
  discussion and interaction required in, 402
  in hospital laboratories, 294
  medical and social impact of, 419
Moltke, Helmut von, 109–110
Monad, hieroglyphic. *See* Hieroglyphic monad
Montaigne, 30
Montessori School (Delft, Netherlands), 362
Monuments, 426
Moran, Bruce, 80
Morel, Jean Marie, 320
Morphology, 475–476

Morris, Charles, 362, 366
Morris, Robert, 519
Motherwell, Robert, 510
Moya, Hidalgo, 172, *173*
Müller, Johannes, 130
Müller, Michael, 41
Mumford, Lewis, 152, 153, 261
Murchison, Sir Roderick, 195, 196, 206n48
Museum of Modern Art "International Style" exhibition, 268, 270
Museum of Natural History (Paris), 191
Museum of Practical Geology (Jermyn Street, London), 193–202
  collections of, 194
  function of, 194
  geographical dimension of displays, 194
  geological theory expressed in, 8, 195–200
  as lesson in stone, 195
  Lower Gallery plan, *198*
  main floor with galleries above, *197*
  ornamental arts collection, 194, 201
  in Papworths' guide, 191
  principal floor, *196, Plate 6*
  removal to South Kensington, 200–201
  as representation of natural historical knowledge, 8, 193
  rock collections, 194–200
  Upper Gallery plan, *199*
Museums. *See also* Anthropological museums; Natural history museums
  architectural meaning in, 190
  authority of, 191
  cupolas associated with, 204n21
  domestic context of early, 36, 47, 50
  in education, 191
  emerging from the study, 32, 33, 34
  expanding beyond a single room, 38, 40
  external constraints in construction of, 176, 178
  of Imperato, 34, *35*
  at juncture of private and public, 36

as lessons in stone, 203
libraries compared with, 192
masculine prerogatives in early modern, 29–57
metaphors for, 192
as monographs, 192, 195
national power expressed by, 8–9
Papworths' guide for, 191, 192
as privileged site of knowledge, 32, 190–191
Scamozzi on design of, 40
as setting for masculine pursuits, 36–37
as sociable settings, 34
women excluded from early modern, 30–31, 47, 48

Namuth, Hans, 510, *511*
Narration, 329, 332n51, 347, 348, 363
Nassau Hall (Princeton University), 393
National Museum (United States), 7, 167–172
Natural history. *See also* Natural history museums
in anthropology's development, 166
architecture and classification in, 312
Fontana's role in Aldrovandi's, 44–45
women excluded from Renaissance, 4, 45–50
Natural History (Imperato), 34
Natural History Museum (London)
and Jermyn Street Museum's collections, 201–202, 206n61
Owen's ideas in display system of, 8, 193
Owen's phylogenetic description of, 192–193
Waterhouse as designer of, 183
Natural history museums. *See also* Museum of Practical Geology; Natural History Museum (London)
American Museum of Natural History, 172
display in, 7–9
Geological Museum (London), 202
Museum of Natural History (Paris), 191
Studio Aldrovandi, 29, 48, *49*
women excluded from Aldrovandi's museum, 4, 30–31, 43
Natural philosophy. *See also* Experimental philosophy
dynamic conception of nature, 319, 320
Newtonian, 340, 341
on openness and empiricism, 93
women excluded from, 32
Natural resources, 347
Nature
artificial versus natural, 324
conversions of force in processes of, 134
dynamic conception of, 319, 320
natural and man-made forms, 480
nautilus shell as natural form, 476, *477*
Navis (Alberti), 87
Neomodernism, 15, 396
Nervi, Pier Luigi, 172, *173*
"Net, the," 519, 530
Neue Museum in Berlin, Das (Freydanck), *131*
Neue Sachlichkeit, 10
Neues Gewandhaus (Leipzig), 258, 259, 273n23
"Neue Welt, Die" (Meyer), 234, *235,* 240, *241*
Neurath, Otto, 10, 370n4
Neutra, Richard, 13, 355
New England mills, 388, *389,* 416
Newman, Barnett, 512, 514, 521
Newman, William R., 4–5, 93
Newton, Isaac, 340, 341
Newton's inverse-square law, 318
New York City Noise Abatement Commission, 266
New York Life Insurance Building (New York City)
as Gothic, 266–267
interior furnishings of, 267
kitchen of, 267, *267*

sound-absorbing materials in, 11, 264, *265*, 266, 270–271
New York School, 510, 530, 531
Nietzsche, Friedrich, 339–340, 344, 348, 349
Niggl, Joseph, 151, 153
Nightingale, Florence, 284
Nihilism, 344, 348, 349
Nobel Prize, 532, 533
Noble, David, 32, 48
Nogarola, Isotta, 36, 51n24
Noise abatement, 11, 264, 266
"Non-sites," 520, 521, 523, 533
Northwest Coast Hall (American Museum of Natural History), 172
Nouvel, Jean, 13, *328,* 329

Oak Ridge Laboratory
    as laboratory-industry hybrid, 18, 530–531
    as largest factory in the world, 504
    Tennessee Eastman Co. involvement in, 504
    wartime factory as template for, 499
Objectivity, 346
Observatory (Paris), 314
Ochsner, Albert, 290
Ohl, Herbert, 357
Oldenburg, Henry, 94n1
Olgyay, Aldar, *356,* 370n7
Olgyay, Victor, 355, *356,* 370n7
Olivetti project (Le Corbusier), 220, *221*
On Growth and Form (Thompson), 316, 359, 475–476
On the Art of Building (Alberti). See *De re aedificatoria*
On the Conservation of Force (as Work or Energy) (Helmholtz), 132, 134, 135
On the Remains of Bloodless Animals (Aldrovandi), 45
On Wifely Duties (Barbaro), 45
"Open lab" concept, 401, *407,* 408, 442, 444
Openness
    Alberti on, 6, 86, 88–89
    empiricism and rhetoric of, 5, 79–103
    in fifteenth-century trades, 85
    Filarete on, 6, 92
    of modern science, 4–6, 59, 72, 79–80
    of museum collections, 38
    Philo of Byzantium on, 81–82, 84
    printing encouraging, 93
    and priority and patronage, 80
    various meanings of, 80
    Vitruvius on, 5–6, 82–85
Oppel, John, 86
Optical correction, 312, 339
Optical Institute, 150–151
    Fraunhofer at, 151
    located at Benediktbeuren, 142, 151–152
    purpose of, 150
    secrecy at, 154–155
Optics. *See also* Optical Institute
    Fraunhofer's contributions to, 141–163
Ordonnance for the Five Kinds of Columns (Perrault), 338
Oren, Moshe, 420–421
Ovitt, George, Jr., 160n36
Owen, Richard, 8, 192–193, 202
Oxford University Museum. *See also* Pitt Rivers Museum Project
    Museum of Practical Geology compared with, 195

Pages paysage (journal), 329
Palazzo Rucellai (Florence), *389*
Paleotti, Ippolita, 31
Palladio, Andrea, 38, 52n40, 238
Pandechion Epistemonicon (Aldrovandi), 44
Panopticon, 248, 300n13, 322, 339
Papua New Guinea Parliament House, 427
Papworth, John, 191, 192
Papworth, Wyatt, 191, 192
Paracelsians, 62, 63, 67, 68, 70, 72–73
Paracelsus, 62, 159n22
Parliament House (Papua New Guinea), 427

Partridge, Loren, 85
Pastore, John, 473n4
Pataphysics, 346–347
Patronage
    Alberti and, 88, 92
    Filarete and, 92, 485
    Filarete on the architect and the patron, 89–90
    institutions replacing, 182
    and openness, 80
    of Vitruvius, 82
Patte, Pierre, 214, 257, 318
Pavilion system, 316
Payette, Thomas, 15, 400, 401
Payette Associates Inc.
    Center for Advanced Biotechnology and Medicine, 431
    Fairchild Biochemistry Building, 400, 401
    Lewis Thomas Laboratory, 398, 400, 413, 431
    Welman Research Building, 401
Paz, Octavio, 347
Peach, C. S., 204n14
Pennethorne, James, 194
Pentagon, the (Washington, D.C.), 504
Penty, Arthur J., 536n41
Peoples, John, 470
Pérez-Gómez, Alberto, 13, 228, 309, 324
Perrault, Claude
    as architect and scientist, 311
    on architecture as search for probable solutions, 13, 339
    engravings for Vitruvius translation and anatomical memoirs, 312
    Louvre Colonnade reinforcement, 312
    on muscle action, 312, *313*
    Observatory of Paris, 314
    Ordonnance for the Five Kinds of Columns, 338
    scientific understanding extrapolated into architecture, 338–339
Perronet, Jean Rodolphe, 319, *320,* 331n31
Persius, Ludwig, 116, *117,* 118, 126, 132

Perspective paradigm, 20n5
Petersschule (Basel), 244–250, *245*
    basic configuration of, 244
    functionalism of, 246
    materiality of, 246
    in Meyer's scientized conception of architecture in, 10, 244
    sectional organization, 247
    site of, 247
Petrarch, 33, 53n42
P53 gene, 420–421
Phenomenology, 342–343, 348
Philadelphia Savings Fund Society (Philadelphia)
    banking room of, 268, *269*
    escalators of, 268, *269*
    functionally differentiated spaces of, 268
    Hitchcock and Johnson on, 270, 276n63
    machined look of, 268
    modernist aesthetic of, *256,* 257, 271
    sound-absorbing materials in, 268, *269,* 270, 276nn59, 61
Phillips, John, 195
Philo of Byzantium, 81–82, 84
Philosophers' stone, 5, 66, 68, 71
Physical (biological) anthropology, 166, 172
Physics. *See also* High-energy physics centers
    Berlin Physikalische Gesellschaft, 132, 134
    as business for Alvarez, 497
    cosmic-ray physics, 505, 507, *508,* 530
    "Discussion on the Design of Physics Laboratories" of 1947, 501
    engineer sharing Nobel Prize for, 533
    Die Fortschritte der Physik, 132
    quantum electrodynamics, 342
    roots of big physics, 504
Physiology, 134
Piano, Renzo
    double-glazed thermal wall used by, 367

Kansai Airport, 360–361, *361,* 371n25
Lowara office building, 371n27
Picon, Antoine, 12–13
Piero della Francesca, 89
Pigeon House (Iran), 478, *478, 479*
Pike, Alexander, 358, 371n20
Pitt Rivers, Augustus Lane Fox, 8, 172
Pitt Rivers Museum Project, 172–176
    archeological gallery, *174*
    current state of, 176, *177*
    ethnological gallery, *175*
    as ideological pastiche, 8
    Nervi, Powell and Moya proposal for, 172, *173*
Plato
    Libavius relying on authority of, 64
    on Socrates, 83
    Timaeus, 13, 338, 342
Pollock, Jackson, 510, *511,* 521
Polyclitus, 96n14
Polyphilo or the Dark Forest Revisited (Pérez-Gómez), 350n19
Pop art, 512, 514
Portrait of the Daughter of Pedro Gonzales (Fontana), *31,* 51n9
Positivism, 318, 342, 346
Post-industrial society, 520, 536n41
Postmodernism
    deconstruction, 326, 343, 363
    dispersal of the postmodern subject, 519–530
    Lyotard on postmodern condition, 363
    postmodern hospitals, 12, 295–299
    sites of artistic and scientific production in, 498
    Smithson in postmodern turn, 521
    and symbolism, 15, 17
Poststructuralism, 363
Potteries Thinkbelt project, 360, *360*
Powell, John Wesley, 167, 169–170, *175*
Powell, Philip, 172, *173*
Pozzo, Modesta, 48
Praxis, 85, 97n25
Précis des leçons d'architecture (Durand), 214

Preece, William Henry, 186
Price, Cedric, 360, *360*
Princeton University. *See also* Lewis Thomas Laboratory
    biochemical sciences department, 452n30
    conservative construction policy of, 400
    Frick Chemistry Complex, 400
    Guyot Hall, 404, *405,* 446, *447,* 448
    Moffett building, 446, *447,* 448
    molecular biology department, 419
    as least specialized university in America, 414
    Nassau Hall, 393
    traditional building materials at, 404
    Wu Hall, 400
Principles of Architectural History (Frankl), 218–220
Printing, 93
Priority, 80
Privacy
    in Libavius's laboratory, 158
    of the patrician study, 37, 46
    in suburban versus high-density housing, 364, *365*
    of the villa, 41
Production, sites of. *See* Sites of production
Produktform, 360, *361*
Professional associations, 182
Progress, 171, 339, 340, 344, 349
Proportion, 81–82, 96n14, 238, 270, 314, 339
Pru-Care, 298
PSFS. *See* Philadelphia Savings Fund Society
Psychology and architectural form, 356–357
Ptolemaic Museum and Library (Alexandria), 81
Puhlmann, Wilhelm, 130
Pupilage, 182–183

Quantum electrodynamics, 342

Radio City complex (New York City), 276n65
Rainer, Roland, 363–364, *365,* 371n31
Ramsay, A. C., 194, 195, 200
Ramsey, Norman, 463
Rapoport, Amos, 426
Rationalization, 322, 366–367, 368, 528
Raytheon, 504
Reform Club (London), 216, *217,* 218
Refractive indices, 143–150
Reichenbach, Georg, 144
Reimer, Hans, 371n31
Renaissance
   construction in economy of, 85
   history in, 344
   masculine prerogatives in early modern museums, 29–57
   perspective paradigm of, 20n5
   Vitruvian principles in, 311
Renaissance and Baroque (Wölfflin), 222
Reverberation time, 11, 259, 262–263
"Revolutionary architecture," 316
RIBA. *See* Royal Institute of British Architects
Richards Medical Laboratory Building (University of Pennsylvania), 16, *395,* 414–415
Richardson, Henry Hobson, 272n11
Ricoeur, Paul, 332n51, 345
Rider, Alan, 470
Riesch, Pater Udalricus, 151–152
Ritchie, Matthew, 537n55
Roadside architecture, 396, *397*
Robert Rathbun Wilson Hall (Fermilab). *See* Wilson, Robert Rathbun, Hall
Robins, Edward Cookworthy, 184, 185, 203n10
Rockefeller, Laurance S., 413, 434
Roger, Jacques, 331n27
Rogers, Richard, 360
Rohrer, E. Parke, 466, 470
Rosarium philosophorum, 68
Rosenberg, Charles, 11, 282
Rosenblatt, M. C., 276n59
Rosenquist, James, 532

Rosner, David, 282
Rossi, Paolo, 80
Rothman, David J., 300n12
Rothschild, Emma, 367
Rousseau, Jean-Jacques, 344
Rowan, David, 406
Rowe, Colin, 9, 222, 224
Royal Academy of Sciences (Paris), 79
Royal Bavarian Academy of Sciences, 154, 158n5
Royal College of Science (London), 200
Royal Ethnographic Museum (Berlin), 167
Royal Institute of British Architects (RIBA)
   acoustics research, 258
   Geological Museum collaboration, 202
   in professionalization of architecture, 182
   promoting the arts and sciences of architecture, 183
   on science in architecture, 184–185
   Standing Committee for Science, 185–190, 258
   standing committees of, 185
Royal Society (London), 79, 80, 81, 94n1
Rucellai Palazzo (Florence), *389*
Rudenstine, Neil, 414
Rudofsky, Bernard, 476
Rule of Saint Benedict, The
   Benedictine architecture as instantiating, 142, 153–154
   on labor, 142, 152–153
   and scientific research, 142
   on secrecy, 142, 153
   on silence, 142, 153
Rumford tile, 262, *262,* 263, *263,* 264, 274n37
Rupke, Nicolaas, 205n37
Russian Constructivism, 396, *397*
Rutgers University. *See also* Center for Advanced Biotechnology and Medicine
   Waksman Institute, 434, *435,* 446, 447, 448, 450

Sabine, Wallace
   acoustical materials developed by, 260–261
   Boston Symphony Hall, 11, 259–260, *260,* 270, 276n64
   Goodhue consulting with, 253–254, 271, 272n2
   and Guastavino, 261–262, 270
   reverberation-time formula, 11, 259
   St. Thomas's Church, 262, *262,* 270
   sound-absorbing properties of materials studied by, 258–259
Safdie, Moshe
   Beyond Habitat, 480
   Fermilab visit, 483–485
   Form and Purpose, 481
   in Merkava battle tank design, 481, *482*
   Salk Institute work of, 485
   Superconductor Super Collider design, 18, 486, 488–491, 529
St. Bartholomew's Church (New York City), 253–254, 272n2
St. Thomas's Church (New York City), *256,* 257, 262, *262,* 263, 270
Salk, Jonas, 410, 485–486
Salk Institute of Biological Studies (La Jolla, California), 401, 410, 415, 485–486, *485,* 536n45
Salle Pleyel (Paris), 275n53
Sanitary science, 184
Sanssouci (estate), 124–128, *125*
Sant'Apollinare Nuovo (Ravenna), 396
Sarrocchi, Margherita, 47–48
Satolas railway station (France), 326, *327*
Scamozzi, Vincenzo, 40, 216
Schiebinger, Londa, 46
Schiegg, Pater Ulrich, 150–151, 152
Schinkel, Karl Friedrich
   The Aesthetic View from Pegasus, 115
   Albrecht's palace, 120, *121*
   Altes Museum, 131, *131*
   Babelsberg, 108, 118
   in Beuth's Technical Deputation, 112
   Charlottenhof, 120, 122–123, 132
   English gardens built by, 107
   Glienicke, 108, 116
   Kreuzberg monument, 120, *121*
   Moabit suburb, 129
   portrait of Beuth, 114, *115*
   tour of English gardens, 112
Schmidt, Hans, 234, 240
Schmieden, Heinrich, 273n23
Schopenhauer, Arthur, 222
Schürmann, Astrid, 82
Schweigger, Johann Salomon Christoph, 154–155
Schwitters, Roy, 486, 490
Science. *See also* Big science; Biology; Buildings of science; Chemistry; Experimentation; Geology; Laboratories; Natural philosophy; Physics; Scientific societies; Scientists
   ancient through early modern science as about wholes, 340
   as architecturally sited, 1
   and architecture and technology, 309–335
   architecture as, 337–351
   and architecture in early modern Europe, 27–103
   and architecture in Victorian Britain, 181–208
   artistry and intuition in, 375
   art separated from, 494, 498
   buildings and the subject of, 1–25
   changed status of, 310–311
   is architecture science, 12–14
   marketability of research, 402
   Meyer's scientization of architecture, 10, 233–252
   mutual limits of architecture and, 353–373
   openness of modern, 4–6, 59, 72, 79–80
   and phenomenology, 342–343
   Plato's *Timaeus* as prototype for, 13, 338
   professionalization of, 181, 182
   public character in Britain, 134

sociology of, 95n2
Thomas's romantic image of, 14
unified science movement, 370n4
Wissenschaft, 10
Science (journal), 167
"Science, Art and Technology" (Morris), 362, 366
Science studies, 19
Scientific revolution, 86
Scientific societies
    Accademia dei Lincei, 46, 47, 54n71
    Accademia del Cimento, 79, 94n1
    Berlin Physikalische Gesellschaft, 132, 134
    on openness, 79
    in professionalization of science, 182
    Royal Academy of Sciences (Paris), 79
    Royal Bavarian Academy of Sciences, 154, 158n5
    Royal Society (London), 79, 80, 81, 94n1
Scientists
    buildings of science affecting, 2–3, 414, 422, 423
    as gentleman amateurs before nineteenth century, 181, 182
    hierarchy of reflected in laboratories, 401
    as solitary creators, 509
    as territorial, 444
    Thomas on behavior of, 377
Scott Brown, Denise, 14
Seaborg, Glenn, 462
Secrecy. *See also* Privacy
    in alchemy, 59, 60, 67
    as Benedictine characteristic, 142, 153
    of collectors, 38
    in glassmaking, 155–156
    of guilds, 6, 7, 85, 155
    at the Optical Institute, 154–155
    secret room in Libavius's laboratory, 70–73
    in twentieth-century biology, 80
Sedgwick, Adam, 195, 206n48
"Self-Portrait" (Meyer), *242*

Selo, Hermann, 132
Sensationalist philosophy, 318
Serialization, 238
Serlio, Sebastiano, 41
Server and served philosophy, 402
Setting and place, 388, 391–393
Sevestre, M., *321*
Sforza, Countess Catarina, 29–30
Sforza, Francesco, 89
Sforzinda (ideal city), 6, 89
Shackelford, Jole, 60
Shapin, Steven
    on Boyle's laboratory, 4, 5, 157–158, 161n90
    on invisible technicians, 447, 453n46
    on Royal Society's openness, 80
    on women in seventeenth-century experimental culture, 32, 51n10
Shatkin, Aaron, 436, 444, 448, 452n30
Shenk, Thomas, 399, 401, 414
Siemens, Werner, 130–131, 132
Silent-Ceal, 276n59
Silver, Ira, 425
Simmel, Georg, 4
Simondon, Georges, 324
Sirleto, Guglielmo, 47
Sites/nonsites, 520, 521, 523
Sites of production, 497–540. *See also* Factories; Laboratories; Museums; Studios
    centralization of, 498–519
    decentralization of, 519–530
    productive spaces, 497–498
Six Lamps Experiment (Fraunhofer), 143–147, *146, 147*
SLAC (Stanford Linear Accelerator Center), 489, 497, 526
Sloane, David C., 11–12, 15
Smith, Christine, 85
Smith, Pamela H., 80
Smith, T. Roger, 186, *187,* 188, 204n17, 258
Smith, Terry, 255
Smithson, Robert
    authorship challenged by, 532–533

centralized studio challenged by, 519, 520
in postmodern turn, 521
as "post-studio" artist, 520, 521, 523, 532
Spiral Jetty, 521–523, *522,* 532–533
Smyth, Henry, 504–505
Snelson, Kenneth, 359
Social (cultural) anthropology, 166
Social space, 157–158
Sociology of science, 95n2
Socrates, 83
Sodium lines, 142
Solar Control and Shading (Olgyay and Olgyay), 370n7
Solar Light Experiment (Fraunhofer), 147–148
Soldner, Johann von, 158n5
Solomon, Nancy, 536n45
Solomon's Temple, 340–341
Sombart, Werner, 2
Sonnblick Observatory (Austria), *508*
Sound-absorbing materials, 261–270
    Acousti-Celotex, 263
    Akoustolith, 11, 262, 263, 264, 274n37
    Celotex Corporation, 11, 263, 264
    Herringbone Rigid Metal Lath, 264, *265*
    Mies on, 268
    in New York Life Insurance Building, 11, 264, *265,* 266, 270–271
    for noise abatement, 11, 264, 266
    in Philadelphia Savings Fund Society, 268, *269,* 270, 276nn59, 61
    Rumford tile, 262, *262,* 263, *263,* 264, 274n37
    Sabine in development of, 260–261
    Silent-Ceal, 276n59
    Sprayo-Flake Acoustical Plaster, 263, 264
Soundproof construction, 264, *265*
Space
    of anthropology, 165–166
    Husserl on knowledge of, 219

and knowledge as intertwined, 3
    modern space, 9–12
    patriarchal authority controlling, 37
    productive spaces, 497–498
    social space, 157–158
Space syntax, 229n10
Space, Time and Architecture (Giedion), 362
Spain, Daphne, 51n15
Spengler, Oswald, 2
Spiral Jetty (Smithson), 521–523, *522,* 532–533
Sprayo-Flake Acoustical Plaster, 263, 264
Squaring the circle, 68, *69,* 75n32
SSC. *See* Superconducting Super Collider
Stam, Mart, 234, 240
Standardization, 238
Stanford Linear Accelerator Center (SLAC), 489, 497, 526
Starn, Randolph, 85
Statham, Henry, 186, 188, 204n19
Steadman, Philip, 354, 357, 370n18
Steam engines, 107–140
    at *Babelsberg,* 118, *119*
    at Borsig villa, 129
    at *Charlottenhof,* *122,* 123
    Cugnot's steam-powered vehicle, 325
    in English gardens, 7
    in first industrial revolution, 325
    in Germany, 112, 116
    at *Glienicke,* 108, *108,* 109, 110, 116, *117,* 119, Plate 3
    heat converted to work in, 134
    house and keeper as integral parts of, 116
    in modernization, 254
    Prussian power associated with, 107
    at *Sanssouci,* 124, 126, *126,* 128
Stein, Stephen, 445
Stella, Frank
    abandoning romantic artist-hero, 531
    Aluminum series, 514, *515*
    on the "executive artist," 512, 514
    painting in his studio, *513*
    painting method of, 514

Stevin, Simon, 311
Stocking, George W., Jr., 7–8, 182, 191, 193, 205n41
Stone and Webster, 499
Strack, Johann Heinrich, 129
Stratigraphic analysis, 195–200
Studies
    Alberti's ideal of, 33–40, *37*
    of Giganti, 38, *39*
    Lotto's depiction of, 34, *34*
    of Machiavelli, 33, *33*
    museums emerging from, 32, 33, 34
    of Nogarola, 36, 51n24
    as places of solitude, 33, 34
    of the scholar, 166
    as setting for masculine pursuits, 36–37
    women excluded from, 37, 46
Studio Aldrovandi, 29, 48, *49*
Studios
    alchemist's lair compared with, 510
    centralization of, 498, 508–519, 530–531
    decentralization of, 519–523
    Delacroix's *Michelangelo in His Studio*, 509, 530
    factory model of, 18, 498, 508–519
    Jackson Pollock in His Studio, 510, *511*
    in loft buildings, 510, 512, 530, 531
    as site of individual contemplation, 510
    Smithson as "post-studio" artist, 520, 521, 523, 532
    of Stella, *513*
    virtual studios, 530, 537n54
    Warhol's Factory, 497, 516–519, *517*, 520, 532
Study (Lotto), 34, *34*
Subjectivity
    Damisch on multiple meanings of, 20n5
    destabilization of the subject, 363
    dispersal of the postmodern subject, 519–530
    Foucault on historicization of, 20n4
    Meyer on architecture as rebuilding, 10
    Meyer on modern, 234–238

Sullivan, Louis, 272n11, 396, 485
Summa perfectionis (Geber), 67, 71, 75n25
Superconducting Super Collider (Waxahachie, Texas), 483–496
    a center for, 18, 490–491, 529
    cooling pond as organizing element for, 491, *492*, Plate 12
    corrugated sheet metal for warehouse buildings, 494, *495*
    cost and staffing of detectors for, 524
    designing, 488–491
    excavation spoils used sculpturally at, 494
    as expandable, 380, 494
    horizontalism of, 18
    landscaping of, 494
    multiple control rooms for, 529
    public and private sectors of, 491, *493*
    Safdie in planning of, 486, 488
    scale as concern for, 489
    site of, 489
    the "street," 491, *493*, 529
    temporary facilities of, 488–489
Sustainable growth, 364
Swensson, Earl, 298
Symbolism
    in Boullée and Ledoux, 322
    Cram on, 261
    in generic architecture, 388, 393–398
    in Lewis Thomas Laboratory, 15
    in postmodernism, 15, 17
    of rulership, 6, 88
    Venturi on, 404
Symphony Hall (Boston), 11, 258, 259–260, *260*, 270, 274n30, 276n64
Synchrotrons, 459, 461, 473n2

Taegio, Bartolomeo, 44, 53n51
Tal, Israel, 481, *482*
Tasso, Torquato, 47
Techne, 85, 97n25
Technical Deputation (Prussia), 112, 113
Technical education
    architectural training in technical universities, 353

Berlin technical schools, 113
at Center for Advanced Biotechnology and Medicine, 438
Ecole des Ponts et Chaussées, 317, 319–320
Hochschule für Gestaltung in Ulm, 357, 360
Imperial College (London), 183–184, 200
pupilage in architecture, 182–183
Robins's contributions to, 184, 203n10
Technical School & College Building (Robins), 184, 203n10
Technical University Delft (Netherlands), 353
Technology. *See also* Engineering; Materials; Steam engines
analysis in, 319
architectural response to, 13
architecture and science and, 309–335
architecture problematized by, 347
beauty of the engineered environment, 267
changed status of, 310–311
crisis of modern, 324–330
electronic, 396
explaining similar inventions, 167–168
and fifteenth-century rulership, 85
as founded on weak truths, 348
glassmaking, 155–156
Magnus as professor of, 133
meaning of the technological artifact as uncertain, 12
medical, 290
Meyer on subjective effects of, 234–238
modernism justifying itself in terms of, 254–255, 361–362
in modernization, 254
new materials of the twentieth century, 261
our world as technological, 347
public character in Britain, 134
the technological sublime, 110
"Technology and Science as Ideology" (Habermas), 366–367

Temple of Solomon, 340–341
Ten Books on Architecture (Vitruvius). *See De architectura*
Tennessee-Eastman Co., 504
Tensegrity masts, 359
Terragni, Giuseppe, 224, *225,* 226, *227*
Territorial planning, 319–320
Tesla, Nikola, 509, *509*
Testing of Materials of Construction (Unwin), 190
Tethering, 448
Teufelsbrücke, 108, *108,* 109, 110, 118, Plate 2
Teyssot, Georges, 317
Thackray, John, 206n61
Theophilus, 159n22
Thin-shelled (timbrel) vaulting, 261
Thomas, Lewis
on biology and values, 419
Lewis Thomas Laboratory named for, 413, 434
Lives of a Cell, 14
as poet-scientist, 380
romantic image of science of, 14
visiting Lewis Thomas Laboratory, 422
on wildness in science, 376–377
Thompson, D'Arcy, 316, 359, 475–476, 486
Thompson, Emily, 10–11, 12
Thompson, John D., 299n2
Thorndike-Shutt Group, 527
Timaeus (Plato), 13, 338, 342
Time Projection Chamber (SLAC), 526–528
Ting, Ann, 475
Tönnies, Ferdinand, 2
Törten Housing, 355
Trades Union School (Bernau), 355
Transmutation, 71
Tratatto di architettura (Filarete)
display in, 89
on openness and empiricism, 81
as remaining in manuscript, 92
technical subjects in, 91
Trust, 158

Tschumi, Bernard, 329
Turte, K. D., 113, 133
Tycho Brahe, 4–5, 60, 64, 70

UMDNJ tower, 434, 448, *449,* 450
Unified science movement, 370n4
Universities Research Association (URA), 459, 463, 473n1
University College London
    architectural courses at, 183
    laboratories for, 186–188, *187*
University of London. *See also* University College London
    architectural training at, 354
University of Medicine and Dentistry of New Jersey. *See also* Center for Advanced Biotechnology and Medicine
    UMDNJ tower, 434, 448, *449,* 450
Unwin, William Cawthorne, 188, 189, 205n28
Urbahn, Max O., 470
Urban planning, 375–376, 378, 379
Uterwer, Johann Cornelius, 45
Utzschneider, Joseph von, 142, 150–151, 154, 155, 158, 159n23

Vale, Lawrence, 427
Valerio, Luca, 47
Vanderweil, R. G., Inc., 408
Vattimo, Gianni, 344, 345
Velten, Olga von, 130, 132
Venturi, Robert
    on abandoning modernist functionalism, 17
    on exterior of Lewis Thomas Laboratory, 15
    Fairfield Building visit, 401
    and Louis Kahn, 16
    on Lewis Thomas Laboratory as generic, 380
    on Lewis Thomas Laboratory design team, 413, 414
    meeting with Levine, 415, 442
    on symbolism, 404
    on teaching spaces of Lewis Thomas Laboratory, 434

Venturi, Rauch and Scott Brown
    Gordon Wu Hall, 400
    Lewis Thomas Laboratory, 400, 413, 431
Venturi, Scott Brown and Associates Inc., 15, 400, 413, 431
Vereinigte Artillerie-und-Ingenieurschule (Berlin), 113, 133
Verein zur Beförderung des Gartenbaues (Prussia), 116, 129, 133
Verein zur Beförderung des Gewerbefleisses (Prussia), 113–114, 124, 132, 133
Vernacular buildings, 476, 478, 480, 484
Vico, Giambattista, 344, 345
Vidler, Anthony, 21n17, 317
Vienna Circle, 10, 370n4
Vignola, Antonio, 41
Villa (Taegio), 44
Villa Jacobs (Potsdam), 116
Villalpando, Juan Bautista, 66, 314
Villa Persius (Potsdam), 116
Villas
    Alberti on, 41, 42, 52n36
    in nineteenth-century Germany, 116
    Palladio on, 52n40
    women in villa life, 43–44, 53n51
Villa Schöningen (Potsdam), 116
Viollet-le-Duc, Eugène
    circulation metaphor in, 9, 214, 216, 218, 220, 337
    deterministic theory of, 339
    on structure and living organisms, 314–316, *315*
Virtual studios, 530, 537n54
Vischer, Robert, 222
Vitruvius. *See also* De architectura
    on acoustics, 273n15
    Alberti's criticism of, 86, 89
    on architectural symbolism of governance, 6
    on authorship, 83–84
    on "commodity, fitness and delight," 483
    European architecture and, 311
    on human body as design concept, 64, 66

on openness, 5–6, 82–85
on optical correction, 312
on progressive development of architecture, 82–83, 96n18
sixteenth-century architects drawing on, 38
on Socrates' wisdom, 83
on three parts of architecture, 81
Vitz, Paul C., 357
Vogelstein, Bert, 421

Wachsmann, Konrad, 359, *359*, 360
Waddy, Patricia, 54n80
Wagner, Pater Josef Maria, 152
Wagner, Rudolph, 134
Waksman Institute (Rutgers University), 434, *435*, 446, 447, 448, 450
Warhol, Andy, 514–519
    abandoning romantic artist-hero, 531
    as "boss," 518, 532
    on the "business art business," 497
    commercial art of, 514
    "everybody should be a machine," 497
    The Factory, 497, 516–519, *517*, 520, 532
    as late modernist, 520
    silk screening by, 516, *516*, 518
Waterhouse, Alfred, 183–184, 190, 193, 204n19
Watt, James, 134
Weber, Max, 152, 160n36
White, William, 185
Wigley, Mark, 36, 54n72
Wilhelm, Prince, 108, 114, 118, 120
Wilkins, William, 186
Wilson, Robert R.
    Broken Symmetry, 465
    in Fermilab construction, 17–18, 461–470
    Safdie on, 483, 486
    as sculptor, 463, *464*
Wilson, Robert Rathbun, Hall (Fermilab)
    aerial view of, *460*, *484*, *Plate 11*
    design process for, 466, 470
    exterior of, *467*

interior of, *468*, *469*
Safdie's criticism of, 18, 484
Wilson, William, 431, 436
Wise, M. Norton, 6
Wissenschaft, 10
Wittkower, Rudolph, 66
Wölfflin, Heinrich, 9, 222
Women
    Alberti on restricting within the home, 43, 44
    at British Association for the Advancement of Science meeting, 134
    in early modern domestic economy, 44
    exclusion from Aldrovandi's collection, 4, 30–31, 43
    exclusion from early modern museums, 30–31, 47, 48
    exclusion from Renaissance natural history, 4, 45–50
    exclusion from the study, 37, 46
    keys of the household, 38
    knowledge segregated from, 32, 51n15
    Meyer on masculinization of, 236
    in molecular biology, 433
    Montaigne on, 30
    and natural philosophy, 32
    Renaissance moralists on, 46
    in villa life, 43–44, 53n51
Women's Worth (Fonte), 48
Woodward, William, 204n22
Wool, Christopher, 537n53
Woolworth Building (New York City), 276n63
Workman, Herbert, 152
Works Progress Administration (WPA), 508
World's Columbian Exposition (1893), 272n11
World War II, 429, 530
World Wide Web, 519, 530, 537n54
Wren, Christopher, 311
Wright, Frank Lloyd, 272n11
Wu, Gordon, Hall (Princeton University), 400

Xenophon, 38

Yale University School of Medicine (sketch), *392*

Zilsel, Edgar, 80
Zoilus, 84
Zymotic theory of disease, 284